Lecture Notes in Production Engineering

For further volumes:
http://www.springer.com/series/10642

Katja Windt
Editor

Robust Manufacturing Control

Proceedings of the CIRP Sponsored Conference RoMaC 2012, Bremen, Germany, 18–20th June 2012

Springer

Editor
Katja Windt
Global Production Logistics
School of Engineering and Science
Jacobs University Bremen
Bremen
Germany

ISSN 2194-0525 ISSN 2194-0533 (electronic)
ISBN 978-3-642-30748-5 ISBN 978-3-642-30749-2 (eBook)
DOI 10.1007/978-3-642-30749-2
Springer Heidelberg New York Dordrecht London

Library of Congress Control Number: 2012952009

Printed on acid-free paper

Springer is part of Springer Science+Business Media (www.springer.com)

Preface

The international conference on "Robust Manufacturing Control—Innovative and Interdisciplinary Approaches for Global Networks" (RoMaC 2012) was held on the campus of Jacobs University in Bremen, Germany. As expressed by the title, one major intention of the conference is to focus on transdisciplinary approaches toward robustness in manufacturing. The conference was sponsored by the International Production Engineering Academy (CIRP) and the Alfried Krupp von Bohlen und Halbach-Foundation, to both of which I am very thankful.

Today, Global Production Networks (i.e., the nexus of interconnected material and information flows through which products and services are manufactured, assembled, and distributed) are confronted with and expected to adapt to:

- sudden and unexpectable large-scale changes of important parameters which occur more frequently.
- event propagation in networks with high degree of interconnectivity which leads to unforeseen fluctuations.
- non-equilibrium states which increasingly characterize daily business.

These multi-scale changes deeply influence logistic target achievement and call for robust planning and control strategies. Therefore, understanding the cause and effects of multi-scale changes in production networks is of major interest in order to achieve robustness in respect of stabilizing and sustaining systems performance. New methodological approaches from different science disciplines are promising to contribute to a new level of comprehension of network processes. Unconventional methods from biology, perturbation ecology, or auditory display are gaining increasing importance as they are confronted with similar challenges. Advancements from the classical disciplines such as mathematics, physics, and engineering are of continuing importance.

This Lecture Notes Volume starts out with Part I "Interdisciplinary Approaches for Robustness in Manufacturing". The contributions presented in Part I cover interdisciplinary work between manufacturing research and a wide range of disciplines, such as systems biology, auditory display, network sciences, or nonlinear dynamics. Especially for today's global manufacturing systems, interdisciplinary research offers a possibility to tackle research questions that arise due to the interplay between a need for robustness and a growing system complexity in manufacturing. As for instance shown in the first paper of Part I by Beber et al., strong parallels exist between manufacturing and metabolic systems. This justifies the application of methods from systems biology, which are designed to cope with the complexity of natural systems (i.e., cells) and offer possibilities to analyze and describe system robustness. Further, it is shown in Part I by Iber et al. that the analysis of manufacturing feedback data with methods from auditory display can identify causes and impacts of certain parameters in complex manufacturing networks which graphical analysis is not able to. This can support and contribute to an increasing robustness of manufacturing processes.

Part II "Robust Manufacturing Control Methods" addresses the issue of how important it is to have novel tools and approaches, which enable manufacturers to keep their high performance in today's unpredictable market conditions. Techniques from three different areas are presented. First, several scheduling methods are described. Scheduling is a well-known problem, which has been extensively studied in the literature. However, manufacturing systems nowadays are highly complex and also often highly automated and therefore further advancements are necessary. Moreover, as production systems have to face sudden changes and fluctuations, innovative robust scheduling procedures are needed. Second, this part also presents methods related to the concept of autonomous control. Granting various logistic objects decision-making abilities could lead to increased robustness of the systems. Third, the part finishes with data mining techniques, which can be used in order to discover knowledge from databases. Such tools are commonly applied in many fields and their use is also growing in manufacturing and logistics. Data mining algorithms can be very beneficial in a complex manufacturing environment, where numerous parameters are involved. For example, they can be utilized to form different product families or to generate production planning rules.

The central topic of the contributions summarized in Part III is "Robustness in Manufacturing Networks and Adaptable Logistics Chains". All contributions focus on the fact that the majority of nowadays manufacturing companies organize their production in a production network: suppliers, manufacturing sites, distribution hubs, and customers are spread around the whole world. Challenges that companies are faced with and solutions to problems that they encounter if they want to keep their production network robust and adaptable are presented here. Within this overarching range, contributions in Part III address several different problems: first, methods to design, configure, or plan robust production networks are presented. This is followed by contributions that deal with the issue of quality management as a means to achieve robustness in global production networks. Part III further includes contributions on collaboration, coordination, and adaptability within global production networks. It concludes with contributions that address questions of decentralized manufacturing, putting also a focus on environmental impacts and issues.

Part IV "Process Optimization and Strategic Approaches toward Robustness" presents a selection of papers, which elaborate on diverse aspects of robust manufacturing control. First, companies should establish adaptable production processes, which are able to operate under changing market conditions. Manufacturers need to ensure that their logistics performance matches the requirements of the customers in terms of delivery reliability, for example. Therefore, concepts such as productivity of the production processes, the level of decentralization of production control, and optimization of the decision-making procedures in production planning and control are of high importance and are addressed in some of the papers in this part. In addition to looking into their processes, manufacturers should also carefully select their strategies. They need to develop manufacturing and strategic flexibilities, which enable them to have strategies of higher robustness. Finally, it is argued that enterprises should also consider the trade-off between robustness and efficiency when making their strategic decisions.

I would like to express my gratitude to all authors contributing to the conference as well as to all participants of the conference making this event successful. Moreover, I would like to thank the members of the program committee for their valuable comments in the respective reviews. In particular, I cordially thank Professor Neil A. Duffie and Professor Hans-Peter Wiendahl acting as editorial committee members for their highly appreciated recommendations and advice on how to prepare and run an international conference as RoMaC 2012 was the first conference ever organized by the Global Production Logistics workgroup at Jacobs University. I am explicitly grateful to Stanislav Chankov and Mirja Meyer who are research associates in my workgroup for their valued assistance in organizing and double checking all paper-relevant processes including the conference preparation. And finally, I thank Silke Tilgner for her high engagement in the conference planning and organization.

I very much hope that with this conference Robust Manufacturing Control was started as a topic on its own and will get further consideration in the future.

Katja Windt
Professor of Global Production Logistics
School of Engineering and Science
Jacobs University
Chair of CIRP sponsored Conference on Robust Manufacturing Control 2012

Acknowledgments

The research of Katja Windt is supported by the Alfried Krupp Prize for Young University Teachers of the Alfried Krupp von Bohlen und Halbach-Foundation.

Contents

Part III Robustness in Manufacturing Networks and Adaptable Logistics Chains

Editorial Committee

Katja Windt Prof. Dr.-Ing. Katja Windt studied mechanical engineering at the University of Hannover. Her studies also involved a stay as a visiting scholar at the Massachusetts Institute of Technology (MIT). She received her diploma in 1995 and continued to work with the Institute of Production Systems and Logistics (IFA) at the Leibniz University of Hannover as a research associate and Ph.D.-student, receiving her Ph.D. in 2000. From 2001 to 2007, she worked as a post-doctoral researcher and manager of the department "Intelligent Planning and Control Methods for Logistics Systems" at the Bremen Institute of Production and Logistics (BIBA). Since 2008 she has been working as a Professor of Global Production Logistics at Jacobs University Bremen.

Hans-Peter Wiendahl Prof. Dr.-Ing. E.h. mult. Dr. sc. h.c. Dr.-Ing. Hans-Peter Wiendahl studied mechanical engineering in Dortmund and Aachen. After one-year research visit at the Massachusetts Institute of Technology (MIT), he started as a research associate and Ph.D.-student at the "Laboratory for Machine Tools and Production Engineering" at RWTH Aachen, from which he graduated as a Ph.D. in 1970. After Post-Doc work in Aachen, Prof. Wiendahl gained extensive industry experience, before he was appointed professor and chair of the Institute of Production Systems and Logistics (IFA) at the Leibniz University of Hannover in 1979. Prof. Wiendahl has been a member of the International Academy for Production Engineering (CIRP) since 1989. He is an honorary doctor of several universities and in 2003 he became an emeritus professor at the Leibniz University of Hannover. In 2010 he received the Society of Manufacturing Engineers (SME) Gold Medal.

Neil Duffie Prof. Neil A. Duffie earned his Bachelor (1972), Master (1974) and Ph.D. (1980) degrees from the University of Wisconsin-Madison, where he now works as a professor at the department of mechanical engineering within the college of engineering. His research and teaching focus on the design and control of manufacturing systems. He is mainly interested in the integration of sensors, actuators, and data sources in highly automated, non-hierarchically controlled production systems. Prof. Duffie became a member of the International Academy for Production Engineering (CIRP) in 1999. He has been a fellow of the American Society of Mechanical Engineers (ASME) since 2006 and has received several awards.

International Program Committee

Dieter Armbruster School of Mathematical and Statistical Sciences, Arizona State University, Tempe, AZ 85287-1804, USA

Werner Bergholz School of Engineering and Sciences, Jacobs University Bremen, Campus Ring 1, 28759 Bremen, Germany

Alexandra Brintrup CABDyN Complexity Centre, Said Business School, University of Oxford, Park End Street, Oxford, OX1 1HP, United Kingdom

George Chryssolouris Laboratory for Manufacturing Systems and Automation, Department of Mechanical Engineering and Aeronautics, University of Patras, Greece

Pedro Cunha Escola Superior der Tecnologia de Setúbal, Instituto Polytécnico de Setúbal, Estefanilha I 2910-761 Setúbal, Portugal

Jochen Deuse Chair of Industrial Engineering, Technische Universität Dortmund, Leonhard-Euler-Str. 5, 44227 Dortmund, Germany

Hoda ElMaraghy Intelligent Manufacturing Systems (IMS) Centre, University of Windsor 401 Sunset Avenue, 204 Odette Building, Windsor, ON Canada N9B 3P4, Canada

Fred van Houten Chair of Design Engineering, University of Twente, Gebouw Horstring Kamer N 202, Postbus 217, 7500 AE Enschede, The Netherlands

Michael Hülsmann Systems Management Workgroup, Jacobs University Bremen, Campus Ring 1, South Hall, Room 140, 28759 Bremen, Germany

Marc-Thorsten Hütt Computational Systems Biology Group, Jacobs University Bremen, Campus Ring 1, 28759 Bremen, Germany

Gisela Lanza Karlsruhe Institute of Technology, Kaiserstaße 12, Karlsruhe, 76131 Germany

Giacomo Liotta Institute of Industrial Technologies and Automation, National Research Council (CNR), Stradadella Neve, via Salaria Km 29.300, 00010 Montelibretti, Roma, Italy

Hermann Lödding Institut für Produktionsmanagement und –technik, Technische Universität Hamburg-Harburg, Denickestraße 17 (L), 21073 Hamburg, Germany

Paul Maropoulos The Innovative Design & Manufacturing Research Centre, The University of Bath, Bath BA2 7AY, United Kingdom

Horst Meier Lehrstuhl für Produktionssysteme, Ruhr-Universität Bochum, Universitätsstraße 150, D-44780 Bochum, Germany

László Monostori MTA SZTAKI, Hungarian Academy of Sciences, H-1111 Kendeutca 13–17, Budapest, Hungary

Peter Nyhuis Institut für Fabrikanlagen und Logistik, Leibniz Universität Hannover, An der Universität 2, 30823 Garbsen, Germany

Marcel Oliver School of Engineering and Science, Jacobs University Bremen, Campus Ring 1, 28759 Bremen, Germany

Bernd Scholz-Reiter BIBA – Bremer Institut für Produktion und Logistik GmbH, Universität Bremen, Hochschulring 20, 28359 Bremen, Germany

Paul Schönsleben Departement Management, Technologie und Ökonomie, Eidgenössische Technische Hochschule (ETH) Zürich, Weinbergsstrasse 56/58, 8092 Zürich, Switzerland

Günther Schuh Laboratory for Machine Tools and Production Engineering (WZL) RWTH Aachen University, Steinbachstr. 19 52074 Aachen, Germany

Wilfried Sihn Institut für Managementwissenschaften, Technische Universität Wien, Theresianumgasse 27, 1040 Wien, Austria

Dieter Spath Fraunhofer-Institut für Arbeitswirtschaft und Organisation (IAO), Nobelstraße 12, 70569 Stuttgart, Germany

Meike Tielebein Institut für Diversity Studies in den Ingenieurwissenschaften, Universität Stuttgart, Pfaffenwaldring 9, 70569 Stuttgart, Germany

Tullio Tolio Institute of Industrial Technologies and Automation, National Research Council (CNR), Via Bassini, 15 - 20133 Milano MI Lombardia, Italy

Kanji Ueda Research into Artifacts Center for Engineering, University of Tokyo, 5-1-5 Kashiwanoha, Kashiwa, Chiba 277-8568, Japan

Anna Valente Dipartimento di Meccanica, Politecnico di Milano, Campus Bovisa Sud - via La Masa 1, 20156 Milano, Italy

József Váncza Research Laboratory of Engineering and Management Intelligence, Hungarian Academy of Sciences, MTA SZTAKI, H-1111 Budapest XI. Kende u. 13–17, Hungary

Contributors

Dieter Armbruster School of Mathematical and Statistical Sciences, Arizona State University, Tempe, AZ 85287-1804, USA, e-mail: armbruster@asu.edu

Fabian Bachmann Laboratory for Machine Tools and Production Engineering (WZL), RWTH Aachen University, Steinbachstr. 19, 52074 Aachen, Germany, e-mail: F.Bachmann@wzl.rwth-aachen.de

Maik Bähr Institute for Transport and Economics, Dresden University of Technology, Würzbuger Str. 35, 01187 Dresden, Germany; Institute of Traffic Telematics, Dresden University of Technology, Andreas-Schubert-Str. 23, 01069 Dresden, Germany

Moritz Emanuel Beber Computational Systems Biology Group, School of Engineering and Science, Jacobs University, Campus Ring 1, D-28759 Bremen, Germany, e-mail: m.berber@jacobs-university.de

Sebastian Beck Institut für Fabrikanlagen und Logistik, Leibniz Universität Hannover, Hannover, Germany, e-mail: beck@ifa.uni-hannover.de

Werner Bergholz Electrical Engineering & Computer Science, School of Engineering and Sciences, Jacobs University Bremen, Campus Ring 1, 28759 Bremen, Germany, e-mail: w.bergholz@jacobs-university.de

Christoph Besenfelder Chair of Factory Organization, Technische Universität Dortmund, Leonard-Euler-Street 5, 44227 Dortmund, Germany, e-mail: besenfelder@lfo.tu-dortmund.de

Fabian Bohnen Chair of Industrial Engineering, Technische Universität Dortmund, Dortmund, Germany, e-mail: fabian.bohnen@tu-dortmund.de

Johannes Book Institute of Production Sciences, Karlsruhe Institute of Technology, Kaiserstaße 12, 76131 Karlsruhe, Germany, e-mail: Johannes.Book@kit.edu

Felix Brambring Laboratory for Machine Tools and Production Engineering (WZL), RWTH Aachen University, Steinbachstr. 19, 52074 Aachen, Germany, e-mail: f.brambring@wzl.rwth-aachen.de

Alexandra Brintrup CABDyN Complexity Centre, Sa Business School, University of Oxford, OX1 1HP Oxford, UK, e-mail: alexandra.brintrup@sbs.ox.ac.uk

Emanuele Carpanzano Institute of Industrial Technologies and Automation (ITIA), National Research Council of Italy (CNR), Via Bassini 15, 20133 Milano, Italy, e-mail: emanuele.carpanzano@itia.cnr.it

Amedeo Cesta Institute of cognitive sciences and technologies (ISTC), National Research Council of Italy (CNR), Via S. Martino della Battaglia 44, 00185 Roma, Italy, e-mail: amedeo.cesta@istc.cnr.it

George Chryssolouris Laboratory for Manufacturing Systems and Automation, Department of Mechanical Engineering and Aeronautics, University of Patras, Patras, Greece, e-mail: xrisol@lms.mech.upatras.gr

Richard Colmorn Systems Management Workgroup, School of Engineering and Sciences, Jacobs University Bremen, Campus Ring 1, 28759 Bremen, Germany, e-mail: r.colmorn@jacobs-university.de

Philip Cordes Systems Management Workgroup, School of Engineering and Sciences, Jacobs University Bremen, Campus Ring 1, 28759 Bremen, Germany, e-mail: p.cordes@jacobs-university.de

Wilhelm Dangelmaier Business Computing, Heinz Nixdorf Institute, University of Paderborn, Fuerstenallee 11, Paderborn 33102, Germany, e-mail: wilhelm.dangelmaier@uni-paderborn.de

Jochen Deuse Chair of Industrial Engineering, Technische Universität Dortmund, Dortmund, Germany, e-mail: jochen.deuse@tu-dortmund.de

Reik V. Donner Research Domain IV-Transdisciplinary Concepts & Methods, Potsdam Institute for Climate Impact Research, Telegrafenberg A31, 14473 Potsdam, Germany, e-mail: reik.donner@pik-potsdam.de

Thomas Dorka Lehrstuhl für Produktionssysteme, Ruhr-Universität Bochum, Bochum, Germany, e-mail: dorka@lps.ruhr-uni-bochum.de

Michalis Doukas Laboratory for Manufacturing Systems and Automation, Department of Mechanical Engineering and Aeronautics, University of Patras, Patras, Greece

Neil Duffie Department of Mechanical Engineering, University of Wisconsin-Madison, Madison, USA, e-mail: duffie@engr.wisc.edu

Konstantinos Efthymiou Laboratory for Manufacturing Systems and Automation, Department of Mechanical Engineering and Aeronautics, University of Patras, Patras, Greece

John Fenske Department of Mechanical Engineering, University of Wisconsin-Madison, Madison, USA, e-mail: jfenske@alumni.iastate.edu

Markus Florian Fraunhofer Austria Research GmbH, Vienna, Austria, e-mail: florian@imw.tuwien.ac.at

Jasper Fonteijn Department of Mechanical Engineering, Eindhoven University of Technology, POB 513, 5600MB Eindhoven, The Netherlands

Thomas Froitzheim Laboratory for Machine Tools and Production Engineering (WZL), RWTH Aachen University, Steinbachstr. 19, 52074 Aachen, Germany, e-mail: T.Froitzheim@wzl.rwth-aachen.de

Sascha Fuchs Laboratory for Machine Tools and Production Engineering (WZL), RWTH Aachen University, Steinbachstr. 19, 52074 Aachen, Germany, e-mail: s.fuchs@wzl.rwth-aachen.de

Friedrich Gehler Gesellschaft für Technologie Transfer mbh, Hannover, Germany, e-mail: friedrich.gehler@gtt-online.de

Konstantinos Georgoulias Laboratory for Manufacturing Systems and Automation, Department of Mechanical Engineering and Aeronautics, University of Patras, Patras, Greece

Henrik Gommel Institut für Managementwissenschaften, Technische Universität Wien, Vienna, Austria

Norbert Gronau Chair of Business Information Systems, University of Potsdam, August-Bebel-Str, 14482 Potsdam, Germany, e-mail: norbert.gronau@wi.uni-potsdam.de

Hans-Christian Haag Institut für Diversity Studies in den Ingenieurwissenschaften, Universität Stuttgart, Stuttgart, Germany, e-mail: hans-christian.haag@ids.uni-stuttgart.de

Annika Hauptvogel Laboratory for Machine Tools and Production Engineering (WZL), RWTH Aachen University, Steinbachstr. 19, 52074 Aachen, Germany, e-mail: a.hauptvogel@wzl.rwth-aachen.de

Carlo Hausberg Laboratory for Machine Tools and Production Engineering (WZL), RWTH Aachen University, Steinbachstr. 19, 52074 Aachen, Germany, e-mail: c.hausberg@wzl.rwth-aachen.de

Jens Heger BIBA—Bremer Institut für Produktion und Logistik GmbH, Universität Bremen, Hochschulring 20, 28359 Bremen, Germany, e-mail: heg@biba.uni-bremen.de

Torsten Hildebrandt BIBA—Bremer Institut für Produktion und Logistik GmbH, Universität Bremen, Hochschulring 20, 28359 Bremen, Germany, e-mail: hil@biba.uni-bremen.de

Manuel Hirsch DITF Denkendorf—Centre for Management Research, Denkendorf, Germany, e-mail: manuel.hirsch@ditf-mr-denkendorf.de

Sho Hosokawa The University of Tokyo, 7-3-1 Hongo, Tokyo 113-8658, Japan, e-mail: hosokawa@css.t.u-tokyo.ac.jp

Michael Hülsmann Systems Management Workgroup, School of Engineering and Sciences, Jacobs University Bremen, Campus Ring 1, 28759 Bremen, Germany, e-mail: m.huelsmann@jacobs-university.de

Marc-Thorsten Hütt Computational Systems Biology Group, School of Engineering and Science, Jacobs University, Campus Ring 1, D-28759 Bremen, Germany, e-mail: m.huett@jacobs-university.de

Michael Iber Global Production Logistics Workgroup, School of Engineering and Sciences, Jacobs University Bremen, Campus Ring 1, 28759 Bremen, Germany, e-mail: m.iber@jacobs-university.de

Christioph Illigen Systems Management Workgroup, School of Engineering and Sciences, Jacobs University Bremen, Campus Ring 1, 28759 Bremen, Germany, e-mail: c.illigen@jacobs-university.de

Sandra Kaczmarek Chair of Factory Organization, Technische Universität Dortmund, Leonard-Euler-Street 5, 44227 Dortmund, Germany, e-mail: kaczmarek@lfo.tu-dortmund.de

Ekaterina Kaganova International Graduate School–Dynamic Intelligent Systems, University of Paderborn, Fuerstenallee 11, Paderborn 33102, Germany, e-mail: kaganova@hni.uni-paderborn.de

Kyle Kippenbrock Institute of Production Sciences, Karlsruhe Institute of Technology, Kaiserstaße 12, 76131 Karlsruhe, Germany

Mathias Knollmann Global Production Logistics Workgroup, School of Engineering and Sciences, Jacobs University Bremen, Campus Ring 1, 28759 Bremen, Germany, e-mail: M.Knollmann@jacobs-university.de

Benedikt Konrad Chair of Industrial Engineering, Technische Universität Dortmund, Dortmund, Germany, e-mail: benedikt.konrad@tu-dortmund.de

Herbert Kopfer Chair of Logistics, University of Bremen, Wilhelm-Herbst-Straße 5, Bremen 28359, Germany, e-mail: kopfer@uni-bremen.de

Benjamin Korsmeier Systems Management Workgroup, School of Engineering and Sciences, Jacobs University Bremen, Campus Ring 1, 28759 Bremen, Germany, e-mail: b.korsmeier@jacobs-university.de

Mirko Kück BIBA—Bremer Institut für Produktion und Logistik GmbH, Universität Bremen, Hochschulring 20, 28359 Bremen, Germany, e-mail: kue@biba.uni-bremen.de

Daniel Kupke Laboratory for Machine Tools and Production Engineering (WZL), RWTH Aachen University, Steinbachstr. 19, 52074 Aachen, Germany, e-mail: D.Kupke@wzl.rwth-aachen.de

Philipp von Lamezan University of Bremen, Bibliothekstraße 1, 28359 Bremen, Germany, e-mail: p.v.lamezan@googlemail.com

Gisela Lanza Institute of Production Sciences, Karlsruhe Institute of Technology, Kaiserstaße 12, 76131 Karlsruhe, Germany

Sander Lass Chair of Business Information Systems, University of Potsdam, August-Bebel-Str, 14482 Potsdam, Germany, e-mail: sander.lass@wi.uni-potsdam.de

Armin Lau DITF Denkendorf—Centre for Management Research, Denkendorf, Germany, e-mail: armin.lau@ditf-mr-denkendorf.de

Daniel Lieber Chair of Industrial Engineering, Technische Universität Dortmund, Dortmund, Germany, e-mail: daniel.lieber@tu-dortmund.de

Klaus M. Liebler Supply Chain Engineering, Fraunhofer-Institute for Material Flow and Logistics, Dortmund, Germany, e-mail: Klaus.Liebler@iml.fraunhofer.de

Giacomo Liotta Institute of Industrial Technologies and Automation (ITIA), National Research Council of Italy (CNR), Via Bassini 15, 20133 Milano, Italy, e-mail: giacomo.liotta@itia.cnr.it

Fernando Marinò Institute of cognitive sciences and technologies (ISTC), National Research Council of Italy (CNR), Via S. Martino della Battaglia 44, 00185 Roma, Italy, e-mail: fernando.marin@istc.cnr.it

Heiko Matheis DITF Denkendorf—Centre for Management Research, Denkendorf, Germany, e-mail: heiko.matheis@ditf-mr-denkendorf.de

Horst Meier Lehrstuhl für Produktionssysteme, Ruhr-Universität Bochum, Bochum, Germany, e-mail: meier@lps.ruhr-uni-bochum.de

Mirja Meyer Global Production Logistics Workgroup, School of Engineering and Sciences, Jacobs University Bremen, Campus Ring 1, 28759 Bremen, Germany, e-mail: mi.meyer@jacobs-university.de

Max Monauni Graduate School of Excellence Advanced Manufacturing Engineering, University of Stuttgart, Stuttgart, Germany, e-mail: max.monauni@gsame.uni-stuttgart.de

Katharina Morik Chair of Artificial Intelligence, Technische Universität Dortmund, Dortmund, Germany, e-mail: katharina.morik@tu-dortmund.de

Dimitris Mourtzis Laboratory for Manufacturing Systems and Automation, Department of Mechanical Engineering and Aeronautics, University of Patras, Patras, Greece, e-mail: mourtzis@lms.mech.upatras.gr

Beata Mrugalska Faculty of Engineering Management, Poznań University of Technology, Poznań, Poland

Nariaki Nishino The Graduate School of Engineering, Department of Technology Management for innovation, The University of Tokyo, 7-3-1 Hongo, Tokyo 113-8658, Japan

Peter Nyhuis Institut für Fabrikanlagen und Logistik, Leibniz Universität Hannover, Hannover, Germany, e-mail: nyhuis@ifa.uni-hannover.de

Andrea Orlandini Institute of Industrial Technologies and Automation (ITIA), National Research Council of Italy (CNR), Via Bassini 15, 20133 Milano, Italy, e-mail: andrea.orlandini@itia.cnr.it

Nikolaos Papakostas Laboratory for Manufacturing Systems and Automation, Department of Mechanical Engineering and Aeronautics, University of Patras, Patras, Greece

Till Potente Laboratory for Machine Tools and Production Engineering (WZL), RWTH Aachen University, Steinbachstr. 19, 52074 Aachen, Germany, e-mail: t.potente@wzl.rwth-aachen.de

Foivos Psarommatis Laboratory for Manufacturing Systems and Automation, Department of Mechanical Engineering and Aeronautics, University of Patras, Patras, Greece

Christian Rainer Department of Economics and Business Management, University of Leoben, Franz-Josef-Straße 18, 8700 Leoben, Austria, e-mail: christian.rainer@wbw.unileoben.ac.at

Riccardo Rasconi Institute of cognitive sciences and technologies (ISTC), National Research Council of Italy (CNR), Via S. Martino della Battaglia 44, 00185 Roma, Italy, e-mail: riccardo.rasconi@istc.cnr.it

Anamika Saxena Indian Institute of Technology, Kharagpur, West Bengal, India

Sebastian Schmitz Laboratory for Machine Tools and Production Engineering (WZL), RWTH Aachen University, Steinbachstr. 19, 52074 Aachen, Germany, e-mail: s.schmitz@wzl.rwth-aachen.de

Bernd Scholz-Reiter BIBA—Bremer Institut für Produktion und Logistik GmbH, Universität Bremen, Hochschulring 20, 28359 Bremen, Germany

Jörn Schönberger Chair of Logistics, University of Bremen, Wilhelm-Herbst-Straße 5, Bremen 28359, Germany, e-mail: jsb@uni-bremen.de

Günther Schuh Laboratory for Machine Tools and Production Engineering (WZL), RWTH Aachen University, Steinbachstr. 19, 52074 Aachen, Germany, e-mail: G.Schuh@wzl.rwth-aachen.de

Thomas Seidel Institute for Transport and Economics, Dresden University of Technology, Würzbuger Str. 35, 01187 Dresden, Germany; AMC Managing Complexity GmbH, An der Tongrube 1-3, 40789 Monheim am Rhein, Germany

Wilfried Sihn Institut für Managementwissenschaften, Technische Universität Wien, Vienna, Austria

Steffen Sowade BIBA—Bremer Institut für Produktion und Logistik GmbH, Universität Bremen, Hochschulring 20, 28359 Bremen, Germany, e-mail: sow@biba.uni-bremen.de

Marco Stolpe Chair of Artificial Intelligence, Technische Universität Dortmund, Dortmund, Germany, e-mail: marco.stolpe@tu-dortmund.de

Hanna Theuer Chair of Business Information Systems, University of Potsdam, August-Bebel-Str, 14482 Potsdam, Germany

Christina Thomas Laboratory for Machine Tools and Production Engineering (WZL), RWTH Aachen University, Steinbachstr. 19, 52074 Aachen, Germany, e-mail: c.thomas@wzl.rwth-aachen.de

Meike Tilebein Institut für Diversity Studies in den Ingenieurwissenschaften, Universität Stuttgart, Stuttgart, Germany, e-mail: meike.tilebein@ids.uni-stuttgart.de

Tullio Tolio Institute of Industrial Technologies and Automation (ITIA), National Research Council of Italy (CNR), Via Bassini 15, 20133 Milano, Italy

Michael Toth Supply Chain Engineering, Fraunhofer-Institute for Material Flow and Logistics, Dortmund, Germany, e-mail: Michael.Toth@iml.fraunhofer.de

Marcello Urgo Manufacturing and Production Systems Division, Mechanical Engineering Department, Politecnico di Milano, Milano, Italy, e-mail: marcello.urgo@mecc.polimi.it

Yilmaz Uygun Chair of Factory Organization, Technische Universität Dortmund, Leonard-Euler-Street 5, 44227 Dortmund, Germany, e-mail: uygun@lfo.tu-dortmund.de

Madhu Vadali Department of Mechanical Engineering, University of Wisconsin-Madison, Madison, USA, e-mail: vadali@wisc.edu

Anna Valente Dipartimento di Meccanica, Politecnico di Milano, Via Bassini 15, 20133 Milano, Italy, e-mail: anna.valente@itia.cnr.it

Rawina Varandani Laboratory for Machine Tools and Production Engineering (WZL), RWTH Aachen University, Steinbachstr. 19, 52074 Aachen, Germany, e-mail: R.Varandani@wzl.rwth-aachen.de

Matt Wienke School of Mathematical and Statistical Sciences, Arizona State University, Tempe, AZ 85287-1804, USA

Katja Windt Global Production Logistics Workgroup, School of Engineering and Sciences, Jacobs University Bremen, Campus Ring 1, 28759 Bremen, Germany, e-mail: K.Windt@jacobs-university.de

Part I
Interdisciplinary Approaches for Robustness in Manufacturing

How Do Production Systems in Biological Cells Maintain Their Function in Changing Environments?

Moritz Emanuel Beber and Marc-Thorsten Hütt

Abstract Metabolism is a fascinating natural production and distribution process. Metabolic systems can be represented as a layered network, where the input layer consists of all the nutrients in the environment (raw materials entering the production process in the cell), subsequently to be processed by a complex network of biochemical reactions (middle layer) and leading to a well-defined output pattern, optimizing, e.g., cell growth. Mathematical frameworks exploiting this layered-network representation of metabolism allow the prediction of metabolic fluxes (the cell's 'material flow') under diverse conditions. In combination with suitable minimal models it is possible to identify fundamental design principles and understand the efficiency and robustness of metabolic systems. Here, we summarize some design principles of metabolic systems from the perspective of production logistics and explore, how these principles can serve as templates for the design of robust manufacturing systems.

Keywords Systems biology · Metabolic networks · Enzymes · Design principles · Simulated evolution

This contribution was previously published in Logistics Research (2012) pp. 79–87. DOI: 10.1007/s12159 012-0090-0.

M. E. Beber (✉) · M.-T. Hütt
School of Engineering and Science, Jacobs University,
Campus Ring 1, D-28759 Bremen, Germany
e-mail: m.berber@jacobs-university.de

M.-T. Hütt
e-mail: m.huett@jacobs-university.de

DOI: 10.1007/978-3-642-30749-2_1, © Springer-Verlag Berlin Heidelberg 2013

1 Introduction

There is a deep intrinsic parallel between the metabolism of biological cells and industrial production. Cells function efficiently under typical environmental conditions. At the same time, they are viable (thus maintaining a certain level of function) across a vast range of atypical environments. It is precisely this robustness with respect to large changes (and significant fluctuations) in the composition of the environment (the 'input pattern') that makes metabolic networks a potentially very interesting role model for technical production and distribution systems (see, e.g., [1]).

The network of metabolic reactions in a cell is responsible for providing a wide range of substances at the right time in the right proportions for a specific purpose of consumption. At the same time, metabolic systems construct complicated chemical substances out of nutrients taken up from the environment. With several thousands of interacting machines (enzymes, catalyzing biochemical reactions) the underlying production network is about as complex as the most involved processes of industrial production. The key challenges are comparable: How do systems in both domains ensure robustness with respect to perturbations? How can these systems react rapidly to important changes in their environment by ensuring the achievement of the logistics targets? For metabolism, the young scientific discipline addressing these questions in a strong interplay between mathematical approaches and experimental efforts is called Systems Biology. It is a 'melting pot' of many scientific fields, contributing to the understanding of the larger-scale organization of living cells and their dynamic behavior in response to external and internal stimuli, including disease development (see, e.g., [2–4]). Systems Biology is situated at the intersection between the Biological Sciences, Mathematics, Statistical Physics, Biophysics, and Computer Science. For the part of Systems Biology discussed here, the principal aim is not the representation of a cell in a computer, but rather it is about understanding function beyond the level of a few elements: How is robustness achieved? How can a system react rapidly to changes in the environment?

The parallel between metabolism and manufacturing has been emphasized by others before (e.g., [5, 6]; see also [7]). Systems Biology has over the last 6–8 years provided a remarkable basis for a more refined, detailed and quantitative comparison of these two realms. In the present paper we repeat some of the arguments from [8] and briefly review two articles exploring abstract model representations of metabolic systems [1, 9].

Our focus here is on metabolism as a potential 'template' for manufacturing systems. Other biological principles, like adaptation, self-organization and aspects of biological evolution have been explored to allow manufacturing systems to deal with environmental variability and internal fluctuations. Two important examples are the framework of Biological Manufacturing Systems (see, e.g., [10, 11]) and the idea of Emergent Synthesis [12], which suggests that regulation on all scales requires integration in a self-organized fashion.

The aim of this paper is to review some material from Systems Biology on the functioning of metabolism and then show, how abstract model representations of metabolic systems can serve as a starting point for transferring metabolic design principles to industrial production. We first describe the general features of metabolic systems that form the basis of a comparison with industrial production (Sect. 2). Next, we discuss a broad range of recently identified metabolic design principles of interest to manufacturing (Sect. 3). In Sect. 4 we then explore the possibility of constructing abstract mode l representations of metabolic systems that are suitable interfaces between Systems Biology and industrial production, helping us to transfer such knowledge into manufacturing contexts. Lastly, in Sect. 5 we discuss, how such biological understanding, in particular of design principles of metabolic systems, can serve as templates for robust technical and industrial systems.

2 Metabolism From a Production Logistics Perspective

Metabolism is at the same time a transportation network, an assembly line, and a storage depot. Substances are taken up from the environment (by exchange reactions) and distributed in the cellular compartments (by transport reactions). Large parts of metabolism are responsible for degrading complex substances into more elementary building blocks (catabolism). These chemical building blocks are used in the formation of more complex compounds (anabolism) that are needed for cellular maintanance, growth or storage. The elementary organizational unit of metabolism is the individual biochemical reaction, often represented by the enzyme (or enzyme complex) serving as catalyst for a reaction. Qualitatively speaking, the exchange reactions can be regarded as an input layer, followed by a complex intracellular processing layer. In many modeling approaches the overall goal of metabolic function is abstracted as a (fictitious) biomass reaction, where each component entering this reaction is known to contribute to cell growth. Figure 1 (left) summarizes this situation.

The flow of substances through the metabolic network is the cell's equivalent of the complex material flows encountered in industrial production. The enzymes represent machines responsible of constructing well-defined products out of a specific set of incoming materials.

The appropriate mathematical tools for analyzing successful configurations of metabolic systems on the scale of a whole cell (rather than an individual metabolic pathway) are constraint-based modeling and, more specifically, flux-balance analysis (FBA), reviewed, e.g., in [14, 15]. FBA can be used to predict metabolic flux distributions (the biological equivalent of material flow) under various nutrient input patterns and for diverse cellular objective functions (serving as the output pattern of the system maximized during flux-balance analysis).

Within the elegant framework of flux-balance analysis, the optimal steady-state distribution of metabolic fluxes can be predicted, given the structure of the

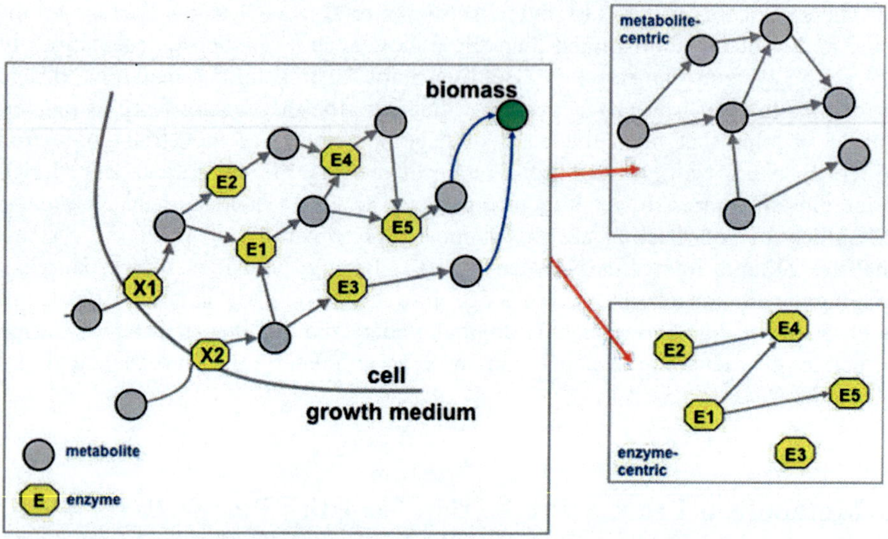

Fig. 1 Network representations of cellular metabolism (schematic view), together with the projection of the bipartite representation of metabolism (*left*) to a metabolite-centric graph (*right*; *top*) and to an enzyme-centric graph (*right*; *bottom*). Figure adapted from [13]

environment (the availability of nutrients) and the cellular objective function (e.g., biomass production or ATP maximization). The objective function, serving as the output pattern of the system, is maximized during flux-balance analysis. FBA has a similar methodological core to many optimization problems in logistics, namely linear programming. It is capable of serving as an interface between the biological and biochemical foundations of metabolic systems and the representation (and conceptual understanding) of metabolism as a complex network.

In virtually all Systems Biology studies, statistical methods play an important role in identifying the intrinsic mechanisms behind the performance of a system. More precisely, the analysis of system-wide information with any statistical methods requires a clear concept of a null hypothesis, a random background, against which the observations can be compared. Expressing levels of cellular organization in terms of networks has turned out to be particularly helpful for the task of formulating the appropriate null models. The strength of graph theory is that it can represent a complex system in a unified formal language of nodes and links.

A suitable network representation of metabolic systems is a bipartite graph, i.e. a graph with two types of nodes (here: metabolites and enzymes) interacting in an alternating fashion. Typically, projections of this bipartite graph are discussed: a metabolite-centric projection, where two metabolites are linked, if an enzyme catalyzes the conversion from one to the other; a enzyme-centric projection, where two enzymes are linked, if they share a common metabolite. Figure 1 illustrates the three network representations of metabolism.

At this point, we wish to emphasize that in spite of the apparent simplicity and autonomy of metabolism as a network converting an input vector (nutrients) into an output vector (biomass), metabolism is also embedded in an intricate system of regulation, the gene-regulatory network and the regulatory action of genome structure (see, e.g., [16]).

How are complex networks characterized? The degree of a node is the number of links entering or leaving this node. It is thus the number of direct neighbors of a node. A network with a degree distribution given by a power law is called 'scalefree', because such networks do not contain a particular scale of reference: There is no typical (e.g., average) degree of a node. The degree is a node property spread over several orders of magnitude. In a scalefree network, the vast majority of nodes only have very few links. At the same time, the network contains nodes, which have several orders of magnitude more links. These hubs are topologically the most important system components. It is surprising that many natural and technical systems seem to follow this scheme (see, e.g., [17]).

Remarkably, the degree distribution of the metabolite-centric graph approximates a power law. Thus, the most appropriate random network representation of this graph is a scalefree graph [18]. Early studies on metabolic network topologies mostly focused on this broad degree distribution of the metabolite-centric graph [19–21]. In contrast, the degree distribution of the enzyme-centric projection has a rather narrow peak with no heavy tail, and thus more resembles an Erdős-Rényi (ER) random graph [22].

All these observations are interesting starting points for a quantitative comparison of metabolic systems and industrial manufacturing systems. While many characteristics of network architecture are similar between the two types of production systems (metabolic and industrial), in [23] striking differences on the level of dynamical quantities (in particular, material flows) are observed. More details on all of the topics listed in this Section can be found in [8].

3 Metabolic Design Principles

Metabolic networks are at the same time scale-free [19], modular [24], layered (one can for example distinguish between the input layer given by the set of uptake reaction, the collection of all reactions directly contributing to biomass as an output layer, and a 'processing layer' consisting of all reactions in between) and bipartite (with metabolites and reactions/enzymes forming the two sets of nodes). Their projections contain two categories of bidirectional links (a bidirectional link can be truly bidirectional or the co-occurrence of two opposite unidirectional links), as well as substantial degree correlations (see, e.g., [25]). These complications all make it a formidable challenge to explore the interrelations between network topology and dynamical function for metabolic systems [26, 27].

Many attempts have been formulated over the last seven years to understand the structure of metabolic networks from first principles using evolutionary or biochemical arguments [28–32]. Robustness of metabolic networks beyond the single-knockout level (i.e., with more severe perturbations than the loss of a single enzyme) has been explored using flux-balance analysis [33] and elementary flux modes [34]. Several works have argued that the network topology of metabolic systems is markedly optimized for robustness. The study by Marr et al. [35] uses binary probes to measure, whether fluctuations are on average dampened out or enhanced on metabolic network architectures.

There seems to be a selection for minimal metabolic pathways, given the environmental conditions (i.e., the set of available nutrients). The accessible nutrients for a species may thus be inferred by analyzing the network topologies [36].

Temperature differences of typical habitats correlate with structural differences in metabolic networks [37], a phenomenon that can be qualitatively reproduced in a simple model involving gene duplications [38]. On a more theoretical side, suitable definitions of a metabolic 'null model' have been formulated recently [39].

For the bacterium, *Escherichia coli*, the remarkable success of flux-balance analysis in predicting growth rates for mutants has been demonstrated by [40]: Even though FBA over-estimates the initial growth rate for most single-gene mutants, in many cases an adaptive evolution over several generations will allow the cells to converge to the computationally predicted growth rates.

A refined method [41], called 'minimization of metabolic adjustment' (MOMA) selects the mutant flux composition relative to the flux composition of the unperturbed system (the 'wildtype'). The view that MOMA rather predicts the initial growth rates of mutants, while FBA predicts the maximally achievable growth rates has been used to establish the concept of 'synthetic rescues' [42] (see also [43]): On the basis these growth-rate differences between MOMA and standard FBA predictions, *compensatory mutations* can be applied selectively to the differences between the underlying flux distributions predicted by MOMA and standard FBA, respectively. These compensatory mutations transform the lower MOMA growth rate into the higher FBA growth rate.

Among the approaches mentioned so far, three seem of particular interest for application to manufacturing systems: (1) elementary flux modes, which is the counting of the number of paths compliant with certain subsidiary conditions; in this way, the importance of a system component is represented by the number of paths it is involved in; this general principle may well be applicable to manufacturing (see [44] for a first attempt in this direction); (2) evaluating network robustness with simple dynamic probes; often, it is unfeasible to perform realistic dynamical simulations due to limited knowledge of the large number of parameters; the organization of binary dynamics on a network may provide a rapid orientation, which sites are prone to large fluctuations; (3) restoring a function lost due to component failure by compensatory perturbations, as in the case of synthetic rescues; the low-performance states a system settles into under component failure may be local optima, and additional targeted perturbations may be necessary to guide the system into a (more) global optimum; for metabolism, the various forms of flux-balance analysis are capable of providing clear practical guidelines for the compensatory perturbations.

4 Abstract Models of Metabolic Systems

On the technical level, in spite of the deep functional parallels between these two systems—material flow in processes of industrial production and metabolic flow in biochemical networks in cells—and the strong similarities in challenges and unsolved problems, methodological exchanges and attempts of quantitative comparison are difficult. They are in particular impeded by the lack of a common terminology and common formal representation of these systems. Identifying design principles in abstract model representations can provide guidelines, what signatures of, e.g., robustness to expect, how to search for them in data and how to transfer the 'structural essence' of an enhanced function (like robustness) into the other realm. Here we will briefly summarize our work on two such abstract model representations at the interface of metabolism and industrial production: generic flow networks and networks of cyclic machines.

4.1 Generic Flow Networks

Recently [9], we analyzed the successful networks in the Kaluza–Mikhailov model of evolved flow networks, exploring their topological properties in more detail than in the previous work [45–47]. The networks consist of three distinct layers: an input layer that may only connect to the intermediate layer of nodes; the middle layer that may interconnect and are also linked to the input and output nodes; output nodes that have only incoming links from the middle layer. The process invoked on these networks is one of flow distribution. A unit flux is applied to each input node. At each node the incoming flux is distributed equally over all outgoing links, before it is finally gathered again at the output nodes. For each network a prescribed *output matrix* (the 'target structure' during the simulated evolution) is determined in a random process, prescribing the proportions, in which each input should reach the outputs. The matrix is thus a set of output vectors, one per unit flux inserted in each input node.

Given a network with random initial links and a prescribed output pattern, two distinct goals are required of the evolutionary algorithm, yet both goals depend on the output pattern. The first goal, and thus the first phase of the evolution, is to adjust the topology of the network in such a way that its output matches the prescribed pattern. This is achieved with a simulated annealing that minimizes a quantity termed 'flow error', a sum of squares over the difference between elements of the actual output and the prescribed pattern. We call the first phase of the evolution 'pattern recognition'.

In a second phase of the evolution, the topology of the network is further altered to maintain an output with a flow error below a threshold in the presence of certain types of damage. The damage applied to these networks is either removal of links, removal of nodes, or small fluctuations of the connectivity of the network, leading

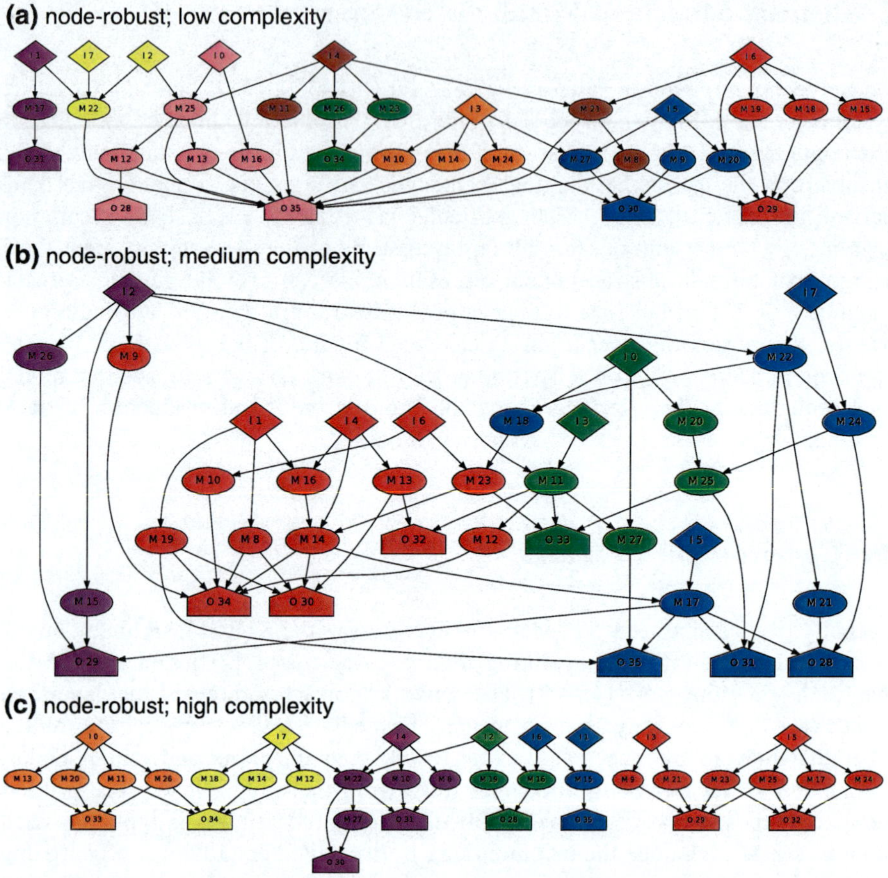

Fig. 2 Example of evolved noise-robust flow networks arising in the Kaluza–Mikhailov model with increasing output pattern complexity. Input nodes, middle nodes and output nodes are diamonds, ellipses and house-shaped polygons, respectively. Figure adapted from [9]

to 'link-robust', 'node-robust' and 'noise-robust' networks. We will refer to this second phase as 'robust pattern recognition'.

Example of evolved node-robust networks at three different 'complexities' of the output pattern (given by the orthogonality of the individual output vectors and thus measuring, how distinct the individual goals are that each output vector provides) are shown in Fig. 2.

The evolved flow networks show strikingly clear topological signatures that we can attribute to function (like modularity) and robustness (like the subgraph composition and degree correlations, depending on the specific type of robustness enhanced during the simulated evolution). In Fig. 2 the strong modularity, as well as the increase in modularity with output pattern complexity, are clearly visible.

From our perspective, the results of [9, 45–47] on the subgraph composition and the modularity of robust evolved flow networks suggest that a specific imprint of robustness and function in the network topology can also be expected for metabolic networks and manufacturing networks. On the basis of these studies we now come to a set of hypotheses relevant to general flow systems and potentially testable in empirical observations: (1) Output pattern complexity regulates the modularity of the successful networks. (2) If the network is robust against link removal, we should see a specific motif signature; if on the other hand the network is robust against node removal, we may expect negative degree correlations (even though we assume that other not yet identified topological features may also help characterizing node and noise robust networks and may even discriminate between node and noise robustness).

As already mentioned in the Introduction, one can think of metabolism as a layered flow system, much like the flow networks in the Kaluza–Mikhailov model. The input layer is given by the available nutrients (or, more specifically, by the list of uptake reactions capable of metabolizing those nutrients), while all reactions directly contributing to the cellular objective function (the required 'output pattern'; e.g., biomass production) can be summarized as output nodes. The robustness of metabolic systems against various forms of perturbations, as well as the modularity of metabolic systems both have been under intense investigation in systems biology (see, e.g., [15, 34, 48]). Across species, the size of the input layer and the diversity of the environments vary substantially. The impact of environmental diversity and other habitat properties on network architecture have been discussed from a variety of perspectives (see, e.g., [49]). The relevant 'pattern recognition' task for metabolic networks is to convert the diverse, given set of input patterns all into optimal outputs. We thus expect the modularity of the metabolic networks to positively correlate with the environmental diversity. Some evidence for this relationship can be seen in [36]. Additionally, we expect that biological evolution has enhanced the robustness of metabolic networks against the loss of enzymes, rather than the loss of metabolites. On the basis of the results from [9, 45–47], we therefore expect a very specific subgraph composition of metabolite-centric metabolic networks. Some evidence for a non-random subgraph signature of metabolic network is found in [25, 50]. Due to the proper selection of a metabolic null model (see also [25, 39]), the computation of a reliable motif signature of metabolic networks is non-trivial and has yet to be done.

4.2 Networks of Cyclic Machines

In [1] first steps towards a framework suitable for modeling scenarios from traffic, metabolism, and production logistics has been presented. In particular, cyclic machines (periodic devices) are taken as the basic constituents of the system and explore how they shape system behavior. Figure 3 summarizes the representation of machines and enzymes as periodic devices. A commonly used production logistics model, the throughput element, describes the different phases of

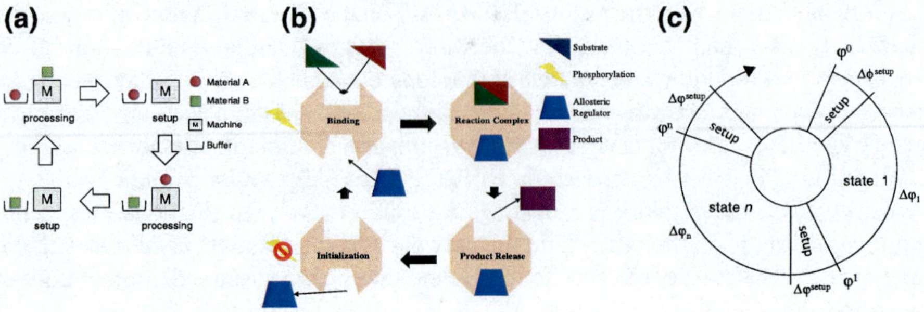

Fig. 3 Representation of production processes (**a**) and metabolic reactions (**b**) as periodic devices (**c**). Figure adapted from [1]

processing that are repeated in a specific operation [51]. The phases are divided into inter-operation time (consisting of transport time and waiting time of the material) and operation time (consisting of setup time and actual processing time of the machine). Therefore, a machine cycles through the two states of setup and operation for each production lot (Fig. 3a). The observation that enzymes can be represented as interacting cyclic machines capable of synchronization and collective behaviors goes back to [52]. Similarly, in order to depict the cyclic performance of allosteric enzymes that bind substrate and regulatory molecule, release product, and resume their initial state, Casagrande et al. [53] used a stochastic phase oscillator model (Fig. 3c). In fact, all catalytic enzymes that are returned to their initial state inherently have a cyclic nature (see Fig. 3b) that can be used for such an abstraction (and the vast majority of enzymes belong to this category). Generally, enzymes (or enzyme complexes) catalyze a specific reaction as long as the substrates are present at favorable concentrations, until there is a regulation event that prevents them from processing the compounds involved, or until they are removed from the cell by, for example, degradation.

What can be achieved with such an abstract model representation? Even though some of the organizational features are comparable, transportation, manufacturing and metabolism all are represented by very different network architectures and are controlled by very different regulatory systems. It is therefore not *a priori* clear that optimization methods from one domain can be applied successfully in another domain. A proof of principle has been given in [1], where a strategy (adaptive control) from one domain (traffic modeling, see [54, 55]) can also be successfully applied to the other domains (industrial production, metabolic systems).

Extending the formal view of periodic devices characterized by a phase variable to (empirical and simulated) data from a large-scale transportation system, in [56] the synchronization of arrival/departure events in the network of long-distance train connections has been analyzed. It has been shown that the performance (in a very general sense) of a given timetable of train connections is essentially determined by its phase pattern and thus the intrinsic levels of synchronization. The results show a clear and surprising negative correlation between the synchronization index of a station and its robustness to delays. This negative correlation between synchronization and robustness that was observed in the data could also be understood in a minimal model of delay propagation [56].

5 Conclusion

The main purpose of the present article is to emphasize the strong parallel between metabolic systems in a biological cell and processes of industrial production. Putting the diverse findings together that have emerged from Systems Biology investigations over the last decade, we start to understand, how production systems in biological cells compute efficient metabolic states under diverse environmental conditions. Specifically, we have made some progress over the last years in understanding some design principles of metabolic systems (e.g., [35, 57]) and making them accessible to industrial production [8, 23]. For addressing such 'transferable' metabolic design principles, we employed abstract model representations of metabolic systems [1, 9] and the analysis of biological data [58] in combination with flux-balance analysis [16, 57] and the exploration of metabolic networks with dynamic probes [35]. The next natural steps are to quantitatively compare material flow in both domains and understand the relationships between fluctuations in supply and demand on the one hand and material flow patterns (or effective network architectures) on the other.

From our perspective, the most important topic to be addressed jointly by the two disciplines, Systems Biology and Production Logistics, is systemic robustness. The balance between the antagonistic pair of requirements, efficiency and robustness, is of broad interest across many disciplines, ranging from industrial production to cell biology and ecology. Lack of robustness due to too high efficiency is related to the notion of systemic risk, which has recently been discussed from a theoretical perspective, for example in the context of complex economical systems (see, e.g., [59]).

For cellular processes this balance between efficiency and robustness has been explored in a multitude of ways resorting to both analysis of experimental data and the mathematical modeling of cellular processes. Motivated by graph theory and nonlinear dynamics, an influential trend in systems biology at the moment is to relate robustness to small regulatory devices [60, 61], serving e.g. as a noise buffer or providing a suitable amount of redundancy for maintaining systemic function even under perturbations.

With these thoughts we hope to contribute to the onset of a rich and stimulating dialogue between the two disciplines, Systems Biology and Production Logistics.

Acknowledgments MEB is supported by a Deutsche Forschungsgemeinschaft grant to MTH (grant HU-937/6). We are indebted to Nikolaus Sonnenschein (San Diego, USA) for providing his expertise on flux-balance analysis. We gratefully acknowledge discussions and close collaboration with Katja Windt (Bremen, Germany) on the parallels of metabolism and manufacturing.

References

1. Becker, T., Beber, M., Windt, K., Hütt, M., Helbing, D.: Flow control by periodic devices: a unifying language for the description of traffic, production, and metabolic systems. J. Stat. Mech. Theory Exp. **2011**, P05004 (2011)
2. Kitano, H.: Computational systems biology. Nature **420**, 206–210 (2002)
3. Palsson, B.: Systems biology: properties of reconstructed networks. Cambridge University Press, Cambridge (2006)
4. Kholodenko, B.: Cell-signalling dynamics in time and space. Nat. Rev. Mol. Cell Biol. **7**, 165–176 (2006)
5. Demeester, L., Eichler, K., Loch, C.H.: Organic production systems: what the biological cell can teach us about manufacturing. Manuf. Serv. Oper. Manage. **6**, 115–132 (2004)
6. Armbruster, D., Mikhailov, A.S., Kaneko, K.: Networks of Interacting Machines. World Scientific, Singapore (2005)
7. Helbing, D., Deutsch, A., Diez, S., Peters, K., Kalaidzidis, Y., Padberg-Gehle, K., Lämmer, S., Johansson, A., Breier, G., Schulze, F., et al.: Biologistics and the struggle for efficiency: concepts and perspectives. Adv. Complex Syst. **12**, 533–548 (2009)
8. Beber, M., Windt, K., Hütt, M.T.: Production research on metabolic systems. In: Spath, D., Ilg, R., Krause, T. (eds.) International Conference on Production Research (ICPR 21): Innovation in Product and Production 31 July–4 August 2011 in Stuttgart. Stuttgart, Germany, Fraunhofer-Verlag, Germany (2011)
9. Beber, M.E., Armbruster, D., Hütt, M.T.: Pattern complexity regulates modularity of flow networks. Phys. Rev. E (2012) (submitted)
10. Ueda, K., Vaario, J., Ohkura, K.: Modelling of biological manufacturing systems for dynamic reconfiguration. CIRP Ann. Manuf. Technol. **46**, 343–346 (1997)
11. Ueda, K., Kito, T., Fujii, N.: Modeling biological manufacturing systems with bounded-rational agents. CIRP Ann. Manuf. Technol. **55**, 469–472 (2006)
12. Ueda, K., Markus, A., Monostori, L., Kals, H.J.J., Arai, T.: Emergent synthesis methodologies for manufacturing. CIRP Ann. Manuf. Technol. **50**, 535–551 (2001)
13. Smith, J., Hütt, M.: Network dynamics as an interface between modeling and experiment in systems biology. In: Tretter, F., Gebicke-Haerter, P.J., Mendoza, E.R., Winterer, G. (eds.) Systems Biology in Psychiatric Research: From High-Throughput Data to Mathematical Modeling, pp. 234–276. Wiley-VCH(2010)
14. Varma, A., Palsson, B.O.: Metabolic flux balancing: basic concepts, scientific and practical use. Nat. Biotech. **12**, 994–998 (1994)
15. Price, N.D., Reed, J.L., Palsson, B.Ø.: Genome-scale models of microbial cells: evaluating the consequences of constraints. Nat. Rev. Microbiol. **2**, 886–897 (2004)
16. Sonnenschein, N., Geertz, M., Muskhelishvili, G., Hütt, M.T.: Analog regulation of metabolic demand. BMC Syst. Biol. **5**, 40 (2011)
17. Barabási, A.L., Oltvai, Z.N.: Network biology: understanding the cell's functional organization. Nat. Rev. Genet. **5**, 101–113 (2004)
18. Barabási, A.L., Albert, R.: Emergence of scaling in random networks. Science **286**, 509–512 (1999)
19. Jeong, H., Tombor, B., Albert, R., Oltvai, Z.N., Barabási, A.L.: The large-scale organization of metabolic networks. Nature **407**, 651–654 (2000)
20. Ma, H., Zeng, A.: The connectivity structure, giant strong component and centrality of metabolic networks. Bioinformatics **19**, 1423–1430 (2003)
21. Arita, M.: The metabolic world of Escherichia coli is not small. Proc. Natl. Acad. Sci. USA **101**, 1543–1547 (2004)
22. Erdős, P., Rényi, A.: On random graphs i. Publ. Math. Debrecen **6**, 290 (1959)
23. Becker, T., Beber, M.E., Meyer, M., Windt, K., Hütt, M.T.: A comparison of network characteristics in metabolic and manufacturing systems. In: 3rd International Conference on Dynamics in Logistics—LDIC 2012, Springer (2012)

24. Ravasz, E., Somera, A.L., Monaru, D.A., Oltvai, Z.N., Barabási, A.: Hierarchical organization of modularity in metabolic networks. Science **297**, 1551–1555 (2002)
25. Beber, M., Fretter, C., Jain, S., Müller-Hannemann, M., Hütt, M.T.: Artefacts in statistical analyses of network motifs. Proc. Roy. Soc. Interface (2012) doi:10.1098/rsif.2012.0490
26. Papp, B., Teusink, B., Notebaart, R.A.: A critical view of metabolic network adaptations. HFSP J. **3**, 24–35 (2009)
27. Basler, G., Grimbs, S., Ebenhöh, O., Selbig, J., Nikoloski, Z.: Evolutionary significance of metabolic network properties. J. The Roy. Soc. Interface **9** (71), 1168–1176 (2011)
28. Handorf, T., Ebenhoh, O., Heinrich, R.: Expanding metabolic networks: scopes of compounds, robustness, and evolution. J. Mol. Evol. **61**, 498–512 (2005)
29. Riehl, W.J., Krapivsky, P.L., Redner, S., Segrè, D.: Signatures of arithmetic simplicity in metabolic network architecture. PLoS Comput. Biol. **6**, e1000725 (2010)
30. Noor, E., Eden, E., Milo, R., Alon, U.: Central carbon metabolism as a minimal biochemical walk between precursors for biomass and energy. Mol. Cell **39**, 809–820 (2010)
31. Maslov, S., Krishna, S., Pang, T., Sneppen, K.: Toolbox model of evolution of prokaryotic metabolic networks and their regulation. Proc. Natl. Acad. Sci. **106**, 9743 (2009)
32. Zhu, Q., Qin, T., Jiang, Y.Y., Ji, C., Kong, D.X., Ma, B.G., Zhang, H.Y.: Chemical basis of metabolic network organization. PLoS Comput. Biol. **7**, e1002214 (2011)
33. Suthers, P.F., Zomorrodi, A., Maranas, C.D.: Genome-scale gene/reaction essentiality and synthetic lethality, analysis. **5** (Dec 2164) 1–17
34. Behre, J., Wilhelm, T., von Kamp, A., Ruppin, E., Schuster, S.: Structural robustness of metabolic networks with respect to multiple knockouts. J. Theoret. Biol. **252**, 433–441 (2008)
35. Marr, C., Müller-Linow, M., Hütt, M.T.: Regularizing capacity of metabolic networks. Phys. Rev. E. Stat. Nonlinear Soft Matter Phys. **75**, 041917 (2007)
36. Borenstein, E., Kupiec, M., Feldman, M.W., Ruppin, E.: Large-scale reconstruction and phylogenetic analysis of metabolic environments. Proc. Natl. Acad. Sci. **105**, 14482–14487 (2008)
37. Takemoto, K., Nacher, J.C., Akutsu, T.: Correlation between structure and temperature in prokaryotic metabolic networks. BMC Bioinf. **8**, 303 (2007)
38. Takemoto, K., Akutsu, T.: Origin of structural difference in metabolic networks with respect to temperature. BMC Syst. Biol. **2**, 82 (2008)
39. Basler, G., Ebenhöh, O., Selbig, J., Nikoloski, Z.: Mass-balanced randomization of metabolic networks. Bioinformatics **27**, 1397–1403 (2011)
40. Fong, S.S., Palsson, B.Ø.: Metabolic gene-deletion strains of *Escherichia coli* evolve to computationally predicted growth phenotypes. Nat. Genet. **36**, 1056–1058 (2004)
41. Segrè, D., Vitkup, D., Church, G.: Analysis of optimality in natural and perturbed metabolic networks. Proc. Natl. Acad. Sci. USA **99**, 15112–15117 (2002)
42. Motter, A.E., Gulbahce, N., Almaas, E., Barabási, A.L.: Predicting synthetic rescues in metabolic networks. Mol. Syst. Biol. **4**, 1–10 (2008)
43. Kim, D.H., Motter, A.E.: Slave nodes and the controllability of metabolic networks. New J. Phys. **11**, 113047 (2009)
44. Windt, K., Hütt, M., Meyer, M.: A modeling approach to analyze redundancy in manufacturing systems. In: ElMaraghy, H.A., (ed.) Enabling Manufacturing Competitiveness and Economic Sustainability: Proceedings of the 4th International Conference on Changeable, Agile, Reconfigurable and Virtual production (CARV2011), pp. 493–498. Springer (2011)
45. Kaluza, P., Mikhailov, A.S.: Evolutionary design of functional networks robust against noise. Europhys. Lett. **79**, 48001 (2007)
46. Kaluza, P., Ipsen, M., Vingron, M., Mikhailov, A.: Design and statistical properties of robust functional networks: a model study of biological signal transduction. Phys. Rev. E **75**, 15101 (2007)
47. Kaluza, P., Vingron, M., Mikhailov, A.: Self-correcting networks: function, robustness, and motif distributions in biological signal processing. Chaos **18**, 026113 (2008)
48. Famili, I., Forster, J., Nielsen, J., Palsson, B.Ø.: Saccharomyces cerevisiae phenotypes can be predicted by using constraint-based analysis of a genome-scale reconstructed metabolic network. Proc. Natl. Acad. Sci. USA **100**, 13134–13139 (2003)

49. Nam, H., Conrad, T.M., Lewis, N.E.: The role of cellular objectives and selective pressures in metabolic pathway evolution. Curr. Opin. Biotechnol. **22** (4), 595–600 (2011)
50. Eom, Y.H., Lee, S., Jeong, H.: Exploring local structural organization of metabolic networks using subgraph patterns. J. Theoret. Biol. **241**, 823–829 (2006)
51. Nyhuis, P., Wiendahl, H.: Fundamentals of Production Logistics: Theory. Springer Verlag, Tools and Applications (2008)
52. Stange, P., Mikhailov, A.S., Hess, B.: Mutual synchronization of molecular turnover cycles in allosteric enzymes. The J. Physi. Chem. B **102**, 6273–6289 (1998)
53. Casagrande, V., Togashi, Y., Mikhailov, A.: Molecular synchronization waves in arrays of allosterically regulated enzymes. Phys. Rev. Lett. **99**, 48301 (2007)
54. Lämmer, S., Kori, H., Peters, K., Helbing, D.: Decentralised control of material or traffic flows in networks using phase-synchronisation. Physica A **363**, 39–47 (2006)
55. Lämmer, S., Helbing, D.: Self-control of traffic lights and vehicle flows in urban road networks. J. Stat. Mech. Theory Exp. (JSTAT) **2008**, P04019 (2008)
56. Fretter, C., Krumov, L., Weihe, K., Müller-Hannemann, M., Hütt, M.: Phase synchronization in railway timetables. Eur. Phys. J. B **77**, 281–289 (2010)
57. Sonnenschein, N., Marr, C., Hütt, M.T.: A topological characterization of medium-dependent essential metabolic reactions. Metabolites (2012) (in print)
58. Marr, C., Theis, F., Liebovitch, L., Hütt, M.: Patterns of subnet usage reveal distinct scales of regulation in the transcriptional regulatory network of *Escherichia coli*. PLoS Comput. Biol. **6**, e1000836 (2010)
59. Lorenz, J., Battiston, S., Schweitzer, F.: Systemic risk in a unifying framework for cascading processes on networks. Eur. Phys. J. B **71**, 441–460 (2009)
60. Alon, U.: Network motifs: theory and experimental approaches. Nat. Rev. Genet. **8**, 450–461 (2007)
61. Brandman, O., Meyer, T.: Feedback loops shape cellular signals in space and time. Science **322**, 390–395 (2008)

Order Related Acoustic Characterization of Production Data

Michael Iber and Katja Windt

Abstract The conductor of an orchestra is able to distinguish not only between different instruments, but even among dozens of string players performing on instruments with similar sound qualities. Trained human ear not only is capable to highly differentiate between pitches and colors of sound, but also to localize the position, where the sound is coming from. This chapter presents a parameter mapping sonification approach on production data, which is based on these human perceptual skills. Representatively for other logistic parameters, throughput times of orders are sonified and allocated in a sonic space. Additionally to auditory representations of the established resource and order oriented views in logistics, a third perspective is introduced, which displays the complete workflow of an order simultaneously as a multi-pitched spatial sound. Thus, causes and impacts of high throughput times in the data set example could be identified.

Keywords Manufacturing · Parameter mapping sonification · Data mining · Logistic analysis

---

This contribution was previously published in Logistics Research (2012) pp. 89–98. DOI:10.1007/s12159-012-0084-y.

M. Iber (✉) · K. Windt
Jacobs University Bremen, Campus Ring 1, 28759 Bremen, Germany
e-mail: m.iber@jacobs-university.de

K. Windt
e-mail: k.windt@jacobs-university.de

K. Windt (ed.), *Robust Manufacturing Control*, Lecture Notes in Production Engineering, 17
DOI: 10.1007/978-3-642-30749-2_2, © Springer-Verlag Berlin Heidelberg 2013

1 Introduction

Profound analysis of actual and planning data and their correlation is an essential requirement for the adjustment of operating levers in production planning and control. Depending on the amount of work systems of a production shop, the number of product variants to be produced, and the quantity of restrictions caused by technical requirements or customer demands, the structure of manufacturing data easily reaches the complexity of NP-hard problems [1]. Whereas traditional methods [2] rely on averaging in order to reduce complexity, more recent approaches include advanced statistics and data mining [3] for a deeper understanding of production data. An important component of both, data mining and traditional statistic approaches as applied in logistic analysis, is exploratory data analysis (EDA). The term [4] comprises the participation of a human analyzer, who interactively explores the structure of data in recursive proceedings between generating and proving hypotheses. Well-established approaches are graphical statistics and data visualizations [5]. In the context of chronologically structured data such as production data, the acoustic equivalent to graphical display, the auditory display of statistical data (as provided by sonifications) is a promising method to gain knowledge about temporal fluctuations of bottlenecks in production workflows.

In natural science, auditory display still is widely disdained in comparison to its visual correspondent [6]. This might be caused by the visual alignment of human thinking per se including written language as the legitimate form to capture thoughts and scientific results. Still, the transfer of auditory cognition to a graphical representation meets a major challenge in most cases. But human ear has qualities that are literally complementary to the ones of the eyes. Wheras the latter tends to focus on singular events, the former is capable to perceive far more complex acoustic information. A conductor of an orchestra e.g. is able to distinguish not only between different instruments, but even among dozens of string players performing on instruments with similar sound qualities. Trained human ear is not only capable to highly differentiate between pitches and colors of sound, but also to localize the position, where the sound is coming from. Since sound only exists in temporal space, sonification of chronologically structured data such as production data is almost self-evident.

This chapter presents an approach on auditory display of production data, which is based on the human perceptual skills described. Exemplary for other logistically relevant parameters (such as work content, setup time, or schedule deviation), throughput times of orders were sonified and allocated in sonic space in order to reveal evident information about fluctuations in the overall workflow.

The following sections of this chapter depict the starting point of this research considering exploratory data analysis in logistic engineering (Sect. 2) as well as parameter mapping sonification in the scientific field of auditory display (Sect. 3). Section 4 describes the methodical approach of this research on the basis of the exploration of a data sample from sheet metal production. Finally, Sect. 5 presents a conclusion and a critical discussion.

2 Data Analysis in Logistics

The contradictoriness of logistic targets (low inventory, low throughput times, high schedule adherence, and appropriate utilization) as described by Gutenberg's scheduling dilemma [7] has comprehensively been treated in literature. Whereas there are practically approved solutions to balance inventory, throughput times, and utilization such as Logistic Operation Curves [2], schedule adherence and its impact on the overall workflow has not equally been investigated. Yu [8] criticizes an insufficient consideration of schedule adherence in production planning and control systems (PPC) and develops a scheduling operation curve in order to quantify the impact and causes of schedule deviations. As an extension of the Logistic Operation Curves, this approach is also based on averaging and not suitable for the identification of characteristic patterns in the chronological sequence of operations.

In order to enhance the level of detail in production planning and control as a first step, analysis methods need to be developed which provide a deeper understanding of the structure of logistic data itself. Therefore, novel approaches in logistic analysis increasingly rely on Knowledge Discovery in Databases (KDD) including artificial neural networks (ANN) and explorative data analysis such as the multi-stage quality information model (MSQIM), which reveals causal factors of quality defects [9]. Windt and Hütt [10] use cluster analysis and methods adapted from gene expression analysis to classify product variants that are the cause of lateness. Although classifications are capable to identify correlations between several qualities of orders and processes, in contrary to e.g. time series analysis they do not consider the serial impact of order sequences. Only [11] combines clustering with dynamically changing data.

Apparently, there have been no explicit researches using time series analysis, which in certain aspects is related to auditory display, for the identification of dynamic bottlenecks on planning and feedback data in manufacturing. The auditory analysis of production data therefore may also be considered as a first step to time series analysis in bottleneck analysis.

3 Auditory Data Analysis

Auditory Display has been established as a scientific discipline at the first Conference on Auditory Display at the Santa Fee Institute in 1992. The initiative, which led to the International Community on Auditory Display (ICAD),[1] aimed to bundle different activities in several scientific fields that examine the potential of information carried by sound. Auditory display therefore embraces a wide range of subcategories between the design of sound signals (e.g. for monitoring in medical environments or human computer interaction) and auditory data analysis.

---

[1] http://www.icad.org

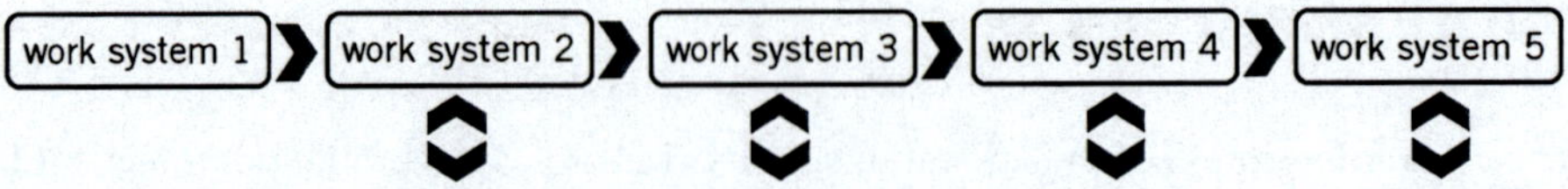

Fig. 1 Five linearly coupled work systems as an extract of a major manufacturing network. For detailed analysis only work systems 3, 4, and 5 were considered

Two key events demonstrated the potential of data sonification essentially. First, the detection of the consistency of the rings of Saturn [12]. Second, the final prove of the assumption that particle currents in weekly coupled macroscopic quantum systems would oscillate between the two systems [13]. Already in 1982, Sara Bly demonstrated in a case study including the sonification of six-dimensional data "that the auditory display was at least as effective as the visual display, and that the combined display outperformed them both" [14]. Numerous researches in fields such as neurology, theoretical physics, sociology, or psychology have refined the methodical approaches toward data sonification [15], including Parameter Mapping Sonification (PMS) and Model-Based Sonification (MBS) for exploratory data analysis [16].

4 Parameter Mapping Sonification of Production Data

In order to keep the information transfer (mapping) from a logistic datum to its representative sound event as immediate as possible, we chose Parameter Mapping Sonification (PMS) as method for the auditory exploration of manufacturing data. The target of this experiment was to identify the cause of high throughput times of production orders and their impact on successive work systems. The used data sample consisted of planning and feedback data of sheet metal production including all processes that had been completed within one year. Only orders in a linear work flow of five work systems (Fig. 1) were regarded.

Any order consisted of one or (mostly) several physically identical material pieces that were processed independently. The throughput time of an order n with k material pieces m consequently was calculated

$$\mathrm{TTP_n} = \mathrm{t_{end\ actual}}(\mathrm{m_k}) - \mathrm{t_{end\ previous}}(\mathrm{m_1}), \tag{1}$$

whereas $\mathrm{t_{end\ actual}}$ is the end of operation at the actual work system and $\mathrm{t_{end\ previous}}$ is the end of operation at the previous work station.

In the following subsections of this chapter, it will be demonstrated how we mapped these orders and material pieces to auditory display. With the sonification software, developed within this research, we explored the data sample from three auditory perspectives in order to gain knowledge about its characteristics. Two of these perspectives represented the resource oriented and order oriented views [17] as established in logistics (Fig. 2). An additional third one (Fig. 3) displayed the

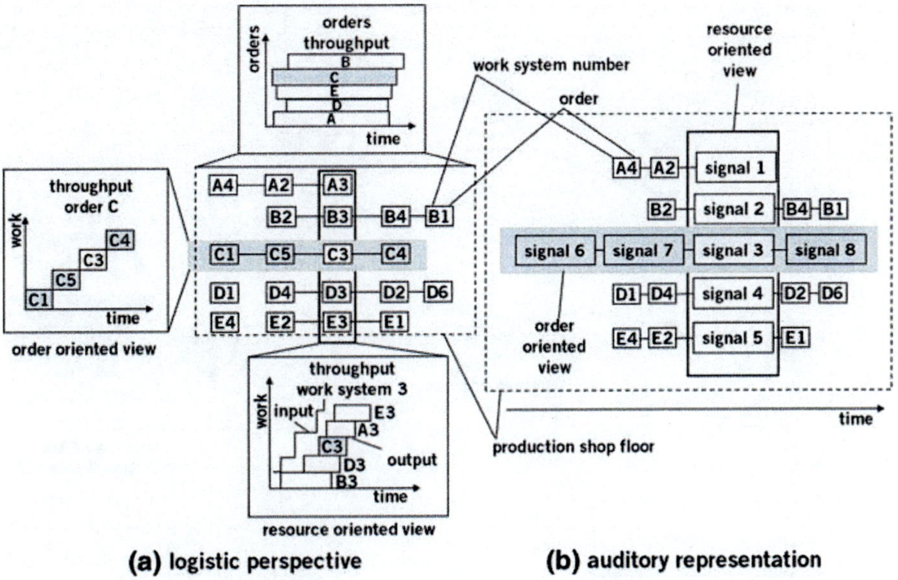

Fig. 2 Auditory representation (**b**) of order and resource oriented views (**a**) in logistics[2]

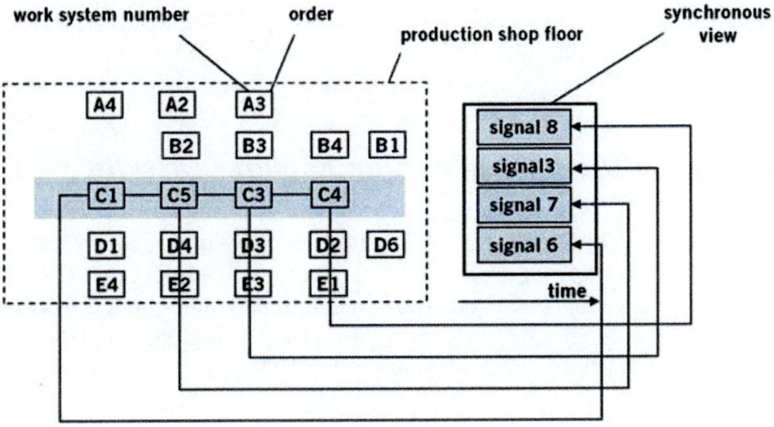

Fig. 3 In synchronous view sequential operations of an order were mapped to a multi-pitched signal

[2] Graphics according to Gläsner, J., Fastabend, H.: lecture presentation material of Institut für Fabrikanlagen (IFA), Leibniz University Hannover

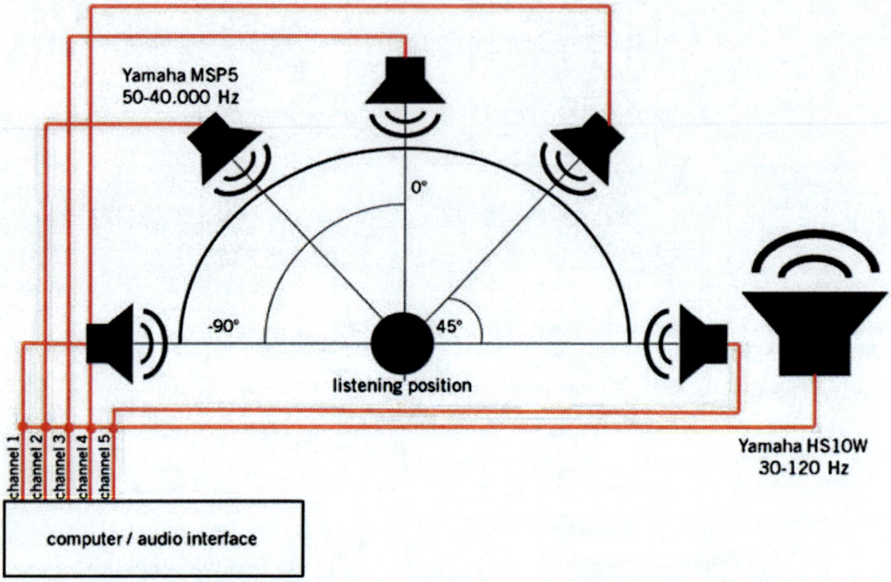

Fig. 4 Wiring of 5.1 surround audio system. In the described experiment, each work system was mapped to a discrete speaker in the upper frequency range (50–40.000 Hz) and merged in the low registers to a subwoofer channel

sequential processes of an order synchronously (synchronous view) in a single multi-pitched sound (Sect. 4.2).

4.1 Overview of Data Sample Using Resource Oriented View

Comparable to throughput diagrams [2], auditory displays in resource oriented view provide a general overview of processing at the monitored work systems. As a start-up of this research, we mapped average and accumulated throughput times ($TTP_{avg, acc}$) of all orders at each work system to sinusoidal sound signals (Fig. 5) with frequency f(t) which is a function of time, e.g. for TTP_{avg} according to the equation

$$f_i(t) = f_{low} \times \left(\frac{f_{\mathrm{high}}}{f_{\mathrm{low}}}\right)^{\frac{\frac{\sum_{n_i=1}^{\mathrm{N}} TTP_{n_i}(t)}{\mathrm{N}} - TTP_{avg}min}{TTP_{avg}max - TTP_{avg}min}} \qquad (2)$$

whereas $f_{\mathrm{low,\ high}}$ is the definable frequency range of the signal, i is the work system, n is the order, N is the number of orders, TTP(t) is the throughput time at the selected time unit, TTP_{avg} *min*, *max* define the minimum and maximum of average throughput times of all orders and systems. This mapping logarithmically scales the TTP of an order to a definable frequency range, in our sonifications between 80 and 8.000 Hz.

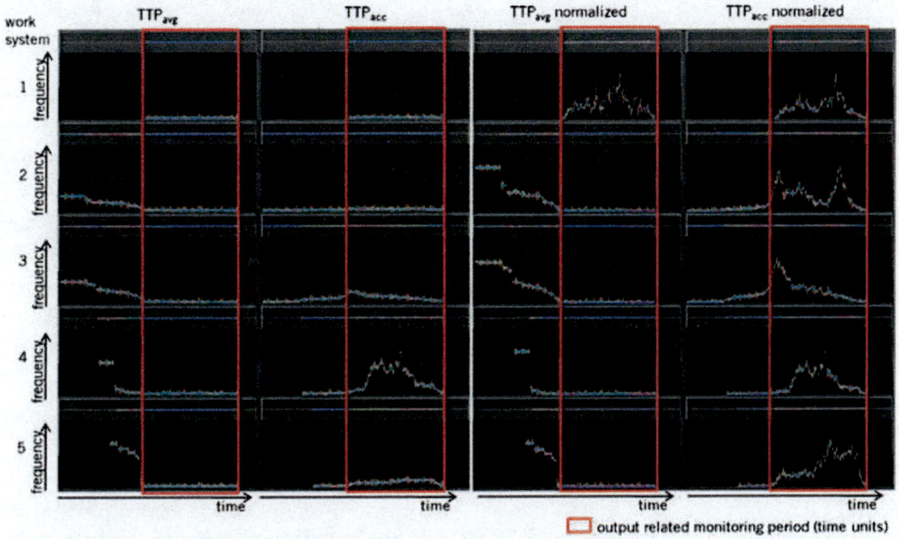

Fig. 5 TTP_{avg}, TTP_{acc} and normalized TTP_{avg} and TTP_{acc} of the five work systems

The mapping of accumulated TTPs was calculated accordingly, summing up TTPs of orders at work systems. The sum of these individual signals $f_i(t)$ representing work stations i resulted in an auditory display with signals:

$$s(t) = \sum\nolimits_{i=1}^{5} f_i(t) \tag{3}$$

The distribution of signal s in a 180° panorama (Fig. 4) facilitated the identification of the work systems.

The left half of Fig. 5 shows the spectrograms of the sonifications of TTP_{avg} and TTP_{acc} at the five work systems over the time span[3] of the data sample. The red rectangles frame time units[4] at which orders exited work systems (output). The auditory display embraced approximately twice this range since some orders had very high throughput times. The spectrograms of TTP_{avg} clearly show steady states of all work systems inside the red rectangle, while TTP_{acc} at work system 4 exhibited major fluctuations, which were subject to further investigations in Sect. 4.2.

To emphasize the individual fluctuations of each work system, we normalized the analyzed parameters (TTP_{avg}, TTP_{acc}) multiplying by:

$$f_{norm} = \frac{100}{TTP_i max}, \tag{4}$$

[3] Playback speed of sonification can be set arbitrarily in the software.

[4] For confidential reasons, time-related information refers to neutral time unit.

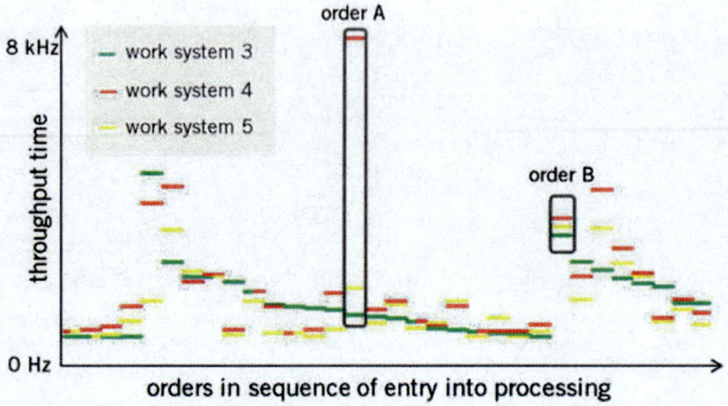

Fig. 6 Synchronous view of an extract of the data sample. Throughput times of all operations of an order were displayed as a synchronous sound. Highlighted orders A and B were subject to further analysis

whereas i is the work system, and TTP_{max} is the maximum average (respectively accumulated) throughput time. Thus, fluctuations of each work system are independently displayed over the complete frequency range defined (80–8.000 Hz).

The normalized sonifications of TTP_{acc} (Fig. 5, right half) indicated potentially mutual (seasonal) impacts of fluctuations, particularly between work systems 3, 4, and 5, which we also further investigated by sonifications based on synchronous and order oriented perspectives.[5]

4.2 Order Characterization in Synchronous View

Contrary to sonifications in resource and order oriented view, which maintain the chronology of the data structure and therefore are related to time series analysis, synchronous view is based on a sorting of orders according to an arbitrary parameter.

As shown in Fig. 6, all sequential processes of an order are displayed as synchronous multi-pitched sound signals s(n) representing the throughput times of operations according to the equation:

$$s(n) = \sum\nolimits_{i=1}^{N} f_{n_i} \tag{5}$$

with

$$f_{n_i} = f_{low} \times \left(\frac{f_{high}}{f_{low}}\right)^{\frac{TTP_{n_i} - TTP_{min}}{TTP_{max} - TTP_{min}}} \tag{6}$$

[5] Fluctuations of work system 1 and 2 at least partly depended on incomplete data and therefore were not further considered.

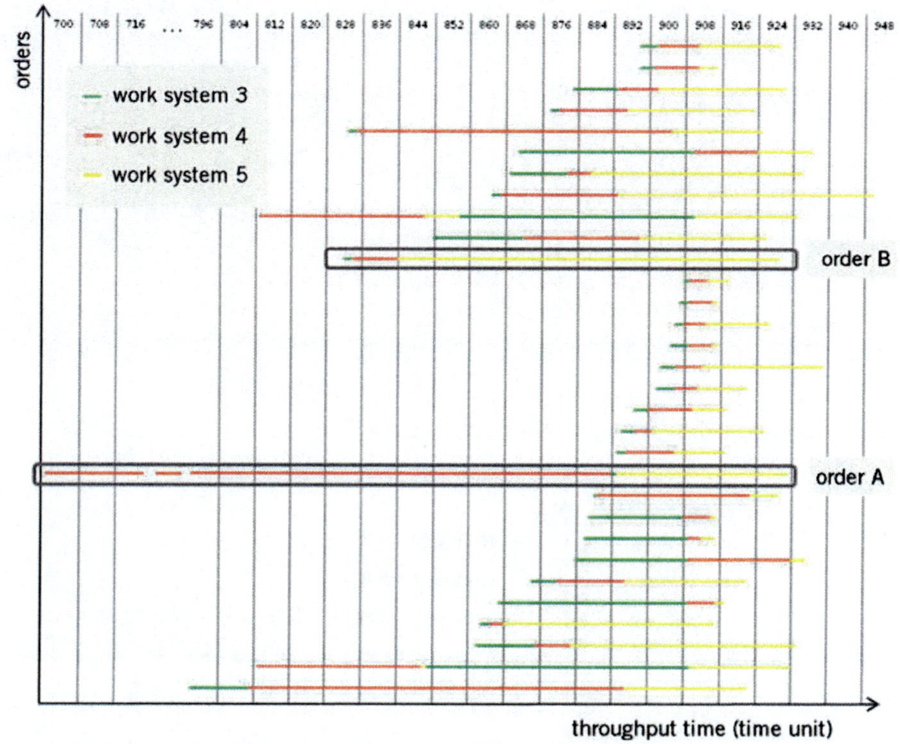

Fig. 7 Screenshot of the schematic graphical display integrated into the sonification software displaying an extract of the data sample. It should be noted that in case of overlapping throughput times in one order the graphical representations will also overlap and mask each other. The sonification displays the situation correctly

whereas $f_{low,\ high}$ is the definable frequency range of the signal, i is the work system, n is the order, TTP is the throughput time, $TTP_{min,\ max}$ is the range of throughput times over all orders and systems.

Signals s representing orders n create a set A, whereas

$$A = \{s(n_1),\ s(n_2),\ s(n_3),\ \ldots\ s(n_n)\} \tag{7}$$

The sonification sequentially displays all elements of A (Fig. 6) in a speed adjustable by the listener. Through the spatial distribution of work systems (Fig. 4), the listener can attribute sound events to the corresponding work system also in synchronous view. Thus, orders are precisely characterized by the frequency distribution of the representing sounds.[6] In the auditory display of TTPs in synchronous view, we found two evident patterns of orders: One, with

6 So to speak "acoustic fingerprints" of orders

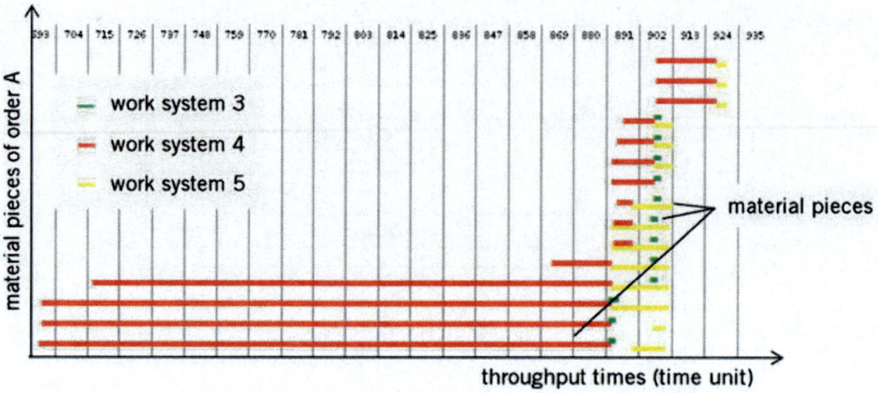

Fig. 8 Throughput times of material pieces of order A

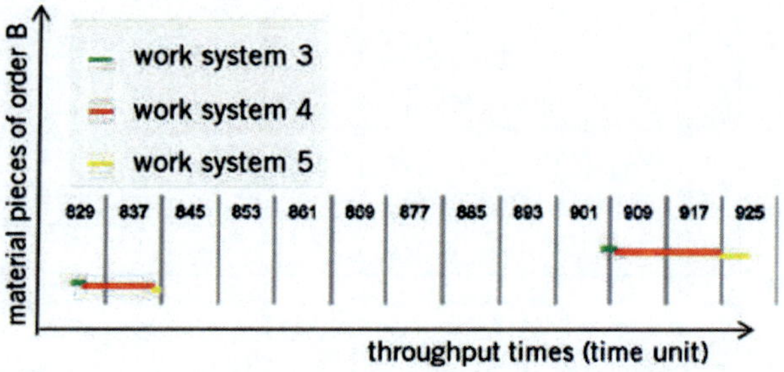

Fig. 9 Throughput times of material pieces of order B

extremely high throughput times at only one of the work systems and another one with above-average throughput times synchronously at work systems 3, 4, and 5.[7]

We selected two orders, order A and order B (Fig. 6) representing either of these two patterns for an exemplary detailed analysis in order oriented view. Order A exhibits extremely high TTP at work system 4. TTPs of order B are not as high, but above average at the three work systems taken into account.

4.3 Detailed Analysis with Order Oriented View

Figure 7 shows order A in the auditory representation of order oriented view (Fig. 2) which was mapped

[7] The operation times (TOP) at the work systems were of comparable length and, given the overall duration of TTPs, negligible.

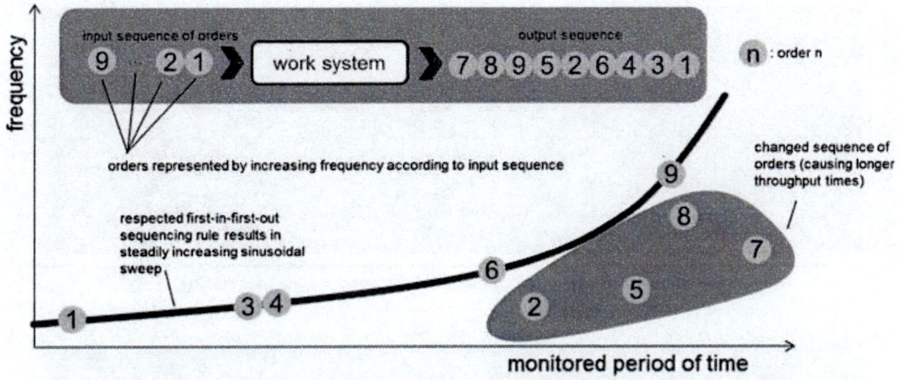

Fig. 10 Deviations of FIFO sequencing rule, represented by an increasing sinusoidal sweep

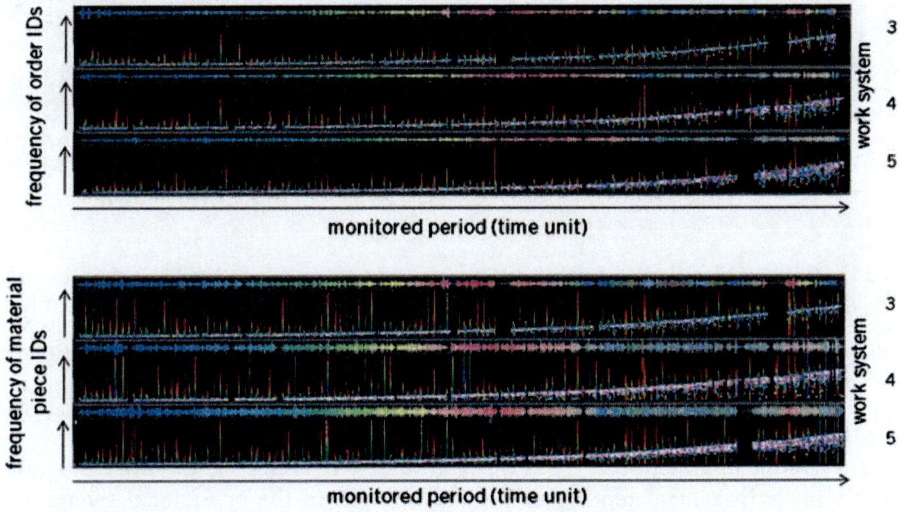

Fig. 11 Sonification of data set using statically attributed frequency to orders or respectively material pieces

$$f_{n_i}(t) = f_{low} \times \left(\frac{f_{high}}{f_{low}}\right)^{\frac{TTP_{n_i}(t) - TTP\,min}{TTP\,max - TTPmin}} \tag{8}$$

whereas $f_{low,\,high}$ is the definable frequency range of the signal, i is the work system, n is order, TTP(t) is the throughput time at the displayed time unit, $TTP_{min,\,max}$ is the range of throughput times of all orders and systems. The resulting signal s equated to the sum of non-stationary signals f of orders n at work systems i (Fig. 7):

$$s(t) = \sum\nolimits_{n_i=1}^{N} f_{n_i}(t), \tag{9}$$

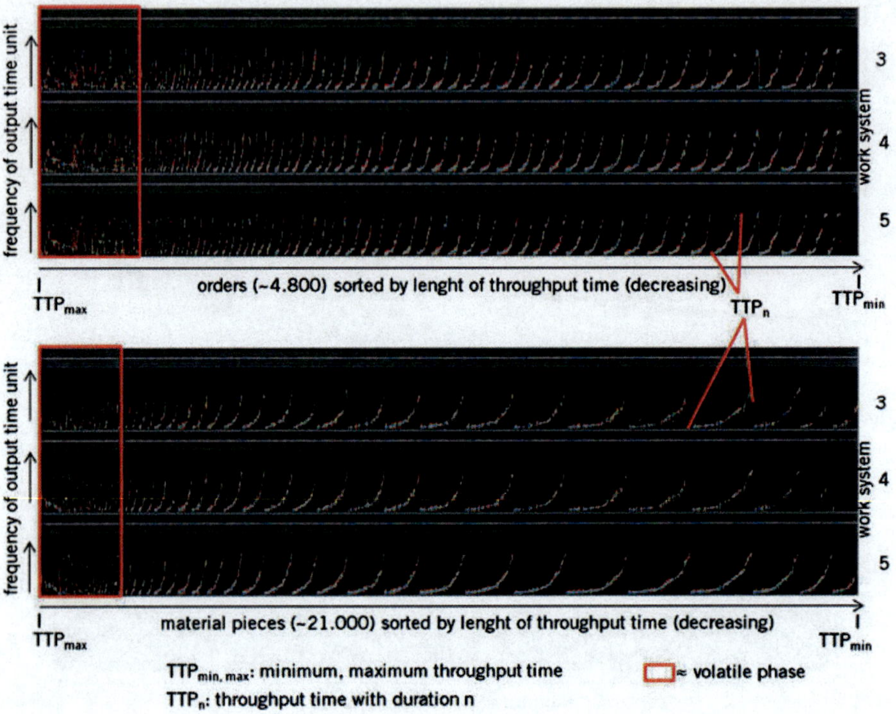

Fig. 12 Orders and material pieces sorted decreasingly (from left to right) by their largest throughput time (TTP) displaying the output time unit (increasingly) as pitch (first time unit ≈ 200 Hz, last time unit ≈ 8.000 Hz). After a short unstructured period (extreme TTPs) both sonifications adjusted to a more or less consistent distribution between output days and TTPs

whereas N is the number of orders.

The sonification of order A (Fig. 7) revealed overlapping of processes not visible in the graphical representation and changes of the operational sequence. Noticeable was the early start (at time unit 699) and the long duration of the throughput time (TTP) at work system 4, which we further tracked down by a sonification of the discrete material pieces, the order consisted of (Fig. 8). The sonification of the material pieces of order A supported our findings revealing major deviations of the expected sequence of operations.

The analysis of order B (Figs. 6, 7, 9) was exemplary for another cause of high TTPs as we confirmed by further spot samples. Individually, both material pieces (Fig. 9), order B consisted of, were produced within average TTPs. Only their combination as an order caused above-average TTPs at the regarded work systems. Our findings resulted from the re-allocation of material pieces to orders, which is characteristic to sheet metal production.

In order to quantify the amount of orders with high TTPs (as an indicator for the degree of re-allocation applied) and to analyze their distribution over the three work

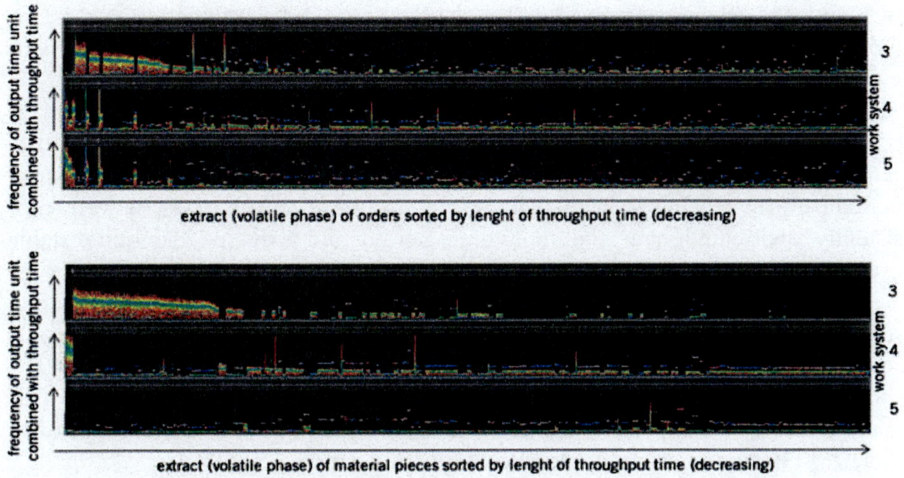

Fig. 13 Auditory display (≈ volatile phase of Fig. 12) of throughput times (represented as band-passed noise) and output days (sinusoidal sounds). The predominance of the band-pass filtered noise in the spectrograms does not reflect the auditory results, where the sinusoidal sounds were in the foreground

systems during the monitored period, we generated auditory displays, where each event was sonified at its respective output time unit only. According to the equation

$$f_n = f_{low} \times \left(\frac{f_{high}}{f_{low}}\right)^{\frac{n}{N}} \tag{10}$$

whereas n is the order (respectively material piece), and N is the number of orders (or material pieces), a static frequency was attributed to each order according to its first entry into the monitored scenario.

For a linear workflow respecting first-in-first-out sequencing rules (FIFO) this mapping would result in a constantly increasing sweep (Fig. 10) at each work system.

The sonifications of orders and material pieces (Fig. 11) indicated an increase of deviations to the main sweep during the course of time. While there was a common tendency in both sonifications, it was noticeable that particularly the material pieces at work system 4 were affected by extended TTPs, which can also visually be identified by the high amount of low frequencies in the spectrogram, but the listening results were much more detailed. The sonification of material pieces at work system 5 displayed a similar trend containing a high amount of dispersing frequencies. However, the amount of low frequencies toward the end of the sonification was clearly less. Considering the large amount of high TTPs of these results, it seemed surprising that the average throughput time (TTP_{avg}) of all work systems was at a more or less constant level (Fig. 5).

In order to get a more refined understanding of the distribution of TTPs over the monitored production period, we applied further sonifications using synchronous view. Orders and material pieces were sorted by their respectively highest TTP. The output time unit at the respective work system was correspondingly mapped as pitch (frequency). Except for orders with extremely high TTPs (about 8 % of data sample, which exhibited a volatile behavior), the spectrograms (Fig. 12) show periodically increasing sweeps. This means that most values of TTPs were consistently distributed over the monitored period and explains the quite stable average TTPs stated before.

For the volatile phase of this sonification (Figs. 12, 13) we found quite different structures of the distribution of TTPs between orders and material pieces. A sonification using band-passed noise to additionally sonify the corresponding TTPs showed that, after a burst of extreme TTPs at work system 4, high TTPs of material pieces mostly were attributed to work system 3, whereas compound to orders, TTPs considerably contributed to work systems 4 and 5.

5 Conclusion and Discussion

For the chosen data sample of sheet metal production, we demonstrated that high throughput times were concealed by a relatively consistent distribution. We also revealed that high throughput times resulted from inappropriate re-allocations of material pieces to orders. Considering that around 8 % of orders were affected by very high throughput times and given the synchronous distribution of orders with long and short throughput times mentioned above, we expect that there is reasonable potential for improvements by reducing re-allocations to a minimum level.

Particularly in combination with traditional and advanced statistical methods [2, 10], auditory data analysis becomes a powerful addition that e.g. distinctly indicates seasonal fluctuations of processes and allows to partition data samples into expedient segments for further analysis.

At this stage of our research on auditory data analysis of production data, it can be said that the questions, which arose during the experiments, differed from the ones usually asked using only established analysis approaches. These questions finally revealed results, which had not been analyzed by traditional methods performed beforehand. Analogue to the function of an engineer as a "hypotheses generator" in a recursive data-mining process [9], in its approach to data itself lies a major benefit using auditory display.

Auditory data analysis requires specialized expertise and experience that make it unsuitable for internal company use. Hence, an application would be well embedded in logistic consultancy projects as an additional analysis tool in order to re-adjust production planning and control strategies in industry.

One of the major problems of the introduced approach is a meaningful graphical transformation of auditory displays in order to fulfill scientific standards. Up to now, graphical representations and written descriptions can only be understood as hints toward the far more detailed information sonifications provide.

Acknowledgments The research of Prof. Dr.-Ing. Katja Windt is supported by the Alfried Krupp Prize for Young University Teachers of the Krupp Foundation. This project was initiated and supported by the research group "Rhythm" of The Young Academy at the Berlin-Brandenburg Academy of Sciences and Humanities and the German Academy of Natural Scientists Leopoldina: www.diejungeakademie.de

References

1. Windt, K., Philipp, T., Böse, F.: Complexity cube for the characterization of complex production systems. Int. J. Comput. Integr. Manuf. **21**(2), 195–200 (2008)
2. Nyhuis, P., Wiendahl, H.-P.: Fundamentals of Production Logistics. Springer, Berlin (2009)
3. Choudhary, A.K., Harding, J.A., Tiwari, M.K.: Data mining in manufacturing: a review based on the kind of knowledge. J. Intell. Manuf. **20**(5), 501–521 (2008)
4. Tukey, J.W.: Exploratory Data Analysis. Addison-Wesley, Boston (1977)
5. Wainer, H.: Graphic Discovery: A Trout in the Milk and Other Visual Adventures. Princeton University Press, Princeton (2005)
6. Iber, M., Klein, J., Windt, K.: Die Grooving Factory, Logistische Datenanalyse im Klanglabor. In: Schoon, A., Volmar, A. (eds.) Das geschulte Ohr: Eine Kulturgeschichte der Sonifikation, 1st edn, pp. 147–163. Transcript, Bielefeld (2012)
7. Gutenberg, E.: Grundlagen der Betriebswirschaftslehre, Erster Band: Die Produktion, 24th edn. Springer, Berlin (1983)
8. Yu, K.-W.: Terminkennlinie, Eine Beschreibungsmethodik für die Terminabweichung im Produktionsbereich. VDI, Düsseldorf (2001)
9. Mizuyama, H.: "Artificial-Neural-Network-based MSQIM for Exploratory Analysis of Manufacturing Data". In: Proceedings of the 7 th Asia Pacific Industrial Engineering and Management Systems Conference pp. 17–20 , Bangkok, Thailand, no. December, pp. 10–19 (2006)
10. Windt, K., Hütt, M.-T.: Exploring due date reliability in production systems using data mining methods adapted from gene expression analysis. CIRP Ann. Manuf. Technol. **60**(1), 473–476 (2011)
11. Liao, T.W., Ting, C.-F., Chang, P.-C.: An adaptive genetic clustering method for exploratory mining of feature vector and time series data. Int. J. Prod. Res. **44**(14), 2731–2748 (2006)
12. Kramer, G.: "Sonification Report: Status of the Field and Research Agenda," (1997). [Online]. Available: http://www.icad.org. [Accessed: 19-March-2012]
13. Pereverzev, S.V., Loshak, A., Backhaus, S., Davis, J.C.: "Quantum oscillations between two weakly coupled reservoirs of superfluid 3 He," Nature, pp. 449-452 (1997)
14. Frysinger S. P.: "A Brief History of Auditory Data Representation to the 1980 s," in First Symposium on Auditory Graphs (2005)
15. de Campo, A., Dayé, C., Frauenberger, C., Vogt, K., Wallscih, A., Eckel, G.: "Sonification as an Interdisciplinary Working Process," in Proceedings of the 12 th International Conference on Auditory Display, pp. 28–35 (2006)
16. Hermann, T.: "Sonification for exploratory data analysis," (2002) [Online]. Available: http://pub.uni-bielefeld.de/pub?func=drec&id=2017263. [Accessed: 24-Feb-2012]
17. Wiendahl, H.-P.: Betriebsorganisation für Ingenieure. Carl Hanser Verlag, München und Wien (2008)

Potentials of Nonlinear Dynamics Methods to Predict Customer Demands in Production Networks

Bernd Scholz-Reiter and Mirko Kück

Abstract Nowadays, markets are characterized by increasing dynamics and complexity. In particular, customer demands are often highly volatile. These conditions complicate demand forecasting and reduce the average accuracy of forecasting data. Nevertheless, manufacturing companies have to predict customer demands precisely, in order to achieve a well-founded production planning and control. The paper at hand deals with methods to predict customer demands in application scenarios of production logistics. Firstly, forecasting methods for smooth customer demand are described with a particular emphasis on nonlinear dynamics methods. Subsequently, a new algorithm to predict intermittent demand is introduced. In both cases of demand evolution, different methods are applied to predict demand data generated by a discrete-event simulation of a production network. Forecasting results are interpreted and the different methods are rated regarding their applicability. The research displays that an application of nonlinear dynamics methods can lead to improved forecasting accuracy.

Keywords Demand forecasting · Intermittent demand · Nonlinear dynamics · Time series analysis

1 Introduction

The prediction of customer demands is an important component of production planning and control. High accuracy of forecasting data is of crucial importance due to its impact on the following planning steps. The medium-term capacity

B. Scholz-Reiter · M. Kück (✉)
BIBA—Bremer Institut für Produktion und Logistik GmbH,
University of Bremen, Bremen, Germany
e-mail: kue@biba.uni-bremen.de

K. Windt (ed.), *Robust Manufacturing Control*, Lecture Notes in Production Engineering,
DOI: 10.1007/978-3-642-30749-2_3, © Springer-Verlag Berlin Heidelberg 2013

planning is influenced by demand data quality as well as the short-term production control [1]. Hence, manufacturing companies need to predict future customer demands as precisely as possible, in order to benefit efficiency. However, an achievement of accurate demand data often turns out to be a difficult task due to highly volatile demand evolutions [2]. This is a result of several static and dynamic effects that impact on customers and their demands. For instance, demand evolutions are determined by the success of marketing campaigns, changing conditions in the economic, political and ecologic environment or the structure of the own production system including inventory policies and procurement strategies [3, 4]. Methods of nonlinear dynamics promise improved applicability compared to common forecasting methods like classical statistical ones or the Box-Jenkins method [5]. Forecasting methods of nonlinear dynamics are able to reconstruct the dynamic properties of a whole dynamic system which comprises the demand evolution as well as further influencing values. In this way, different impacts on the demand evolution can be considered. Moreover, qualitative means are incorporated in addition to quantitative means of demand data. Hence, nonlinear dynamics methods are able to identify possible deterministic structures of demand evolutions which can improve the accuracy of forecasts.

In general, demand evolutions are separated into the two classes of smooth and intermittent demand. Intermittent, sporadic or lumpy demand evolutions are characterized by a high amount of periods without any demand on the one hand and few periods of high demand on the other hand [6]. For example, demand evolutions of service parts are often intermittent [7]. By contrast, smooth demand evolutions do not have such intermittent structure. In this paper, statistical methods, the Box-Jenkins method as well as nonlinear dynamics methods are applied to predict smooth demand. In order to predict intermittent demand, two quantities have to be predicted for every point of time. Firstly, the amount of periods until the next expected demand occurs has to be predicted. Secondly, the demand size at this period has to be predicted. In the paper at hand, a new forecasting algorithm for intermittent demand is introduced. Here, the different methods for smooth demand evolutions can be used to predict one of the two required quantities. Applying this algorithm, different combinations are compared to the commonly used Croston method which predicts both quantities by exponential smoothing.

For the application of the different forecasting methods, the availability of a successive equidistant stationary time series of past customer orders

$$\boldsymbol{y} = \{y_1, y_2, ..., y_N\} \tag{1}$$

until the present time point N is assumed within the whole paper. In the following, different forecasting methods are described and applied to predict demand evolutions in production networks. After outlining different forecasting methods for smooth demand evolutions, the new forecasting algorithm to predict intermittent demand is described and compared to the often used Croston method [8]. Subsequently, the methods are applied and evaluated. In particular, potentials of nonlinear dynamics methods are pointed out.

2 Forecasting Methods for Smooth Demand

This section gives an overview of different methods to predict smooth demand evolutions. On this basis, advantages and disadvantages of the different methods are briefly discussed. Classical statistical methods, the Box-Jenkins method and nonlinear dynamics methods are considered. The section closes with a description of measures to evaluate forecasting methods for smooth demand by accuracy.

2.1 Statistical Methods

An often used approach to forecast customer demands is the application of classical statistical methods like moving average or exponential smoothing [9, 10]. Given a time series of past customer orders (1), an application of the moving average method to predict a future demand value y_{N+1} leads to

$$\hat{y}_{N,1}^{\mathrm{MA},r} = \frac{1}{r} \sum_{i=N-r+1}^{N} y_i. \tag{2}$$

The parameter r determines the number of past customer orders that are used in averaging. A weighted average is computed by exponential smoothing. This method predicts a future value y_{N+1} of the demand time series by

$$\hat{y}_{N,1}^{\mathrm{ES},\alpha} = \alpha y_N + (1-\alpha)\hat{y}_{N-1,1}^{\mathrm{ES},\alpha}. \tag{3}$$

This method computes a future value as a weighted sum of the current time series value y_N and the predicted value for the current period $\hat{y}_{N-1,1}^{\mathrm{ES},\alpha}$. The parameter α is called the smoothing constant. It determines to which extend both values are weighted. The algorithms (2) and (3) are applicable to predict one step into the future. For a recursive prediction of h steps, the unknown future values $y_{N+1},\ldots,y_{N+h-1}$ are replaced by the one step predictions $\hat{y}_{N,1},\ldots,\hat{y}_{N+h-2,1}$.

The basic reason for the frequent application of classical statistical forecasting methods is their simplicity in application and comprehension. These methods base on intuitive algorithms. Indeed, because of their simplicity, these methods are not able to predict complex dynamic demand evolutions. In this case, an application of the Box-Jenkins method or nonlinear dynamics methods can perform better forecasts.

2.2 Box-Jenkins Method

A more sophisticated forecasting approach is the Box-Jenkins method [5, 9, 11]. It bases on autoregressive moving average models (ARMA) which use linear stochastic processes. An ARMA(p,q)-model can be displayed as

$$y_t = \sum_{i=1}^{p} \alpha_i y_{t-i} + \varepsilon_t - \sum_{j=1}^{q} \beta_j \varepsilon_{t-j}. \tag{4}$$

Here, the first term represents an autoregressive process of model order p, the second term is an error process and the third term constitutes a moving average process of model order q. In order to build up a prediction model using (4), three steps have to be passed iteratively: model identification, parameter adjustment and model validation. In the model identification step, the model orders p and q are identified by using the autocorrelation plot and the partial autocorrelation plot. Subsequently, the model parameters α_i, $i = 1, \ldots, p$, and $\beta_j, j = 1, \ldots, q$, have to be estimated. Here, different approaches exist. In this paper, the method of least squares is used [9]. For given model orders p and q as well as given parameters α_i and β_j, the ARMA(p, q) model is uniquely defined. Now, the model is validated by reference to the original time series (1) and by computing model errors. An iteration of the three modeling steps assures a well-adjusted ARMA-model (4) of the real time series (1). Based on this model, future customer orders can be predicted by extrapolation. A one-step prediction is achieved by

$$\hat{y}_{N,1}^{\text{ARMA},p,q} = \sum_{i=1}^{p} \alpha_i y_{N+1-i} - \sum_{j=1}^{q} \beta_j \varepsilon_{N+1-j}. \tag{5}$$

For a h-step prediction

$$\hat{y}_{N,h}^{\text{ARMA,p,q}} = \sum_{i=1}^{p} \alpha_i y_{N+h-i} - \sum_{j=1}^{q} \beta_j \varepsilon_{N+h-j} \tag{6}$$

the unknown values $y_{N+h-1}, \ldots, y_{N+1}$ are replaced by the predicted values $\hat{y}_{N,h-1}, \ldots, \hat{y}_{N,1}$, the values $\varepsilon_N, \varepsilon_{N-1}, \ldots$ are given by the prediction errors $y_N - \hat{y}_{N-1,1}, y_{N-1} - \hat{y}_{N-2,1}, \ldots$ and the values $\varepsilon_{N+h}, \ldots, \varepsilon_{N+1}$ are set zero. For a detailed description of the algorithm, see [5] or [11].

The Box-Jenkins method approximates real processes by a linear model and includes a noise term for nonlinear processes. Thus, it is able to deal with more complex demand evolutions than classical statistical methods. However, real processes are nonlinear, in general. For a consideration of deterministic nonlinear processes within the prediction model, nonlinear dynamics methods can be used.

2.3 Methods of Nonlinear Dynamics

Recently, nonlinear dynamics methods have been applied to model several aspects of production and logistics systems. For example, these methods are appropriate to characterise the complexity and model the dynamic evolution of production systems [12, 13]. In this paper, these methods are applied to predict customer demands in production networks. Prediction methods of nonlinear dynamics base on a modeling by dynamic systems [14]. The evolution of a dynamic system can be described by a system of ordinary differential equations:

$$\frac{\mathrm{d}}{\mathrm{d}t}\boldsymbol{x}(t) = \boldsymbol{F}(\boldsymbol{x}(t)), \text{ with } \boldsymbol{x} = \begin{bmatrix} x_1 \\ \vdots \\ x_m \end{bmatrix}, \boldsymbol{F} = \begin{bmatrix} F_1 \\ \vdots \\ F_m \end{bmatrix}, t \in \mathbb{R}. \tag{7}$$

Here, the vector $\boldsymbol{x}$ denotes the state of the dynamic system at time t and $\boldsymbol{F}$ denotes a slope vector. The m-dimensional space that is spanned by the components of $\boldsymbol{x}$ is called the phase space $M \subset \mathbb{R}^{>}$. A state of the dynamic system is represented by a point $\boldsymbol{x}$ in phase space M. A mapping $\varphi : \mathbb{R} \rightarrow \mathbb{M}$ that illustrates the state's evolution in phase space is called a trajectory. In this paper, the dynamic system is defined to be deterministic which means that a trajectory is uniquely defined by a state vector $\boldsymbol{x}$ and through every point $\boldsymbol{x}$ in phase space M there exists a unique trajectory with $\boldsymbol{x}$ as initial condition. Moreover, dissipative dynamic systems are considered because they are more reasonable to model evolutions in accordance to production systems than non-dissipative [12]. Dissipation implies that after a certain time a set of initial conditions will be attracted to a subset of the phase space. This subset $A \subset M$ is called the attractor of the system. The attractor contains the relevant dynamic properties of the whole system. Normally, it has a smaller dimension than the phase space.

The dynamic system considers various components that impact on its evolution. In order to illustrate this evolution, all dependencies have to be known which generally cannot be assured. Often, only measurements on one component of the dynamic system are known. In this paper, only a scalar time series of past customer orders (1) is given. The demand evolution constitutes one component of the dynamic system which describes evolutions of a whole production and delivery system. Here, the embedding theorem of Takens [15] states that the dynamic evolution of a whole deterministic dynamic system can be reconstructed out of a scalar time series of one component of the system. The method of delay coordinate embedding is used to embed the attractor of the system in a so called embedding room E. By this approach, the system's dynamic properties as well as the topological and differential properties of the attractor are reconstructed in E.

The method of delay-coordinate embedding uses the successive and equidistant time series

$$\begin{gathered} \boldsymbol{y} = \{y_1, y_2, \ldots, y_N\} \\ y_k = x_j(t_k), \quad t_k = t_1 k\tau_S, \quad y_k \in \mathbb{R} \end{gathered} \tag{8}$$

of measurements on the component x_j of the state vector $\boldsymbol{x} = [x_1, \ldots, x_m]^{\mathrm{T}}$. The component x_j describes the customer orders. The parameter τ_S is called the sampling time. A vector

$$\boldsymbol{v}_{k,y}^{n,\tau_L} = \left[y_{k-(n-1)\tau_L}, \ldots, y_{k-\tau_L}, y_k\right]^{\mathrm{T}} \tag{9}$$

is called delay coordinate vector of length n corresponding to time point k. The delay time τ_L is a multiple of the sampling time τ_S. A delay coordinate vector $\boldsymbol{v}_{k,y}^{n,\tau_L}$ is a segment of the original time series $\boldsymbol{y}$. While the original time series $\boldsymbol{y}$ involves N measurements with successive time distance τ_S, the delay coordinate vector contains n of the N components with successive time distance $\tau_L = c\tau_S$, $c \in \mathbb{N}$. The length n of the delay coordinate vector is called the embedding dimension. By a delay coordinate vector, the dynamic properties of the original state vector $\boldsymbol{x}$ can be reconstructed. In order to build a delay coordinate vector, the embedding dimension n and the delay time τ_L have to be chosen appropriately for the demand evolution described by $\boldsymbol{y}$. In this paper, the method of false nearest neighbors is applied to compute a reasonable embedding dimension n [16]. The delay time is calculated as the first minimum of the average mutual information [17]. Further methods to choose these parameters can be found in [14] or [18].

Using the method of delay coordinate embedding, the dynamic properties of a state $\boldsymbol{x}(k)$ in an unknown phase space M can be reconstructed by a vector $\boldsymbol{v}_{k,y}^{n,\tau_L}$ in an embedding space E, if a scalar time series of equidistant measurements of one component x_j of $\boldsymbol{x}$ is available. Based on this nonlinear system reconstruction, a prediction model for future values of the given time series of customer orders can be build. In this paper, a local linear prediction model is applied. For this purpose, a delay coordinate vector $\boldsymbol{v}_N$ corresponding to the present state is build and a prediction function is needed that extrapolates into the future. In order to achieve this function, all delay coordinate vectors of length n with successive distance τ_L are build out of the given time series $\boldsymbol{y}$. These $N-(n-1)\tau_L$ vectors are sorted by their Euclidean distances to the delay coordinate vector $\boldsymbol{v}_N$ corresponding to the last point of time N within the given time series $\boldsymbol{y}$. Now, the z nearest neighbors $\boldsymbol{v}_{\mathrm{NN},1}, \ldots, \boldsymbol{v}_{\mathrm{NN},z}$ of $\boldsymbol{v}_N$ are chosen to build up the prediction model where the qth nearest neighbor is the delay coordinate vector that has the qth shortest distance to $\boldsymbol{v}_N$. After mapping the z nearest neighbors h steps into the future in the reconstructed phase space to get the values $P_h(\boldsymbol{v}_{\mathrm{NN},1}), \ldots, P_h(\boldsymbol{v}_{\mathrm{NN},z})$, the affine model

$$P_h(\boldsymbol{v}_N) = \boldsymbol{a} \cdot \boldsymbol{v}_N + d \tag{10}$$

which best fits the points $(\boldsymbol{v}_{\mathrm{NN},1}, P_h(\boldsymbol{v}_{\mathrm{NN},1})), \ldots, (\boldsymbol{v}_{\mathrm{NN},z}, P_h(\boldsymbol{v}_{\mathrm{NN},z}))$ is computed. Here, the unknown coefficients $\boldsymbol{a}$ and d are calculated by a least-squares method. Finally, a h-step prediction by the nonlinear dynamics prediction method is established by

$$\hat{y}_{N,h}^{\mathrm{NLD}} = P_h(\boldsymbol{v}_N). \tag{11}$$

Methods of nonlinear dynamics approximate a given time series by a nonlinear deterministic dynamic system. These methods are able to model complex dynamic evolutions. The dynamic properties of the demand evolution as well as additional influencing values are reconstructed and considered within prediction. Hence, these methods incorporate qualitative and quantitative information of the given time series. Thus, they offer potential to forecast demand evolutions in production networks. For a detailed description of the applied prediction algorithm of nonlinear dynamics, see [4] and [19]. Within these papers, additional interpolation and filtration steps as well as a second projection are proposed. For the paper at hand, these tasks are neglected. In this way, a simplified prediction algorithm is achieved and tested for general applicability in production networks.

2.4 Accuracy Measures

For the evaluation of forecasting methods for smooth demand in terms of their accuracy, different measures can be used. In this paper, the mean absolute percentage error $\mathrm{MAPE} = \frac{100}{h}\sum_{i=1}^{h}\left|\frac{y_i - \hat{y}_i}{y_i}\right|$ is computed [9]. This is a non-scale dependent variability measure that states prediction errors in percent of the demand size.

3 Forecasting Methods for Intermittent Demand

In contrast to smooth demand evolutions, intermittent demand is characterized by a high number of periods without demand and averagely high demand sizes in the few other periods [6]. For an exemplary intermittent demand evolution, see Fig. 1a. Prediction methods for smooth demand are poorly able to predict intermittent demand evolutions. On that account, a commonly used method is the one of Croston [8]. This method predicts two different quantities. At first, the number of periods until the next expected demand occurs is predicted. Subsequently, the demand size for this period is predicted. Within the Croston method, both values are predicted by exponential smoothing (3).

Based on the Croston method, a new prediction algorithm for intermittent demand is introduced in the paper at hand. This algorithm also predicts the required values in two steps. In comparison to the method of Croston, different prediction methods can be used for the two steps. Here, each of the two steps can be predicted by one of the displayed methods for smooth demand, moving average (2), exponential smoothing (3), the Box-Jenkins method (6) or the nonlinear dynamics prediction algorithm (11). The application scheme of the new prediction algorithm is illustrated in Fig. 1b.

In order to evaluate the forecasting accuracy of methods for intermittent demand, the common measures for smooth demand are not appropriate because they get biased by the high number of periods without demand [20]. Hence, the paper at hand uses the cumulated forecast error $\mathrm{CFE}_h = \sum_{i=1}^{h}(y_i - \hat{y}_i)$ that sums the forecast errors over all periods until present. In addition, the greatest shortage $\mathrm{CFE}_{\max} = \max_{i\in\{1,\ldots,h\}}(\mathrm{CFE}_i)$ and the greatest surplus $\mathrm{CFE}_{\min} = \min_{i\in\{1,\ldots,h\}}(\mathrm{CFE}_i)$ identify

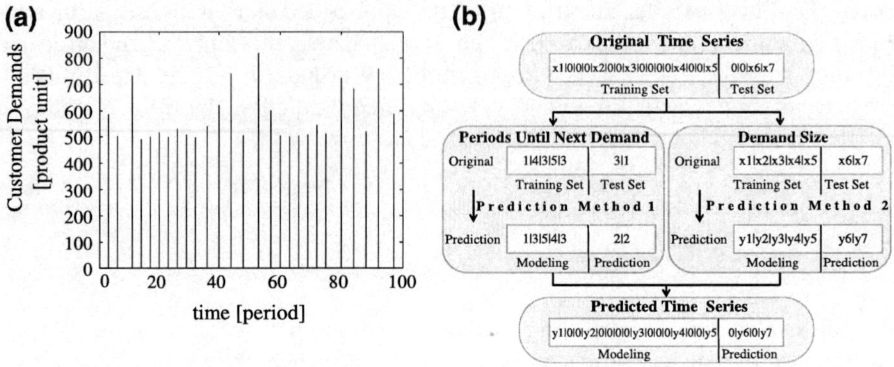

Fig. 1 **a** Exemplary intermittent demand evolution. **b** Prediction algorithm for intermittent demand

the maximum and minimum values of CFE. Moreover the number of shortages $\text{NOS} = \{\#i : \text{CFE}_i > 0, i = 1, \ldots, h\}$ indicates how often a method predicts too small values.

4 Application and Results

In this section, the prediction accuracy results of the different methods are displayed after their application to scenarios of production networks. Firstly, the application scenario is described. Subsequently, the results of smooth as well as intermittent demand prediction are illustrated.

4.1 Application Scenario

In order to generate representative customer demand data, a discrete-event simulation model of a production network is applied, Fig. 2. This network consists of three suppliers that deliver raw material to a customer on the first level which supplies ten customers on the second level with products. Model input data are orders of the 2nd-level customers to the 1st-level customer. As input data, sine functions around a mean value $40 + i \cdot \sin(kx)$ with different cycle lengths k are used. For the creation of different values of external dynamics, different amounts of intensity i are applied. When the stock of a product decreases below a defined reproduction level, new products are manufactured. Here, different values of the reproduction level are adjustable. The production of the 1st-level customer is simulated as a job shop system with different work shops whose internal dynamic effects impact on the model output demand data. For the production, raw material

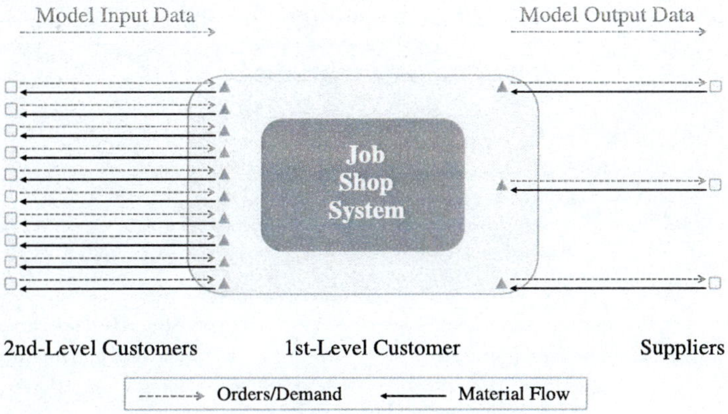

Fig. 2 Application scenario

is taken from the raw material stock. A decrease of material stock below the reorder level generates customer orders which constitute model output data. For the considerations of this paper, the orders from the 1st-level customer to the first supplier are used. By the configuration of reproduction and reorder levels, smooth as well as intermittent demand evolutions can be mapped. For a detailed description of the simulation model, see [21].

4.2 Forecasting Smooth Demand Data

After generating smooth demand data, now, the different methods of Sect. 2 are applied to predict the demand evolutions. Moving average (MA) (2) orders of $r = 1, 2, \ldots, 10$ are applied. Exponential smoothing (ES) (3) parameters $\alpha = 0.1, 0.2, \ldots, 1$ are adapted. Within the Box-Jenkins method (BJ) (6), model orders of $p = 2, \ldots, 6$ and $q = 1, \ldots, 7$ are applied. For the nonlinear dynamics prediction method (NLD) (11), embedding parameters are computed like described in Sect. 2.3 and additional s redundant neighbors, $s = 0, 1, \ldots, 10$, are considered for modeling. Figure 3 displays the accuracies of the best parameter configurations of the different prediction methods as mean absolute percentage error over the prediction horizon. Different cases of external dynamic influence intensity $i = 1, 5, 10$ within the sine input data functions are considered. In the case of low external dynamics $i = 1$, all methods show similar low error and hence high accuracy values. For increased dynamics ($i = 5$), for short and middle-term predictions, BJ shows slightly better accuracy than NLD and the statistical methods. For long-term predictions, NLD shows better results than the other methods. In the case of high dynamics intensity $i = 10$, the statistical methods have very poor prediction accuracy, in contrast to BJ and NLD which perform better. This is due to the internal dynamic induced by the production policy and the order policy of the

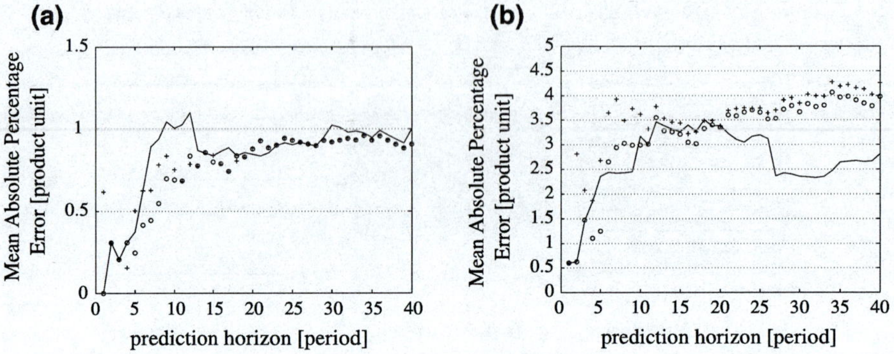

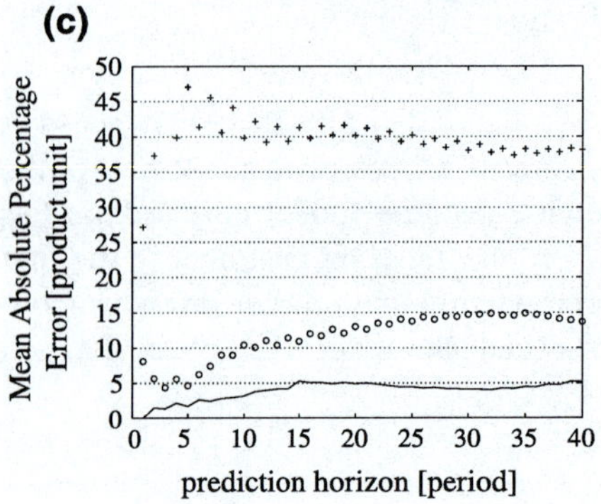

(b)

		Intensity		
		1	5	10
Horizon	1	NLD	STA	NLD
	5	BJ	BJ	NLD
	10	BJ	BJ	NLD
	40	BJ	NLD	NLD

Fig. 3 Accuracies of the prediction methods for smooth demand (+ best statistical method [STA], o Box-Jenkins method [BJ], —nonlinear dynamics method [NLD]), **a** intensity $i = 1$, **b** intensity $i = 5$, **c** intensity $i = 10$, **d** best methods for different prediction horizons (h) and different intensities (i)

1st-level customer in the simulation model (Fig. 2). The specific configuration leads to alternating high and low orders. The statistical methods predict by averaging. Hence, their accuracy is low. The best predictions for this case and all prediction horizons are achieved by NLD which performs about two to three times better than BJ and eight times better than MA and ES. Concluding, one can say that NLD shows improved potential in forecasting dynamic smooth demand evolutions compared to statistical methods and BJ.

4.3 Forecasting Intermittent Demand Data

In addition to smooth demand evolutions, intermittent demand is investigated and predicted. Here, the new prediction algorithm for intermittent demand of Sect. 3 is applied (see Fig. 1). In the paper at hand, a first exemplary application of the algorithm is described. In this process, the different methods for smooth demand

Table 1 Accuracy measures of the prediction methods for intermittent demand

	CFEmax	CFEmin	NOS	Mean	Std
Croston method (CM)	919	−660	26	189.72	296.13
2x Moving average (2MA)	917	−626	26	209.97	290.67
2x Box-Jenkins method (2BJ)	687	−421	16	−34.39	303.20
2x Nonlinear dynamics method (2NLD)	270	−730	10	−125.81	189.97

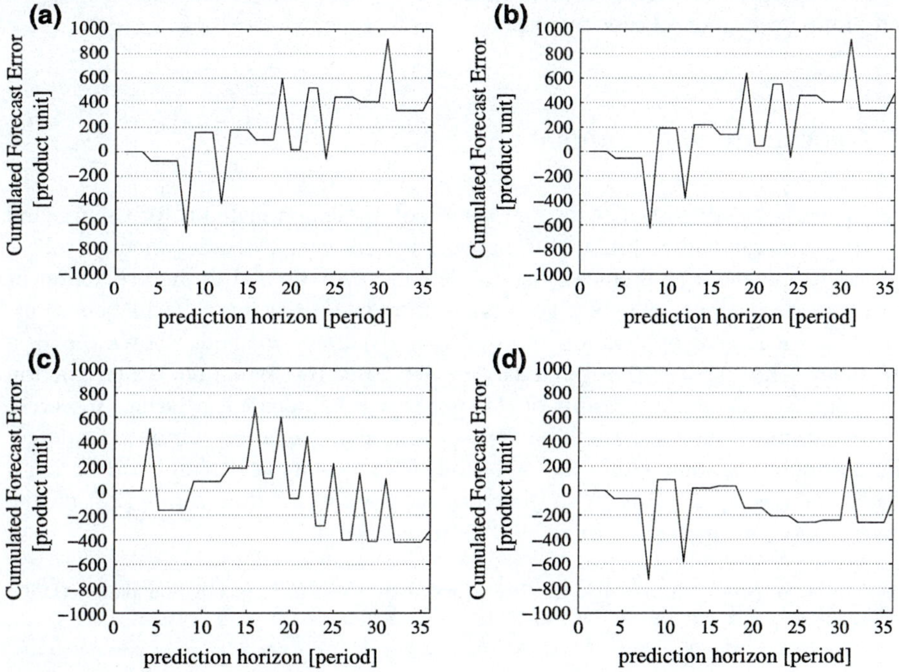

Fig. 4 Cumulated forecast errors of the prediction methods for intermittent demand. **a** Croston method (CM). **b** 2x moving average (2MA). **c** 2x Box-Jenkins method (2BJ). **d** 2x nonlinear dynamics method (2NLD)

prediction are applied to predict the amount of intervals until the next expected demand and the demand size at this period subsequently. The prediction accuracies of the Croston method (CM) which uses exponential smoothing in both steps as well as double applications of moving average (2MA), the Box-Jenkins method (2BJ) and the nonlinear dynamics method (2NLD) are compared. Table 1 shows the accuracies of the different methods. Figure 4 illustrates the evolutions of the cumulated forecast errors over the prediction horizon. Here, ten successive demand events are predicted which results in a prediction horizon of 36 periods. 2MA shows almost the same accuracy as CM. Especially for long prediction horizons, these methods perform poorly compared to 2BJ and 2NLD. A value of the cumulated forecast error above zero means that too low demand sizes have been predicted. In contrast, a value below zero implies too high predictions. The number of shortages (NOS) states in how many periods the cumulated value of all demands until this period is higher than the cumulated value of predicted demands. By usage of 2MA or CM, too few products would be available in 26 of the 36 considered periods. This deficiency would occur 16 times for 2BJ and 10 times for 2NLD. 2BJ shows the best mean value of −34.39 but the highest standard deviation of 303.20. On average, 2NLD predicts higher demands than the actual values of customer orders. However, this method shows the best performance of the different methods due to the lowest value of NOS and the closest proximity to zero of CFE on average.

5 Conclusion and Outlook

The paper at hand, describes the application of prediction methods for smooth and intermittent demand in production networks. Nonlinear dynamics methods show particular potential to predict dynamic demand evolutions. For the prediction of intermittent demand, a new algorithm is introduced that predicts in two steps. Here, a successive application of nonlinear dynamics methods shows the best accuracy. The considered demand data are generated by simulation of a production network. The exemplary results of this paper will be detailed in further research. Different model input data will be applied and other scenarios will be considered. Further combinations within the algorithm for intermittent demand will be evaluated. Moreover, the impact of general predictability of time series [22] on the accuracy of the applied methods will be investigated.

Acknowledgments This research has been funded by German Research Foundation (DFG) under the reference number SCHO 540/21-1.

References

1. Wiendahl, H.P.: Betriebsorganisation für Ingenieure. Carl Hanser, München (2010)
2. Günther, H.O., Tempelmeier, H.: Produktion und Logistik. Springer, Berlin (2005)
3. Scholz-Reiter, B., Kück, M.: Auswahl von Prognoseverfahren für Kundenbedarfe - Erstellung einer Datenbank mit Handlungsempfehlungen zur Auswahl geeigneter Prognoseverfahren. In: Industrie Management, vol. 28, pp. 61–65. GITO, Berlin (2012)
4. Scholz-Reiter, B., Kück, M., Toonen, C.: Improved demand forecasting using local models based on delay time embedding. In: International Journal of Systems Applications Engineering& Development, vol. 6, pp. 17–27. University Press (2012)
5. Box, G.E.P., Jenkins, G.M.: Time Series Analysis: Forecasting and Control. Holden-Day, San Francisco (1976)
6. Wedekind, H.: Ein Vorhersagemodell für sporadische Nachfragemengen bei der Lagerhaltung. In: Ablauf- und Planungsforschung, vol. 9, pp. 1–11. Oldenbourg, München (1968)
7. Altay, N., Litteral, L.A.: Service Parts Management—Demand Forecasting and Inventory Control. Springer, London (2011)

8. Croston, J.D.: Forecasting and stock control for intermittent demands. Oper. Res. Q. **23**, 289–303 (1972)
9. Mertens, P., Rässler, S.: Prognoserechnung. Physica, Heidelberg (2005)
10. Alicke, K.: Planung und Betrieb von Logistiknetzwerken: Unternehmensübergreifendes Supply Chain Management. Springer, Berlin (2005)
11. Schlittgen, R., Streitberg, B.H.J.: Zeitreihenanalyse. Oldenbourg, München (1995)
12. Scholz-Reiter, B., Freitag, M., Schmieder, A., Müller, S.: A dynamical concept for production planning & control. In: Proceedings of 16th International Conference Production Research, pp. 27–40. Prague (2001)
13. Papakostas, N., Efthymiou, K., Mourtzis, D., Chryssolouris, G.: Modeling the complexity of manufacturing systems using non-linear dynamics approaches. CIRP Ann. Manuf. Technol. **58**, 437–440 (2009)
14. Kantz, H., Schreiber, T.: Nonlinear Time Series Analysis. University Press, Cambridge (2004)
15. Takens, F.: Detecting strange attractors in turbulence. In: Rand, D.A., Young, L.-S. (eds.) Dynamical Systems and Turbulence, Lecture Notes in Mathematics. vol. 898, pp. 366–381. Springer, Warwick (1981)
16. Kennel, M., Brown, R., Abarbanel, H.D.I.: Determining embedding dimension for phase-space reconstruction using a geometrical construction. Phys. Rev. A **45**, 3403–3411 (1992)
17. Fraser, A.M., Swinney, H.L.: Independent coordinates for strange attractors from mutual information. Phys. Rev. A **33**, 1134–1140 (1986)
18. Buzug, T., Pfister, G.: Optimal delay time and embedding dimension for delay-time coordinates by analysis of the global static and local dynamical behavior of strange attractors. Phys. Rev. Lett. A **45**, 7073–7084 (1992)
19. Sauer, T.: Time series prediction by using delay coordinate embedding. In: Weigend, A.S., Gershenfeld, N.A. (eds) Time Series Prediction: Forecasting the Future and Understanding the Past. pp. 175–193. Addison-Wesley, Harlow (1994)
20. Wallström, P., Segerstedt, A.: Evaluation of forecasting error measurements and techniques for intermittent demand. Int. J. Prod. Econ. **128**, 625–636 (2010)
21. Scholz-Reiter, B., Kück, M., Toonen, C.: Simulation-based generation of time series representing customer demands in networked manufacturing systems. In: Proceedings of 16th Annual International Conference Industrial Engineering Theory, Applications and Practice, pp. 488–493, Stuttgart (2011)
22. Sugihara, G., May, R.M.: Nonlinear forecasting as a way of distinguishing chaos from measurement error in time series. Nature **344**, 734–741 (1990)

The Structure of the Value Creation Network for the Production of Electric Vehicles

Lessons to be Learned from Complex Network Science

Richard Colmorn, Michael Hülsmann and Alexandra Brintrup

Abstract The production of electric mobility can be considered as a network of networks, because of the internal linkages of industry-wide value creation chains, the international linkages of suppliers with several car manufacturers and the international linkages between car manufacturers due to different forms of cooperation. As a result, a classification approach for the structural relations is needed to better analyze e.g. the robustness of efficiency of cooperative partnerships, because these are not adequately formalized in the logistics and supply chain management literature. For this reason, the paper introduces a systematic classification approach based on the methodology of complex network sciences and a three-dimensional matrix notation for the production of electric vehicles.

Keywords Complex network science · Network architecture · Electric mobility · Automotive industry · Structural complexity

R. Colmorn (✉) · M. Hülsmann
Jacobs University Bremen gGmbH, Campus Ring 1,
28759, Bremen, Germany
e-mail: r.colmorn@jacobs-university.de

M. Hülsmann
e-mail: m.huelsmann@jacobs-university.de

A. Brintrup
Said Business School, University of Oxford,
OX1 1HP, Oxford, UK
e-mail: alexandra.brintrup@sbs.ox.ac.uk

K. Windt (ed.), *Robust Manufacturing Control*, Lecture Notes in Production Engineering,
DOI: 10.1007/978-3-642-30749-2_4, © Springer-Verlag Berlin Heidelberg 2013

1 Introduction

Our central question is how to analyze the network structure as an influencing factor on the supply chain target achievement? It is now widely accepted that companies compete with one another through their value creation networks—a trend that will continue increasing in the future [1]. Hence, it becomes possible to gain and maintain a competitive advantage through the whole network through the cooperation of companies in the value creation process. Among the vast literature of various types and intricacies of organizational networks, particularly the concept of Supply Chain Management (SCM) takes the structure and coordination of autonomously acting organizational units explicitly into account from an inter-organizational perspective (cf. e.g. [2, 3]). A vast academic enterprise has been built on the question of supply chain cooperation, with the dominant theoretical positions on as for example Transaction Cost Economics, Resource Based View and Agency theory (for an overview please see: [4]). Despite these foundations, the actual the inter-organizational network structure that arises from dyadic cooperation is largely unknown [5, 6]. For this reason, before the effects of the network structure on the supply chain target achievement can be investigated, a conceptual study on the application of network theory to supply chains needs to be conducted.

In this paper we address this gap by developing a systematic classification approach for the application of network metrics on the supply chain value creation system. From a methodological point of view we use data from the German automotive industry for the production of electric mobility, a topical value-creation challenge facing the industry today.

2 The Problem

2.1 *Cooperative Strategic Partnerships as a Success Factor for the Competitiveness of Electric Vehicles*

The car manufacturers of the German automotive industry have announced the market introduction of several electric vehicles for this year and the year 2013, e.g. the R8 e-tron of the Audi AG in 2012, the Megacity Vehicle of the BMW AG in 2013, the E-Cell of Mercedes-Benz in 2013 or the Golf blue e-motion and the e-Up of the Volkswagen AG in 2013 [7]. The reasons for this can be traced back to increased ecological customer requirements, binding legal regulations and a decreasing availability of fossil fuels from few—politically instable—regions [8]. Therefore, electric vehicles—consisting of an electric motor that is powered from electric energy from renewable energy sources stored in lithium-ion-batteries—present a promising alternative. The program is backed by the German government and aims to bring the German economy as the leading market for electric mobility

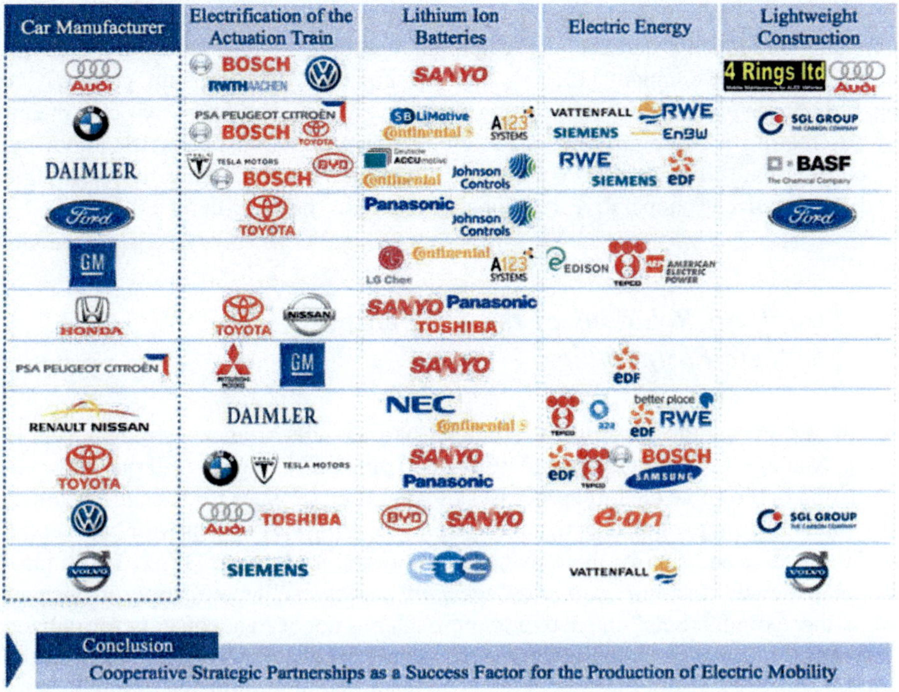

Fig. 1 Cooperative strategic partnerships as a success factor for the production of electric vehicles

technologies. Several other national promotional programs compete with the initiatives (an overview can be found in [9] and [10]). Therefore, increasing the competitiveness of electric vehicles is of particular importance both for the strategic management of the car manufacturers and the German government. While we previously investigated the specific product characteristics of electric vehicles in order to meet the customer preferences by developing a customer value-based model [11], this paper focuses on the supply chain side as a key driver for the competitiveness of electric vehicles.

As outlined above, for the production of electric vehicles the areas of the electrification of the actuation train, of lithium-ion-batteries for the storage of electric energy from external—and renewable—energy sources are the central areas of electric vehicles. Furthermore, lightweight materials will also play an integral part for the production of electric mobility due to the fact that the range is still the limiting factor [8]. For this reason, car manufacturers have agreed to cooperate due to several advantages. For example, the opportunities of a cooperation exceed often the risks and comprise, e.g. the access to new markets and technologies, the concentration on core competencies, the reduction of costs and capital commitment or an increase of the productivity and capacity for innovation

[8]. As such, goods and services can be brought together that could not be offered by a single company, because of restrictions regarding capital, time, competencies or size of a single company [12, 13]. As a consequence, the car manufacturers have agreed upon new cooperation targeting the above-mentioned four areas as a means of co-creating a strategic success factor for the production of electric mobility. According data that has been collected through newspaper articles between 2008 and 2012, current cooperative relationships include the ones depicted on Fig. 1.

2.2 *Structural Relations of the Cooperative Strategic Partnerships*

According to Schulte cooperative strategic partnerships are considered in Supply Chain Management in order to effectively and efficiently integrate all partners participating in the value creation processes of goods and services for the end-customer [14]. Hence, despite of the fact that the systems thinking approach is widely accepted in the logistics and supply chain management literature (cf. e.g. [15, 2, 16]) understanding logistics as a system of interdependent processes, "where actions in one part affect those of all others" the network approach has not been adequately formalized [17]. For this purpose and based on the data that has been collected for studying the current cooperative relationships depicted in the previous section, the following subsections aim to conceptualize these relationships for enabling the interdisciplinary solution approach of complex network science.

Firstly, it can be argued that the production of electric vehicles can be considered as a network of multiple value creation chains as previously unrelated industries are increasingly incorporated to the value creation network to play pivotal roles, e.g. electric power companies [18]. In addition, the value added chains for the lithium-ion-batteries, modules and components for efficient energy management, appropriation of electricity from—renewable—energy sources all need to be effectively and efficiently integrated so that a network of industry-wide integrated value creation chains arises (cf. [19, 20, 8]).

Secondly, it can be argued that the production of electric vehicles is a global network of cooperative linkages of suppliers in the four above-mentioned core areas, because by considering Fig. 1, it can be recognized that some companies cooperate with several car manufacturers. For example, Sanyo as a producer of lithium-ion-batteries cooperates with Audi, PSA Peugeot Citroen, Honda, Toyota and Volkswagen so that some suppliers cooperate with different car manufacturers, while the car manufacturers cooperate with different suppliers. As a result, a network of cooperative international linkages arises.

Finally, it can be argued that the production of electric mobility is a heterogeneous network consisting of different types of links, as car manufacturers forms several types of relationships for horizontal cooperation. For example, in the special edition of the industry-sector-specific journal "Automobil Produktion" [4] have shown the international connections due to six different forms of cooperation—investments, joint venture, exchange of components, marketing agreement, production agreement and R&D agreement.

Consequently, the production of electric vehicles can be considered as a network of networks with heterogeneous links, given multiple industry-wide value creation chains, international linkages of the suppliers with several car manufacturers and the heterogeneous linkages between car manufacturers.

The "network of networks" viewpoint is not adequately considered in the logistics and supply chain management literature, because, firstly, definitions about the supply chain management frequently implies a linear understanding of the value creation processes from the raw materials to the end-customer. For example, widely accepted definitions of SCM include "Planning and controlling all of the business processes – from end-customer to raw material suppliers – that link together partners in a supply chain in order to serve the needs of the end-customer [17]" or "concept of Supply Chain Management defines the company-wide coordination and optimization of the flow of materials, information and capital along the whole value creation process from the raw materials to end-customer focusing on the needs of the customers. [2]". Hence, some authors refer to the network aspect by talking about so-called International Supply Networks, but the analytical formalization of this network of cooperative relationships has not yet taken place (cf. [28]), [21]).

Secondly, for example, the supply chain operations (SCOR)-approach is a widely accepted concept for the description, analysis and evaluation of supply chains from a company-wide perspective [cf. [22]]. Because of the fact that the SCOR-approach only considers the efficiency of the flow of materials in the traditional logistic processes of transportation, transaction and storage, cooperative connections with their intended contributions are not considered so that the widely accepted SCOR-approach in the logistics discipline cannot be used for the analysis of cooperative linkages.

Thirdly, for example, [17] distinguish three groups to describe characteristics of cooperative strategic partnerships (cf. for example [23] who use a similar approach by using the three groups of cooperation, coordination and composition):

1. Cooperation, i.e. the description of the character and forming of the collaboration;
2. Coordination, i.e. the description of the character and forming of the daily collaboration based on company-wide processes and methods in "terms of establishing 'rules of the road'";
3. Composition, i.e. the description of the character and forming of "long-term commitments to technology sharing and to closely integrated planning and control systems".

Hence, Schulte points out that these characteristic groups have to concretized through surveys, but their overarching aim is to create a common logistic understanding and knowledge base so that the approach only intends to describe the characteristics of cooperative strategic partnerships [14, p. 525]. Therefore, the causal interrelations of these connections are not analyzed so that possible network phenomena as a result of these causal interplays are neglected.

Network phenomena are the result of the characteristics of the elements of a system and their interplay and these emergent phenomena cannot simply reduced to the behavior on the individual level [24] and comprise, e.g. the efficient and/or transportation of goods in the Supply Chains as a result of the complex and dynamic interplay between institutions, products, technologies and markets [25]. Other network phenomena are known as Co-evolution (e.g. the performance of a company depends on the performance of another company [26]) or the aspect of self-organization (e.g. the appearance of new technologies lead to new organizational arrangements [27]). Consequently, the formalization of a network of networks brings the need for a systematic classification approach for the structural relations in the automotive industry can be derived.

3 Solution Approach

3.1 On Complex Network Science

Several authors argue that Complex Network Science "is rapidly becoming a lingua franca across virtually all of the sciences from anthropology to physics" [28] (cf. also [29, 30, 31, 32, 33]). Boccaletti et al. point out that in order to investigate global properties of such systems like efficiency or robustness one has "to model them as graphs whose nodes represent the dynamical units, and whose links stand for the interactions between them" [31]. Therefore, following Borgatti's and Li's generic discussion on Supply Chains, it is the intention to formalize his approach to the value creation network of networks for the production of electric vehicles [28] in order to consecutively conceptualize this interdisciplinary approach of complex network science applied in a supply chain management context for future research.

3.2 On Complex Network Variables

Selecting Nodes and Ties. By defining a supply network as a network "in which a tie exists from A to B if A supplies to B", Borgatti and Li point to the aspect that one also has to define what the relationship "supplies" mean [28]. It is highlighted that network scientists would be generally comfortable with the idea to "study whatever relations" one might "choose to define", because it depends on one is interested in studying. Furthermore, they also refer to the concept of multiplexity, i.e. the "property of given pair of actors having ties of many kinds simultaneously" (Fig. 2).

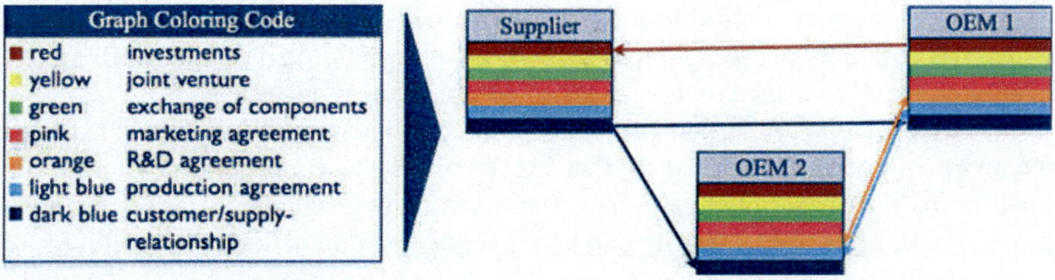

Fig. 2 The graph coloring code for the cooperative relationships between companies

Against this background and by considering the definitions of Supply Chain Management the companies shall be considered as the nodes in the network. The ties between the nodes can be one of the introduced different forms of cooperation (including the supply relationship) so that both the aspect of multiplexity according to Borgatti and Li and the network of networks as outlined can be represented for the production of electric mobility.

Ego network Composition. Borgatti and Li consider the concept of an ego network as "probably the closest to a supply chain theorist's intuitive understanding of an international supply network" [28]. Hereby, an ego network comprises a focal actor, known as the ego, the "set of actors with any kind of tie to ego", known as alters, and "all ties among alters and between the alters and the ego". As a result, such a network can be mathematically represented as a matrix X in which $x_{ij} = 1$ indicates that the firm $_i$ supplies to firm $_j$, while $x_{ij} = 0$ indicates that no supply relationship exists. In this conjunction, "the row sums of this matrix give the number of outgoing ties for each node", i.e. the out-degree centrality (cf. below) that could indicate the number of different customers for each firm, while the column sums represents the number of incoming ties, i.e. the in-degree centrality that could indicate the number of different suppliers. Furthermore, Borgatti and Li point out that the "quality" of ego's alters as a "useful ego-network property, because the strength of the focal firm can be derived from the strength of its trading partners" [28]. As a result, this measure can be defined as the average of the "attribute values for a given ego's alters":

$$q_i = \sum_j x_{ji} \cdot a_j \tag{1}$$

When we consider the concept of the ego network composition in the context of electric mobility, each company is characterized through r matrices X^r, where $r = 1,\ldots,7$ represents each one of the introduced different forms of cooperation. Herewith, it becomes possible to refer a set of quality vector q_i^k for each form of cooperation. Hereby, one has to distinguish between a quality of a company that can be referred to each form cooperation, e.g. the revenue, the number of employees or the profit of the company, and cooperation-specific qualities, e.g. the stock share for the cooperation form of investments, the number of patents for the R&D agreement or the value of the good for the customer-supply-relationship.

However, bear in mind that the simple ego network measures are only useful, when the full network—e.g. comprising the ego networks of the German car manufacturers—is assembled in order to compare these measures.

Node Centrality. A further key concept of complex network analysis "has been the notion of node centrality" [28] that tries to define the "importance of a node" in relation to "its structural position in the network as a whole". Hence, different characteristics of node centrality can be distinguished and defined as it follows:

- The characteristic of **node centrality of a node** is the number of incoming (in-degree) or outgoing links (out-degree) (cf. [32, p. 18]).
- The characteristic of **node centrality of a network** is the statistical distribution of the degrees of the nodes over the whole network (cf. [32, p. 29]).
- The characteristic of **closeness centrality** is defined as the inverse sum of the shortest paths to all other nodes (cf. [34, p. 20]). Borgatti and Li [28] further state that the closeness centrality defined as the sum of—graph-theoretically—distances to or from all other nodes enables the indication of the number of steps that "a focal firm's raw materials had to go through to get to the focal firm". In the case, that everything else is kept equal it can be hypothesized that "longer chains provide greater opportunities for disruption and cost increases" [5].
- The characteristic of **eigenvector centrality** is defined to measure the influence of a node in a network by assigning a relative value to all nodes so that connections with highly valued nodes contribute more (cf. [35]). Therefore, the notion of eigenvector centrality expresses the idea that "a node that is connected to nodes that are themselves well connected should be considered more central than a node that is connected to an equal number of less connected nodes" so that indirect and direct influences are both considered.
- The characteristic of **alpha centrality** is an adaptation of the notion of eigenvector centrality and the importance of a node is represented through the implementation of a vector of exogenous importance (cf. [36]). Borgatti and Li [28] express the eigenvector concept in the form of alpha centrality according to [37] for directed ties in the following form.

$$c_i = \alpha \sum_j x_{ij} c_j + e_i \tag{2}$$

$$c = (I - \alpha X)^{-1} e \tag{3}$$

(c_i: alpha centrality of node *i*, α: attenuation factor for the importance of length of a chain, *e*: a vector of exogenous importance, I: Identity matrix)

Applying the concept of alpha centrality to the production of electric mobility the vector *e* could represent the extent of the value creation in order to indicate to power of a tier supplier.

Structural Holes. Borgatti and Li use the concept of structural holes to explain the difference for Toyota's network to be able bring production back to normal levels within a couple of days, after a fire at a key supplier, A is in [28]. Here, the lack of structural holes made it possible that affiliated supplier was able to create an alternative configuration to produce the missing component. In the classical theory of Burt, structural holes increase the "amount of nonredundant information available to ego" so that in the case of ties as information exchanges structural holes should be negatively related to firm performance. However, the types of links being considered here is key. As Borgatti points out, in the Toyota context, the links mentioned are tight, cooperative relationships, increasing trust among suppliers; which caused the network to quickly reconfigure itself as suppliers were not hesitant to share know-how and blueprints. Therefore Borgatti sounds a cautionary note to the use of generic network concepts in supply chains and urges the practitioner to think in context.

When we consider structural holes to the production of electric mobility, especially in the situation of research and development of technologies for electric mobility (in the four areas) a lack of structural holes should make a positive difference. As a result, a denser ego network with few holes in the cooperative network of R&D agreements should be positively related with a successful production of electric mobility.

Hubs and Authorities: By introducing the measures μ for the extent if a company supplies firms with many suppliers and v for the extent if a company is supplied by firms with many customers, the following equations with a scaling component λ can provide an "inverse measure of the firm's bargaining power" [28].

$$\mu_i = \lambda \sum_j x_{ij} v_j \tag{4}$$

$$v_i = \lambda \sum_j x_{ij} \mu_j \tag{5}$$

When we consider the concept of hubs and authorities in the context of electric mobility, it would help to analyze the bargaining power of the suppliers indicated in Fig. 1 that have higher levels of cooperation with different car manufacturers.

3.3 Analytical Extensions for the Network of Networks for the Production of Electric Vehicles

To consider the above-outlined network of networks for the production of electric mobility, an analytical extension of [28]'s approach consists in adding a third dimension to indicate the different forms of cooperation (cf. Fig. 3). Herewith, it becomes possible to extend the introduced Eqs. (1–5) with the new dimension $r = 1, \ldots, 7$ in order to indicate specific kinds of relationships. For example, Eq. (1) for the representation of an alter's attributes can be re-written as Eq. (6) so that different values for each form of cooperation can be considered.

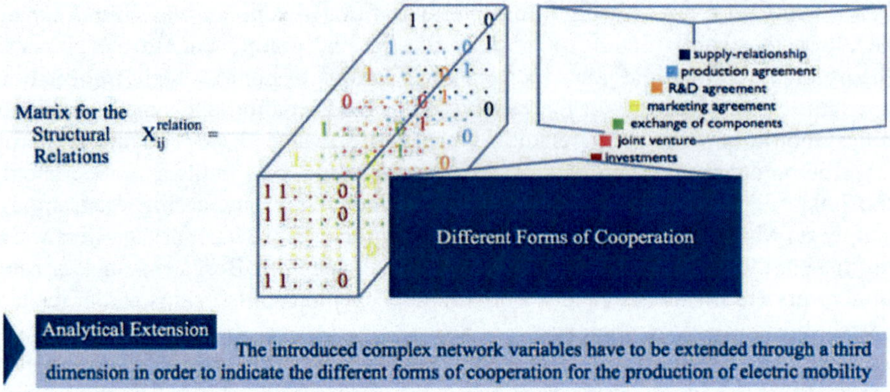

Fig. 3 Illustration of the extension of the approach of [28] for the production of electric mobility by adding a new dimension for the different forms of cooperation

$$q_i^r = \sum_j x_{ji}^r \cdot a_j^r \tag{6}$$

An analytical extension for each of the above-mentioned complex network variables is approached, because of the need to better analyze the causal interrelations between the companies. For example, by only considering the "whom-supplies-to-whom"-structure, the robustness as part of the risk management could only be simplified calculated, because a big supplier that has only a R&D agreement with several car manufacturers might have a bigger influence on the robustness (or efficiency) of the whole value creation network. For this reason, the number and diversity of company's structural relations should also be taken into account for investigating it's influence on the whole network as an indicator e.g. for its bargaining power for the analysis of the power shift in the value creation network.

4 Discussion and Concluding Remarks

4.1 Empirical Validation

Our initial point was to conduct a conceptual study on the application of network theory to supply chains before our overarching research question about the effects of the network structure on the supply chain target achievement can be investigated. Therefore, a systematic classification approach for the application of network metrics on the supply chain value creation system has been developed in the previous sections. Hence, the question arises about the validity of the solution approach as a quality criterion for scientific work (cf. e.g. [38]). For this reason, in [39] the validation of the developed framework is investigated by using an empirical dataset about the German automotive industry. Hereby, the dataset consists of the supply

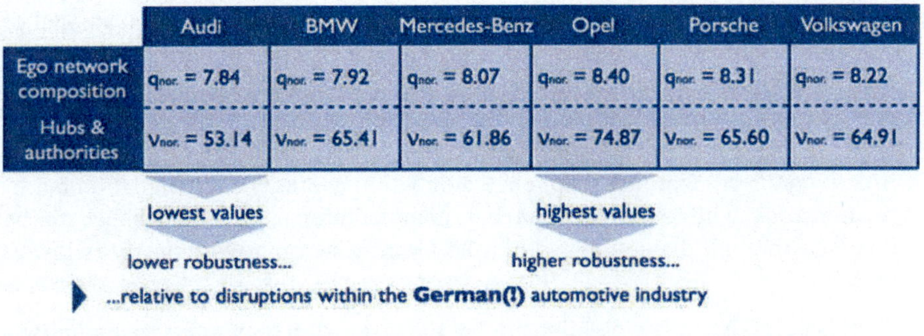

Fig. 4 Illustration of the goal of the network research with respect to robustness relative to disruptions within the German automotive industry

relationships between *i* German or German-speaking 1st tier suppliers and *j* car manufacturers, whereat the delivered component *c* on each supply relationship is also known, i.e. $X_{ij}^{r=7}$; 7 : *supply relationship*, $j = 1\ldots498$, $i = 1\ldots10$, $c = 1\ldots579$.

Figure 4 shows an extraction of the empirical results for the *Ego network composition* and *Hubs & authorities* for six German car manufacturers. At this, the normalized quality of $q_{i,nor.}$—i.e. *q* divided through the total amount of supply relationships of a car manufacturer *i*—represents the average value of potential suppliers in the German automotive industry for the ego network of the car manufacturer *i*, because the attribute *a* [cf. Eq. (6)] represents the number of potential—competitive—German supplier for a given component. The value ϑ_i goes one step further in order to identify *hubs* by representing if a car manufacturer is delivered from a German supplier that supplies to many other German car manufacturers. Therefore, the normalized amount of $\vartheta_{i,nor.}$ represents the average number of average of relationships with German car manufacturers (*note*: *the analysis in is done with* $i = 10$ *and the high features can be traced back that* 1st *tier suppliers have several relationships with a car manufacturers because of several components*). Against this background, the following implications can be derived as it follows:

- If $q_{i,nor.}$ is low/high, then the average amount of potential supply relationships with German suppliers for a given component *c* is low/high for the car manufacturer *i*.
- If $\vartheta_{i,nor.}$ is low/high, then a car manufacturer *i* is delivered from German suppliers that deliver in average to a low/high number of relationships with German car manufacturers.

In order to address the above-mentioned overarching research question, these implications shall be shortly discussed with respect to robustness. Hereby, a network (of one German car manufacturer) would be robust relative to disruptions within the German automotive industry, e.g. through environmental, technological, economic or political influences, that may lead for example to a breakdown of one or more German suppliers (for details, please consider [39]). In this conjunction, a low/high…

- $\ldots q_{i,nor.}$ might indicate a lower/higher robustness, because a low/high amount of potential relationships (in average) within the German automotive industry (for a given component c) might decrease/increase the chance to be involved through disruptions within the German automotive industry;
- $\ldots \vartheta_{i,nor.}$ might indicate a lower/higher robustness, because if a car manufacturer is delivered from German suppliers that have (in average) a low/high number of relationships with other German car manufacturers, then the chance to be involved through disruptions within the German automotive industry is lower/higher.

At this point, it shall be highlighted that the discussion only refers to disruptions within the German automotive industry. Hence, as illustrated in Fig. 4 these results imply to empirically validate the developed framework, because of the lowest and highest results for Audi and Opel. Due to the fact that Audi AG is a premium car manufacturer with its two main German production plants (Ingolstadt and Neckarsulm) and part of the Volkswagen AG (cf. [40]) it could be expected that most of the relationships would have been established with German suppliers. On the other side, despite of the fact that the Opel AG is a German car manufacturer, it is an affiliated company of General Motors Company [41] and an integration into foreign supply chains could be expected.

Consequently, with respect to the overarching research question about the application of network theory to supply chains and their contributions to the logistics target achievement the first results of the empirical investigation seem to validate the developed framework.

4.2 Contributions and Limitations

We aim to close a gap of knowledge in the logistics and supply chain management literature on to the formalization of cooperative strategic partnerships as a success factor through the production of electric mobility, as the emerging industrial landscape challenges us with the analysis of different forms of cooperative linkages in the automotive industry. A first empirical validation supports potential contribution of network theory to the logistics target achievement by referring to robustness relative to disruptions within the German automotive industry. In [39] the developed framework is further empirically validated and implications of the ratios for the competitiveness of the German automotive industry are discussed by pointing to a multi-dimensional analysis of the statistical correlations between the linkages and supply chain performance.

A number of limitations exist primarily because of the resistance towards a network perspective in the field of managerial economics, sensitivity of commercial relationships, and the consequent lack of adequately large-scale data that would make the application of complex network measures to network performance measures possible. As outlined in Sect. 3.2, taking the perspective of the ego network of the focal firm, the introduced complex network measurements have only limited explanatory power, because most of the measures gain importance by comparing them with the whole network.

4.3 Outlook

Future research includes the collection of empirical data to map the three different networks and investigate network properties. Furthermore, a multivariate statistical analysis will be carried out to connect certain complex network metrics with performance measurements of the focal firm. Therewith, future research aims to consecutively extend the applicability of methods from complex network science in a supply chain management context in order to systematically close the gap of knowledge about the complex causal interrelations of cooperative strategic partnerships for the production of electric vehicles.

Acknowledgments This research was supported by the Federal Ministry of Transport, Building and Urban Development as part of the "Model Regions Electric Mobility".

References

1. Bach, N., Buchholz, W., Eichler, B.: Geschäftsmodelle für Wertschöpfungsnetzwerke - Begriffliche und konzeptionelle Grundlagen. In Bach, N., Buchholz, W., Eichler, B. (eds.) *Geschäftsmodelle für Wertschöpfungsnetzwerke*, pp. 1–20. Ilmedia, Ilmenau (2010)
2. Arndt, H.: Supply Chain Management: Optimierung Logistischer Prozesse. Gabler, Wiesbanden (2010)
3. Sydow, J.: Management von Netzwerkorganisationen - Zum Stand der Forschung. In: Sydow, J. (ed.) Management von Netzwerkorganisationen, pp. 387–472 (2006)
4. Ketchen, D., Hult, T.: Bridging organisation theory and supply chain management—The case of best value supply chains. J. Oper. Manag. **25**(2), 573–580 (2007)
5. Brintrup, A., Kito, T., New, S., Reed-Tsochas, F.: From transaction cost economics to food webs: a multi-disciplinary discussion on the length of supply chains. EUROMA, Cambridge UK, July (2011)
6. Göpfert, I.: Logistik: Führungskonzeption. Gegenstand, Aufgaben und Instrumente des Logistikmanagements und -controllings. Vahlen, München (2005)
7. BEM: Elektroautos in Deutschland. Neue Mobilität - Das Magazin vom Bundesverband für eMobilität **2**, 24 (2011)
8. Wallentowitz, H., Freialdenhoven, A., Olschewski, I.: Strategien zur Elektrifizierung des Antriebsstranges: Technologien, Märkte und Implikationen. Vieweg + Teubner Verlag - GWV Fachverlage GmbH, Wiesbaden (2011)
9. BEM e.V.: Elektromobilität in Europa. *Neue Mobilität - Das Magazin vom Bundesverband eMobilität*, pp. 88–89, Jan 2011
10. Schröder, A.: Staatliche Förderung der eMobilität im internationalen Vergleich. Neue Mobilität - Das Magazin vom Bundesverband eMobilität, pp. 92–93, Jan (2011)
11. Colmorn, M., Hülsmann, M., Ergin, M., Chavez, T.: Are battery electric vehicles competitive?—The development of a customer value-based model. In: Proceedings of the 3rd European Conference Smart Grids and E-Mobility, pp. 60–67. Ostbayerisches Technologie-Transfer-Institut e.V. (OTTI), Regensburg (2011)

12. Colmorn, R., Hülsmann, M.: Wertschöpfungsnetzwerke - Berücksichtigung der Komplexität für einen erfolgreichen Systemwechsel. *Neue Mobilität - Das Magazin vom Bundesverband eMobilität*, 103 (2011)
13. Rösler, O.; Gestaltung von kooperativen Logistiknetzwerken: Bewertung unter ökonomischen und ökologischen Aspekten. Dt. Univ.-Verl (2003)
14. Schulte, C.: Logistik - Wege zur Optimierung der Supply Chain. Vahlen, München (2009)
15. Alicke, K.: Planung und Betrieb von Logistiknetzwerken - Unternehmensübergreifendes Supply Chain Management. Springer, Berlin (2005)
16. Schönsleben, P.: Integrales Logistikmanagement: Planung und Steuerung von umfassenden Geschäftsprozessen. Springer, Berlin (1998)
17. Harrison, A., van Hoek, R.: Logistics Management and Strategy—Competing Through the Supply Chain. Pearson Education Limited, Essex (2008)
18. Sigl, K.: Uns steht ein Systemwechsel bevor. Reflex Verlag, Berlin (2011)
19. Kreutzer, K.: Elektromobilität - Chancen und Risiken für Energieversorger. KREUTZER Consulting UG, München (2009)
20. Roland Berger Consultants: Powertrain 2020—The Future Drives Electric. Roland Berger Strategy Consultants (2008)
21. Brintrup A.K.T.-T.: Mapping the Toyota supply network: the emergence of resilience. In: Said Business School Working Paper 2011-05-012 (2012)
22. Schönsleben, P., Hieber, R.: Gestaltung von effizienten Wertschöpfungspartnerschaften im Supply Chain Management. In: Busch, A., Dangelmaier, W. (eds.) Integriertes Supply Chain Management - Theorie und Praxis effektiver unternehmensübergreifender Geschäftsprozesse, 1st edn, pp. 45 – 62. Gabler, Wiesbaden, (2002)
23. Supply Chain Council: Supply Chain Operations Reference Model. 6th Version, Pittsburgh (2003)
24. Ludwig, B.: Management Komplexer Systeme - Der Umgang mit Komplexität bei unvollkommener Information: Methoden, Prinzipien, Potentiale. Edition sigma, Berlin (2001)
25. Bruce, K.: The network as knowledge—generative rules and the emergence of structure. Strateg. Manag. J. **21**(3), 405–425 (2000)
26. Li, G., Ji, P., Sun, L., Lee, W.: Modelling and simulation of supply network evolution based on complex adaptive system and fitness landscape. Comput. Ind. Eng. **124**, 839–853 (2009)
27. Wycisk, C., McKelvey, B., Hülsmann, M.: 'Smart parts' logistics systems as complex adaptive systems. Int. J. Phys. Distrib. Logistics Manage. **38**(2), 108–125 (2008)
28. Borgatti, S., Li, X.: On social network analysis in a supply chain context. J. Supply Chain Manage. **45**(2), 5–22 (2009)
29. Albert, R., Barabasi, A.: Statistical mechanics of complex networks. Rev. Mod. Phys. **74**, 47–97 (2002)
30. Amaral, L., Ottino, J.: Complex networks: augmenting the framework for the study of complex systems. Eur. Phys. J. B **38**(2), 147–162 (2004)
31. Bocaletti, S., Latora, V., Moreno, Y., Chavez, M., Hwang, D.: Complex networks: structure and dynamics. Phys. Rep. **424**, 175–308 (2006)
32. Dorogovtsev, S.: Lectures on Complex Networks. Oxford University Press, Oxford (2010)
33. Kim, Y., Choi, T., Yan, T., Dooley, K.: Structural investigation of supply networks: a social network analysis approach. J. Oper. Manage. (2010)
34. Newman, M.: The Structure and Function of Complex Networks (2002)
35. Borgatti, S., Everett, M.: A graph-theoretic perspective on centrality. Soc. Netw. **28**(4), 464–484 (2006)
36. Bonacich, P., Lloyd, P.: Eigenvector-like measures of centrality for asymmetric relations. Soc. Netw. **23**(3), 191–201 (2001)
37. Bonacich, P.: Factoring and weighting approaches to status scores and clique identification. J. Math. Sociol. **110**(1), 113–120 (1972)
38. Schnell, R., Hill, P. B., Esser, E.: Methoden der empirischen Sozialforschung. Oldenbourg Wissenschaftsverlag (2011)

39. Colmorn, R., Hülsmann, M.: Eine Untersuchung der strukturellen Komplexität des Netzwerkes an Wertschöpfungsnetzwerken der deutschen Automobilindustrie - Implikationen zur Bestimmung ihrer Wettbewerbsfähigkeit. *4. Wissenschaftsforum Mobilität "Steps to Future Mobility"* p. 12. Springer Gabler, Duisburg (2012)
40. Volkswagen, A.G.: Vielfalt erfahren - Geschäftsbericht 2011. Volkswagen Finanzpublizität, Wolfsburg (2012)
41. General Motors Company: This is the new GM—annual report 2010. Detroit (2011)

Network Configuration in Presence of Synchronization Requirements

Jörn Schönberger and Herbert Kopfer

Abstract This article investigates a multi-commodity network flow problem. The generated network flows represent processes running in this network. Redundant processes are installed in order to increase the robustness of the transportation system. A product can be served by two or even more of these processes so that ad-hoc re-assignments from one mode to another mode can be applied. Special attention is paid to the temporal and spatial synchronization of alternative processes. We propose a mathematical model for the investigated problem and evaluate this model within computational experiments.

Keywords Network configuration · Mathematical programming · Transport network · Synchronization · Multi-commodity network flow

1 Introduction

This contribution investigates the consideration of transport network robustness issues during the network configuration phase. In this phase, explicit transport tasks are still unknown but origin-destination pair relations to be served are specified. Ad hoc load peaks on some origin-to-destination connections occur frequently so that some customer demand is in danger to remain unfulfilled.

J. Schönberger (✉) · H. Kopfer
Chair of Logistics, University of Bremen,
Wilhelm-Herbst-Straße 5,
Bremen 28359, Germany
e-mail: jsb@uni-bremen.de

H. Kopfer
e-mail: kopfer@uni-bremen.de

K. Windt (ed.), *Robust Manufacturing Control*, Lecture Notes in Production Engineering,
DOI: 10.1007/978-3-642-30749-2_5,

The basic idea to increase the system's robustness against such unexpected demand peaks is to connect each origin with its destination by two or even more services that offer transport capacity on different paths through the network (redundant services).

Transport processes in networks follow strict schedules. Customers adapt their internal processes to these schedules to ensure a smooth material flow throughout their value creation systems. For this reason, the arrival times of redundant services must be synchronized. The consideration of coordinated arrival times of redundant transport services in a network enables switching from one transport service to a redundant service at short notice without significant delays or temporary storage of goods within the network. Such a coordination of processes of different entities (e.g. vehicles) is referred to as synchronization [4].

Within this contribution, a transport network configuration problem with inherent synchronization requirements of vehicle routes and vehicle arrival times at nodes is investigated. It is analyzed if (and to which extent) the consideration of synchronizing constraints within network flow optimization models impacts the process execution costs and/or the compilation of the generated processes. An informal description of the network configuration problem at hand is given in Sect. 2. Next, a network flow optimization model as starting point of a computational evaluation is presented in Sect. 3. Finally, initial evaluation results are presented and discussed in Sect. 4.

2 Network Services Generation

2.1 Related Literature and Examples

The configuration of a network comprises the linking of network nodes by inserting connections (arcs) between selected pairs of nodes. Each node represents an activity in the network, or a resource or a network state. Logical or temporal precedence relations between the connected activities are reflected in the network representation by the inserted arcs. An alternating sequence of nodes and arcs can be interpreted as a network state transformation or as a value creation process. Examples of processes include routes of vehicles visiting a sequence of customer sites [9] or machine scheduling tasks [2]. Each process is setup to fulfill a certain task and might be evaluated by determining used capacities, occurred costs or gained revenues. The selection or construction of such a task is often guided by one or more planning goal(s) like the minimization of the process execution costs or the maximization of the throughput. Network configuration problems that aim at generating one process are subsumed under the term *shortest path problem* [10]. In a *multi-commodity network flow problem* [17] several processes must be determined in parallel. Often, diverse requirements restrict the process generation.

A process in a network is referred to be *robust* if it is able to manage exogenous task variations without significant service output quality variation (punctuality, quality, reliability, ...) at reasonable input variations (costs, man power, ...) [14]. In order to install processes that are robust in the operational (short-term) scope it is necessary to consider robustness issues already in the network configuration phase. Arrangements to protect network processes against compromising external impacts must be installed during the network configuration phase. They will become then effective during the process deployment. A well known strategy to achieve robustness is to install redundant (parallel) processes within the network [12] offering alternative request fulfillment modes. The installation of redundant processes enables the value creation system to switch from one fulfillment mode to another mode in an ad-hoc fashion in the event that the first mode is not available due to a failure or if it is exhausted. Costly emergency activities necessary to extend manpower at short hand (overtime working or stand-by-machines) can be avoided and also costly short-term rentals of additional equipments becomes unnecessary.

The synchronization of processes aims at establishing connections between several processes defined in a given network [6]. Such connections are established in order to enable the transfer of information and/or (partly-finished) goods among independent processes. In (in-house) manufacturing logistics, synchronization is mainly used to feed core assembling processes by supply processes that provide the required materials . In transportation logistics hierarchical consolidation and de-consolidation of flows of individual goods to/from bundled transportation flows are used. Furthermore, transfer opportunities for passengers must be installed in order to offer comfortable passenger transport chains [7]. Another topic is the achievement of "green waves" in traffic signal control [11]. In addition, it is tried to connect staff rosters with machine and/or routes in order to enable staff replacements [3].

Drexl [4] classifies the synchronization requirements that appear in transport logistics into five categories. *Task synchronization* is required in the event that several resources (e.g. vehicles) must fulfill a demand cooperatively (like in the split delivery vehicle routing problem [5]). If loading or unloading operations must be coordinated by time and/or location then this coordination is referred to as *operation synchronization*. *Movement synchronization* means that two or even more resources (vehicles) must use the same path like in truck-trailer applications [3]. If cargo is interchanged between vehicles then *load synchronization* is needed and *resource synchronization* addresses the situation when two or more vehicles use ("share") the same (scarce) resource(s).

2.2 Basic Notation

A **product** p requires the transformation of a system from an origin state p^+ into a destination state p^-. Thus, in a given system a product p is described by the ordered pair $p = (p^+;p^-)$. Resources are deployed to vary a system from one state a into another state b. A **resource** res is therefore determined by the ordered pair $res = (a;b)$. Two resources $(a;b)$ and $(c;d)$ are called **concatenated resources** if and only if $b = c$. The **demand** of resource res is denoted by the real-numbered value $U(res)$

A finite sequence m of concatenated and pair wise different resources $m := ((a;b),(c;d),,(w;x),(y;z))$ so that $a = p^+$, $b = c$, $d = e$,... ,$x = y$ and $z = p^-$ is called **mode** m of product $p = (p^+,p^-)$. The ordered pair $S := (m,c)$ consisting of the mode m (of a certain product) and of a real-valued number c is called a **service** and the real-valued number $C(m)$ represents the **capacity of the service** m. A service can be interpreted as a routed vehicle. The capacity c is the maximal sum of demand of several resources that can be executed simultaneously by the service S. If a service s executes the demand of a resource then the service capacity is reduced by $U(res)$. The consecutive execution of concatenated resources varies the remaining capacity c step-by-step. Since also negative resource demand values $U(res)$ are allowed it is possible that the remaining capacity re-increases after a resource demand has been executed.

The **load** of service S on a resource (i,j) is defined as the sum of executed demand of the so far processed resources including resource (i,j) according to the mode m of S. A **service** $S := (m,c)$ **offers product** $p = (p^+,p^-)$ if and only if there exists at least one mode $m(p)$ of product p so that $m(p)$ is a subsequence of m.

All resources donated by the aforementioned modes are indexed with the donating mode and the index resources are then put into the set $\mathcal{A}$. The ordered pair $\mathcal{G} := (\mathcal{N};\mathcal{A})$ is then a mathematical graph with the vertices set $\mathcal{N}$ and the arc set $\mathcal{A}$. Since the arcs are indexed by the donating mode it is possible to maintain parallel arcs in $\mathcal{G}$.

Of course it is possible to maintain more than one mode for a product in the graph representation. A product from the set $\mathcal{P}$ that has two or more modes in the derived graph $\mathcal{G}$ is called a **flexible product** [8]. In contrast, if there is only one mode available for a product then such a product is called a **specific product**.

A flexible product is served by two or more services. Each service determines a process that is used to execute and fulfill the requests that have been assigned to this service. In the event that one of these services (processes) becomes unavailable, another service (process) can be selected to serve the request for a flexible product. Thus, the existence of flexible products determines the existences of redundancies within the considered value creation system (network). These redundancies enable a system to keep its performance on an acceptable level even if components of the system are unavailable [15, 18]. Therefore, the setup and maintenance of a collection of services should take the definition of a large number of flexible products into account. However, maintaining too much services and too high (regularly unused) capacities is costly. Thus, a reasonable trade off between the number of maintained services and the number of implied flexible products must be identified.

2.3 Informal Challenge Description

Flexible products offer alternative production modes. In the event that a certain mode fails then tasks related to flexible products are able to be re-assigned to another mode. However, the installation and the maintenance of any alternative mode imply costs as well as the absorption of man power or machines are achieved. Consequently, a reasonable trade off between the positive impacts of offering redundant services (e.g. reliability) and the associated negative impacts (e.g. costs) must be identified. From the system's perspective and with the intention to offer a sufficient high reliability level it is reasonable to specify the number of offered flexible products. However, from the cost perspective the number of installed and maintained services in the production network should be kept as small as possible. The challenge is now to find out the minimal cost services required in the production network so that all products are served and so that the desired number of flexible products is achieved. In this investigation, the number of the flexible products is given because we want to compare the network performance and appearance with respect to the number of flexible products that must be served.

Although the minimal number of required services to meet the desired number of flexible products is not known in advance there is a maximal number of available services given (derived from a given budget or from available capacities). The set $\mathcal{S}$ contains all available services. Each service $s_i \in \mathcal{S}$ starts from an initial state s_i^+ and terminates in a final state s_i^- after having been executed. All initial states are collect in the set $\mathcal{N}^+$ and all final states are put in the set $\mathcal{N}^-$. We integrate these additional nodes into the existing graph $\mathcal{G}$ and define the extended graph $\mathcal{G}^* := (\mathcal{N}^*; \mathcal{A}^*)$. The set of nodes of graph $\mathcal{G}^*$ is updated to $\mathcal{N}^* := \mathcal{N} \cup \mathcal{N}^+ \cup \mathcal{N}^-$. Arcs originating from $\mathcal{N}^+$ and terminating in $\mathcal{N}$ as well as arcs originating from $\mathcal{N}$ and terminating in $\mathcal{N}^-$ are added to the set $\mathcal{N}$ so that the extended arc set $\mathcal{A}^*$ is formed.

Using terminology from graph theory it is necessary to determine an origin-destination-path for each service $s \in \mathcal{S}$ through $\mathcal{G}$ that starts at the source node s_i^+ and terminates in the sink node s_i^-. If such a path does not cover any product then the path of such service is not deployed. An origin-destination-path (od-path) s (associated with service $s \in \mathcal{S}$) is feasible if the following conditions are fulfilled: It starts at the given starting node $s^+ \in \mathcal{N}^+$ (C1) it terminates at the given terminating node $s^- \in \mathcal{N}^-$ (C2) all intermediately visited nodes are product-related nodes contained in $\mathcal{N}$ (C3) every node is visited only once by a service s (C4) if s

serves product p then p^+ is visited before p^- is visited (C5) the capacity of a service s is not exceeded at any node contained in s (C6). Beside the service-specific requirements there is the requirement that at least N^{flex} of the available products must be served by at least two services (C7). If two or more services serve a flexible product then their arrival times at the associated pickup and at the associated delivery location must not differ more than DT^{max} time units (C8).

Following the classification of Drexl [4], the outlined problem comes along with three types of synchronization constraints. First, task synchronization must be achieved because several vehicles must serve a flexible product. Second, operation synchronization is needed due to the need for similar arrival time of all the vehicles serving a flexible product and, third, resource synchronization is required because all products share the limited number of vehicles.

To ease the upcoming informal and the formal problem statement we assume that the involved four nodes p_1^+, p_1^-, p_2^+, p_2^- of each two products (p_1^+, p_1^-) and (p_2^+, p_2^-) are pair wise different. The sum of process costs that is determined by the sum of traveled distances of all services must be minimized.

3 Mixed-Integer Linear Program

Decision Variables. The number of services serving product p is stored in the non-negative integer-valued decision variable N_p. In addition, the binary decision variable v_{ps} ($p \in \mathcal{P}, s \in \mathcal{S}$) indicates whether service s is selected to serve product p ($v_{ps} = 1$) or not (**product service assignment variables**).

For the determination of the services, we follow the commonly used path coding idea and deploy the family of binary decision variables x_{ijs} ($i,j \in \mathcal{N}^*, s \in \mathcal{S}$). If and only if service s uses the resource (i,j) then x_{ijs} equals 1 (**routing variables**).

The **arrival time** of a service s at the node i within its path is stored in the continuous and non-negative decision variable t_{si}. Arrival times are updated recursively along the path of a service. Similarly, the **inbound load** ω_{si} of a service s going to node i is determined and stored.

In order to simplify the model the following dependent decision variables are employed in the model. The binary indicator variable y_p is 1 if and only if $N_p \geq 2$ (p is a flexible product if and only if $N_p \geq 2$). Finally, the load variation of service s at node i is stored in the continuous non-restricted decision variable ${}_{si}$, which is ≥ 0 if i is a loading node and which is ≤ 0 if i is an unloading node in the path of service s. Its absolute value gives the loaded (or unloaded) quantity.

Service Assignment to Products. In order to implement the requirement to get at least N^{flex} flexible products (C7) it is necessary to count the products served by two or more services. Therefore, the number of services assigned to a certain product p is counted and the achieved number is stored in the decision variable N_p (1). Second, the binary indicator variable y_p is set to 1 if and only if $N_p \geq 2$ (2), (3).

Finally, the y_p-indicator variables are summed up at it is ensured that this sum is at least N^{flex} (4). The constraints (1)–(4) implemented the required task synchronicity.

$$N_p = \sum_{s \in \mathcal{S}} v_{ps} \ \forall p \in \mathcal{P} \tag{1}$$

$$N_p \geq y_p + 1 \ \forall p \in \mathcal{P} \tag{2}$$

$$N_p \leq y_p \cdot \mathcal{M} + 1 \ \forall p \in \mathcal{P} \tag{3}$$

$$N^{flex} \leq \sum_{p \in \mathcal{P}} y_p \tag{4}$$

Service Path Determination. The basic idea for coding the requirements of a feasible service path in linear constraints is to re-use common flow-preserving and flow directing constraints so that a service path connects an (artificial) origin node with an (artificial) destination node. The linear constraints (5)–(13) are required to ensure the feasibility of the generated od-paths of the available services $s \in \mathcal{S}$ as described by (C1)–(C4) in Sect. 2.3.

$$x_{iis} 38; = 0 \ \forall s \in \mathcal{S} \ \forall i \in \mathcal{N}^* \tag{5}$$

$$\sum_{j \in \mathcal{N}} x_{jis} 38; = 0 \ \forall s \in \mathcal{S} \ \forall i \in \mathcal{N}^+ \tag{6}$$

$$\sum_{j \in \mathcal{N} \setminus \mathcal{N}^+} x_{i_s^+ js} = 1 \ \forall s \in \mathcal{S} \tag{7}$$

$$\sum_{i \in \mathcal{N}^+} \sum_{j \in \mathcal{N}} x_{ijs} = 1 \ \forall s \in \mathcal{S} \tag{8}$$

$$\sum_{j \in \mathcal{N} \cup \mathcal{N}^+} x_{j i_s^- s} = 1 \ \forall s \in \mathcal{S} \tag{9}$$

$$\sum_{j \in \mathcal{N}} x_{ijs} = 0 \ \forall s \in \mathcal{S}, \forall i \in \mathcal{N}^- \tag{10}$$

$$\sum_{i \in \mathcal{N}} x_{ijs} \leq 1 \ \forall s \in \mathcal{S}, \forall j \in \mathcal{N} \tag{11}$$

$$\sum_{i \in \mathcal{N}} x_{jis} \leq 1 \ \forall s \in \mathcal{S}, \forall j \in \mathcal{N} \tag{12}$$

$$\sum_{j \in \mathcal{N}} x_{jis} = \sum_{j \in \mathcal{N}} x_{ijs} \ \forall s \in \mathcal{S}, \forall i \in \mathcal{N} \tag{13}$$

The constraint (5) prevents self-loops while constraint (6) hinders the travel to a service starting node after a service has left its initial node. The second node in each service is a non-starting node (7). Each service originating from its starting node goes either to a product node or to a terminating node next (8). Furthermore, each service visits its terminating node exactly once (9) and a terminating node $i \in \mathcal{N}^-$ is never left (10). In addition, each node it targeted at most one time (11) and left not more than once (12). Finally, a service that leaves a product node has to go to this node before (13).

Scheduling. The following constraints (14)–(17) are defined to preserve the necessary temporal precedence condition that a pickup node of a product must be visited earlier than the associated delivery node. The defined decision variables only consider the direct successor node of a pickup node. In order to ensure that the delivery node p^- is visited after p^+ the arrival time of a service at a node is determined and it is enforced that a service s goes to p^+ earlier than to node p^-.

$$t_{i_s^+} 38; = 0 \ \forall s \in \mathcal{S} \tag{14}$$

$$t_{si} + d_{ij} \leq t_{sj} + (1 - x_{ijs}) \cdot \mathcal{M} \ \forall s \in \mathcal{S}, \forall i \in \mathcal{N}, \forall j \in \mathcal{N} \tag{15}$$

$$t_{i_s^-} 38; \leq 15 \ \forall s \in \mathcal{S} \tag{16}$$

$$t_{sp^+} \leq t_{sp^-} + (1 - v_{ps}) \cdot \mathcal{M} \ \forall s \in \mathcal{S}, \forall p \in \mathcal{P} \tag{17}$$

All services start at time 0 (14). In the event that service s uses the arc (i,j) constraint (15) is activated: The Big-M-term $\mathcal{M}$ on the right side disappears then, so that the temporal difference between the visiting time at node i and at its successor node j is at least the transformation time d_{ij}. In the event that service s does not employs the arc (i,j) then the requested inequality is fulfilled independently from the values of t_{si} and t_{sj}. We restrict the arrival time at the final node of a service to be not later than 15 (16). This is a technical constraint to enforce the definition of an arrival time at the finale node of a mode. Finally, (17) ensures that the pickup node p^+ of a product p is visited by service s before the associated delivery node p^- in the event that service s is selected to serve p.

Consideration of Limited Service Capacity. In the event that the service load decreases (increases) monotonically along a service it is sufficient to restrict the initial (final) load as required in the traditional vehicle routing application [9] (in a collecting vehicle routing problem [16]). In the here investigated application, a initial service capacity is reduced at product starting nodes but it re-increases at product terminating nodes similar as done in the well-known pickup-and-delivery routing applications [13]. Therefore, it is necessary to explicitly restrict the en-route service utilization at every intermediate node included in a service s in order to fulfill the feasibility condition (C6).

$$\omega_{si_s^+} = 0 \; \forall s \in \mathcal{S} \tag{18}$$

$$\Delta_{si} = \sum_{p \in \mathcal{P}} p^+(p,i) v_{ps} v(p) - \sum_{p \in \mathcal{P}} p^-(p,i) v_{ps} v(p) \; \forall s \in \mathcal{S}, \forall i \in \mathcal{N} \tag{19}$$

$$\omega_{si} + \Delta_{sj} \leq \omega_{sj} + (1 - x_{ijs}) \cdot \mathcal{M} \; \forall s \in \mathcal{S}, \forall i,j \in \mathcal{N} \tag{20}$$

$$\omega_{sj} \leq C(s) + (1 - \sum_{i \in \mathcal{N}} x_{ijs}) \cdot \mathcal{M} \forall j \in \mathcal{N} \; \forall s \in \mathcal{S} \tag{21}$$

$$\omega_{si_s^-} = 0 \; \forall s \in \mathcal{S} \tag{22}$$

The restriction (18) initializes the service load $\omega_{si_s^+}$ at the service start node i_s^+ to 0. Parameters are introduced in order to describe if a certain node i is a loading node or an unloading node associated with a product p. The binary parameter $p^+(p,i)$ is set to 1 if and only if node i is the starting node of product $v(p)$. Similarly, the binary parameter $p^-(p,i)$ is set to 1 if and only if node i is the terminating node of product p. Using these parameters, the restriction (19) determines the load variation of service s at node i. The calculated load variation is used to update the service load along the service path (20) and (21) is the necessary capacity restriction. Finally, constraint (22) ensures that a service terminates empty. The capacity update constraints as well as the load limitation constraints (20) and (21) are only activated, if the service s uses arc (i,j). Following the classification in [4] (18)–(22) are the resource synchronization constraints that control the usage of the vehicle's capacities.

Coupling of Services and Service-to-Product Assignments. After the constraints for determining the services have been stated and after the assignment of products to services has been addressed it is necessary to couple both decision subproblems so that the determined services can cover those products that are assigned to each service. Therefore, (23) ensures that sufficient services leave the loading node p^+ of product p and (24) forces at least N_p services to go the unloading node p^- of product p. Furthermore, constraints (25) and (26) ensure that a service s visits both the loading as well as the unloading node of product p in the event that p is assigned to s. In addition, the constraint (17) compels that the loading node is visited before the associated unloading node.

$$\sum_{s \in \mathcal{S}} \sum_{j \in \mathcal{N}} x_{p^+js} \geq N_p \ \forall p \in \mathcal{P} \tag{23}$$

$$\sum_{s \in \mathcal{S}} \sum_{j \in \mathcal{N}} x_{jp^-s} \geq N_p \ \forall p \in \mathcal{P} \tag{24}$$

$$(1 - v_{ps}) \cdot \mathcal{M} + \sum_{j \in \mathcal{N}} x_{p^+js} \geq v_{ps} \ \forall s \in \mathcal{S}, \forall p \in \mathcal{P} \tag{25}$$

$$(1 - v_{ps}) \cdot \mathcal{M} + \sum_{j \in \mathcal{N}} x_{jp^-s} \geq v_{ps} \ \forall s \in \mathcal{S}, \forall p \in \mathcal{P} \tag{26}$$

Operation Synchronization for Flexible Products. Let p be a flexible product that can be served by the modes s_1 as well as s_2. In order to ensure that both services start and complete the execution of a request for p in a comparable fashion (making both modes exchangeable from the customers perspective) it is necessary that the starting times of the operations at p^+ and p^- are similar, e.g. they differ not more than DT^{max} time units. This condition is described in (C8). In order to model this condition in terms of linear constraints, the four constraint (27)–(30) are setup. Again, the $\mathcal{M}$-factor ensures that any of these four constraints is only activated if (and only if) product p is assigned to both services s_1 as well as s_2. These four constraints are the required operation synchronization constraints.

$$t_{s_1p^+} - t_{s_2p^+} \leq DT^{max} + (2 - v_{ps_1} - v_{ps_2}) \cdot \mathcal{M} \; \forall p \in \mathcal{P} \; \forall s_1, s_2 \in \mathcal{S} \tag{27}$$

$$t_{s_2p^+} - t_{s_1p^+} \leq DT^{max} + (2 - v_{ps_1} - v_{ps_2}) \cdot \mathcal{M} \; \forall p \in \mathcal{P} \; \forall s_1, s_2 \in \mathcal{S} \tag{28}$$

$$t_{s_1p^-} - t_{s_2p^-} \leq DT^{max} + (2 - v_{ps_1} - v_{ps_2}) \cdot \mathcal{M} \; \forall p \in \mathcal{P} \; \forall s_1, s_2 \in \mathcal{S} \tag{29}$$

$$t_{s_2p^-} - t_{s_1p^-} \leq DT^{max} + (2 - v_{ps_1} - v_{ps_2}) \cdot \mathcal{M} \; \forall p \in \mathcal{P} \; \forall s_1, s_2 \in \mathcal{S} \tag{30}$$

Minimization of the overall service costs. The overall sum of costs for the multi-commodity network flow (31) is minimized (c_{ijs} are the costs resulting from the usage of (i,j) by service s (proportional to the length d_{ij} of (i,j)).

$$\sum_{s \in \mathcal{S}} \sum_{i \in \mathcal{N}} \sum_{j \in \mathcal{N}} c_{ijs} \cdot x_{ijs} \tag{31}$$

The mixed-integer linear program (5)–(26), (27)–(31) is a representation of the task to set up a least cost collection of services in a given network so that all products are served and the synchronization requirements are considered.

4 Experimental Evaluation

The proposed model is evaluated in computational experiments. All experiments are executed within a network build up on a set $\mathcal{N}$ comprising 10 nodes. The maximal number of services available is 5, so that the set $\mathcal{N}^+$ ($\mathcal{N}^-$) contains 5 artificial starting (terminating) nodes leading to a node set $\mathcal{N}^*$ consisting of 20 nodes. Artificial test cases are setup for this configuration of the proposed model. Each test case is defined by the triple $(N^{flex}, DT^{max}, \alpha)$. The first component N^{flex} defines the requested number of flexible products. The second component is the maximal allowed arrival time difference DT^{max}. The third component α seeds the random pairing of nodes for the definition of 10 products in the used network. The travel distance matrix $(d_{ij})_{i,j \in \mathcal{N}^*}$ is randomly drawn (with uniform distribution) from the interval $[1;2]$ using also the seeding α. Test cases are generated for $N^{flex} \in \{0;2;4,6,8;10\}$, $DT^{max} \in \{0;0.25;0.5;0.75;1;3;5;25\}$ and $\alpha \in \{0,1,2\}$. Overall, $1 + 5 \cdot 8 \cdot 3 = 121$ test cases are set up. Each test case parameterizes an instance of the mixed-integer linear program proposed in Sect. 3, which is solved by CPLEX 12.3 on a Core2 DUO CPU T7500 2.2 GHz with 2 GB RAM computer. The maximal solver processing time is set to 6 minutes. The variation of the total travel length caused by the variation of N^{flex} and DT^{max} is observed. For given

Table 1 Relative travel distance variation $\Delta L(N^{flex}, DT^{max})$ (averaged)

N^{flex}	DT^{max}							
	25	5	3	1	0.75	0.50	0.25	0
2	0.20	0.20	0.20	0.20	0.20	0.20	0.20	0.21
4	0.38	0.40	0.40	0.40	0.38	0.38	0.38	0.38
6	0.58	0.61	0.58	0.58	0.60	0.60	0.58	0.58
8	0.78	0.80	0.79	0.78	0.78	0.84	0.82	0.82
10	1.01	1.03	1.03	1.03	1.03	0.81*	1.01	1.05

* lower bound

values of N^{flex} and DT^{max} let $L(N^{flex}, DT^{max})$ be the averagely observed sum on the travel lengths of all involved services. The increase L of L in relation to the actually unconstrained cases with $N^{flex} = 0$ and $DT^{max} = 25$ is calculated by $L(N^{flex}, DT^{max}) := L(N^{flex}, DT^{max})/L(0, 25)$.

The similarity of two service collections contained in the generated service plans P and Q is evaluated by means of a pseudo Hamming Distance function $H^1(P, Q)$ [1]. For a given service plan P, the binary parameter $a_{ij}(P)$ is 1 if and only if both nodes i as well as j are assigned to the same service according to P. Given two nodes $i, j \in \mathcal{N}$ and two service plans P and Q, the binary parameter $b_{ij}(P, Q)$ is 1 if and only if $a_{ij}(P) = 1$ as well as $a_{ij}(Q) = 1$, which means that i and j can be found together in at least one service in both service plans. These two parameters enable the definition of the pseudo Hamming Distance $H^1(P, Q)$ between two service plans: $H^1(P, Q) = 1 - \frac{\sum_{i,j \in \mathcal{N}} b_{ij}(P,Q)}{\sum_{i,j \in \mathcal{N}} a_{ij}(P)}$. In contrast to a regular Hamming Distance, H^1 is not reflexive, which means that $H(P, Q) = H(Q, P)$ cannot be guaranteed for all pairs of service plans P and Q. Therefore, we calculate $H^1(P, Q)$ as well as $H^1(Q, P)$ and define $H(P, Q) := \frac{H^1(P,Q)+H^1(Q,P)}{2}$. Furthermore, $\bar{H}(a, b)$ is the average of the $H(P, Q)$-values observed for plans P generated with $DT^{max} = a$ and for plans Q generated with $DT^{max} = b$.

In Table 1 the averagely observed travel length variations $L(N^{flex}, DT^{max})$ are summarized. In general, the specification of two additional flexible products implies an increase of the traveled distance of around 20% compared to the case with $N^{flex} = 0$. Furthermore, it can be seen that the variation of the strictness of the temporal synchronization DT^{max} of different modes belonging to a certain flexible product influences the travel length only slightly. In the event that all products are flexible then the total travelled distance is more than doubled compared to the case with $N^{flex} = 0$. The highest increase is achieved for $DT^{max} = 0$.

It has been shown that tightening the maximal allowed difference between departure and arrival times of different modes within a flexible product does not significantly affect the overall travel length. In order to find out why the overall travel length reacts insensitively with respect to the variation of the maximal allowed time variation DT^{max}, we investigate the similarity of the generating service plans. Table 2 contains average H-values that quantify variations of the

Table 2 Similarity: $\bar{H}(\cdot,\cdot)$-values

N^{flex}	(25;5)	(5;3)	(3;1)	(1;0.75)	(0.75;0.5)	(0.5;0.25)	(0.25;0)
2	0.00	0.09	0.00	0.00	0.00	0.00	0.08
4	0.27	0.00	0.00	0.27	0.00	0.00	0.27
6	0.11	0.11	0.00	0.00	0.00	0.00	0.00
8	0.38	0.38	0.00	0.00	0.18	0.19	0.16
10	0.00	0.00	0.02	0.00	0.00	0.00	0.04

clustering of the request portfolio in services in the event that the maximal allowed arrival and departure time variation is stepwise increased. Surprisingly, no general trend is observed. The highest variations are observed in the events that (i) DT^{max} is tightened for the first time from 25 to 5 and (ii) DT^{max} is reduced from 0.25 to 0. For intermediate reductions, modifications of the clusters of requests are considered only for isolated configurations (e.g. variation of DT^{max} form 25 down to 5 and $N^{flex} = 4$). There is no obvious trend for the development of the $\bar{H}$-values with respect to the value of N^{flex}.

5 Summary & Conclusions

A straight-forward approach to establish synchronized processes in a network has been proposed. The process synchronization covers both spatial as well as temporal issues and allows the specification of flexible products in a given network. Robustness is addressed in the sense that two or even more redundant execution modes are available for fulfilling network demand. An initial evaluation of the proposed mathematical optimization model has been carried out. From the observed simulation results it is learnt that the crucial issue is the extend of robustness (here: the number of requested flexible products). It mainly determines the process costs. The dependency between costs and the desired strictness of the synchronization (here expressed in the maximal allowed arrival/departure time variation) is not strong. In order to find out the reason for this surprising observation we have analyzed the optimized processes and we have found that the assignment of nodes to processes is changed only if extreme maximal allowed arrival/departure time variations are involved.

Future research will address alternative model solving approaches which might include the development of heuristics or even metaheuristic algorithms in order to master the very high complexity of the model that restrict the application of optimizing solver software to small instances like the ones used here. Applications of heuristics enable the implementation of the synchronization issues even in large networks. Another planned investigation covers the analysis of the impacts of determining the flexible product in advanced so that redundant processes are generated for pre-specified products in a given network.

References

1. Bäck, T.: Binary strings. In: Bäck, T. et al. (Eds.) Evolutionary Computation 1—Basic Algorithms and Operators, pp. 132–135. IoP, Bristol (2000)
2. Cheng, R., Gen, M., Tsujimura, Y.: A tutorial survey of job-shop scheduling problems using genetic algorithms—representation. Comput. Ind. Eng. **30**(4), 983–997 (1996)
3. Drexl, M., Rieck, J., Sigl, T., Berning, B.: Simultaneous Vehicle and Crew Routing and Scheduling for Partial and Full Load Long-Distance Road Transport, Technical report LM-2011-05, University of Mainz, Lehrstuhl für Logistikmanagement (2011)
4. Drexl, M.: Synchronization in Vehicle Routing—A Survey of VRPs with Multiple Synchronization Constraints, Technical report LM-2011-02. University of Mainz, Lehrstuhl für Logistikmanagement (2011)
5. Dror, M., Laporte, G., Trudeau, P.: Vehicle routing with split deliveries. Discrete Appl. Math. **50**(3), 239–254 (1994)
6. Fiedler, C.: Integration by synchronization: logistics planning and control based on petri nets. In: Proceeding of 1st Annual IEEE Systems Conference (2007)
7. Fretter, C., Krumov, L., Weihe, K., Müller-Hannemann, M., Hütt, M.-T.: Phase synchronization in railway timetables. Eur. Phys. J. B. **77**(2), 281–289 (2010)
8. Gallego, G., Phillips, R.: Revenue management of flexible products. M&SOM **6**(4), 321–337 (2004)
9. Golden, B.L., Raghavan, S., Wassil, E.A.: The Vehicle Routing Problem: Latest Advances and New Challenges. Springer, New York (2008)
10. Irnich, S., Desaulniers, G.: Shortest path problems with resource constraints. In: Desaulniers, G., Desrosiers, J. (eds.) Column Generation, pp. 33–65. Springer, Heidelberg (2005)
11. Lämmer, S., Donner, R., Helbing, D.: Anticipative control of switched queueing systems. Eur. Phys. J. B. **63**(3), 341–347 (2007)
12. Laporte, G., Marin, A., Mesa, J.A., Perea, F.: Designing robust rapid transit networks with alternative routes. J. Adv. Transp. **45**(1), 54–65 (2011)
13. Savelsbergh, M.W.P., Sol, M.: The general pickup and delivery problem. Transp. Sci. **29**(1), 17–29 (1995)
14. Schönberger, J.: Model-Based Control of Logistics Processes in Volatile Environments. Springer, New York (2011)
15. Schönberger, J., Kopfer, H.: A general approach to robustness in logistics—basic concepts, quantification approaches and experimental evaluations, In: Voss, S. et al. (Eds.) Logistik Management, pp. 299–323, Springer, Berlin (2009)
16. Tang, X., Xianpeng, W.: Iterated local search algorithm based on very large-scale neighborhood for prize-collecting vehicle routing problem. Int. J. Adv. Manufact. Technol. **29**, 1246–1258 (2006)
17. Tomlin, J.A.: Minimum-cost multicommodity network flows. Oper. Res. **14**(1), 45–51 (1966)
18. Ziha, K.: Redundancy and robustness of systems of events. Probab. Eng. Mech. **15**, 347–357 (2000)

Modeling Production Planning and Transient Clearing Functions

Dieter Armbruster, Jasper Fonteijn and Matt Wienke

Abstract The production planning problem, i.e. to determine the production rate of a factory in the future, requires an aggregate model for the production flow through a factory. The canonical model is the clearing function model based on the assumption that the local production rate instantaneously adjusts to the one given by the equilibrium relationship between production rate (flux) and work in progress (wip), e.g. characterized by queueing theory. We will extend current theory and modeling for transient clearing functions by introducing a continuum description of the flow of product through the factory based on a partial differential equation model for the time evolution of the wip-density and the production velocity. It is shown that such a model improves the mismatch between models for transient production flows and discrete event simulations significantly compared to other clearing function approaches.

Keywords Production planning · Transient clearing functions · Continuum models

---

This contribution was previously published in Logistics Research (2012) pp. 133–139. DOI:10.1007/s12159-012-0087-8.

D. Armbruster (✉) · M. Wienke
School of Mathematical and Statistical Sciences, Arizona State University,
Tempe, AZ 85287-1804, USA
e-mail: armbruster@asu.edu

J. Fonteijn
Department of Mechanical Engineering, Eindhoven University of Technology,
POB 513, 5600 MB Eindhoven, Netherlands

K. Windt (ed.), *Robust Manufacturing Control*, Lecture Notes in Production Engineering,
DOI: 10.1007/978-3-642-30749-2_6,

1 Introduction

The production planning problem is a well studied problem in industrial engineering. Fundamentally it involves finding the correct starts into a factory such that production meets demands. The problem is complicated by two different major issues: Stochasticity and nonlinearity. Stochasticity manifests itself through the uncertainty of the demand and the variation of any demand realization. In addition, variations in the production speed and quality introduce other fundamental stochastic processes. While demand fluctuations are covered via suitably sized and placed inventories, stochasticity in the production process leads to variable lead times to refill these inventories.

Note that stochasticity is the more fundamental issue than nonlinearity, since the latter is generated by the former via queueing: Nonlinearity is generated by the fact that the variable lead times do not only depend on the stochastic processes that impact the production, it is mostly generated by waiting in queues. Such waiting depends crucially on the amount of material produced concurrently, i.e. the wip. Specifically, the lead times increase dramatically together with the lengths of the queues, if the flux through the factory approaches the capacity limit of the factory. A typical scenario goes like this: Demand is projected to increase at a certain time in the future. Meeting demand requires increasing the start rate into the factory by a lead time earlier than the requested delivery time. However, increasing the start rate will increase the wip in the factory and as a result increases the cycle time—the time a product needs to completely go through the factory increases. The resulting nonlinear optimization is at the core of the production planning problem.

1.1 Clearing Function

The baseline for all modeling in production systems is given by discrete event simulation (DES) models where every part, every machine and every production step are modeled with (different) probability distributions characterizing the specific stochastic process responsible for uncertain steps. As the characterization of these stochastic processes is non-trivial and as such simulations are very expensive, aggregate deterministic models that represent *average* behavior have been developed.

The canonical aggregate model has become known as a clearing function, first introduced by Karmarkar [9], and as Betriebskennlinen in German by Wiendahl [12]. The clearing function can be defined for any size of production unit, i.e. a group of machines, a production line, a full factory or even a supply chain. It is a state equation that defines the outflux F of the production unit as a function of the wip W in steady state in that unit, i.e.

$$F = \Phi(W). \tag{1}$$

The functional form of the clearing function Φ has been determined in many different ways: Measured in real factories, modeled via an $M/M/1$ queue (i.e. a queue with exponentially distributed arrivals and exponentially distributed machine processing times), modeled after the fundamental diagram of a traffic model [10] etc. (see e.g. [5, 6]),

$$\begin{aligned} \Phi &= \frac{\mu_0 W}{1+W} \qquad M/M/1 \\ \Phi &= \mu_0 W - W^2 \text{ fundamental diagram of traffic.} \end{aligned} \tag{2}$$

Aouam et al. [2] notice that the clearing function can be approximated by piecewise linear functions, making the production planning problem an Integer-LP optimization problem.

Notice that the clearing function is used with a wip level that is a function of time and hence models the outflux as a function of time. The fundamental assumption here is known as the adiabatic or quasi-steady assumption: The wip level changes slowly relative to the damping time of the underlying stochastic process. Hence the outflux is never transient and instantaneously relaxes to its steady state behavior.

Missbauer [11] extends the clearing function concept to capture transient phenomena in a three parameter clearing function. He shows that the outflux of a system depends on the initial wip of the system, the expected number of arriving lots and the probability distribution for sampling the initial wip. He studies an M/M/1 single server queue with infinite buffer, a mean arrival rate λ and a mean machine process rate $\mu = 1$. The number of lots in the system (queue plus machine) at the beginning of period t is denoted by $W(t)$ the wip in the workstation. Missbauer studies a version of the clearing function that characterizes the expected outflux at time t over a time interval $[t, t+T]$ denoted by X_t. The expected outflux is a function of the expected load $E[L_t]$ in the system given by the initial wip at time t and the new arrivals over the time interval

$$E[L_t] = W_t + A_t, \tag{3}$$

$$A_t = \int_t^{t+T} \lambda(s)ds. \tag{4}$$

For constant arrival rate $\lambda = \lambda_c$ we get $\lambda_c = \frac{A_t}{T}$. For time varying influx a DES was used. For further reference we define an decreasing and a increasing influx

$$\lambda_D(t) = \frac{\lambda_c}{(t/T + 1/2)ln(3)}, \tag{5}$$

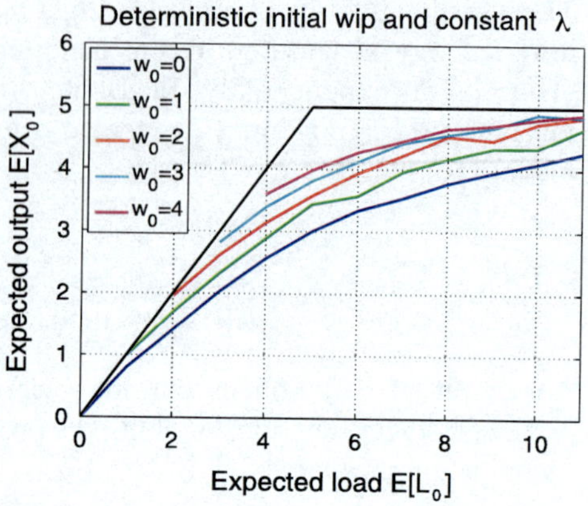

Fig. 1 Expected total ouflux for an M/M/1 queue for a time interval of five mean cycle times as a function of the expected total load

$$\lambda_I(t) = \frac{\lambda_c}{(t/T - 3/2)ln(3)}, \tag{6}$$

corresponding to a linear interpolation of the inter-arrival times between the two steady states related to the initial queue and the queue associated with $E[L_t]$. Missbauer's experiments were done for a time interval of $T = 5$. Figure 1 shows the expected output generated as averages of DES simulations for constant λ with 5 different initial wips. The dependence on the initial wip is obvious.

2 Transient Clearing Functions

While Missbauer restricted himself to a constant arrival rate, we have studied arrival rates that vary over the simulated time period but that lead to the same expected total load. We investigated five cases and generated clearing functions as a function of the initial wip like Fig. 1 for five different influx protocols: (i) constant influx, (ii) instantaneous influx at the beginning of the time period, (iii) instantaneous influx at the end of the time period, (iv) a monotonic decreasing influx rate (Eq. 5) and (v) a monotonic increasing influx rate (Eq. 6). Figures 2a, b show the expected outflux for an increasing and a decreasing influx. Since a decreasing influx introduces more of the load into the system early in the time period, it is not surprising that the variance in the outflux due to the initial wip becomes smaller than for increasing influx. The extremes of these cases are instantaneous influx at the very beginning of the time interval leading to high outflux almost independent of initial wip, and instantaneous influx at the end of the time interval leading to outflux based entirely on the initial wip. The conclusion of these DES is that, instead of extending the clearing function concept to three parameters as suggested in [11] the influx-outflux relationship in a transient setting in much more complicated: in addition to the total load, the functional form of the influx over the time interval of interest is

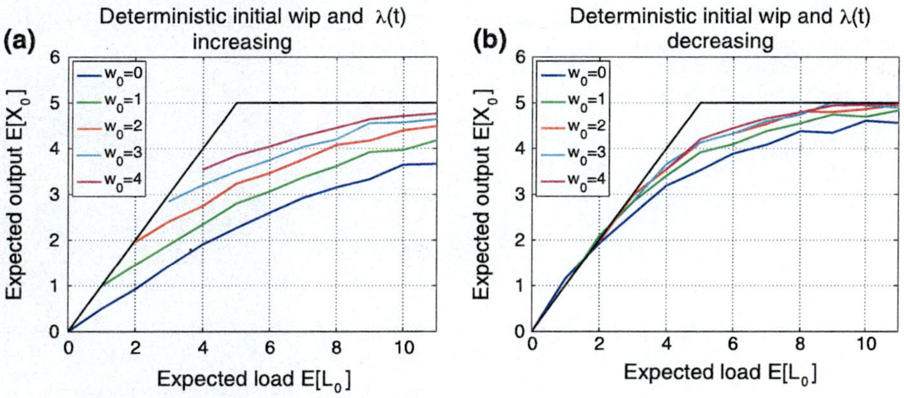

Fig. 2 Clearing functions for an $M/M/1$ queue with different initial wips and (**a**) increasing, (**b**) decreasing influx protocols

highly important and therefore the clearing function cannot be just a parametric relationship between input and output.

The applicability of Missbauer's result [11] that the probability distribution of the initial wip has a big influence on the clearing function needs to be clarified further: We can imagine two fundamentally different scenarios for the experiments described above:

1. Production has been halted and the state of the system can be examined. Hence the wip in the system is known exactly. When production is resumed, the initial wip is known deterministically.
2. Alternatively, one might want to plan a transition of the state of the factory from a steady state to another steady state and initial wip may only be known in the mean but no specific sample will be taken to determine the actual initial wip at the beginning of the planning period. In that case, the initial wip follows the geometric probability distribution associated with the probability of finding an M/M/1 queue at a particular level for a given arrival process and a given exit process.

Figure 3 shows the clearing functions for different mean initial wips for constant influxes. Figures 1 and 3 report on the same experiment—the difference is that in the former the initial wip is known whereas in the latter the clearing function is an average over many samples taken from the steady state distribution associated with the mean initial wip. We confirm that increasing variance of the initial wip leads to a lower outflux. However, the striking result of Fig. 3 is the fact that the dependence of the clearing function on the mean initial wip is almost completely gone.

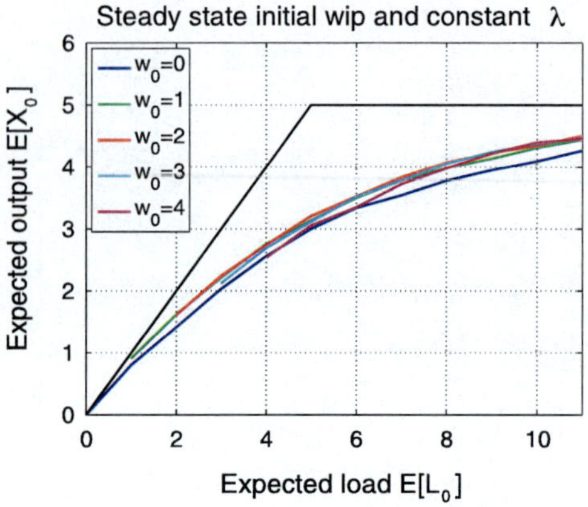

Fig. 3 Expected ouflux for an M/M/1 queue as in Fig. 1. Here, the outflux represents an ensemble average over a steady state probability distribution for the initial wips with a mean as indicated

3 Continuum Models

3.1 Transport Equations

A clearing function model gives an instantaneous relationship between outflux and wip in steady state. Since it is used to model an influx that changes in time, it will not be able to model the delay associated with the production time and waiting in the factory [5]. This is the fundamental reason why the clearing function cannot be parameterized by a finite number of parameters but depends on the complete history of the influx function. Attempts for transient clearing functions will therefore always be restricted to the special experimental setups.

In an attempt to design a complete time dependent theory of production flows we have therefore in recent years developed an aggregate theory of production flows based on standard transport equations studied in physics, especially in fluid mechanics and in some traffic models [3, 4]. Transport equations are partial differential equations that describe the time and space evolution of a density under an influx. In our case the spatial variable is given by the degree of completion of the part or the stage of the production. We scale the stage or completion variable $x \in [0,1]$ and define density of parts at stage x at time t by $\rho(x,t)$. If the fluid moves with a velocity field $v(x,t)$ then the flux is described as $F(x,t) = v(x,t)\rho(x,t)$. Mass conservation then is given by the partial differential equation

$$\frac{\partial \rho}{\partial t} + \frac{\partial F}{\partial x} = 0. \tag{7}$$

Since $v(x,t) \geq 0$, the fluid moves from left to right, allowing a boundary condition to be imposed at $x = 0$. Typically the boundary condition is $F(0,t) = \lambda(t)$, i.e. the local flux at zero is the arrival rate of the parts into the factory. Together with an initial wip profile $\rho(x,0) = \rho_0(x)$ this sets up a well defined hyperbolic problem.

Notice that we are describing a flow that is continuous in its parts and continuous in its spatial direction. This should be distinguished from the so-called fluid equation models of queueing theory [8] which are continuous in its parts but describes a flow through a finite and distinct number of queues, leading to a set of Ordinary Differential Equations (ODEs).

In [4] we extended the fluid analogy even further and derived macroscopic transport equations from kinetic models leading to Boltzmann equations which is akin to deriving the Euler equations of fluid dynamics from first principles based on Newton's law. Defining a particle density $f(x, v, t)$ describing the number of parts at state x at time t moving with a velocity v in completion space, we derive equations for the first moments of this density of the form:

$$\frac{\partial \rho(x,t)}{\partial t} + \frac{\partial v(x,t)\rho(x,t)}{\partial x} = 0, \tag{8}$$

$$\frac{\partial v(x,t)}{\partial t} + v(x,t)\frac{\partial v(x,t)}{\partial x} = 0. \tag{9}$$

An initial value problem appropriate for the DES experiments described in Sect. 1.1 can be defined by setting $\rho(x,0) = w_0$ and $v(x,0) = v_0$ with w_0 and v_0 constants.

3.2 Boundary Conditions

Determining the right boundary conditions to describe the DES experiments is the major modeling issue here. Equations (8, 9) are a set of hyperbolic partial differential equations whose solution travel from left to right as long as $v(x,t) > 0$. Hence boundary conditions have to be imposed at the boundary $x = 0$ and the outflux at the other boundary $x = 1$ is a result of the transport. Clearly the flux has to be given by the production start rate, hence

$$\rho(0,t)v(0,t) = \lambda(t). \tag{10}$$

The other boundary condition is based on the relationship between queuing theory and Equation (9): Equation (9) is Burgers equation and can be solved via characteristics. Hence ignoring the intitial conditions, after a while the solution $v(x,t)$ is determined by the value of the velocity at the boundary. As a result, a mass ρdx arrives at the boundary and travels downstream with the velocity it acquires at the moment of arrival at the boundary. Translating this into the M/M/1 setting and defining the velocity at the boundary as $v(0,t) = \frac{1}{cycle\ time}$ we see that the velocity $v(0,t)$ should depend on the queue length $w(t) = \int_0^1 \rho(x,t)dx$ at the moment a part arrives at the end of the queue.

The problem therefore reduces to finding the expected cycle time, conditioned on the length of the queue. For a steady state queue, the cycle time is determined by the PASTA (Poisson Arrivals See Time Averages) property of M/M/1 queues: In steady state a part arriving at the end of queue will find an average queue length w_0 and the resulting cycle time for this part will be $\tau = \frac{1}{\mu}(1 + w_0)$. Hence

$$v_{ss}(t) = \frac{\mu}{1 + w(t)} \tag{11}$$

is the velocity related to the well known M/M/1 clearing function (cf Eq. 2).

The same PASTA property gives us the initial condition: At the beginning of the experiment we have an initial wip w_0 that we assume is a known deterministic quantity. This initial wip is the length of the queue. For a Markov process, the history of arrivals—whether they arrived in packets or spaced out—is not important. Hence we can use the average cycle time formula for a M/M/1 queue or the heuristic extensions discussed below to determine the initial condition for the velocity in the factory.

To improve on the steady state result requires significantly more queuing theory machinery: Since we are describing transient phenomena, the system is not ergodic any more and hence ensemble averages and time averages are not the same. We therefore need to be more specific about the "expected" cycle time. The natural setup following the experiments in Sect. 2 is to determine the probability distribution of the cycle times given that we restart a factory with a given initial wip w_0 over many instances of this scenario—i.e. we are interested in the ensemble average, conditioned on the initial wip.

Our current models for the expected value of the cycle time are preliminary and based on fitting heuristic boundary condition models to the DES. We distinguish two regimes:

1. If the production start rate $\lambda(t)$ is less than the mean production rate μ, then we expect that any initial wip distribution exponentially fast decays to the wip distribution associated with the steady state related to the arrival rate. Hence the boundary condition is determined by the solution to an ordinary differential equation

$$\frac{dv(0,t)}{dt} = -\sigma(v(0,t) - v_{ss}(t)) = -\sigma\left(v(0,t) - \frac{\mu}{1 + \int_0^1 \rho(x,t)dt}\right) \tag{12}$$

   where the decay constant σ will be determined experimentally.
2. If the production start rate $\lambda(t)$ is bigger than the mean production rate there is no associated steady state since the queue length will become unbounded. In this case the cycle time at arrival of a part at a queue length of $w(t)$ will become just $\tau = \frac{1}{\mu}w(t)$ which would lead to a velocity equation of

$$v_{hw}(t) = \frac{\mu}{w(t)}. \tag{13}$$

It turns out that for small wip and for $\lambda - \mu << 1$ this model creates a velocity that is too high and hence the production in the PDE simulations is overestimated relative to the DES. This is due to the fact that basing the ensemble average only on the stochastic properties of the exit process it not a good model in these cases since for small wips machines do occasionally idle as a result of missing arrivals. We settled for a model that averages between the steady state Eq. (11) and the high wip model (13) of the form

$$v(t) = \frac{\mu}{0.5 + w(t)}. \tag{14}$$

Hence the full boundary conditions for Eq. (9) become

$$\begin{aligned} v(0,t) &= \frac{\mu}{0.5 + \int_0^1 \rho(x,t)dx} \qquad \text{for} \quad \lambda \geq \mu, \\ \frac{dv(0,t)}{dt} &= -\sigma\left(v(0,t) - \frac{\mu}{1 + \int_0^1 \rho(x,t)dt}\right) \qquad \text{for} \quad \lambda < \mu, \\ v(0,0) &= \frac{\mu}{0.5 + \int_0^1 \rho(x,t)dx}. \end{aligned} \tag{15}$$

The last equation describes the initial condition for the ordinary differential equation. It is based on the assumption of a deterministic initial condition, i.e. the initial wip is exactly known and hence the ensemble average will be mostly affected by the stochasticity of the machine process and little affected by the stochasticity of the arrival process.

4 Numerical Results

We have been reproducing the DES of Sect. 2. Since there are only a small number of lots involved in these simulations, the discretization error between the DES and the partial differential equation becomes an issue. In the discrete case wip measures whole lots whereas the continuous model registers infinitesimally small lots. This is not a problem for large wips but for these experiments partial lots in the PDE are counted earlier than they really appear in the DES and hence they lead to lower velocities than in the DES. We compensate for this by calculating wip with a floor function i.e. $w(t) = \frac{1}{2}\lfloor 2\int_0^1 \rho(x,t)dx \rfloor$. In that way the PDE system observes partial lots only after half of the lot has already appeared. Figures 4a, b compare the outflux of the PDE simulations for different constant influxes and initial wip of $w_0 = 0$ and $w_0 = 3$ with the corresponding DES. The clearing functions of the DES for these wips as well as others for different initial wips are very well reproduced by the PDE simulations. The decay constant σ has been adjusted to

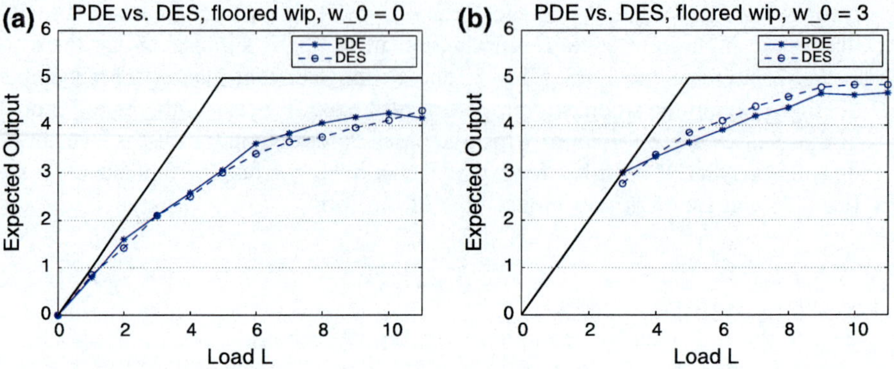

Fig. 4 Outflux over five time intervals as a function of the total expected load for the PDE model (8, 9) with boundary conditions (10) and (15) and constant influx. **a** shows the DES simulation for an initial wip of $w_0 = 0$ and the corresponding PDE simulation, **b** initial wip $w_0 = 3$

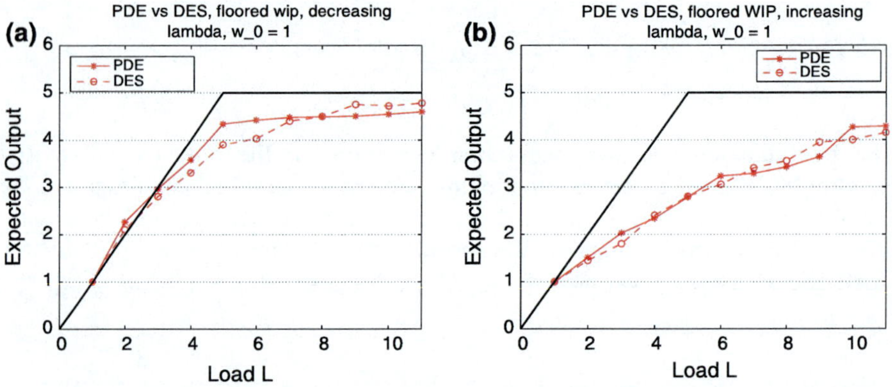

Fig. 5 As in Fig. 4 with initial wip $w_0 = 1$. **a** shows a decreasing influx rate, **b** an increasing influx rate

give the best fit of the two curves over all the data points. Notice that the best fit depends on the initial wip: For $w_0 = 0$ the best fit is $\sigma = 2.3$ indicating fast relaxation to the steady state and for $w_0 = 3$ the best fit is $\sigma = 0.3$ indicating a very slow relaxation that has not yet equilibrated after the five time intervals used in the experiment.

The advantage of a PDE simulation becomes apparent in the Fig. 5a, b which show the clearing functions for an influx that corresponds to a linearly increasing inter-arrival time and a linearly decreasing inter-arrival time for an initial wip of $w_0 = 1$. Although the decreasing case 5a shows a slight overproduction of the PDE compared to the DES, the overall trend of the PDE simulations captures the DES simulations very well.

5 Conclusion

We have developed a PDE model for a transient M/M/1 queuing experiment representing the most simplified case of a production model. We used a coupled system of evolution equations for the part density and the velocity to describe the production system as transport equation and showed that the crucial modeling aspect of the problem is the boundary condition of the velocity equation.

We showed that a heuristic model based on exponential relaxation of initial queue lengths to their steady state values given by M/M/1 queuing theory in addition to a high wip limiting model of queueing behavior leads to very good agreement between the PDE simulations and the DES. Specifically we compare the two approaches visually via plots presenting the expected throughput over five time units for an ensemble average of repeated experiments as a function of the average total load in the factory similar to clearing function models. The mean relative error for each of the clearing functions experiment is of the order of 6%, which is far better than any other modeling approach for these experiments. Overall the heuristics requires data fitting of a single decay parameter (σ) and a global choice of the functional dependence of the ensemble average of the velocity for the case when there is no steady state.

The current state of the project to model the ensemble average of transient behavior of production systems is clearly unsatisfactory. While the heuristic model presented here is a clear improvement for any practical considerations of the production planning problem that can easily be implemented for a practical code, the theoretical state of the model is very unsatisfactory. Research based on exact and approximate solutions of transient queueing theory [1, 7] is currently under way to bridge the gap between theoretical and heuristic models.

Acknowledgments This project was started when D.A. was a part time professor at the Department of Mechanical Engineering, Eindhoven University of Technology. Helpful discussions with Erjen Lefeber and Ivo Adan on transient queueing models are gratefully acknowledged. Discussions with Hubert Missbauer and Reha Uzsoy helped to clarify the experiments on transient queuing and the relationship between clearing function models and the production planning problem. This research was supported by a grant from the Volkswagen Foundation under the program on Complex Networks.

References

1. Abate, J., Whitt, W.: Transient behavior of the M/G/1 workload process. Operations Research **42**(4), 750–764 (1991)
2. Aouam T., Uzsoy R.: An exploratory analysis of production planning in the face of stochastic demand and workload-dependent lead times. In: Armbruster D., Kempf K. (eds.) Decision Policies for Production Systems, pp. 173–209. Springer, London (2012)
3. Armbruster, D., Marthaler, D., Ringhofer, C., Kempf, K., Jo, T-C.: A continuum model for a re-entrant factory. Operations Research **54**(5), 933–950 (2006)

4. Armbruster, D., Marthaler, D., Ringhofer, C.: Kinetic and fluid model hierarchies for supply chains. SIAM Multiscale Model. Simul. **2**(1), 43–61 (2004)
5. Armbruster, D.: The production planning problem: clearing functions, variable leads times, delay equations and partial differential equations. In: Armbruster, D., Kempf K. (eds.) Decision Policies for Production Systems, pp. 289–302. Springer, London (2012)
6. Asmundsson, J.M., Rardin, R.L., et al.: Production Planning Models with Resources Subject to Congestion. Naval Research Logistics **56**, 142–157 (2009)
7. Berstimas, D., Mourtzinou, G.: Transient laws of non-stationary queueing systems and their applications. Queueing Systems **25**, 115–155 (1997)
8. Bramson, M.: Stability of Queueing Networks, Lecture Notes in Mathematics, 1950. Springer Verlag, Berlin (2008)
9. Karmarkar, U.S.: Capacity Loading and Release Planning with Work-in- Progress (wip) and Lead-times. J. Manuf. Oper. Manag. **2**, 105–123 (1989)
10. Lighthill, M.J., Whitham, G.B.: On kinematic waves. II. A theory of traffic flow on long crowded roads. Proceedings of the Royal Society of London, **A229**, 317–345 (1955)
11. Missbauer H.: Order release planning with clearing functions: A queueing-theoretical analysis of the clearing function concept. Int. J. Prod. Econ. (2010). 10.1016/j.ijpe.2009.09.003
12. Wiendahl H.-P.: Load oriented manufacturing control. Springer Verlag, Berlin (1995)

Part II
Robust Manufacturing Control Methods

Switching Dispatching Rules with Gaussian Processes

Jens Heger, Torsten Hildebrandt and Bernd Scholz-Reiter

Abstract Decentralized scheduling with dispatching rules is applied in many fields of production and logistics, especially in highly complex manufacturing systems, e.g. semiconductor manufacturing. Nevertheless, no dispatching rule outperforms other rules across various objectives, scenarios and system conditions. In this paper we present an approach to dynamically select the most suitable rule for the current system conditions in real time. We calculate Gaussian process (GP) regression models to estimate each rule's performance and select the most promising one. The data needed to create these models is gained by a few preliminary simulation runs of the selected job shop scenario from the literature. The approach to use global information to create the Gaussian process models leads to better local decision at the machine level. Using a dynamic job shop scenario we demonstrate, that our approach is capable of significantly reducing the mean tardiness of jobs.

Keywords Simulation · Gaussian process regression · Scheduling · Dispatching rules

J. Heger (✉) · T. Hildebrandt · B. Scholz-Reiter
Bremer Institut für Produktion und Logistik GmbH at the University of Bremen, Hochschulring 20, 28359 Bremen, Germany
e-mail: heg@biba.uni-bremen.de

T. Hildebrandt
e-mail: hil@biba.uni-bremen.de

B. Scholz-Reiter
e-mail: bsr@biba.uni-bremen.de

K. Windt (ed.), *Robust Manufacturing Control*, Lecture Notes in Production Engineering,
DOI: 10.1007/978-3-642-30749-2_7,

1 Introduction

In today's highly competitive, globalized markets, manufacturing companies have to use their production resources as efficiently as possible. Therefore, especially capital-intensive industries like semi-conductor manufacturing spend considerable effort to optimize their production processes. Improvements in scheduling lead to a better achievement of objectives (e.g., tardiness or flow time of jobs). Scheduling in most practical settings, such as job shops or flexible flow shops, is a combinatorial, NP-hard optimization problem. These problems have attracted researchers and practitioners for many decades now and are still of considerable interest, because of their high relevance and difficulty. Many heuristics, which calculate schedules in a centralized manner, have been introduced, since optimal solutions can only be calculated for small scenarios. New schedules are calculated regularly with a rolling time horizon. If the production scenarios are facing high variability like continuously arriving new jobs, job changes, break-downs etc. decentralized scheduling methods are advantageous compared to central methods.

One class of decentralized scheduling heuristics are dispatching rules [2, 4], which are widely used to schedule even very complex shop floors. Their popularity derives from the fact that they perform reasonably well in a wide range of environments, and they are relatively easy to understand. Furthermore, they only require minimal computational time, which qualifies them to be used in real-time, online scheduling. They can therefore always take the latest information available from the shop-floor into account. Many dispatching rules are proposed in the literature, which perform well on specific scenarios. However, no rule is known to consistently outperform all other rules [16]. One approach to meet this challenge and improve scheduling performance is to select and switch dispatching rules depending on current system conditions. For this task machine learning techniques (e.g., artificial neural networks [11]) are frequently used.

In this paper we present the first simulation study using Gaussian process regression for this task [17, 22]. Analysis from Scholz-Reiter et al. [19] indicates that Gaussian processes can predict dispatching rule performance better than neural networks in many settings. Additionally, a single Gaussian Process model can easily provide a measure of prediction quality. This is in contrast to many other machine learning techniques.

This paper is organized as follows: in Sect. 2 we give a review of previous work of dispatching rules, machine learning in scheduling and Gaussian processes. In Sect. 3 our chosen scenario and the experimental designs are described. Section 4 presents the results of our experiments. The paper concludes with a short summary and provides directions towards future research.

2 State of the Art

2.1 Scheduling with Dispatching Rules

Scheduling is defined by Haupt [4] as "the determination of the order in which a set of jobs (tasks) {i I i = 1,..., n) is to be processed through a set of machines (processors, work stations) (k I k = 1...m)." Since the problem is np-hard, non-optimal heuristics are used. Especially in extremely complex scenarios with high variability priority rules are often employed. Dispatching rules as a special kind of priority rules are applied to assign a job to a machine. This is done each time the machine becomes idle and there are jobs waiting. The dispatching rule assigns a priority to each job. This priority can be based on attributes of the job, the machines or the system. The job with the highest priority is chosen to be processed next. Dispatching rules have been developed and analyzed in the scientific literature for many years; see e.g. [2, 4, 14]. The most well know rules are Shortest Processing Time first (SPT), Earliest Due Date (EDD) and First Come First Served (FCFS).

Since the development of sophisticated and specialized dispatching rules is a tedious and time-consuming task, concepts to generate them automatically have been proposed e.g. [3, 5]. One drawback for sophisticated rules as well as for automatically generated rules remains: depending on the manufacturing system and the various objectives (e.g. mean flow time or mean tardiness etc.) no single rule, which outperforms all others, can be found [11, 16].

2.2 Machine Learning in Scheduling

2.2.1 Machine Learning

Alpaydin [1] stated: "The goal of machine learning is to program computers to use example data or experience to solve a given problem". In this study, we are interested in a system that can predict the value of an objective function from system characteristics, which otherwise would have to be obtained by a costly simulation. As we are interested in a high due date-adherence of jobs, we chose the objective of minimizing the mean tardiness of jobs. Inputs are system attributes (for example the utilization etc.) affecting the tardiness. The output is the estimated tardiness for the dispatching rule chosen to make scheduling decisions in our manufacturing system. Let X denote the (vector of) system attributes and Y the tardiness. Surveying past production processes (or using simulations) we can collect training data and the machine learning program fits a function to these data to learn Y as a function of X.

2.2.2 Machine Learning and Scheduling: Related Research

Kotsiantis [8] gives an overview of a few supervised machine learning techniques, like artificial neural networks, decision trees, Naïve Bayes, support vector machines etc. Priore et al. [15] present a review of machine learning in dynamic scheduling of flexible manufacturing systems. Most approaches are based on artificial neural networks and are described in the following.

A neural network based controller, consisting of an adjustment module and the equipment level controllers, was proposed for scheduling and controlling a manufacturing cell by Sun and Yih [20]. The adjustment module considers the user objectives and the current performance levels to determine the relative importance of performance measures. Based on these importance, values and current machine status, the equipment level controller, implemented by a neural network, selects a proper dispatching rule and the jobs are processed accordingly. The training samples for each equipment level controller are calculated by a one-machine simulation and modified to reflect the impacts of different dispatching rules on the system performance.

El-Bouri et al. [7] used a neural network to select dispatching rule in a job shop. They chose small scenarios with 5 machines and investigated 3 rules. To train the neural network they calculated optimal solutions for 10, 15 and 20 jobs. The neural network was used to select one rule for every machine. With this approach they were able to get better results than just using one of the rules on every machine. The drawback of this approach is that it is limited to scenarios with only a few machines and jobs, otherwise no optimal solutions for learning could be generated.

Mouelhi-Chibani and Pierreval [11] use a neural network to dynamically switch dispatching rules on every machine depending on the current system state. They have selected four system parameters (e.g. shop load) and 22 system state variables (e.g. average slack time of jobs in the first queue), which the neural network uses to decide which rule should be applied. They train the neural network with preliminary simulation runs. The scenario they selected consists of only two machines and the set of dispatching rules consists of SPT and EDD. They outperform the static use of rules, but not that clearly, which might be due to the small scenario.

These are interesting approaches, but the results seem to have potential for improvement. It is not clear if this is due to the selected scenario or the learning technique. There has been no study of Gaussian processes for selecting dispatching rules dynamically until now. Since they have shown good results compared to other machine learning techniques [18], this seems to be a promising approach, which is investigated in the remainder of this paper.

2.3 *Gaussian processes*

2.3.1 Introduction

O'Hagan [13] represents an early reference from the statistics community for the use of a Gaussian processes as a prior over functions, an idea which was only introduced to the machine learning community by Williams and Rasmussen [22].

As stated before we have a simulation model implicitly implementing a (noisy) mapping between a vector of state variable (in our case containing, e.g. utilization) and the objective function (mean tardiness) $y = f(x) + \varepsilon$. The learning consists of finding a good approximation f*(x) of f(x) to make predictions at new points x.

To learn such a model using GP requires some learning data as well as a so-called covariance function. This covariance function, sometimes called kernel, specifies the covariance between pairs of random variables and influences the possible form of the function f* learned. For our study the squared exponential (SE) covariance function is selected, because it should be able to fit to our data well and is a common choice in applications of GP. It is depicted in Eq. (1):

$$k_y(x_p, x_q) = \sigma_f^2 \exp\left(-\frac{1}{2l^2}(x_p - x_q)^2\right) + \sigma_n^2 \delta_{pq} \qquad (1)$$

The squared exponential covariance function used in our experiments has three hyper parameters. There is the length-scale l, the signal variance $\sigma_f^2$and the noise variance σ_n^2. These parameters of a covariance function can be used to fine-tune the GP-model, thus learning of a Gaussian Processes model requires having some learning data, choosing an appropriate covariance function and choosing a good set of hyper parameters. For further information see [17, Chaps. 2 and 4].

2.3.2 Application and Example

Gaussian processes provide a quality estimate of their predicted value. This is denoted by the shaded area in Fig. 1. Ten noisy training points are given and since there is noise the standard deviation close to the training points is small, but not exactly zero. In between two points as well as at the beginning and the end the quality of the estimates decreases.

Learning with Gaussian processes is done by selecting a covariance function and setting its free hyper parameters. For our study the squared exponential covariance function is chosen. To learn, or optimize the hyper parameters, the marginal likelihood should be maximized. Details and mathematical background can be found in [1 Chaps. 5, especially Eq. (5.9), p. 114]. Basically, the hyper parameters are chosen in a way that the generalization error, which is the average error on unseen test examples, is minimized. This is done with cross-evaluation by splitting the training data in learning and test data. The training error is not optimized, because this may lead to over-fitting the data.

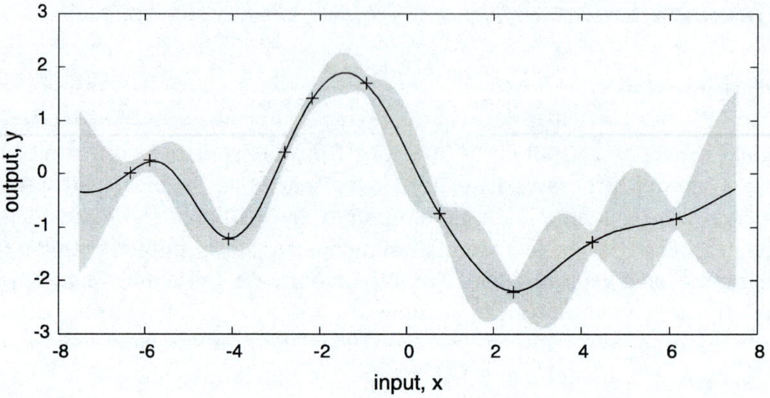

Fig. 1 Example of a Gaussian process regression function with 10 noisy training points observed. The mean prediction is shown as a black line and the shaded area denotes twice the prediction uncertainty

Additionally, since hyperparameters can be interpreted as length-scale parameters in the case of the *squared exponential* covariance function further optimizations can be performed. Rasmussen and Williams [17] describe the hyper parameters informally like this: "how far do you need to move (along a particular axis) in input space for the function values to become uncorrelated". Thus, the squared exponential covariance function implements automatic relevance determination (ARD) [12], since the inverse of the length-scale determines how relevant an input is. A very large length-scale value means that the covariance will become almost independent of that input. ARD has been used successfully for removing irrelevant input by several authors, e.g. Williams and Rasmussen [22].

3 Approach and Experimental Setup

3.1 Approach: Switching Dispatching Rules

The main focus of our research is to develop a new scheduling method, which uses local and global information to make scheduling decisions. Therefore, we suggest learning performance models of the dispatching rules with Gaussian process regression to select the best rule for the current conditions. This is a promising approach, since the major drawback of dispatching rules is their lack of a global view of the problem. Rules approach the overall scheduling problem by taking independent scheduling decisions based on the current, local conditions at the particular machine without consideration of the negative effects they might have on future decisions and on the overall objective function value.

The performance models are learned by preliminary simulation runs, which add a global perspective on the system behavior to the local decision rules.

3.2 Scenario Description

The type of problems we address, are dynamic shop scenarios. Our computational experiments are based on the dynamic job-shop scenarios from Rajendran and Holthaus [16]. In total there are 10 machines on the shop floor, each job entering the system has to visit each machine once, using a random routing, i.e., machine visitation order is random with no machine being revisited. Processing times are drawn from a uniform discrete distribution ranging from 1 to 49 min. The due dates of the jobs are determined by a due date tightness factor, a job's due date is set to x-times the job's total processing time + release time. Job arrival is a Poisson process, i.e., inter-arrival times of jobs follow an exponential distribution. The arrival rate is set to yield a desired long term utilization level of each machine.

The dynamic experiments simulate the system for a duration of 12 months, (about 18,000 jobs), using changing utilization rates and due date factors. All results in Sect. 4 are based on these dynamic setting. The utilization rate of the shop, i.e., arrival rate, is oscillating between 0.75 and 0.99 following a sine function with a period length of 30 days. The sine function to generate due date factors has a period length of 15 days and oscillates between 2 and 7. Performance figures are calculated averaging the tardiness of all jobs started within the simulation length of 12 month.

To generate the learning data we are only interested in the performance for a specific setting of utilization and due date tightness. We therefore closely follow the procedure from Rajendran and Holthaus [16]. We start with an empty shop and simulate the system until we collected data from jobs numbering from 501 to 2,500. The shop is further loaded with jobs, until the completion of these 2,000 jobs [14]. Data on the first 500 jobs is disregarded to focus on the shop's steady state behavior.

3.3 Investigated Dispatching Rules

To have a set of dispatching rules, out of which the best for each system condition can be selected, we have selected several dispatching rules from the literature. The first four are standard rules being used for decades; the fifth rule was developed by Holthaus and Rajendran [6] especially for their scenarios. If the rules calculate the same priority for more than one job, we use ERD (*earliest release date*) as a tiebreaker.

SPT—*Shortest Processing Time First*: SPT breaks ties by choosing the job with the shortest processing time for its imminent operation. Although this rule primarily aims to reduce the flow time of jobs (the difference between its completion and release time), SPT has shown to effectively minimize total tardiness when most jobs cannot meet their due dates, because of a tight due date settings and/or a high shop utilization.

EDD—*Earliest Due Date*: EDD resolves ties among equally weighted jobs by prioritizing the job with the earliest due date, therefore tending to decrease the maximum tardiness of all jobs. Contrary to SPT, the EDD rule is known to perform well for total tardiness when the shop is not congested and most jobs can be completed in time.

FCFS—*First Come First Served*: FCFS selects the job that has been in the buffer for the longest time. Though FCFS generally exhibits a modest tardiness performance, it is easy to implement and often serves as a benchmark.

MOD—*Modified Operation Due Date*: MOD orders the queue of waiting jobs by the larger of each job's operation due date ($d_{i,imt}$) minus the current time (t) or each job's operation processing time ($p_{i,imt}$). Therefore, if all jobs in the queue have positive slack (no job is in danger of missing its due date) then MOD dispatches them in earliest operational due-date (ODD) order. If all jobs have negative slack (all jobs are in danger of missing their due dates), then MOD works like SPT to reduce shop congestion. MOD is defined as follows:

$$MOD_i = \max(p_{i,imt}, d_{i,imt} - t) \quad (2)$$

2PTPlusWINQPlusNPT – *2Processing Time + Work in Next Queue + Next Processing Time:* This rule was suggested by Holthaus and Rajendran [6] and consists of three parts. First, the processing time on the current machine is considered. Secondly, the *Work in Next Queue* is added: **WINQ**—jobs are ranked in the order of a (rather worst case) estimation of their waiting time before processing on the next machine can start. This estimation includes the time needed by a machine m to finish its current job plus the sum of processing times of all jobs currently waiting in front of m. The job where this sum is least has the highest priority. Thirdly, the processing time of a job's next operation (**NPT**—*Next Processing Time*) is added.

$$2PTPlusWINQPlusNPT_i = 2p_{i,imt} + WINQ_i + p_{i,imt+1} \quad (3)$$

3.4 System Architecture

For the simulation experiments of this paper we use Jasima,[1] a self-implemented discrete-event simulation. Jasima is very roughly based on a Java-port of the SIMLIB library [9], as described in (Huffman 2001). It is an efficient simulation

[1] http://code.google.com/p/jasima/

which can utilize multiple cores, and offers a variety of dispatching rules and the flexibility to implement complex scheduling scenarios such as the one used in this paper.

For the Gaussian processes, we have used the software examples provided by Williams [21] and adapted them for our scenarios. The calculations have been performed with MatLab from MathWorks. The MatLab Builder JA is used as an interface between the simulation software and the Gaussian processes.

4 Experiments and Results

4.1 Simulation Runs and Rule Performance

At first we performed simulation experiments with each of the dispatching rules in our scenario with oscillating utilization rates and due date factors as described in 3.2. The results are shown in Table 1 and Fig. 2. The 2PTPlusWINQPlusNPT and the MOD rule performed similarly and significantly outperformed the three other rules. Thus, we decided to take these two for our dynamic switching experiments.

The confidence interval has been calculated with the paired-t method, described in Law [9]. Therefore, the differences in each of the 30 replications between one rule and the MOD rule are calculated. Let Z_j be the differences with j = 1…30. The confidence interval can be constructed as follows:

$$\overline{Z}(n) = \frac{\sum_{j=1}^{n} Z_j}{n} \tag{4}$$

With

$$\widehat{Var}\left[\overline{Z}(n)\right] = \frac{\sum_{j=1}^{n} \left[Z_j - \overline{Z}(n)\right]^2}{n(n-1)} \tag{5}$$

and

we can the form of the (approximate) 100(1 − α) percent confidence interval:

$$\overline{Z}(n) \pm t_{n-1,1-\frac{\alpha}{2}} \sqrt{\widehat{Var}\left[\overline{Z}(n)\right]} \tag{6}$$

If the interval does not contain zero the two perform rules are significantly different.

Table 1 Simulation results with 30 replications and paired-t confidence to MOD

Rule	Mean tardiness in min; (standard error)	Difference to MOD in min	Paired-t confidence interval to MOD (approx. 99 %)
FCFS	532.8 (11.3)	310.3	[298.2–322.4]
EDD	340.7 (7.1)	118.2	[113.1–123.3]
SPT	282.2 (7.8)	60.4	[57.4–63.4]
2PTPlusWINQPlusNPT	222.9 (6.3)	0.4	[−1.9–2.7] (non-significant)
MOD	222.5 (7.0)		

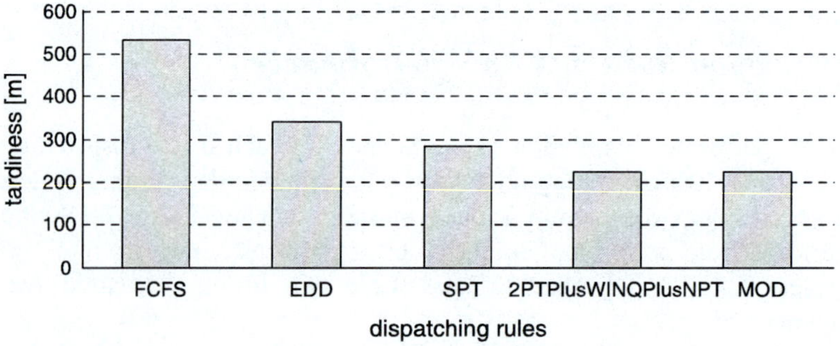

Fig. 2 Simulation results in mean tardiness with 30 replications; 99 % confidence level to MOD (paired-t confidence interval)

4.2 Preliminary Simulation Runs and Gaussian Process Regression

To learn performance models of the two selected dispatching rules, we performed preliminary simulations runs with both rules and different system conditions. For the system conditions we selected two parameters, which are the input for the Gaussian processes. The first is the system's utilization and the second is the due date factor, which defines the job's due date tightness. These two system parameters have been combined in 650 combinations. We have performed simulation runs with system utilizations from 75 % till 99 % and have combined each of these with due date factors from 2 to 7 (in 0.2 steps). The five selected dispatching rules described in Sect. 3.3 have been evaluated for all these parameter combinations. Our performance criterion is mean tardiness, but the general approach is applicable to other objective functions as well. Each result for each combination of utilization, due date factor and dispatching rule is the average of 20 independent replications to get reliable estimates of the performance of our stochastic simulation.

If the interval does not contain zero the two perform rules are significantly different.

If simulation runs are expensive and more system parameters are considered, not all parameter combinations can be simulated in advance. This is where

Table 2 Simulation results with 30 replications and paired-t confidence to MOD

Rule	Mean tardiness in min	Difference to MOD in min	Difference to MOD in %	Standard error	Paired-t confidence to MOD (99 %) [interval delta]
MOD	222.47	0.00	0.00	6.96	0.00
GP 15	212.00	−10.47	−4.71	6.40	2.19
GP 45	210.28	−12.19	−5.48	6.33	2.24
GP 90	209.01	−13.46	−6.05	6.52	2.00
GP 360	207.99	−14.49	−6.51	6.17	4.62
GP 650	207.50	−14.97	−6.73	6.46	4.22

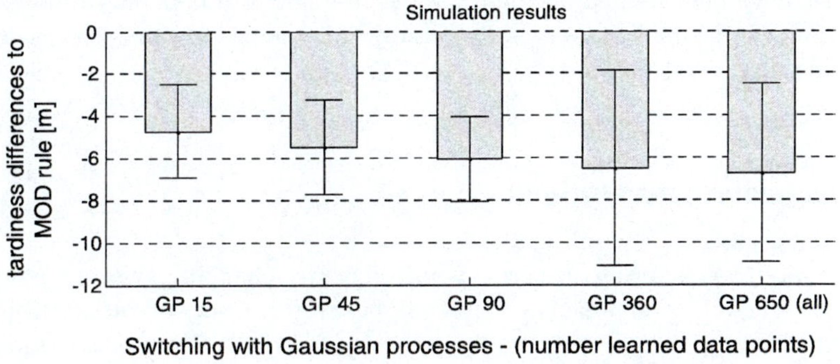

Fig. 3 Simulation results in mean tardiness; 99 % confidence level to MOD (paired-t confidence interval); GP x, with x being the number of data points used for the regression model

supervised machine learning techniques can play an important role, helping to select the best dispatching rule with only a few simulations runs as a learning data set. Therefore, we also investigated how the number of learning data points affects the quality of the learned models and the scheduling results in the end for our chosen scenario. For the selection of data points, i.e. a combination of utilization rate and due date factor, we used latin hypercube designs (e.g. [10]).

4.3 *Simulation of the Dynamic Scenario with Switching Between Rules*

With the help of the preliminary calculated regression models, we are now able to run the dynamic simulation studies. The scenario described in Sect. 3.2 is characterized by oscillating utilization rates and due date factors. These utilization rates and due date factors are unknown to our scheduling rules, thus they need to be estimated. Both values are calculated locally at each machine. The mean utilization rate of the last 24 h at the machine is calculated as an estimate for the utilization rate. The mean due date factor of the waiting jobs are calculated simply by averaging the due date factors the jobs started with when they entered the system.

The results in Table 2 and Fig. 3 show that the mean tardiness in our scenario can be reduced by 6.73 % if all data points are used for learning the static regression models. Even with only 15 preliminary simulation runs the mean tardiness can be reduced by almost 5 %. Compared to less sophisticated rules savings are much higher: SPT (26.5 %), EDD (39.1 %) and FCFS (61.1 %)

For our simulation runs we considered 5 different Gaussian process regression models with 15, 45, 90, 360 and 650 data points randomly selected with a latin hypercube design from our preliminary simulation runs. The results show that more data points lead to better models and thus to lower mean tardiness levels. Since we have only studied one model for each number of data points, no general conclusion can be made, however, how many data points are generally needed for satisfying regression models.

5 Conclusion and Outlook

In dynamic manufacturing scenarios with frequently changing system parameters adaptive scheduling approaches improve the performance of dispatching rule based scheduling. In our study we have shown, that Gaussian process regression models can be used to learn dispatching rule behavior under different system conditions. These models, which are gained with by few preliminary simulation runs, add global knowledge to each local decision unit. Before a rule decides which job to select next, a machine selects the best dispatching rule for the current system conditions first. Our results have shown that improvements of more than 6 % can be achieved.

To our knowledge this is the first time, that Gaussian process regression models have been used to switch between dispatching rules in a dynamical scenario. Their advantage compared to other machine learning techniques is, that they provide solid regression models, even with only a few data points and they provide an estimate of the quality of their predictions. In further studies we want to explore how these estimates can be used to optimize the selection of necessary learning data. Especially, when more system parameters are used and models get more complex this becomes an important factor.

In further studies the underlying scenario could be extended e.g. to semiconductor manufacturing, which is more complicated (i.e. sequence depending setup costs, batch machines etc.). Switching rules in such a scenario might increase the performance strongly, e.g. when the product mix changes and a batch machine becomes the bottleneck, the effect of different rules on the objective can be severe.

Acknowledgments The authors are grateful to the generous support by the German Research Foundation (DFG), grant SCHO 540/17-2.

References

1. Alpaydin, E.: Introduction to Machine Learning (Adaptive Computation and Machine Learning Series), vol. 14. The MIT Press, Cambridge (2004)
2. Blackstone, J.H., Phillips, D.T., Hogg, G.L.: A state-of-the-art survey of dispatching rules for manufacturing job shop operations. Int. J. Prod. Res. **20**(1), 27–45 (1982)
3. Geiger, C.D., Uzsoy, R., Aytug, H.: Rapid modeling and discovery of priority dispatching rules: an autonomous learning approach. J. Sched. **9**(1), 7–34 (2006)
4. Haupt, R.: A survey of priority rule-based scheduling. OR Spektrum **11**(1), 3–16 (1989)
5. Hildebrandt, T., Heger, J., Scholz-Reiter, B.: Towards improved dispatching rules for complex shop floor scenarios—a genetic programming approach. In: GECCO'10: Proceedings of the 12th Annual Conference on Genetic and Evolutionary Computation, pp. 257–264. ACM Press, New York (2010)
6. Holthaus, O., Rajendran, C.: Efficient job shop dispatching rules: further developments. Prod. Plan. Control **11**(2), 171–178 (2000)
7. Huffman, B.J.: An object-oriented version of SIMLIB (a simple simulation package). Inf. Trans. Educ. **2**(1), 1–15 (2001)
8. Kotsiantis, S.B.: Supervised machine learning: A review of classification techniques. Informatica **31**, 249–268 (2007)
9. Law, A.M.: Simulation Modeling and Analysis, 4th edn. McGraw-Hill, New York (2007)
10. McKay, M.D., Beckman, R.J., Conover, W.J.: A comparison of three methods for selecting values of input variables in the analysis of output from a computer code. Technometrics **21**(2), 239–245 (1979)
11. Mouelhi-Chibani, W., Pierreval, H.: Training a neural network to select dispatching rules in real time. Comput. Ind. Eng. **58**(2), 249–256 (2010)
12. Neal, R.M.: Bayesian Learning For Neural Networks (Lecture Notes in Statistics), 1st edn. Springer, Berlin, Aug (1996)
13. O'Hagan, A.: Curve fitting and optimal design. J. Roy. Stat. Soc. **40**(1), 1–42 (1978)
14. Panwalkar, S.S., Iskander, W.: A survey of scheduling rules. Operations Res. **25**(1), 45–61 (1977)
15. Priore, P., de la Fuente, D., Gomez, A., Puente, J.: A review of machine learning in dynamic scheduling of flexible manufacturing systems. AI EDAM **15**(03), 251–263 (2001)
16. Rajendran, C., Holthaus, O.: A comparative study of dispatching rules in dynamic flow shops and job shops. Eur. J. Oper. Res. **116**(1), 156–170 (1999)
17. Rasmussen, C.E., Williams, C.K.I.: Gaussian processes for machine learning (Adaptive Computation and Machine Learning). The MIT Press, Dec 2006
18. Rasmussen, C.E.: Evaluation of gaussian processes and other methods for non-linear regression. PhD thesis, Department of Computer Science, University of Toronto (1996)
19. Scholz-Reiter, B., Heger, J., Hildebrandt, T.: Gaussian processes for dispatching rule selection in production scheduling. In: Procceeding of the International Workshop on Data Mining Application in Government and Industry 2010 (DMAGI10) As Part of The 10th IEEE International Conference on Data Mining, pp. 631–638 (2010)
20. Sun, Y.L., Yih, Y.: An intelligent controller for manufacturing cells. Int. J. Prod. Res. **34**(8), 2353–2373 (1996)
21. Williams, C.K.I.: Gaussian processes for machine learning—software examples. http://www.gaussianprocess.org/gpml/code/matlab/doc (2006)
22. Williams, C.K.I., Rasmussen, C.E.: Gaussian processes for regression. Adv. Neural Inf. Process. Syst. **8**, 514–520 (1996)

An AI Based Online Scheduling Controller for Highly Automated Production Systems

Emanuele Carpanzano, Amedeo Cesta, Fernando Marinò, Andrea Orlandini, Riccardo Rasconi and Anna Valente

Abstract Highly automated production systems are conceived to efficiently handle evolving production requirements. This concerns any level of the system from the configuration and control to the management of production. The proposed work deals with the production scheduling level. The authors present an AI-based online scheduling controller for Reconfigurable Manufacturing Systems (RMSs) whose main advantage is its capacity of dynamically interpreting and adapting any production anomaly or system misbehavior by regenerating on-line a new schedule. The performance of the controller has been assessed by running a set of closed-loop experiments based on a real-world industrial case study. Results demonstrate that the capability of automatically synthesizing plans together with recovery actions severely contribute to ensure a high and continuous production rate.

E. Carpanzano · A. Orlandini · A. Valente (✉)
Via Bassini 15, 20133 Milano, Italy
e-mail: anna.valente@itia.cnr.it

E. Carpanzano
e-mail: emanuele.carpanzano@itia.cnr.it

A. Orlandini
e-mail: andrea.orlandini@itia.cnr.it

Amedeo. Cesta · F. Marinò · R. Rasconi
CNR-National Research Council of Italy, ISTC,
Via S. Martino della Battaglia 44, 00185 Roma, Italy
e-mail: amedeo.cesta@istc.cnr.it

F. Marinò
e-mail: fernando.marino@istc.cnr.it

R. Rasconi
e-mail: riccardo.rasconi@istc.cnr.it

K. Windt (ed.), *Robust Manufacturing Control*, Lecture Notes in Production Engineering, 105
DOI: 10.1007/978-3-642-30749-2_8,

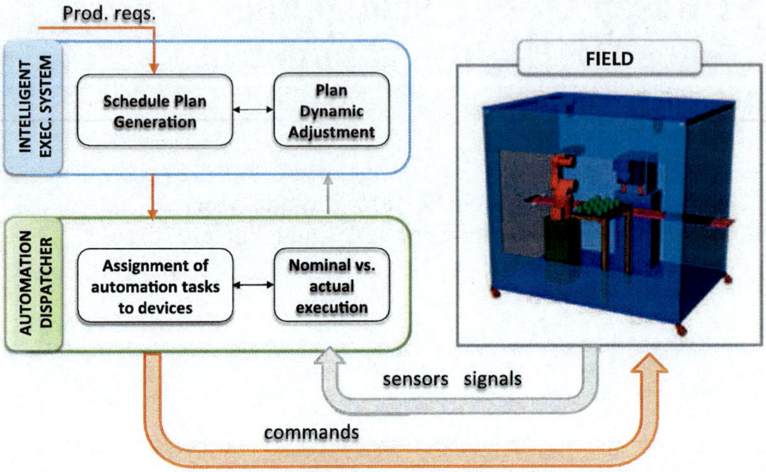

Fig. 1 Production scheduler and automation dispatcher closed-loop

1 Introduction

Highly automated production systems can efficiently compete in evolving production environments if they implement at various levels the capability to adapt or anticipate a change in the production requirements [1–3]. This for Reconfigurable Manufacturing Systems (RMSs) means to have a set of reconfigurability enablers related either to the single component of the system such as the mechatronic device, the spindle axes, or related to the entire production cell and the system layout so that any change of the production demand can be accomplished by implementing the required enablers [4, 5]. For Focused Flexibility Manufacturing Systems (FFMSs) the responsiveness towards the changes relies on the production evolution forecasting. The production system is preliminarily equipped with a set of flexibility degrees designed on the basis of the foreseen events which results available at the moment the change occurs [6, 7]. Differently from RMSs and FFMSs, Flexible Manufacturing Systems (FMSs) represent the most flexible family of system solutions. This allows developing a system that is robust to any production requirement evolution [8–10].

Besides the specific system architecture, any highly automated system is expected to be structured in hardly coupled levels so that any adaptation and recovery action would be comprehensively and persistently streamed across the various layers. Together with an interoperable system infrastructure, this also requires that approaches and methodologies, traditionally developed for each production level in a myopic way such as production and automation, would be conceived exploiting the integration.

A particularly sensitive case concerns the integration of production and automation layers whose scarcely efficient integration severely affects the system global performance [11, 12]. A production schedule module designed for highly automated systems must be able to manage both exogenous (e.g. change of volumes or machining features) and endogenous events (e.g. machine failures or anomalous behavior). Concurrently, it must close the loop with the automation dispatching module that is responsible for mapping production tasks to the related automation tasks that are assigned to the devices coherently to the sequence of jobs to be processed. Based on the actual execution of the tasks, the dispatching module feeds back the current status to the production schedule module that can decide to eventually correct the plan (Fig. 1).

The current work addresses the production and automation synchronization problem from the scheduling perspective. The paper is structured as follows: Sect. 2 presents the dynamic production scheduling approach; Sect. 3 introduces the industrial application to a case study; Sect. 4 describes the formulation of the scheduling model; Sect. 5 outlines the major benefits of the approach while a final section closed the paper.

2 The Proposed Approach

With specific focus on the RMS management aspects, the production scheduling problem implies the capability to develop a short term production plan based on the inputs generated by the capacity planning problem that can be easily and efficiently adjusted and regenerated once a production or system change occurs. There is a number of production scheduling approaches considering changes, both static [13] and dynamic [14].

In [12] we proposed to address the production scheduling problem using the Constraint Satisfaction Problem (CSP) formalism, as it allows to naturally express the features needed to model scheduling problems under uncertainty [14] (e.g., it allows to easily provide the search algorithms with domain-specific heuristic, and to naturally represent *flexible* solutions). Synthesizing a production plan basically entails assigning the available resources to the jobs that are to be processed in the plant with a temporal horizon of the shift; once jobs are allocated to the resources, the schedule is passed to the automation layer that translates the production scheduling in automation plans.

Should disrupting events occur during execution (e.g., machine breakdowns), representational flexibility provides the schedule with strong reconfiguration capabilities, as alternative plans can be quickly produced that use the undisrupted resources, therefore minimizing bottlenecks.

2.1 Modeling the Scheduling Features

The base scheduling problem model employed in this work conforms to the *Resource Constrained Project Scheduling Problem with Time Lags* (RCPSP/max), this is to open the possibility to import a robust algorithmic experience on the problem [14, 15]. The RCPSP/max can be formalized as follows: (i) a set V of n activities must be executed, where each activity a_j has a fixed duration d_j. Each activity has a start-time S_j and a completion-time C_j that satisfies the constraint $S_j + d_j = C_j$; (ii) a set E of temporal constraints exists between various activity pairs $<a_i, a_j>$ of the form $S_j - S_i \in [T_{ij}^{min}, T_{ij}^{max}]$, called start-to-start constraints (time lags or generalized precedence relations between activities); (iii) a set R of renewable resources are available, where each resource r_k has a integer capacity $c_k \geq 1$. The execution of an activity a_j requires capacity from one or more resources; for each resource r_k the integer $rc_{j,k}$ represents the required capacity (or size) of activity a_j. A schedule S is said to be *time-feasible* if all temporal constraints are satisfied, while it is *resource-feasible* if all resource constraints are satisfied (let $A(S,t) = i \in V | S_i \leq t < S_i + d_i$ be the set of activities which are in progress at time t and $r_k(S,t) = \sum_{j \in A(S,t)} rc_{j,k}$ the usage of resource r_k at that same time; for each t the constraint $r_k(S,t) \leq c_k$ must hold). The solving process is performed exploiting a makespan optimization scheduling algorithm called ISES (Iterative Sampling Earliest Solutions) [15]. The ISES solving algorithm basically proceeds by detecting the sets of schedule activities that compete for the same resource beyond the resource maximum capacity (*conflict sets*) and deciding the order of the activities in each set, through the insertion of further temporal constraints between the end time of one activity and the start time of the other, to eliminate conflicting overlaps.

2.2 The Dynamic Scheduling Control Architecture

In this work, we present a real-time control architecture (see Fig. 2) endowed with the flexible production scheduling capabilities discussed above in order to dynamically synthesize updated scheduling solutions as required by the continuously changing environmental conditions.

As shown in Fig. 2, the proposed intelligent execution system architecture (i.e., see the upper-left box in Fig. 1) is designed to provide/receive data to/from the automation layer, and is composed of three different modules, each one holding different responsibilities. The *Controller* is the main component of the architecture and is in charge of: (i) invoking the Scheduler in order to ask for new solutions whenever a new job is entering the system (*find solution* command, see also the following point iv); (ii) updating the internal model of the system according to the observations received by the Dispatcher (*modify model* command); (iii) detecting any possible cause (e.g., anomalous behaviors, failures, etc.) leading to plan unfeasibility; (iv) invoking the Scheduler in order to reschedule the current solution and possibly produce a new feasible solution; (v) disposing completed tasks from the current model. Whenever invoked by the Controller, the *Scheduler* is responsible for (i) producing the initial solution needed to initiate the production process starting from a given problem, and (ii) rescheduling the current solution when it becomes unfeasible due to the onset of some exogenous event. Finally, the

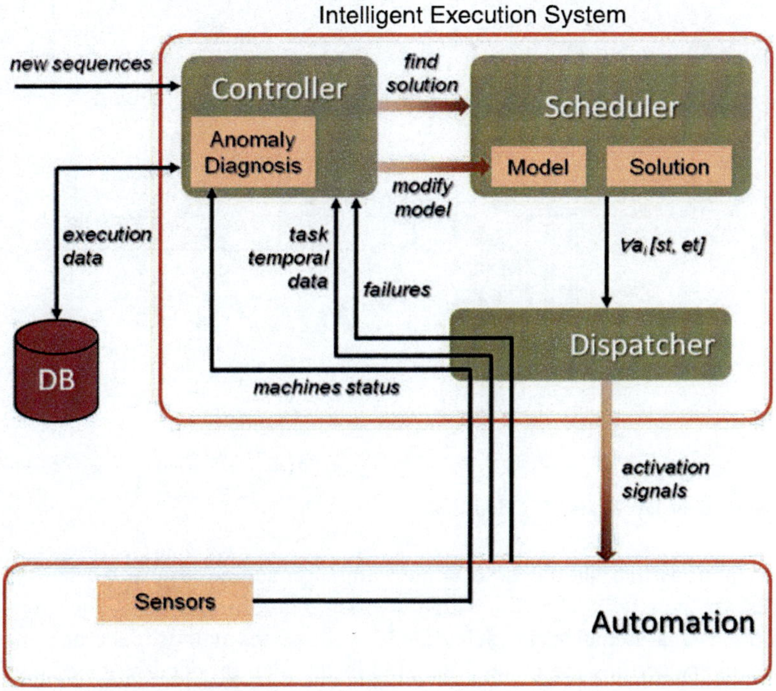

Fig. 2 The overall control architecture

Dispatcher is responsible for (i) realizing the communication from the automation level to the rest of the architecture (all messages coming from the field are pre-processed by the Dispatcher and the related data are forwarded to the Controller), and (ii) dispatching solution-related plan activation signals to the automation layer.

The overall architecture is implemented in Java as a composition of three concurrent and asynchronous processes that interact in a coordinated way to control the production process. In addition, one additional component has been implemented in order to record and store in a database the information flowing within the control system and to provide a human operator with a graphical view of the collected data. Finally, the communication between the control architecture and the automation level has been implemented through the use of the OPC protocol. According to the ISA95 standard, such protocol is fully compatible for Supervisory Control And Data Acquisition (SCADA) connection.

2.3 *Representing Maintenances and Recovery Actions*

In order to make the execution domain as close as possible to the real production system environments, besides the ordinary production tasks the system is able to accommodate *maintenance* activities (ordinary and extraordinary) as well as

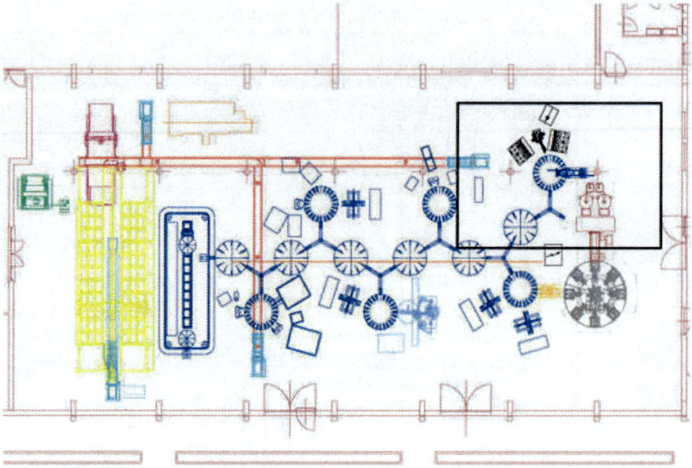

Fig. 3 Shop-floor layout and FRC location

recovery actions that should be executed after a machine failure. Ordinary maintenances are generally scheduled in the plan according to their due frequency, extraordinary maintenances are scheduled in case of anomalous machine behaviors, while recovery actions are instead inserted in the plan on occurrence of particular machine failures. The urgency (i.e., the execution immediacy) of the extra-maintenance will be decided on the basis of the gravity of the occurred anomaly, which is assessed by the Controller's *Anomaly Diagnosis* module (see Fig. 2). It should be noted that as opposed to anomalies, which entail a degraded machine performance and therefore cause a limited impact on the productive process, we assume failures entail the complete inoperability of the affected resource until the failure is resolved (see Sect. 3.1 for details related to the use case considered in this work).

3 Industrial Case Application

The proposed scheduling approach has been applied to an industrial case pertaining to a reconfigurable production line for the manufacturing of customized shoes, representing the European Best Practice in mass customization. The production system is composed by 5 manufacturing cells connected by a flexible transport system composed by rotating tables. The last automated manufacturing island in the shop-floor (Fig. 3) is the Finishing Robotic Cell (FRC), responsible for the shoe finishing before packaging and delivery [16].

As illustrated in Fig. 4, the FRC consists of four machine units, respectively an ABB robot (R1), the island from/to which parts are un/loaded (R2), a controlled brushing machine (R3), a creaming machine (R4) and a spraying machine (R5). The robot operates as *pick and place* and fixturing system; it loads the semi-finished shoe from the island (or rotary table) and, according to the part program,

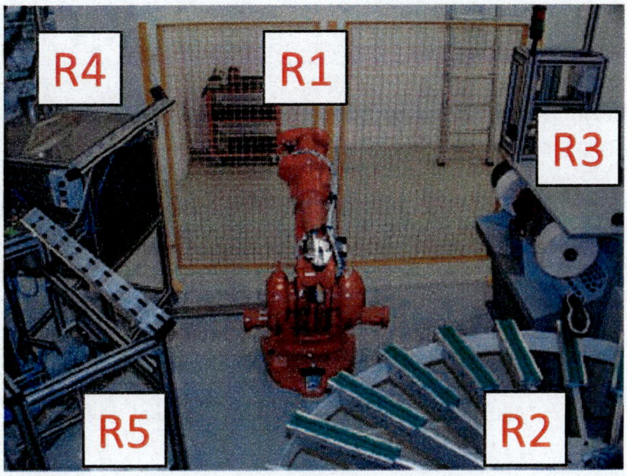

Fig. 4 Resource composing the FRC

transports the part to the related machines, holding the part while the machine is processing it, as a proper fixturing system. Creaming and spraying machines are equipped with two inter-operational buffers with 9 slots each.

As far as the FRC automated system is concerned, the FRC controller is connected with the transportation system Programmable Logic Controller (PLC), the SCADA of the entire line and the low lever cell controller modules. Three types of activities are achieved by means of the existing control architecture: Communication-synchronization with production line controller; Synchronization of tasks in the finishing cell; Control of finishing operations such as rotation speed of the felt rollers, check of spray pressure and drying time, tracking of actual operation execution times compared to nominal expected ones.

3.1 Production and Management Features of the FRC

The FRC finishing process can be clustered in three main families: creaming processes, spraying processes and brushing processes. A typical process sequence is structured in the following steps: part loading; brushing for cleaning the raw piece of dust; finishing by spraying or creaming operations; drying in the buffer; brushing; unloading the finished part.

As highlighted in [12], the considered family of products consists of 8 different part types (i.e., 4 woman models and 4 male models). The processing of each part is to be further divided into the left and right subparts of each shoe model. The production of all parts can be described in terms of the task sequences presented in Table 1. Given a specific shoe model, the left and right part of the model can be produced by means of the same sequence type for both female and male items. However, the durations of the sequence tasks can vary depending on the product type, resulting in 16 different process sequences in total.

Table 1 Description of operation sequences

Sequence #1	Sequence #2	Sequence #3	Sequence #4
Load	Load	Load	Load
Brushing	Brushing	Spraying	Creaming
Spraying	Creaming	Unload	Unload
Unload	Unload	Buffering	Buffering
Buffering	Buffering	Load	Load
Load	Load	Brushing	Brushing
Brushing	Brushing	Unload	Unload
Unload	Unload		

Table 2 Maintenance operation time matrix [sec]

Maintenance task [Rate]	R1	R2	R3	R4	R5	Cell stop	Frequency
Fill cream tank					90	No	1/day
Creaming machine cleaning					60	No	12/day
Creaming machine nozzle cleaning					3	No	2/hour
Fill spray tank				60		No	1/day
Spraying machine cleaning				60		No	12/day
Spraying machine nozzle cleaning				3		No	2/hour
Fill wax in brushing machine			60			No	1/day
Gripper calibration	15					No	1/day

Table 3 Failure modes

Failure types	R3	R4	R5	Duration (mins)	Cell stop
Wax is not reaching	x			2	No
Brush slider is not moving	x			2	No
Brush does not rotate	x			2	No
Dosage is not working			x	5,15	No, Yes
Cream does not arise from sponge			x	15,25	No, Yes
Spray pistol is not responding to control signals		x		10,20	No, Yes
Only air emerges from the spray pistol		x		5	No
Anomalous spray pistol jet		x		10	No

As stated earlier, besides the production tasks a number of maintenance operations need to be foreseen and scheduled to ensure the FRC health. Table 2 synthesizes a few examples of maintenance tasks for FRC resources, considered in this work; in the table, the listed maintenance activities are associated to the related resource, and it is specified whether a stop of the cell is required. The table reports the average expected time (in seconds) for carrying out each maintenance activity as well as the maintenance rate indicated in brackets.

Besides the maintenance tasks, a set of FRC failures have also been systemized and clustered by type in this work (see Table 3). Each failure type mapped upon resources is associated to a number of suitable troubleshooting strategies.

Table 4 Maintenance tasks required from assessment of sensor signals

Maintenance type	Physical source of measured data	Orange %	Red %
Fill cream tank	Level sensor	10–20	0–10
Fill spray tank	Level sensor	10–20	0–10
Fill wax in brushing machine	Level sensor	10–20	0–10
Gripper calibration	Force sensor	10–20	0–10
Creaming machine cleaning	Visual sensor + filter + product quality	15–30	0–15
Spraying machine cleaning	Visual sensor + filter + product quality	15–30	0–15
Creaming machine nozzle cleaning	Cream consumption + valve + product quality	15–30	0–15
Spraying machine nozzle cleaning	Spray consumption + valve + product quality	15–30	0–15

An efficient execution of maintenance and/or recovery tasks relies on a persistent signal interpretation to assess the system status. This evaluation is crucial to identify the gap between actual and nominal system behavior and consequently the related actions to be implemented. Table 4 outlines few examples of signal information associated to the need to undertake specific maintenance tasks. For each considered machine maintenance, the table shows: (i) the polled sensors, and (ii) the predefined signal threshold values beyond which anomalies of different gravity are recognized (e.g., severe (*red*) anomalies are detected when the weighted sum of the anomalous readings obtained from sensors goes below 10 %).

4 The Scheduling-Based Controller

As explained in [12], the FRC scheduling problem is modeled in CSP terms adopting a combination of modeling strategies that allows to capture all the significant aspects of the problem that the solving process must reason upon.

4.1 Modeling in the Static Case

The reader interested in the base model details can refer to [12]; in that work, we focused on a *model abstraction* suitable for the *static* problem solving case, which has allowed us to: (1) decrease the number of involved tasks guaranteeing no loss of expressiveness, and (2) re-use partially modified, if at all, off-the-shelf scheduling algorithms for the solving process.

The solution provided in [12] was taking advantage of the robot acting as a *critical* resource, which allowed the two task subsequences immediately preceding and following the buffering operation to be grouped in two single blocks (the dashed boxes in Fig. 5). In order to allow for a finer treatment of machine faults

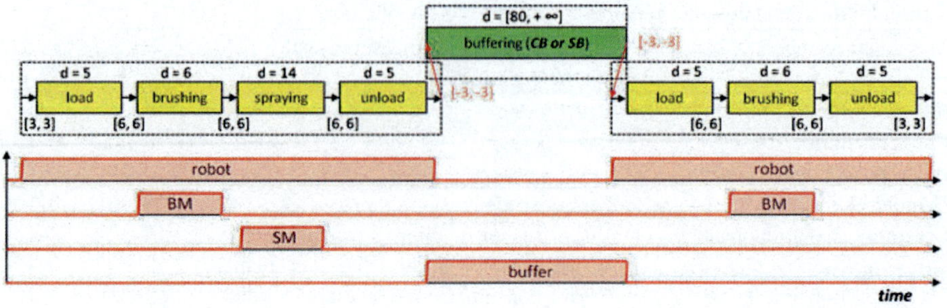

Fig. 5 An example of FRC task sequence (BM = Brushing Machine, SM = Spraying Machine)

and maintenance operations, in the present work it is necessary to abandon such *aggregated* model and keep each individual sequence task separated. Figure 5 depicts a typical sequence that entails the utilization of a subset of FRC machines and tools, e.g., the brushing machine and the spraying machine, as well as one of the two available buffers. Each sequence task is characterized by a nominal duration d, and consecutive tasks are separated by temporal constraints $[a, b]$ where a and b are the lower and the upper bound of the separation constraint. The actual constraint values depicted in Fig. 5 are consistent with the real robot transition times (e.g., the 6 value between the brushing and the spraying tasks represents the time that the robot takes to go from the brushing machine to the spraying machine passing through the *home* position), while the negative constraint values shown in red characterize the fact that the buffering operation actually starts 3 seconds *in advance* with respect to the end of the first dashed box, because the robot must however return to its home position before commencing any other action.

4.2 The Dynamic Model

In order to allow the management of the schedule in a dynamic context, i.e., continuously absorbing all the modifications that pertain to the occurrence of exogenous events as well as to the simple passing of time, it has been necessary to extend the previous model with online knowledge-capturing and management features. In our framework, such features are added using an asynchronous event-based model.

All the information about the environmental uncertainty (e.g., endogenous and/or exogenous events) is organized through an asynchronous message exchange mechanism among the system modules. These messages convey all the information relatively to the deviations between the nominal schedule currently under execution and the *real data* coming from the automation side of the plant. The Controller is in charge of acquiring such information, adapting the plan accordingly, and calling for the necessary rescheduling actions. However, applying a rescheduling to an executing plan generally presents some technical difficulties, mainly stemming from the following circumstances: (i) during the solving phases, all dispatching and/or event-listening features are temporarily disabled, which brings up the requirement to keep the rescheduling steps as fast as possible; (ii) the Scheduler does not have any internal chronological model of the schedule with respect to the passing of time. In other words, it has no knowledge of *past*, *present* and *future* relatively its own activities (i.e., it may decide to reschedule one activity into the past, or postpone the start time of an activity that has already started).

The former issue is taken into account by keeping the number of activities in the current schedule as low as possible, i.e., by eliminating the activities from the plan as they terminate their execution, in order to establish a sort of *dynamic equilibrium* between incoming and outgoing sequences, after an initial transient. The latter issue is solved by introducing a number of constraint-based pre-processing procedures whose objective is to impose new constraints to the executing schedules prior to the solving process, so as to *force* the Scheduler to produce solutions that reflect the temporal reality of execution. Such procedures are the following: (i) *fixActivity()* when the Dispatcher acknowledges from the plant that an activity has started, the Controller must *fix* the activity's start time in the model, so that it is not shifted by the rescheduling process; (ii) *fixActivityDuration()* when the Dispatcher acknowledges from the plant that an activity has terminated, the Controller must fix the activity's end time, so that the latter is not modified by any possible rescheduling process before the activity is eliminated from the current plan; (iii) *disposeCompletedActivity()* this procedure eliminates a completed activity from the model; (iv) *prepareRescheduling()* this procedure performs the very important task of inserting in the plan a set of new *release constraints* relatively to all the activities that will participate to the rescheduling, so as to avoid that such activities will be scheduled in the past w.r.t. to the current execution time. Once all previous preparatory actions are performed, the rescheduling procedure can be safely called by the Controller. The Scheduler will therefore produce an alternative solution that (i) is temporally and resource feasible, (ii) satisfies all problem-related and execution-related constraints, and (iii) complies with the chronological physical requirements.

5 Experimental Results

In this section, we analyze the dynamic scheduling performances of our architecture by deploying it to control the execution of a series of typical production tasks relatively to the FRC case study. In particular, we will test the dynamic scheduling capabilities of our system by simulating the execution of a determined number of production sequences, which entails the online scheduling of the continuously incoming production tasks (equally distributed among the different process types) and ordinary maintenances (defined in Table 2). Both the temporal flexibility of the employed model and the rescheduling efficacy of the solver will be assessed by simulating the onset of perturbing events of random extent during each execution. More specifically, we analyze the performances of our architecture by varying the following settings: (i) we consider randomly variable start and end times for each incoming task, which affects the overall stability of the solution and requires the controller to continuously invoke the scheduler in order to adjust the current solution; (ii) we introduce a number of anomalies on the basis of the values (described in Table 4) detected by the automation layer sensors, and processed by the Diagnosis module. Each time an anomaly is detected, the control architecture reacts by scheduling an extraordinary maintenance activity whose urgency depends on the severity of the anomaly (*orange, red*). Maintenance activities may even cause the complete stop of the cell, and affect in any case the overall makespan; (iii) according to Table 3, we consider a set of possible failures for each machine, that may occur during execution. In this cases, the control architecture is in charge of scheduling the proper recovery task aimed at restoring full machine operability. As for anomalies, failures may introduce idle production periods, thus reducing production capability.

The experiments are organized in two different settings, both entailing the execution of 130 uniformly distributed production sequences. In the *First Setting*, 5 runs are executed for each resource R_i of the FRC. Each run requires the dynamic scheduling of the continuously incoming production tasks, including the periodic maintenances. Temporal uncertainty is introduced by considering an average 10 % randomic misalignment between the nominal (i.e., dispatched) and the real (i.e., acknowledged) start/end times of the production activities. Each run is characterized by the onset of a number of anomalies and failures that depends on the affected machine R_i: in particular, every brushing machine will undergo 5 anomalies and 3 failures, every creaming machine will undergo 3 anomalies and 2 failures, and every spraying machine will undergo 3 anomalies and 3 failures (such numbers are decided on the basis of the available maintenance and recovery operations for each machine as well as of their durations, as per Tables 2 and 3). In order to appreciate the benefits of a controller that allows the concurrent scheduling and execution of both maintenance and production tasks, a second experimental setting is developed (*Second Setting*) where all previous runs are performed anew under the assumption that each maintenance and each failure recovery action entails a full FRC cell stop. All runs are performed on a MacBook Pro with a 64-bit Intel Core i5 CPU (2.4 GHz) and 4 GB RAM. In the following, we illustrate the collected empirical results.

Table 5 summarizes the obtained results; the table is horizontally organized so as to provide the data related to every machine. In particular, for each machine row the table lists data obtained in the first and second experimental settings (first and second row) together with the plain value difference and related percentage (third row). For each setting, the table provides the average values obtained from the five runs executed on each machine of: (i) the final makespan (i.e., the completion time of all 130 production sequences), (ii) the overall average time spent in reschedulings, (iii) the total number of reschedulings. The obtained results show the advantage of deploying an online reasoner that allows to continue execution during

Table 5 Results from the experimental runs

	Makespan (mins)	Rescheduling time (mins)	# of Reschedulings
Brushing Machine			
First setting	251	27	129
Second setting	269	30	157
Δ (Δ %)	18 (7.2 %)	3 (11.1 %)	28 (21.7 %)
Spraying Machine			
First setting	256	26	130
Second setting	279	28	155
Δ (Δ %)	23 (9 %)	2 (7.7 %)	25 (19.2 %)
Creaming Machine			
First setting	250	25	127
Second setting	278	28	154
Δ (Δ %)	28 (11.2 %)	3 (12 %)	27 (21.2 %)

maintenances and recovery actions. Regardless of the machine involved in the performed runs, a significant reduction in makespan can be observed between the two experimental settings, meaning that the cell succeeds in executing all sequences in less time. In the table, makespan gains ranging from 18 up to 28 min are observable, which represent a significant improvement when measured against a total run time of 4 h. Such gains are more evident for the machines that are characterized by longer maintenance and recovery actions (i.e., spraying and creaming). In case of long maintenances or recoveries, the capability to continue the execution of the tasks already scheduled on the unaffected machines is of great importance. Another interesting aspect can be observed by analyzing the higher number of reschedulings necessary in the *Second Setting* w.r.t. to *First Setting* runs; the reason of this stems from the fact that in order to simulate the absence of the execution controller (*Second Setting* runs) we have modeled the cell-blocking condition by considering all maintenances and recoveries as tasks that require the whole cell; this causes a resource conflict that has to be solved by means of a rescheduling each time a maintenance or a recovery must be executed. As a last observation, the table also confirms that the chosen number of failures and anomalies injected during all runs for the different machines was well balanced, as the average total time spent for reschedulings is equally subdivided in all cases of the same type, despite the durations of the recoveries and maintenances varied significantly among the machines (see Tables 2 and 3), the reason being that the longer the recovery/maintenance operation, the higher the possibility of a rescheduling when it is added to the plan.

6 Conclusions

This work has presented an AI-based online scheduling controller capable of dynamically manage a production plan under execution in uncertain environmental conditions. The capabilities of the proposed scheduling controller have been tested with reference to a real-world industrial application case study. The series of closed-loop experimental tests concerning the execution of reality-inspired production plans (i.e., complete with regular maintenances, as well as random failures and anomalies), demonstrate that thanks to the adopted flexible model, the proposed controller enhances the current production system with the robustness necessary to face a subset of typical real-world production requirement evolutions. The current results confirm that the deployment of continuous rescheduling capabilities on a temporally flexible plan model positively contribute to the overall efficiency of the production plant, by allowing the execution of the planned number of jobs in less time. The authors work is currently ongoing with the further objectives of (i) improving the controller's rescheduling optimization capabilities in environments characterized by a higher number of tasks, and (ii) expanding the controller's uncertainty management capabilities to the whole actual set of FRC exogenous events, which represents a necessary step before commencing any experimentation on the real field.

Acknowledgments The research presented in the current work has been partially funded under the Regional Project "CNR - Lombardy Region Agreement: Project 3". Cesta and Rasconi acknowledge the partial support of MIUR under the PRIN project 20089M932N (funds 2008).

References

1. Smith, T., Waterman, M.: Identification of common molecular subsequences. J. Mol. Biol. **147**, 195–197 (1981)
2. Wiendahl, H.P., ElMaraghy, H., Nyhuis, P., Zah, M., Wiendahl, H.H., Duffie, N., Brieke, M.: Changeable manufacturing - classification, design and operation. CIRP Ann. - Manufact. Technol. **56**(2), 783–809 (2007)
3. Terkaj, W., Tolio, T., Valente, A.: Designing Manufacturing Flexibility in Dynamic Production Contexts. Des. Flex. Prod. Syst. Methodol. Tools 1–18 (2009)
4. Koren, Y., Heisel, U., Jovane, F., Moriwaki, T., Pritschow, G., Ulsoy, G., Van Brussel, H.: Reconfigurable manufacturing systems. CIRP Ann. Manufact. Technol. **48**(2), 527–540 (1999)
5. Landers, R., Min, B.K., Koren, Y.: Reconfigurable machine tools. CIRP Ann. Manufact. Technol. **50**(1), 269–274 (2001)
6. Terkaj, W., Tolio, T., Valente, A.: Design of focused flexibility manufacturing systems (FFMSs). Des. Flex. Prod. Syst. Methodol. Tools 137–190 (2009)
7. Terkaj, W., Tolio, T., Valente, A.: A stochastic programming approach to support the machine tool builder in designing focused flexibility manufacturing systems - FFMSs. Int. J. Manufact. Res. **5**(2), 199–229 (2010)
8. Stecke, K.: Design, planning, scheduling and control problem of flexible manufacturing systems. Ann. Oper. Res. **3**, 1–13 (1985)
9. Park, T., Lee, H., Lee, H.: FMS design model with multiple objectives using compromised programming. Int. J. Prod. Res. **39**, 3513–3528 (2001)
10. Shin, H., Park, J., Lee, C., Park, J.: A decision support model for the initial design of FMS. Comput. Ind. Eng. **33**, 549–552 (1997)
11. Valente, A., Carpanzano, E.: Development of multi-level adaptive control and scheduling solutions for shop-floor automation in reconfigurable manufacturing systems. CIRP Ann. Manufact. Technol. **60**(1), 449–452 (2011)

12. Carpanzano, E., Cesta, A., Orlandini, A., Rasconi, R., Valente, A.: Closed-loop production and automation scheduling in RMSs. In: ETFA. International Conference on Emergent Technologies and Factory Automation (2011)
13. Tolio, T., Urgo, M.: A rolling horizon approach to plan outsourcing in manufacturing-to-order environments affected by uncertainty. CIRP Ann. Manufact. Technol. **56**(1), 487–490 (2007)
14. Rasconi, R., Policella, N., Cesta, A.: Fix the schedule or solve again? Comparing constraint-based approaches to schedule execution. In: COPLAS-06. Proceedings of the ICAPS Workshop on Constraint Satisfaction Techniques for Planning and Scheduling Problems (2006)
15. Cesta, A., Oddi, A., Smith, S.: A constraint-based method for project scheduling with time windows. J. Heuristics **8**(1), 109–136 (2002)
16. Valente, A., Carpanzano, E., Brusaferri, M.: Design and implementation of distributed and adaptive control solutions for reconfigurable manufacturing systems. In: CIRP Sponsored ICMS. International Conference on Manufacturing Systems (2011)

Stochastic Scheduling of Machining Centers Production, Estimating the Makespan Distribution

Tullio Tolio and Marcello Urgo

Abstract In the scheduling of manufacturing systems, uncertain events are rather the rule than the exception and are the main responsible of cost increase due to missed due dates, resource idleness, higher work-in-process inventory. Robust scheduling approaches aim at devising schedules insensitive, at least to some degree, to the occurrence of uncertain events. However, robust scheduling must always deal with finding a balanced compromise between expected profit and the protection against extremely unfavorable events having a low occurrence probability. Tackling this problem implies being able of estimating the distribution probability associated with a scheduling objective function, or at least some of its quantiles. In this paper we propose a Markovian approach to estimate the distribution of the completion time of a general network of activities. Grounding on this estimation, an estimation of the objective function distribution can be easily calculated. To demonstrate its viability, the proposed approach is applied to a real industrial case in the machining tool sector.

Keywords Stochastic scheduling · Markovian activity networks

T. Tolio
ITIA-CNR, Institute of Industrial Technologies and Automation National Research Council, Milano, Italy

M. Urgo (✉)
Manufacturing and Production Systems Division, Mechanical Engineering Department, Politecnico di Milano, Milano, Italy
e-mail: marcello.urgo@mecc.polimi.it

K. Windt (ed.), *Robust Manufacturing Control*, Lecture Notes in Production Engineering, 121
DOI: 10.1007/978-3-642-30749-2_9,

1 Introduction

In production, differences between projected future and actual execution of the schedules are rather the rule than the exception. Uncertainty may stem from a number of possible sources, both internal and external. Activities may last more or less than originally estimated, resources may be unavailable, materials may arrive behind schedule, ready times and due dates may change, new activities like reworks could be inserted in the plan. A disrupted schedule incurs high costs due to missed due dates and deadlines, resource idleness, or higher work-in-process inventory [1].

Academic research has recently investigated robust scheduling approaches to make schedules insensitive, at least to some degree, to disruptions. Robust approaches aim at reacting to the occurrence of uncertain events (reactive approaches) or at protecting the performance of the plan by anticipating, to some extent, the occurrence of uncertain events (proactive approaches) [2]. However, the vast majority of robust approaches rely on the minimization of the expected value of a performance indicator (i.e., the expected tardiness). Although providing a significant improvement with respect to pure deterministic approaches, such objective functions do not entirely model the concept of robustness. In fact, minimizing the expected tardiness aims at assuring an average good performance in terms of due date satisfaction, but does not protect against the worst cases if their probability is low.

Protection against worst cases is a basic attitude in management decisions. Plant managers facing uncertainty always try to maximize some profit measure but, at the same time, want to avoid the rare occurrence of unfavorable situations causing heavy losses. Hence, a balanced compromise has to be found between expected profit and acceptable risk. Translating the risk concept into the scheduling area implies being able to estimate the distribution probability associated with a scheduling objective function, or at least some of its quantiles [3]. However, for many scheduling problems, such estimation could be difficult.

In this paper we present an approach to estimate the distribution of the makespan of an activity network with general distributed activity durations by means of the time to absorption of a Continuous Time Markov Chain (CTMC) using a Phase Type (PH) approximation for non exponential distributions. In Sect. 2 the analysis of the related literature is provided while Sect. 3 describes both the traditional markov chain approach and the extension to generally distributed activity durations using Phase Types. In Sect. 4 the proposed approach is applied to a real industrial case related to machining centers production while Sect. 5 comments the results and outlines possible improvement paths.

2 State of the Art

A classic problem in the analysis of stochastic project networks is the calculation of the distribution function of the project completion time (makespan) and also its moments. However, these problems have been demonstrated to be hard to solve in general [4]. Researchers, hence, focused on computing proper bounds to estimate the distribution function. Methods for computing bounds of the makespan distribution have been proposed first in [5–8]. A further approach has been proposed in [9] allowing the use of some dependency between the distributions of the activity durations. [10] introduced a unified model for such bounding results in terms of a chain minor notion for project networks that covers and generalizes all approaches which are based on the transformation of the given project network into series-parallel networks. A wide class of approaches give up in estimating the whole distribution of the objective function and, hence, concentrate on the calculation of a specific indicator (i.e., the mean or the variance) used to characterize the schedule [11, 12]. Other approaches rely on stochastic programming to address the stochastic characteristics of the scheduling problem [13–16].

A different class of approaches are grouped under the name Markov Activity Network (MAN). MAN has been proposed in the literature as a way of modeling the execution of a set of activities in a network defining the precedence relations among them using a markov model. This approach has been first proposed in [17]. Given that (i) the durations of the activities are mutually independent and (ii) exponentially distributed, the execution of the activity network can be represented through a *continuous time markov chain* (CTMC). Being able to model the execution of the network of activities as a CTMC provides the capability of exploiting the wide set of tools devoted to CTMC to analyze the stochastic scheduling problem described by the network of activities. In particular, the completion of all the activities in the network can be considered as a state of the markov model, but this can be also seen as the absorbing state of the markov chain. Analyzing the distribution of the completion time of the network of activities, hence, correspond to the transient analysis of the underlying CTMC. In addition, the CTMC also provides definite algorithms to compute the moments of the completion time. However, the restriction to exponentially distributed activity durations represents a limiting hypotheses, since the exponential distribution is quite seldom applicable to real industrial problems.

In the field of markov models, Phase Type (*PH*) distributions can be used as a tool to provide an approximation of general distributions. Basically, a set of one or more inter-related Poisson processes are put together to form a new CTMC. Hence, a *PH* distribution is the distribution of the time until absorption of a markov chain with one absorbing state. A proper design of the set of considered Poisson processes and their relationships provides a tool to fit a general distribution with a certain degree of accuracy. In fact, each MAN is a *PH* model, since its time to absorption, that does not need to be exponentially distributed, is represented through a system of Poisson processes (the execution of each activity) with mutual interactions (precedence relations among the activities).

Fig. 1 AoA activity network

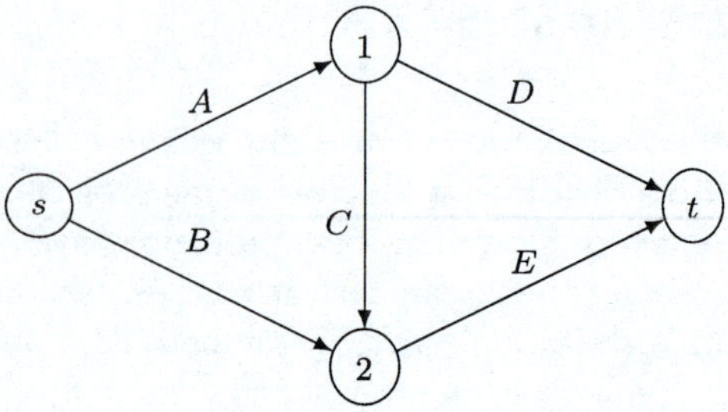

Let us consider a continuous-time markov process with m states plus a state ***0*** that is an absorbing state and, in addition, an initial probability vector α being the probability of the process to start in any of the $m+1$ states (phases), The continuous *PH* distribution is the distribution of the time from the process's start to its absorption in the absorbing state ***0***

Different approaches have been proposed in the literature to define a proper *PH* approximation aiming at best fitting a given general distribution shape [18] or concentrating on matching a certain number of moments with the minimal number of phases [19, 20].

3 Description of the Approach

Let us consider an acyclic directed graph $G = (V, A)$ with a set of nodes $V = v_1, v_2, \ldots, v_m$ and a set of arcs $A = a_1, a_2, \ldots, a_n$. Each arc in V represents the execution of an activity, hence, the graph is an Activity on Arc (*AoA*) network and we can use the term arc or activity as synonymous. The *AoA* network has a single source s representing the start of the execution and a single sink t representing the completion of all the activities in the network. In the model proposed by [17], at each instant time t, an activity could be either idle (= unstarted), or active (= in process) or finished (= completed). Basically, during the execution of the network of activities each activity can be in one and only one of the previous defined states and all the feasible combinations of states of the activities in the network provides the support for a CTMC. It must be noticed that in [17], the authors also provide a way of reducing the states to be considered in the CTMC selecting only a subset of all the possible states that guarantee the modeling of the execution of the CTMC. This set of states is obtained through the definition of all the *uniformly directed cuts* (s, t) in G [21].

As an example we consider the activity of network in Fig. 1 representing the execution of the activities A, B, C, D and E and the set of precedence relations among them. The duration of the activities are exponentially distributed with parameters $\lambda_A, \lambda_B, \lambda_C, \lambda_D, \lambda_E$, respectively.

With the approach proposed in [17] the following set of states can be defined:

(A,B)	both activity A and B are in process;
(A,B^*)	activity A is in process while activity B is completed and *dormant*, i.e. it waits for the completion of C to allow activity E to start;
(B,C,D)	activity A is completed, hence, activity C and D are in process while activity B is still in process;
(B,C^*,D)	activity B and D are in process while activity C is completed and *dormant*, waiting for the completion of B to let E start;
(B,C,D^*)	activity B and C are in process while activity D is completed and *dormant*, waiting for the completion of E to terminate the execution of the whole network;
(B^*,C,D)	activity C and D are in process while activity B is completed and *dormant*, waiting for the completion of C to let E start;
(B,C^*,D^*)	activity B is in process while both activity C and D are completed. Activity C is *dormant*, waiting for the completion of B to let E start while D is *dormant*, waiting for the completion of E to terminate the execution of the whole network;
(B^*,C,D^*)	activity C is in process while both activity B and D are completed. Activity B is *dormant*, waiting for the completion of C to let E start while D is *dormant*, waiting for the completion of E to terminate the execution of the whole network;
(D,E)	bot activity D and E are in process;
(D,E^*)	activity D is in process while activity E is completed and *dormant*, i.e. it wait for the completion of D to terminate the execution of the whole network;
(D^*,E)	activity E is in process while activity D is completed and *dormant*, i.e. it wait for the completion of E to terminate the execution of the whole network;
$(\emptyset)$	all the activities have been processed, the network has been completely executed.

Since all the activites in the network are exponentially distributed, the above described states are the support for the CTMC in Fig. 2.

As stated before, when activity durations are not exponential distributed, the CTMC only provides an approximation of the real scheduling problem. To address this issue, *PH* distributions can be used to approximate non exponential distributions through the time to absorption of a CTMC. *PH* approximations are provided for each activity and then joined together to represent the whole scheduling problem with the associated precedence constraints.

As a brief simple example, let us consider the execution of two activities A and B in series. Activity A can be approximated through a *PH* distribution with n states $A_1, A_2, \ldots, A_n$ while B is approximated through a *PH* distribution with m states $B_1, B_2, \ldots, B_m$ (Fig. 3).

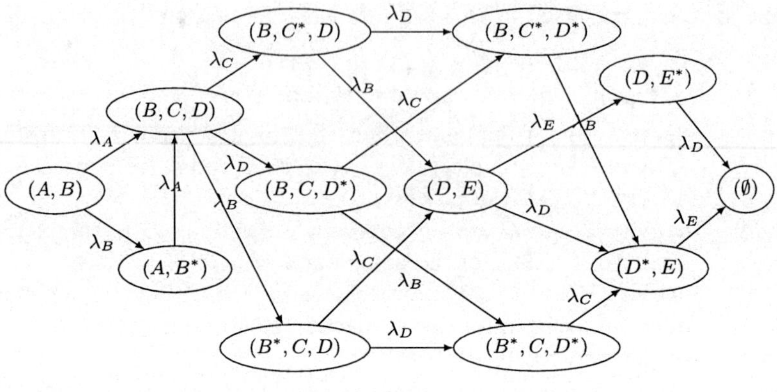

Fig. 2 Continuous time Markov chain for the AoA network in Fig. 1

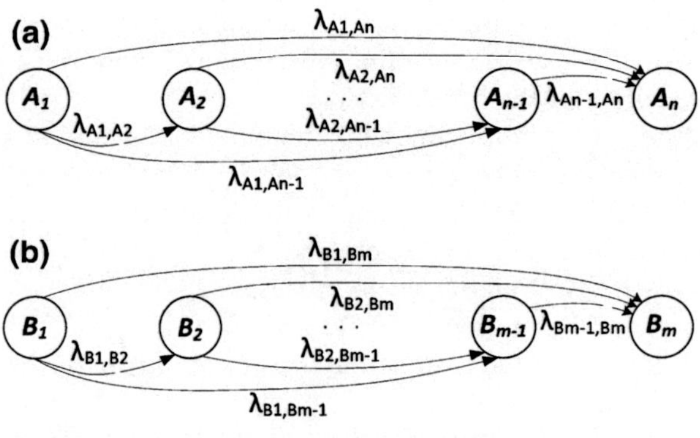

Fig. 3 Phase Type approximation for activity A(**a**) and B(**b**)

Given the fact that activity B can start only after activity A has been completed, the set of possible states is $A_1, A_2, \ldots, A_n, B_1, B_2, \ldots, B_n$. The states representing the transition from the execution of A to the execution of B and the related transition rates can be defined as described in Fig. 4.

Using the same approach for all the activities in the original network, a new CTMC considering the Phase Type approximation for all the activities can be defined. Hence, the distribution of the time to absorption of the obtained CTMC can be calculated as

$$F(t) = 1 - \alpha e^{St}\mathbf{1} \tag{1}$$

where S represents the transition rates matrix and α is the probability vector associated to the Phase Type model. This distribution is also the distribution of the makespan of the underlying stochastic scheduling problem.

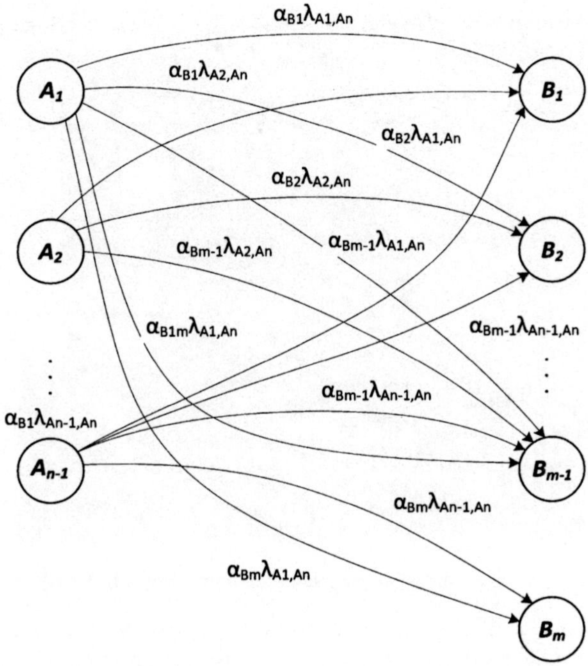

Fig. 4 Transition rates passing from the execution of activity *A* to activity *B* using Phase Types approximations

4 Industrial Case

A machining center is a CNC (Computer Numerical Controlled) machine integrated with an automatic tool changer, and it often has equipment for pallets or parts handling (Fig. 5).

The assembling of a machining center is a rather complex activity. A great number of components must be assembled together onto the machining center structure. Some components are the main components of the machining center, e.g., the spindle, the turning table, the controlled axes and their activation equipment. Other components, like small mechanical components (screws, nuts, clamps, lids, and so on), or components of the pneumatic and hydraulic systems are considered ancillary. Moreover, most of the assembled components must be electrically wired.

Even if a standard configuration for a machining center type exists, most of the machining centers are frequently specifically designed to satisfy customers needs. This can be considered a typical characteristics for European (and in particular Italian) machining center manufacturers. High levels of customization and long flow times results in production plans to be defined before information on product customizations and detailed production activities are completely disclosed. Hence,

Fig. 5 Machining center structure with preassembled components installed

at the planning phase, the duration of the production activities is considered uncertain. To model the production of a machining center, the following production activities are considered: Structure Scraping (*SS*), Structure Painting (*PA*), Pallets Scraping (*PS*), Autonomous Components Assembling (*CA*), Machining Center Assembling (*MA*), Wiring (*WI*), Testing (*TE*).

The precedence relations among the production activities are represented in the AoA network of activities in Fig. 6. The production activities are executed by teams of workers but the resource constraints due to the limited availability of workers are not considered in this analysis.

If the activity durations are exponentially distributed, the state space associated to the CTMC can be easily calculated and is reported in Table 1. However, the considered activities are not exponential distributed but are modeled through *Uniform* and *Weibull* distributions.

The durations of the aggregated activities are defined in terms of hours and their probabilistic distribution is defined grounding on historical data. For those activities with enough historical data, a Weibull distribution is fitted while, for those activities with not enough data, a uniform distribution is used considering the minimum and maximum observed value. In the analysis, the following distribution are used: $SS \approx Uniform(100, 150)$, $PA \approx Uniform(26, 34)$, $PS \approx Weibull(1.99, 63.39)$, $CA \approx Weibull(1.85, 150)$, $MA \approx Weibull(2.56, 913)$, $WI \approx Weibull(8.15, 275)$, $TE \approx Weibull(5.15, 282)$.

The fitted distributions are approximated through 8-phases PH distributions (Figs. 7 and 8).

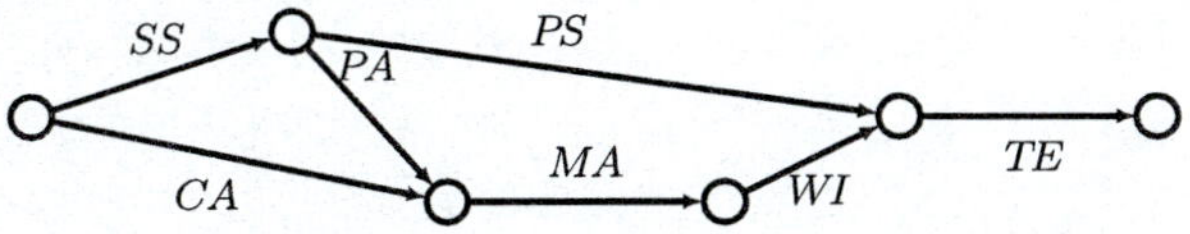

Fig. 6 Activity network for the production of a machining center

Table 1 State space for the machining center production network under the hypothesis of exponential distributed activity durations

0. (CA, SS)	5. (CA, PA^*, PS^*)	10. (WI, PS^*)
1. (CA, PA, PS)	6. (MA, RP)	11.(TE)
2. (CA^*, PA, PS)	7. $(SA, RP*)$	12. $(\emptyset, \emptyset)$
3. (CA^*, PA, PS^*)	8. (WI, PS)	
4. (CA, PA^*, PS)	9. (WI^*, PS)	

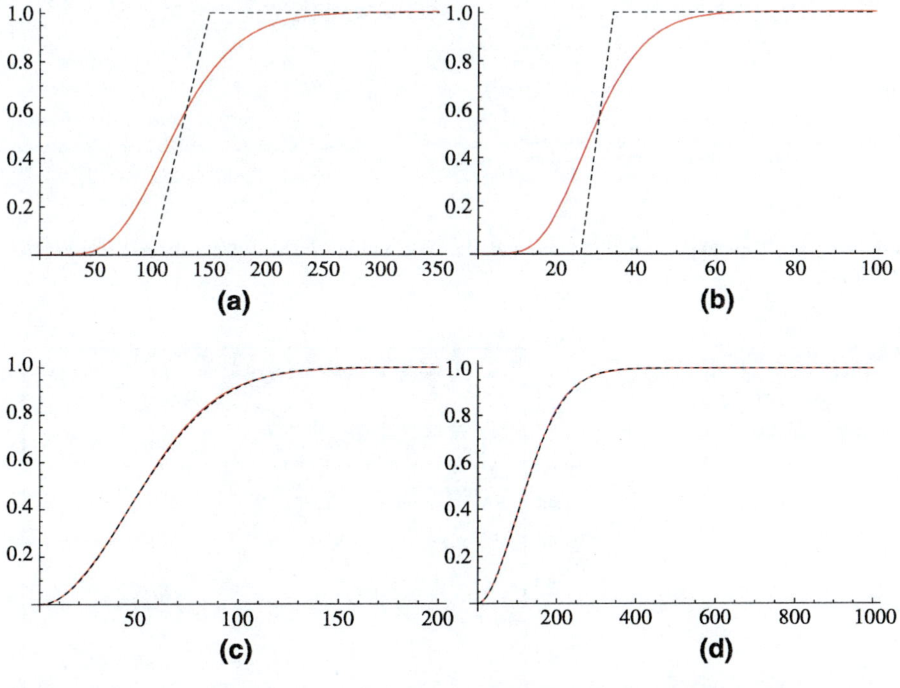

Fig. 7 Phase Type approximation (*red*) and real distribution (*dashed black*) for *SS* (**a**), *PA* (**b**), *PS* (**c**), *CA*(**d**).

Hence, the associated CTMCs are considered together with the state space in Table 1 to compute the different combinations of states and the relative transition rates. The resulting complete state space contains about 78 thousands states. However, considering the constraints induced by precedence relations, the state space is reduced to 638 states. The complete CTMC and the associated transition

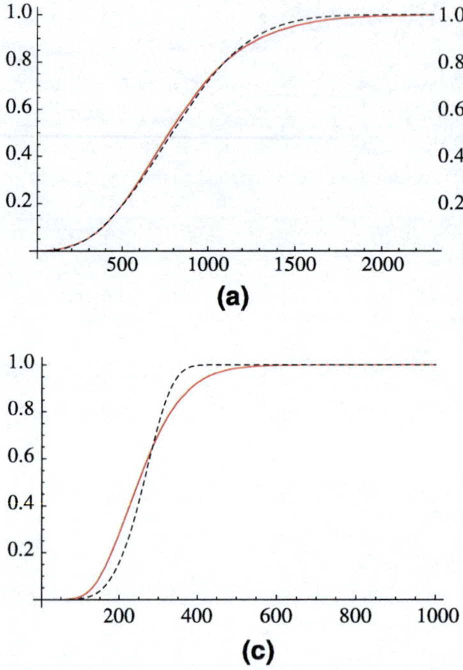

Fig. 8 Phase Type approximation (*red*) and real distribution (*dashed black*) for *MA* (**a**), *WI* (**b**), *TE* (**c**).

Fig. 9 Phase Type approximation cdf (*red*) and the sampled cdf (*blue*) of the makespan distribution

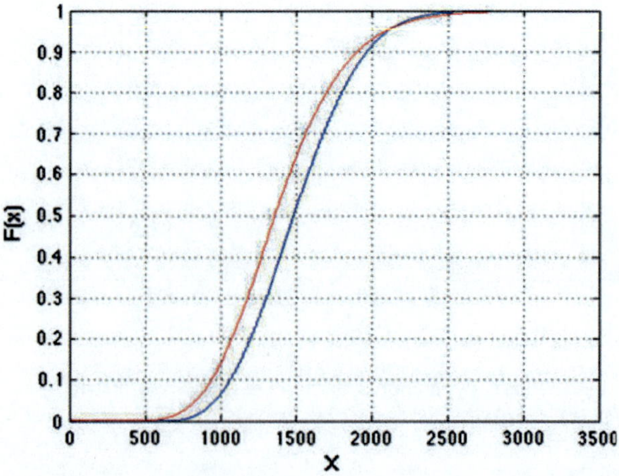

rates are used to calculate the *cumulative distribution function* (cdf) of the time to absorption. To test the accuracy of the Phase Type approximation, this distribution is compared with the real cdf obtained with a Monte Carlo sampling.

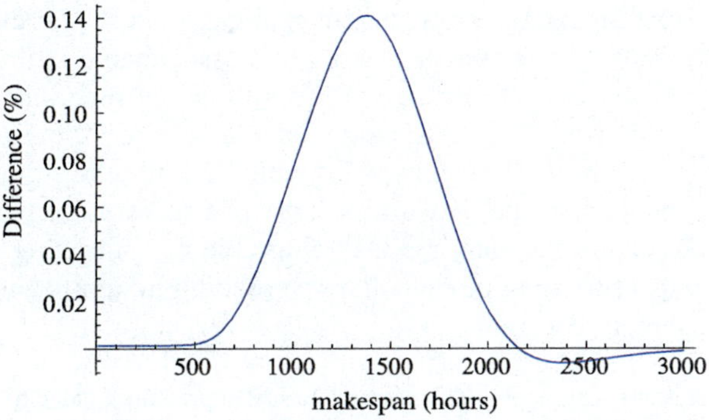

Fig. 10 Difference between the Phase Type approximation cdf and the sampled cdf of the makespan distribution

Figure 9 reports the cdf calculated using the Phase Type approximation and the sampled one. In addition, Fig. 10 reports the plot of the difference between the two cdfs, showing that the approximation is rather rough in some parts of the distribution, reaching a maximum error of about 14 %. The poor quality of the approximation is mostly due to the impossibility of adequately approximate the uniform distribution with a Phase Type distribution with so small a number of phases (Fig. 10).

5 Conclusions

In this paper we presented a markovian approach to estimate the distribution of the completion time (makespan) of a set of activities with generally distributed stochastic durations. The proposed approach extends the Markov Activity Network approach to generally distributed activity durations grounding on Phase Type approximations.

The viability of the approach has been demonstrated on a real industrial case with seven activities whose duration distributions has been approximated using 8-phases PH distributions. The proposed approach was able to provide a reasonable bound for the makespan distribution, although the maximum error of 14 % can be considered a poor performance.

The quality of the *PH* approximation could be easily improved at the price of further increasing the dimension of the state space. However, the capability of decomposing the activity network and/or to specifically selecting the desired level of approximation varying the number of states (phases) for the different *PH* approximations provides a promising research path.

In addition, the capability of formally modeling the execution of the activity network through a markov process provides a promising path towards embedding the proposed estimation approach in scheduling algorithms to optimize a function of the makespan distribution.

As an example of concrete application in the production management area, risk measures are a function of a given associated distribution. In fact, a common risk measure like the *Value at Risk* is basically the quantile of a distribution.

Being able of estimating the value α of the cumulative distribution of the makespan in C^* means being able of estimating the probability $(1 - \alpha)$ that the execution of the network of activities takes more than C^*. In addition, if C^* is a due date, this also implies being able of estimating the probability of being late respect to this due date. Thus providing a significant support to production management decisions.

Acknowledgments This research has been partially funded by the EU FP7 Project VISIONAIR-Vision and Advanced Infrastructure for Research, Grant no. 262044.

References

1. Aytug, H., Lawley, M.A., McKay, K., Mohan, S., Uzsoy, R.: Executing production schedules in the face of uncertainties: a review and some future directions. Eur. J. Oper. Res. **161**, 86–110 (2005)
2. Herroelen, W., Leus, R.: Robust and reactive project scheduling: a review and classification of procedures. Int. J. Prod. Res. **42**(8), 1599–1620 (2004)
3. Tolio, T., Urgo, M., Váncza, J.: Robust production control against propagation of disruptions. CIRP Ann. Manuf. Technol. **60**(1), 489–492 (2011)
4. Hagstrom, J.N.: Computational complexity of PERT problems. Networks **18**, 139–147 (1988)
5. Dodin, B.: Bounding the project completion time distribution in PERT networks. Oper. Res. **33**, 862–881 (1985)
6. Kleindorfer, G.B.: Bounding distributions for a stochastic acyclic network. Oper. Res. **19**, 586–601 (1971)
7. Meilijson, I., Nadas, A.: Convex majorization with an application to the length of critical paths. J. Appl. Probab. **16**, 671–677 (1979)
8. Robillard, P., Trahan, M.: Expected completion time in PERT networks. Oper. Res. **24**, 177–182 (1976)
9. Shogan, A.W.: Bounding distributions for a stochastic PERT network. Networks **7**, 359–381 (1977)
10. Moehring, R.: Scheduling under uncertainty: bounding the makespan distribution. In: Alt, H. (ed.) Computational Discrete Mathematics, pp. 79–97. Springer, Berlin (2001)
11. Golenko-Ginzburg, D., Gonik, A.: Stochastic network project scheduling with non-consumable limited resources. Int. J. Prod. Econ. **48**(1), 29–37 (1997)
12. Golenko-Ginzburg, D., Gonik, A.: A heuristic for network project scheduling with random activity durations depending on the resource allocation. Int. J. Prod. Econ. **55**(2), 149–162 (1998)
13. Alfieri, A., Tolio, T., Urgo, M.: A two-stage stochastic programming project scheduling approach to production planning. Int. J. Adv. Manuf. Technol. (online first) (2012)
14. Keller, B., Bayraksan, G.: Scheduling jobs sharing multiple resources under uncertainty: a stochastic programming approach. IIE Trans. **42**(1), 16–30 (2010)
15. Tolio, T., Urgo, M.: A rolling horizon approach to plan outsourcing in manufacturing-to-order environments affected by uncertainty. CIRP Ann. Manuf. Technol. **56**(1), 487–490 (2007)
16. Zhu, G., Bard, J., Yu, G.: A two-stage stochastic programming approach for project planning with uncertain activity durations. J. Sched. **10**(3), 167–180 (2007)

17. Kulkarni, V.G., Adlakha, V.G.: Markov and markov-regenerative pert networks. Oper. Res. **34**(5), 769–781 (1986)
18. Horváth, A.: Approximating non-Markovian behavior by Markovian models. Ph.D. thesis, Department of Telecommunications, Budapest University of Technology and Economics (2003)
19. Bobbio, A., Horváth, A., Telek, M.: Matching three moments with minimal acyclic phase type distributions. Stoch. Models **21**(2), 303–326 (2005)
20. Osogami, T., Harchol-Balter, M.: Closed form solutions for mapping general distributions to quasi-minimal ph distributions. Perform. Evaluation **63**(6), 524–552 (2006)
21. Ford, L.R., Fulkerson, D.R.: Flows in Networks. Princeton University Press, Princeton (1962)

Coordination of Capacity Adjustment Modes in Work Systems with Autonomous WIP Regulation

Neil Duffie, John Fenske and Madhu Vadali

Abstract A method is presented in this paper for coordinating multiple modes of capacity adjustment in work systems with autonomous WIP regulation with the goal of maintaining desired fundamental dynamic behavior. To prevent overcorrection of capacity, adjustments involving floaters, temporary workers, overtime, etc. need to be coordinated, and it is shown that control-theoretic analysis can be used to develop algorithms for determining combinations of adjustments that result in WIP regulation that is as fast-acting as possible yet non-oscillatory. Results of discrete event simulations in Arena, driven by industrial data, are used to illustrate the dynamic behavior of WIP regulation in an autonomous work system that incorporates such an algorithm and multiple modes of capacity adjustment.

Keywords Capacity · Control · Dynamics

This contribution was previously published in Logistics Research (2012) pp. 99–104. DOI:10.1007/s12159-012-0088-7.

N. Duffie (✉) · J. Fenske · M. Vadali
Department of Mechanical Engineering,
University of Wisconsin-Madison, Madison, USA
e-mail: duffie@engr.wisc.edu

J. Fenske
e-mail: jfenske@alumni.iastate.edu

M. Vadali
e-mail: vadali@wisc.edu

K. Windt (ed.), *Robust Manufacturing Control*, Lecture Notes in Production Engineering,
DOI: 10.1007/978-3-642-30749-2_10,

1 Introduction

With the increasing complexity and uncertainty in demand, as well as the rise of global competition that modern manufacturing industries face, superior control of internal processes is an attribute that companies strive for in order to maintain a competitive "edge." Work-In-Progress (WIP) regulation is an important aspect of this, with objectives of high utilization and keeping lead times short. It has been suggested that optimization of these conflicting objectives can be approached using the concept of ideal minimum WIP [1]. If WIP deviates from this ideal, or some multiple of it, then loss of performance occurs in the form of lower utilization or higher lead times. Regulation of WIP, in the presence of turbulence in demand, requires flexible capacity; the agility with which capacity can be adjusted is a crucial factor.

Beyond simply hiring or laying off permanent employees, several modes of capacity adjustment may be available to manufacturing industries. Each has its positive and negative aspects as well as specific constraints. The use of "floaters" is common. Floaters are cross-trained workers who are able to perform a variety of tasks within a company. In a manufacturing environment, they may be assigned to a different department each day, or assignment changes may occur even more frequently. These personnel are useful for filling in for absent workers or increasing the capacity of a work system that has accumulated backlog. However, training costs are generally higher for these higher-skilled workers and depend on the quantity and difficulty of the tasks they are expected to perform. In addition, the number of these workers available is often limited because they come from within the company. Wild and Schneeweiss presented a hierarchical approach for capacity planning with decisions made at long-, medium- and short- term levels [2]. In their model, every seventh worker is "highly qualified" (i.e., a floater).

The utilization of temporary workers is a second potential mode of capacity adjustment. Temporary workers are extra personnel recruited to increase the available amount of labor and are not part of the primary workforce. Foote and Folta described an appropriate situation for their employment: "heavy use of temporary workers for uncertain expansion projects allows the firm to quickly adjust its workforce in response to worsening or improving economic conditions at negligible cost relative to adjustments involving permanent employees" [3]. Temporary workers may be hired directly by the company in need or dispatched from an agency to a client organization where the work is carried out. Companies that work closely with temporary employment agencies may be able to obtain a large number of workers very quickly depending on the size and experience of the agency. However, these temporary workers are not part of the primary workforce, and additional coordination by management or experienced employees may be needed to achieve effective performance; these temporary workers may require training on specific company processes and policies. Although this mode of capacity adjustment often has only a short delay, floaters usually allow for a quicker response.

In many companies, overtime is an important mode of adjusting capacity, which is achieved by adjusting the work hours of existing employees [4]. By having permanent employees perform their known tasks, the costs of training new employees can be avoided, along with fringe benefits and other costs of hiring and layoffs. However, hourly workers often must be paid at a multiple of their usual rate for time worked above a normal work day or work week. If overtime is occurring frequently, it may be preferable to hire additional permanent employees. Also, requiring workers to work longer hours can cause physical and psychological strain, and union rules may place constraints on who can work overtime and for how long.

For companies that need larger capacity adjustments, employees can be hired or laid off, and the number of working shifts can be changed. Hiring additional permanent workers typically implies higher hiring and training costs, and the company may be "locked-in" for a longer period of time with this mode of capacity adjustment. Hence, careful planning is required, and there usually is a longer delay in implementing such capacity adjustments than with the other modes discussed above. It is important to consider the economics of each mode of capacity adjustment: training costs may be high for companies that employ many floaters, but companies that utilize overtime face other expenses and potential negative effects on workers.

More than one mode of capacity adjustment can be used by a company to regulate WIP. For example, if WIP is higher than desired, then floaters could be immediately assigned to increase work system capacity up to a limit; then, if additional capacity is required, overtime could be used, up to a limit, with some delay in implementation. To prevent overcorrection, such capacity adjustments cannot be made independently, and an appropriate decision-making algorithm must be employed by managers with responsibility for control of internal processes. Delays in implementation associated with the various modes of capacity adjustment, and limits on the magnitudes of the adjustments that can be made, complicate these algorithms and can significantly affect the dynamic behavior of the production system employing them. The tools of control theory can assist in devising algorithms for determining combinations of capacity adjustments that result in WIP regulation that has desirable fundamental dynamic behavior, for example, responses to turbulence that are as fast-acting as possible, yet non-oscillatory.

In the following sections of this paper, an example of a capacity adjustment algorithm will be presented that coordinates capacity adjustments between two adjustment modes in work systems with autonomous WIP regulation [5]. The coordination algorithm is based on the characteristic equation obtained using control-theoretic analysis of WIP regulation. It specifies how the control parameters (gains) for each mode of capacity adjustment are varied as limits in capacity adjustment are reached. The method used in the example can be applied to work systems with various capacity adjustment modes, various combinations of adjustment frequencies and various delays in implementing adjustments. Results of discrete event simulations in Arena, driven by industrial data, are presented that illustrate the dynamic behavior of WIP regulation in an autonomous work system that incorporates the two modes of capacity adjustment and the coordination algorithm. Measures of variation in capacity and WIP are used to compare results and justify presented conclusions.

2 Control-Theoretic Coordination of Capacity Adjustment

Control-theoretic dynamic models have been previously developed for work systems with autonomous WIP control [6–8] that incorporate various capacity adjustment periods and delays. While it is outside the scope of this paper to review these general methods, results are presented for a capacity adjustment scenario that combines two modes: no delay in adjustment; and 1-day delay in adjustment. These could be implemented, for example, by the combination of same-day adjustment using floaters and next-day adjustment using temporary workers. It is assumed that WIP in the work system is measured at the beginning of each work day, and capacity adjustments are implemented at the beginning of each work day as determined using present and past deviations of WIP from what is planned.

In the no-delay mode, capacity adjustment decisions (by what quantity to increase or decrease the capacity) are made each work day and there is no delay in implementation of these decisions. For example, WIP can be measured each morning and appropriate capacity adjustments are immediately implemented. In the 1-day delay mode, capacity adjustment decisions are made each work day, but implementation is delayed by one work day. The equations used for adjusting capacity at time nT, where n is a positive integer and T is the period of time between capacity adjustments (one day in this example), are

$$c_a(nT) = c_p(nT) + \Delta c(nT) \tag{1}$$

$$\begin{aligned}\Delta c(nT) =& k_0(nT)\big(WIP_a(nT) - WIP_p(nT)\big)\\ &+ k_1(nT)\big(WIP_a((n-1)T) - WIP_p((n-1)T)\big)\end{aligned} \tag{2}$$

where $c_a(nT)$ is the adjusted capacity, $c_p(nT)$ is the planned capacity, $\Delta c(nT)$ is the capacity adjustment, $WIP_a(nT)$ and $WIP_a((n\text{-}1)T)$ are the current and previous measured WIP, $WIP_p(nT)$ and $WIP_p((n\text{-}1)T)$ are the current and previous planned WIP, and $k_0(nT)$ and $k_1(nT)$ are WIP-regulation parameters (units time^{-1}) selected to maintain desirable fundamental dynamic behavior. The following discrete characteristic equation describes the fundamental dynamic properties of the work system with WIP regulation:

$$z^2 - (1 - k_0T)z + k_1T = 0 \tag{3}$$

Figure 1 shows the relationship between k_0T and k_1T and the percent overcorrection in capacity adjustment that is represented by this characteristic equation, while Fig. 2 shows the relationship between these two parameters and normalized settling time T_s/T in response to turbulence. (Here, settling time has been calculated using the equivalent damping ratio and natural frequency.) The line of equivalent constant damping ratio $\zeta = 1$ is shown on both figures. For example, when $k_0T = 0$, $k_1T = 0.25$ there is no overcorrection, whereas when $k_0T = 1$, $k_1T = 0.25$ the overcorrection is approximately 20 %. This line indicates the combinations of k_0T and k_1T in Eqs. (2) and (3) that produce response to

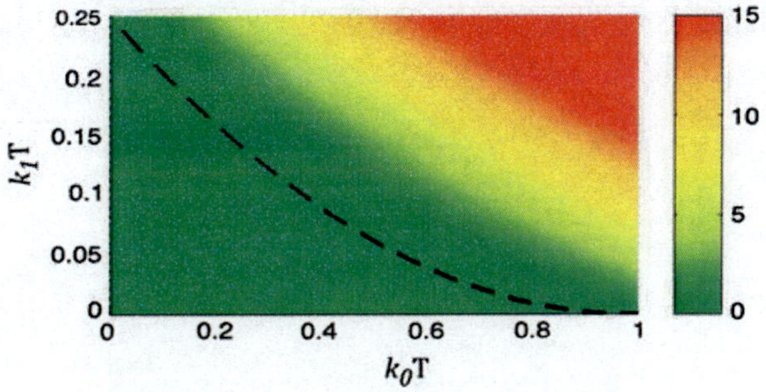

Fig. 1 Percent overcorrection in capacity adjustment versus k_0T and k_1T

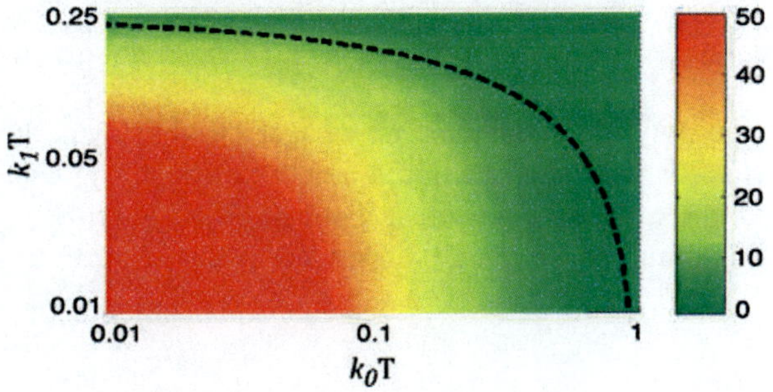

Fig. 2 Normalized settling time T_s/T in response to turbulence versus k_0T and k_1T

turbulence that is as rapid as possible without producing overcorrection (overshoot) in capacity adjustments.

The combination $k_0T = 1$, $k_1T = 0$ produces the most desirable response; however, the amount of capacity adjustment in the no-delay mode (for example, the number of floaters that can be added or removed from the work system) is often limited. In this case, the following algorithm can be used to determine k_0 and k_1 given capacity adjustment period T:

$$k_0(nT) = \frac{1}{T} \tag{4}$$

$$k_1(nT) = 0 \tag{5}$$

$$\Delta c_0(nT) = k_0(nT)\big(WIP_a(nT) - WIP_p(nT)\big) \tag{6}$$

If

$$|\Delta c_0(nT)| > \Delta c_{0_{max}} \tag{7}$$

then

$$k_0(nT) = \frac{\Delta c_{0_{max}}}{|\Delta c_0(nT)|T} \tag{8}$$

and

$$k_1(nT) = \frac{(1 - -k_0(nT)T)^2}{4T} \tag{9}$$

If

$$k_1(nT) > \frac{0.25}{T} \tag{10}$$

then

$$k_1(nT) = \frac{0.25}{T} \tag{11}$$

$$\Delta c_1(nT) = k_1(nT)\big(WIP_a((n-1)T) - WIP_p((n-1)T)\big) \tag{12}$$

where $\Delta c_{0\max}$ is the maximum capacity adjustment that can be made with no delay (for example, the maximum number of floaters). There also can be a limit $\Delta c_{1\max}$ on the capacity adjustment that can be made with 1 day delay, and application of this limit can be readily appended to this algorithm.'

3 Simulation of Coordinated of Capacity Adjustment

The algorithm for coordination of capacity adjustment modes described in the previous section was studied using a discrete event simulation driven by input from a real-world industry dataset. These data were from a supplier to the automotive industry. The dataset contains the orders received and processed over a period of approximately three months. Details include order numbers, machines, order start dates, target order times, actual order times and lot sizes. A significant fraction of the documented orders were processed on shearing and sawing machines as the first step in their production. For the purpose of the research reported here, these machines were grouped as a Shearing/Sawing work system and the data associated with them were examined. Some of the work in this work system was done on weekends, but the amount of work was quite small, and to simplify simulations and clarify results this work was shifted to the following Monday. No setup times for orders or machines were provided in the dataset, nor was failure and preventative maintenance information. Because there were large variations in both work content from order to order and orders arriving day to day, there were large daily variations in work input to the work system. These variations represented turbulence to which the work system was required to react by making capacity adjustments for the purpose of regulating WIP.

In addition to the provided data, several key parameters were needed for simulating autonomous WIP regulation in the Shearing/Sawing work system. The planned capacity c_p for the work system was assumed to be constant and was

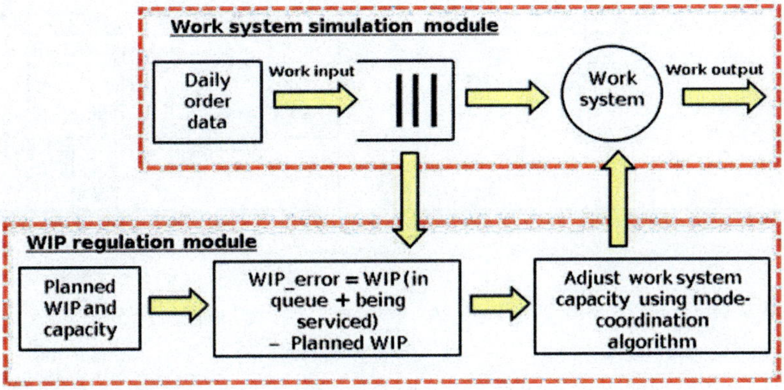

Fig. 3 Discrete event simulation of work system with autonomous WIP regulation

calculated as the average daily work input, which was 49.95 h/day. The planned WIP for the work system was assumed to be the average of the WIP in the data, which was 384 h. Investigation of the effects of planned WIP on utilization and work system dynamic behavior was outside the scope of this work. (see Toshniwal [9] for more information on this production system, its behavior as a function of WIP, and the characteristics of the work input data).

The discrete event simulation model of the Shearing/Sawing work system was constructed using Arena. As indicated in Fig. 3, there were two main modules in the model: a work system simulation module; and a WIP regulation module [9]. At 8:00 am each workday, the (current) WIP (the sum of work in a single work-system queue and work remaining in orders being serviced on machines in the work system) was measured and capacity adjustments were calculated. The work system had six machines and one input queue. There were no limits on queue size, and set up and transportation times were neglected. In the following subsections, simulation results are presented that first illustrate the behavior of the individual modes of capacity adjustment, and then illustrate the behavior of the coordination of the two modes using the algorithm described in the previous section.

3.1 No Delay in Capacity Adjustment

Figure 4 shows simulation results for WIP and work system capacity when there is no delay in capacity adjustment ($T = 1$ day, $k_0 = 1$ day^{-1}, $k_1 = 0$ day^{-1}) and no limit on the magnitude of adjustment. The initial "ramp up" and final "ramp down" portions of the simulation results are not included in performance measurements. In this case, WIP is well regulated, but capacity adjustment magnitudes are large. The standard deviation of capacity and WIP are shown in Table 1.

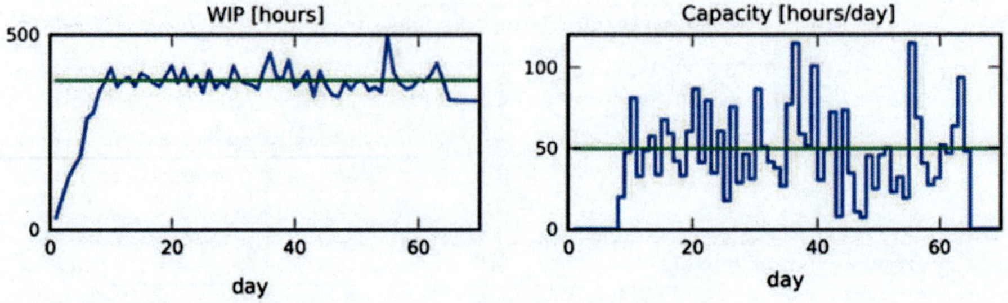

Fig. 4 WIP and capacity with $T = 1$ day, $k_0 = 1$ day^{-1}, $k_1 = 0$ day^{-1}

Table 1 Standard deviation of capacity and WIP obtained from discrete simulations with no limits on capacity adjustments and no coordination between modes

Mode	No delay	1 day delay	No delay + 1 day delay
k_0 [day^{-1}]	1	0	1
k_1 [day^{-1}]	0	0.25	0.25
σ_{cap} [h/day]	4.19	1.80	4.57
σ_{WIP} [h]	28.27	44.32	29.22

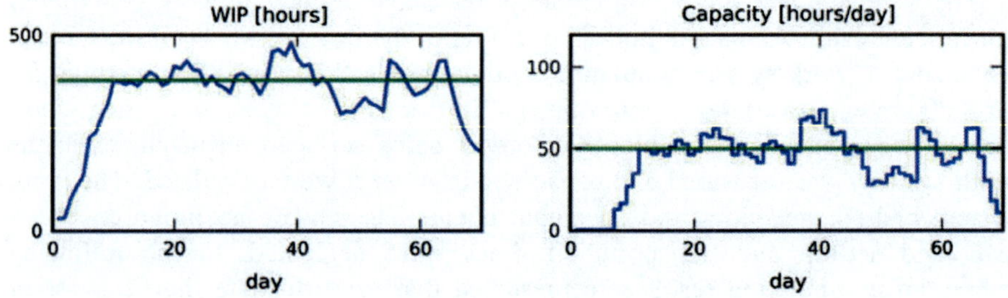

Fig. 5 WIP and capacity with $T = 1$ day, $k_0 = 0$ day^{-1}, $k_1 = 0.25$ day^{-1}

3.2 1 day Delay in Capacity Adjustment

Figure 5 shows the simulation results for WIP and work system capacity when there is a 1 day delay in capacity adjustment ($T = 1$ day, $k_0 = 0$ day^{-1}, $k_1 = 0.25$ day^{-1}) and no limit on the magnitude of adjustment. In this case, WIP regulation is not as effective as in the no-delay case, but capacity adjustment magnitudes are reduced. Again, the standard deviation of capacity and WIP are shown in Table 1.

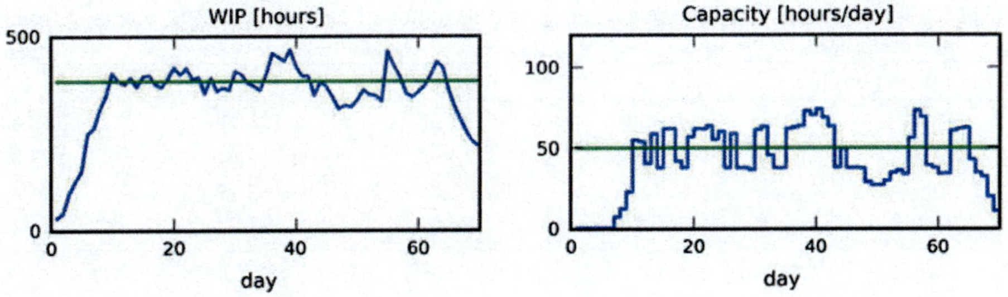

Fig. 6 WIP and capacity with $T = 1$, $\Delta c_{0max} = 12$ h/day

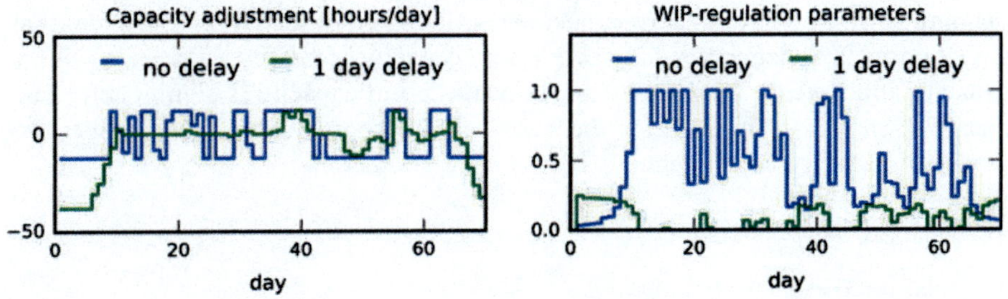

Fig. 7 Capacity adjustments and WIP-regulation parameter adjustments with $T = 1$, $\Delta c_{0max} = 12$ h/day

3.3 *Combination of Capacity Adjustment Modes without Coordination*

When the no-delay and 1 day-delay modes are combined without using the algorithm described in Sect. 2 ($k_0 = 1$ day^{-1}, $k_1 = 0.25$ day^{-1}), the standard deviations of capacity adjustment and deviation of WIP from planned WIP that result are shown in Table 1. Deviations in both capacity and WIP are increased with respect to the no-delay case because there is overcorrection in capacity adjustments as predicted by Fig. 1.

3.4 *Coordination of No-Delay and 1 Day Delay in Capacity Adjustment*

In reality, there are limits on the magnitude of capacity adjustment that are possible in each mode. Therefore, WIP-regulation parameters k_0 and k_1 can be adjusted according to the algorithm described in Sect. 2, which incorporates limits while avoiding overcorrection of capacity. Figure 6 shows the simulation results for WIP and work system capacity when there is a 12 h/day limit on no-delay

Table 2 Mean capacity adjustments and standard deviation of capacity and WIP obtained from discrete simulations for various values of capacity adjustment limit $\Delta c_{0_{max}}$

$\Delta c_{0_{max}}$ [hours/day]	6	12	24	36
$\overline{\Delta c_0}$ [h/day]	5.76	11.38	18.58	22.64
$\overline{\Delta c_1}$ (h/day)	7.34	4.55	1.47	0.80
σ_{cap} (h/day)	1.92	2.40	3.22	3.78
σ_{WIP} (h)	40.10	35.96	31.14	29.34

capacity adjustment ($T = 1$ day, $\Delta c_{0max} = 12$ h/day), and Fig. 7 shows the capacity adjustments and WIP-regulation parameters generated by the algorithm.

Table 2 shows the results of applying this algorithm with various limits on magnitude of no-delay capacity adjustments. It can be observed, as expected, that deviation in WIP decreases and deviation in capacity increases as larger no-delay capacity adjustments are permitted. The variation in capacity is significantly less than that shown in Table 1 for the case without coordination between the two modes of capacity adjustment.

4 Conclusions

Consideration of dynamic behavior is important in designing agility into production systems that must respond effectively to turbulence in demand and capacity. A method for capacity adjustment coordination between various modes of capacity allocation adjustment has been described that maintains constant dynamic damping while using faster-acting modes first, up to their capacity adjustment limit, and then using slower-acting modes. The algorithm is based on results of control-theoretic analysis of WIP regulation. The algorithm for no delay paired with 1 day delay was presented, but similar capacity adjustment algorithms can be obtained for more complex combinations using similar analytical methods: an algorithm for coordinating no delay, 2 day delay, and 1 week delay capacity adjustments for example. Economic factors have not been incorporated into the algorithms, which are designed to eliminate overcorrection of capacity and accommodate limits on the magnitudes of capacity adjustments that can be implemented. The trade-off between variation in WIP and variation in capacity is not optimized, but overcorrection of capacity is prevented.

Results of discrete event simulations in Arena, driven by industrial data, were used to illustrate the dynamic behavior of WIP regulation in an autonomous work system that incorporates two modes of capacity adjustment. The results show that the approach that has been presented produces adaptive WIP regulation that avoids both overcorrection of velocity and sluggish response in work input that has significant turbulence. The results confirm the desirability of coordination of modes of capacity adjustment and confirm the fundamental dynamic behavior predicted by control theory.

References

1. Nyhuis, P., Weindahl, H.-P.: Fundamentals of Production Logistics: Theory, Tools and Applications. Springer, Berlin (2009)
2. Wild, B., Schneeweiss, C.: Manpower capacity planning—A hierarchical approach. Int. J. Prod. Econom. **30–31**, 95–106 (1993)
3. Foote, D.A., Folta, T.B.: Temporary workers as real options. Human Resour. Manage. Rev. **12**, 579–597 (2002)
4. Delarue, A., Gryp, S., Hootegem, G.V.: The quest for a balanced manpower capacity: different flexibiliy strageies examined. Enterprise Work Innovation Stud. **2**, 69–86 (2006)
5. Windt, K., Hülsmann, M.: Changing paradigms in logistics—understanding the shift from conventional control to autonomous cooperation and control. In: Hülsmann, M., Windt, K. (eds.) Understanding Autonomous Cooperation and Control in Logistics, pp. 1–16. Springer, Berlin (2007)
6. Duffie, N.A., Shi, L.: Maintaining constant WIP-regulation dynamics in production networks with autonomous work systems. CIRP Ann. Manuf. Technol. **58**(1), 399–402 (2009)
7. Duffie, N., Shi, L.: Dynamics of WIP regulation in large production networks of autonomous work systems. IEEE Trans. Autom. Sci. Eng. **7**(3), 665–670 (2010)
8. Toshniwal, V., Duffie, N., Jagalski, T., Rekersbrink, H., Scholz-Reiter, B.: Assessment of fidelity of control-theoretic models of wip regulation in networks of autonomous work systems. CIRP Ann. Manuf. Technol. **60**(1), 485–488 (2011)
9. Toshniwal, V.: Assessment of WIP regulation in networks of autonomous work systems. Madison, M.S. Thesis, University of Wisconsin-Madison (2011)

Evaluating the Effects of Embedded Control Devices in Autonomous Logistic Processes

Steffen Sowade, Philipp von Lamezan and Bernd Scholz-Reiter

Abstract Embedded control devices enhance logistic objects with the ability of decision making and execution and can increase the flexibility and robustness of logistic processes. While the influence of different autonomous control strategies on logistic performance has been investigated, an evaluation of a control devices' influence is missing. This work analyses the behaviour of these control devices by means of discrete event simulations for an autonomously controlled production logistics process and a centralised controlled process. The investigation follows the idea that the impact of such devices can be described as additional decision making time. The larger the decision making time is and the smaller the actual processing time is, the higher is the loss of manufacturing time for control purposes. The investigated autonomously controlled process compensates the additional decision making time for specific levels of uncertainties in terms of machine failures.

Keywords Autonomous control · Embedded control device · Benefit analysis · Event discrete simulation

S. Sowade (✉) · B. Scholz-Reiter
BIBA—Bremer Institut für Produktion und Logistik,
Hochschulring 20, 28359 Bremen, Germany
e-mail: sow@biba.uni-bremen.de

B. Scholz-Reiter
e-mail: bsr@biba.uni-bremen.de

P. von Lamezan
University of Bremen, Bibliothekstraße 1,
28359 Bremen, Germany
e-mail: p.v.lamezan@googlemail.com

K. Windt (ed.), *Robust Manufacturing Control*, Lecture Notes in Production Engineering,
DOI: 10.1007/978-3-642-30749-2_11,

1 Introduction

Logistic processes can be evaluated by means of performance and costs for internal and external objectives [1]. The performance of autonomous logistic processes can be described by a structured vector using, e.g. on-time delivery, lead time, utilisation, and inventory [2]. On-time delivery and lead time are time-related [1, 3], while utilisation and inventory are resource-related objectives. Lead time sums up setup time, processing time, process-related waiting time, as well as idle and waiting times at processing stations. Process disturbances may affect lead time negatively [4].

Autonomously controlled logistic processes can cope with such problems. Individual logistic objects are equipped with control device infrastructure and run jointly with a logistic process [5]. On the one hand, these devices automatically compensate process errors and reduce lead time [6]. On the other hand, autonomous decision making takes time and increases lead time. Assuming decision making in centralised controlled production logistics processes takes place preceding the production process, the time required for scheduling does not belong to lead time. In contrast, the time of autonomous decision making is settled in a production process and must be added to the other components of lead time.

Although the additional time for decision making is relatively small (see next section), it becomes relevant for production processes with very short processing times that are in the range of the decision time. For instance, assembly steps of mobile phones may take 1.6 s [7] and processing times in automated bottling lines can be lower than 1 s [8]. The smaller the portion of the processing time is compared to the decision time, the bigger is the risk of changes in a production process state during decision making. Thus, calculated decisions might be inappropriate for the new system state. Short decision times help to reduce this problem.

In addition to these time-related issues, equipping logistic processes with autonomous control devices causes costs, e.g. for development, acquisition and operation of infrastructure components [9]. They can be estimated during the design phase but the estimated costs might differ from an implemented specific autonomous logistic process. Compared to efforts that are measureable technical values, e.g. time, costs aggregate a set of valuated effort types in a single value and depend on a valuators perspective as well as on the market situation for the component in question. Hence, costs are highly volatile variables representing rather a market based valuation process result than a technical characteristic. For this reason and in order to focus on reliable measures for technical systems, this study focuses on a time-related discussion.

While the performance of autonomously controlled processes has been investigated for various control strategies, an evaluation of the impact of specific control device types is still missing. Hence, this paper analyses the behaviour of embedded control devices in autonomously controlled production logistic processes by means of a discrete event simulation and compares the resulting performance of the autonomously controlled process to a centrally controlled process, which performs the controlling of the material flow predetermined without delay. The study is based on the assumption that planning and control tasks of centralised controlled processes take place preceding the process, while these tasks add additional decision processing steps in autonomously controlled processes. Consequently, the processing time of autonomous processes is higher and their capacity is lower than in centralised controlled processes. The bigger the decision time is compared to the processing time; the lower is the capacity of autonomous processes. However, the advantage of the centralised controlled processes vanishes, if e.g. resources fail to work. Since resource downtimes do happen in the real world, autonomously controlled processes can achieve a better performance than centralised controlled processes [6].

The question is: under what circumstances is this the case? Further, the magnitude of the influence of the decision time on the performance of a logistic process is unknown. This work aims to discuss these issues. The results of simulation experiments show, at what downtime rate an autonomously controlled logistic process outperforms a centralised one in terms of lead time. Due to the assumed effects of planning and control, the decision time is modelled in detail. It is believed that autonomously controlled processes compensate for the additional decision time by a shorter waiting time at processing stations for a specific proportion of resource' s downtime rates. The shorter waiting time results from the adaptability of autonomously controlled logistic processes. On the contrary, centralised controlled production processes do not respond to downtimes and thus increase the lead time of dispatched commodities.

The remainder is structured as follows. Section 2 discusses the decision making time. Section 3 presents the experiment design. Sections 5 and 6 illustrate and discuss the results. The paper closes with a conclusion and outlooks for future work.

2 Components of the Decision Making Time

Autonomously controlled logistic processes require various infrastructure components to perform their tasks [10]. Functions and properties of the components influence the decision making time which consists of time to calculate a decision, to gather necessary information, and to execute a decision. The first two elements focus on information technology, while the third refers to actuators as rather mechanical components. Actuators are used in centralized and in autonomously controlled logistic processes, but differ in the way they are controlled. Thus, they do not characterise the decision process itself. In order to focus on the added decision making process, this study investigates on the information technological components and excludes the decision execution system in this study. Hence, the decision making time sums up the time for communication and for decision calculation purposes.

The communication time consists of time to initialise a communication link and time to transmit data. The initialisation time depends on the communications technology and the network topology. Mobile logistic objects cause rapid changes in the network topology in autonomously controlled production processes. Communication technologies like Bluetooth are able to deal with these dynamics [11, 12]. Its energy consumption is low, range and data rate fits small factories and it is relatively inexpensive [13]. Thus, Bluetooth is chosen as the communication technology is this study. Bluetooth networks consist of one master and up to seven slaves being affiliated with the master [12]. When master and slave are connected a data exchange lasts for 1.25 ms (milliseconds) for a request and a response [13]. If a master has to activate a slave from standby mode, switching to inquiry state takes up to 2 and to connection state up to 6 s. The mean initialisation time is 640 ms [12].

Control method and communication technology determine the transmission time. It is calculated as message length divided by data rate [14]. The message length depends on the selected control method, while the communication technology defines the data rate. This study uses the forecast control method QLE (Queue Length Estimation), as it responds faster to changes in a production process than past value methods do. Control devices utilizing QLE query all processing stations for the number of waiting commodities and route them to the station with the least waiting time [15]. A single value is transmitted per request, representing the queue length [15]. If this value is of type integer, it is 16 bits long.

Communication technology and transmission distance determine the available data rate [12]. The transmission distance on shop floor level is assumed to be only a few meters. Hence, the shop floor is sufficiently small to conclude a Bluetooth data rate of 2 Mbps (megabit per second) [13]. The signal propagation time can be neglected, as the signal spreads in magnitude of the speed of light in a small shop floor [16]. The resulting propagation time is many times smaller than that caused by the amount of transmitted data. Hence, the transmission time of a Bluetooth-QLE-integer-message on the assumed small shop floor is 8 ns and scales linearly with the number of responding processing stations.

Microcontrollers are suitable for processing data into decisions by using their computing abilities [17]. The decision calculation time and depends on a microcontroller's performance measured in MIPS (mega instructions per second) [17] and on the efficiency of a chosen control method [18], i.e. given by the complexity of the input data, of the processing algorithm, and of the communication protocol [19]. With an increasing number of input data elements, e.g. to sort n numbers, the importance of fast algorithms rises. The QLE method returns the minimum values out of a list. The task is similar to a sorting procedure and of linear complexity [18]. Hence, calculation lasts 3 μs for $n = 3$ data elements to be sorted by a microcontroller with a performance of 1 MIPS. More complex algorithms may need more than a second for $n = 12$, while linear scaling algorithms require less than 20 ms.

In conclusion, the initialisation time dominates the communication time as well as the processing time by several orders of magnitude when using the QLE method and Bluetooth technology in a small shop floor. The initialisation time take several hundred milliseconds, while the others cause delays of few nano or microseconds. The initialisation time is responsible for over 99.9 per cent of the decision time. Hence, the decision time is reduced to the initialisation time in the subsequent experiments.

3 Experiment Design

The effects of a control device on an autonomously logistic process can be analysed in simulation experiments. Therefore, a production process is modelled and simulated in the software Tecnomatix Plant Simulation 9 (version 9.0.3). For the comparison of an autonomously controlled and a centralised controlled production process a respective control method can be applied in each case. Despite its simplicity, this production process is sufficient to demonstrate the effects of the decision time on the lead time of autonomously controlled processes.

3.1 *Analysed Processes*

Figure 1 shows the centralised controlled production process. After a commodity entered the process through the source on the left, a switch forwards it round robin without loss of time to one of three succeeding processing lines. Each line consists of a buffer and a working station offering the same processing step. The buffers keep commodities temporarily in the event processing stations suffer downtimes or fluctuations in processing time. The commodities pass buffers without delay, when the processing stations are idle. The sink removes the processed commodities from the production process. Figure 2 shows the model of an autonomously controlled production process. Incoming commodities pass a buffer and a decision station forwarding them to a processing line. These elements represent the control device being mounted on a commodity. QLE is used as a control method. The buffer element Buffer0 is used to indicate delayed decisions. If the decision station is jammed Buffer0 keeps the commodities.

3.2 *Selected Parameters*

3.2.1 Decision Making Time

The connection setup between communication participants makes up the largest share of the decision making time. Hence, the simulation study focuses on this parameter. As Bluetooth is used, participants can connect with each other on

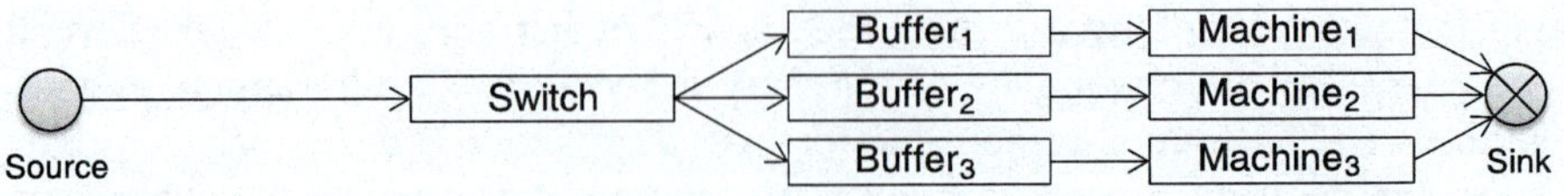

Fig. 1 Centralised controlled logistic process

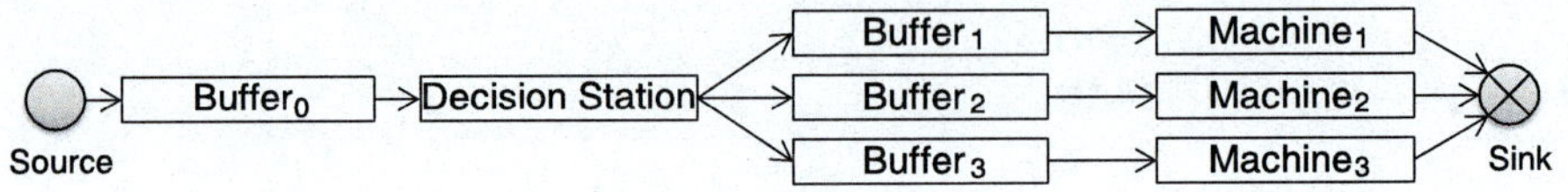

Fig. 2 Autonomously controlled logistic process

average in 0.64 s and to a maximum in 2.56 s [12]. The experiments use a logarithmic normal distribution in order to model varying connection setup times [20], because it's skewness is suitable for the left positioned mean value. The values of this distribution fluctuate unsymmetrical around the mean and can reach large values to the right of the mean. By setting a standard deviation of 0.5 s, 68.3 per cent of all values are in the range 0.44–0.84 s. A larger standard deviation would lead to a broader dispersion of decision time. Hence, the simulation uses a logarithmic normal distribution with parameters $\mu = 0.64$ s, and $\sigma = 0.5$ s to map variations of the decision making time.

3.2.2 Processing Time and Failure Behaviour

The time a machine needs to process a commodity is referred as processing time. Their expected value is constant. The processing time varies to the expected value due to the technical characteristics of a processing station. These fluctuations can be approximated with a normal distribution [21]. The ratio of decision making time to processing time is set to one for the experiments, so that the production process operates at its limit where the capacity of the processing station and the decision station are equal. Hence, variations in processing time can lead to selective over- or under-utilisation of the processing stations. The cumulative average capacity of each of the three processing stations is set to 0.64 s, so that the mean ratio of decision making and processing time is one. Each of the three processing stations needs three times as much time for processing a commodity as the decision processing station needs. Their expected processing time is 1.92 s. The three work stations together have a theoretical capacity of 5625 commodities per hour. Transportation times are neglected. Under the assumption that processing times vary mainly in the vicinity of the expected value, the standard deviation $\sigma = 0.5$ s is specified. In conjunction with the expected value $\mu = 1.92$ s for 68.3 % of all generated are processing times in the interval 1.42–2.42 s.

The share of a processing station's operating time when there are no faults is known as technical availability [22]. It is composed of the MTBF (mean time

Table 1 Dispatch rate and pitch

Level	90 %	92 %	95 %	97 %	99 %	100 %
Pitch(s)	0.711111	0.695652	0.673684	0.659738	0.646465	0.640000

Table 2 Mean lead time of the centrally controlled logistic process

Centralised control		Technical availability of manufacturing stations					
		90 %	92 %	95 %	97 %	99 %	100 %
Load	90 %	1317.90	472.00	151.70	167.20	19.62	2.26
	92 %		711.50	235.68	168.90	18.20	2.41
	95 %			943.40	295.20	31.40	2.88
	97 %				481.20	74.00	3.77
	99 %					213.40	8.37

between failures) and the MTTR (mean time to repair). Plant Simulation needs the technical availability and the MTTR to calculate downtimes of processing stations. Downtimes are calculated independently for each processing station. Nearly 55 per cent of the capital goods industry has a technical availability of 90 % or more of [23]. Therefore, the technical availability in the simulation experiments varies in steps of 90, 92, 95, 97, 99 and 100 %. The MTTR is made up of the repair period, the duration between the occurrence of a fault and the beginning of trouble-shooting by the staff, the duration to detect the fault reason, and the preparation of repair [24]. With the assumption of less complex processing steps and reasons for faults, the MTTR is set to 5 min.

3.2.3 Dispatching

The source dispatches commodities as percentage of the theoretical maximum capacity of the production system. Dispatching is done with constant pitch to limit the fluctuating variables to the decision making time and the processing time and to limit the complexity of the simulations. The pitch is adjusted to the percentage utilisation in each simulation experiment. A pitch of 0.64 s leads to a 100 % utilisation in the centralised controlled manufacturing process (Table 1).

3.3 Procedure

The mean lead time is determined for each combination of the gradually varied parameter dispatching and technical availability in a total of 40 simulation experiments. Tables 2 and 3 show the considered combinations that are examined for both types of control. Dispatching is less than or equal to the technical availability, so that the production system retains the ability to cope with the

Table 3 Mean lead time of the autonomously controlled logistic process

Autonomous control		Technical availability of manufacturing stations					
		90 %	92 %	95 %	97 %	99 %	100 %
Load	90 %	510.70	139.00	31.90	19.40	21.40	3.74
	92 %		181.70	130.70	28.20	15.20	4.10
	95 %			294.40	89.60	22.10	4.02
	97 %				120.30	63.30	5.54
	99 %					29.94	6.99

situation in the long run. Otherwise, the buffers would fill up and the processing time would go to infinity. The simulation duration is set to 24 h for each experiment run, i.e. about 100,000 commodities, which is regarded as a sufficient amount of data to calculate the mean lead time.

4 Experimental Results

4.1 Mean Lead Time Centralised Controlled Logistic Process

Table 2 shows the mean lead times of the centralised controlled production process. Figure 3 presents the results in a three-dimensional diagram. The X-axis indicates the technical availability (unit: per cent), the Y-axis represents the dispatching (unit: per cent) and the Z-axis shows the mean lead time (unit: seconds). Linear interpolation is used between the individual results to create the illustrated surface. Mean lead times increase, where the dispatching rate equals the technical availability. Lead times increase for a given technical availability, when dispatching increases. An exception is the combination of dispatching = 92 % and technical availability = 99 %, where the mean lead time 1.42 s lower than the mean lead time of the combination of dispatching = 90 % and technical availability = 99 %. Lead time decreases for a given dispatching and increasing technical availability. An exception is the combination of dispatching = 90 % and technical availability = 95 %, where the lead time is 15.5 s bigger, compared to technical availability = 97 %.

4.2 Mean Lead Time Autonomously Controlled Logistic Process

Table 3 and Fig. 4 show the simulation results of the autonomously controlled production process. The Z-axis is scaled as in Fig. 3 in order to facilitate a comparison of both figures. Although Fig. 4 displays lower mean lead times, the surfaces in Figs. 3 and 4 show a similar orientation: At dispatching = technical

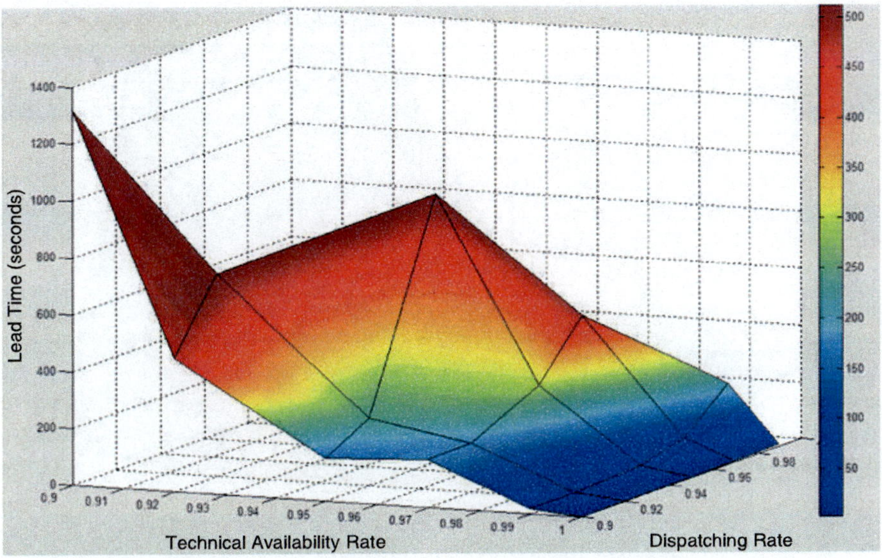

Fig. 3 Mean lead time of the centralised controlled logistic process

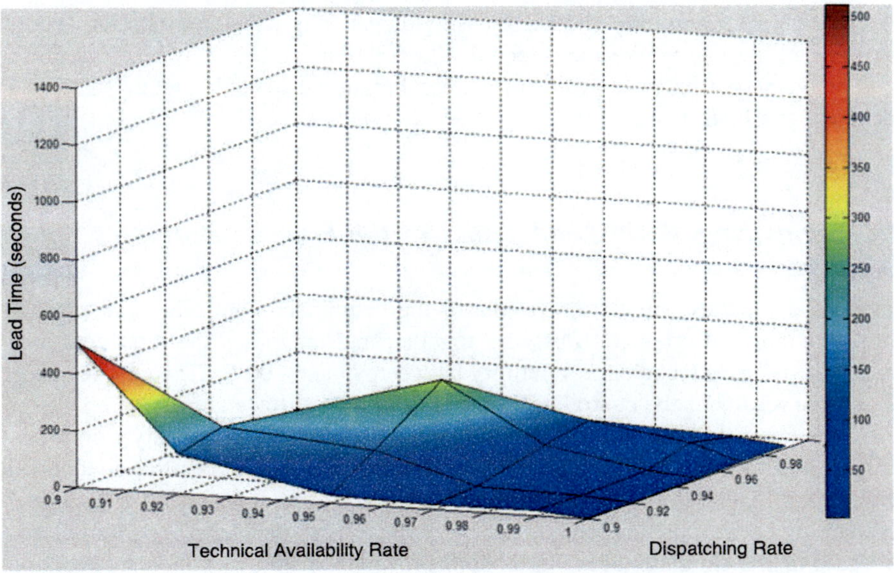

Fig. 4 Mean lead time of the autonomously controlled logistic process

availability occur large lead times. The maximum mean lead time occurs where dispatching = 90 % and technical availability = 90 %. Lead times decrease for a given dispatching rate and an increasing technical availability as well as for a

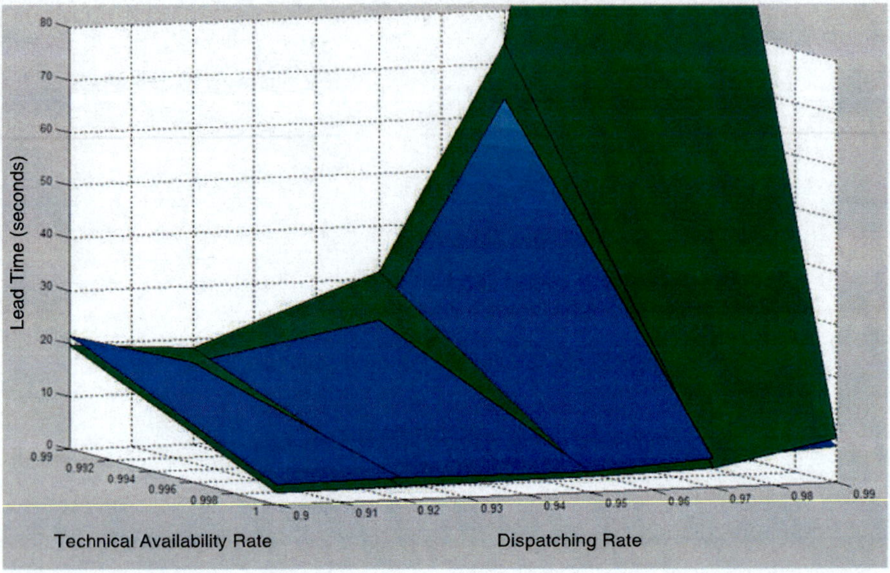

Fig. 5 Comparison of mean lead times: *blue* = autonomous control, *green* = centralised control

given technical availability and a decreasing dispatching rate. An exception occurs for dispatching rate = 95 % and technical availability = 100 %, where the mean lead time is 0.08 s under the value of the combination dispatching rate = 92 % and technical availability = 100 %.

4.3 Comparing Mean Lead Times of the Logistic Processes

Figure 5 depicts the simulation results of the centralised controlled and autonomously controlled production process together and magnifies the intersection of both surfaces for a technical availability between 99 and 100 %. The green surface shows the results of the centralised control. The blue surface shows the results of the autonomously controlled production process. The autonomously controlled process has lower lead times than the centralised controlled process for a technical availability below ca. 99 %, but this result turns to the opposite for a technical availability between 99 and 100 %, where both surfaces intersect. The interpolated surfaces of the simulation results of both processes have a common intersection line. The upper surface is visible. The partial triangular faces are a result of the intersection. Almost all lead times of the autonomously controlled process are higher than for the centralised controlled process at technical availability = 100 % (exception at dispatching = 99 %). At a technical availability of 99 % almost all lead times of the autonomously controlled process are below the centralised controlled process (exception at dispatching = 90 %).

5 Discussion

The experiments show the behaviour of autonomously and centralised controlled logistic processes under specific conditions. Lead times are relative high for particular parameter combinations. Especially, the centralised controlled production process has lead times of up to 686-times higher than the minimum processing time. Lead times increased significantly in the region were dispatching equals technical availability, in particular in the centralised controlled process, due to its static control character.

When considering a single fault, the QLE method reduced buffer stocks faster than the switch in the centralised controlled process. Both production processes reduce buffer stocks at the ratio of dispatching rate to mean processing time. Since autonomous control changes the production process in case of machine breakdowns, the buffers to the processing stations increase slower than in the centralised controlled production process. Thus, the mean lead times remains lower in the autonomously controlled production process. Only when the technical availability is above 99 %, the centralised controlled process outperforms the autonomous process in mean lead time.

This behaviour might occur for two reasons: First, the decision time increases the lead time in autonomously controlled processes. In the experiments, lead time grows on average by 0.64 s compared to the centralised controlled process. Second, the experiment used a rather simple implementation of the QLE method. If the technical availability is high, the first two processing stations exhibit higher load than the third station does because QLE queries the buffer utilisation and neglects a station's processing state. If the first buffer is empty but its processing machine is busy and if the other buffers and their corresponding are empty or idle, this QLE implementation selects the first processing line although others are a better choice. In consequence, commodities wait in buffers although they could be processed by other processing stations without delay. However, querying only buffer stocks keeps the communication efforts low. Decision time is saved at the expense of utilisation.

The experiments confirmed the hypotheses that autonomous control can be beneficial compared to centralised control. The impact of the additional decision time is marginal, if the technical availability of processing stations is only a little lower than 100 %. The higher the technical availability of processing stations is, the smaller are the lead times of the centralised controlled production process. The advantage of an autonomous control decreases with increasing technical availability and reduces the benefit of an autonomous control device. The threshold region for this benefit is between a technical availability of 99 to 100 % for the analysed production process. If a benefit analysis of autonomous control devices in logistic processes solely uses the dimension of lead times, autonomous control is only useful if the technical availability is below 99 %. This area limits the application of autonomous control.

Possible sources of errors are: First, the statistical distribution of the parameters technical availability, decision time and processing time may be different for a real implementation of the analysed process. Second, the implementation of the QLE method offers the advantage of small communication and computing efforts. However, it neglects useful information about the status of processing stations. Third, a possible transient behaviour was not considered in the experiments. In order to avoid effects caused by settling times, the measurement should start with delay after the start of the simulation. Then lead time is determined for the steady case only. Fourth, the linear interpolation of the results in the figures suggests a linear behaviour of the lead times between the measured points. However, the real process behaviour might be different. An increased number of experiments with parameters set to the currently interpolated areas can help to determine what type of interpolation fits well.

6 Conclusion and Outlook

This work analysed the influence of autonomous control devices in an autonomously controlled logistic process on the lead time and compared the results with a centralised controlled process. The analysis focused on the decision time as the main difference between both approaches. The components of the decision time have been discussed and the time to initialise a communication link has been identified as most time consuming part. Thus, it has been used as decision time in simulation experiments. A big influence of the control devices was triggered by using a decision time in the range of the succeeding processing time. The results show the advantage of autonomous control for the given process if the technical availability of its processing stations is below 99 %. Then, the added decision time in an autonomously controlled process is less bad to lead time than a centralised controlled process. The results help to select appropriate components for autonomous control devices for applications with processing times being in the range of decision times, e.g. in electronics and beverage industry.

Future research could examine the benefits of autonomous control devices for interconnected processes to evaluate the dynamics of entire logistics systems. More complex process models and real production processes should be analysed to understand the influence of autonomous control devices on the dynamics of lead time better. Further, the area of the advantageousness shift should be examined at a finer resolution to gain more insights on the performance and the value of autonomous control devices. Follow-up studies could seek to explain the transition process from a stationary state to a non-steady state in detail, i.e. to define and compare the stability of centralised controlled and autonomously controlled processes.

Acknowledgments This research is funded by the German Research Foundation (DFG) as part of the Subproject B2 of the Collaborative Research Centre 637 "Autonomous Cooperation Logistic Processes—A Paradigm Shift and its Limitations" (CRC 637).

References

1. Herman Lödding. Verfahren der Fertigungssteuerung: Grundlagen, Beschreibung, Konfiguration. Springer, Berlin, Heidelberg, 2 edn (2008)
2. Bernd Scholz-Reiter, Torsten Philipp, Christoph de Beer, Katja Windt, and Michael Freitag. Einfluss der strukturellen Komplexität auf den Einsatz von selbststeuernden logistischen Prozessen. In Hans-Christian Pfohl and Thomas Wimmer, editors, Steuerung von Logistiksystemen—auf dem Weg zur Selbststeuerung. Deutscher Verkehrs-Verlag (2006)
3. Michael, E., Porter. *Wettbewerbsvorteile: Spitzenleistungen erreichen und behaupten.* Campus Strategie. Campus-Verlag, Frankfurt/Main, 7 edition, 2010
4. Kletti, J., Schumacher, J.: Die perfekte Produktion: Manufacturing Excellence durch Short Interval Technology (SIT). Springer, Berlin (2011)
5. Bernd Scholz-Reiter, Hartmut Höhns. Selbststeuerung logistischer Prozesse mit Agentensystemen. In Günther Schuh, editor, Produktionsplanung und -steuerung: Grundlagen, Gestaltung, Konzepte, p. 745–780, Berlin, Springer (2006)
6. Windt, K.: Ermittlung des angemessenen Selbststeuerungsgrades in der Logistik—Grenzen der Selbststeuerung. In: Nyhuis, Peter. (ed.) Beiträge zu einer Theorie der Logistik, pp. 349–372. Springer, Berlin (2008)
7. DEPRAG SCHULZ GmbH u. Co, Handarbeitsplätze in der Wirtschaftskrise. online. URL=http://www.deprag.com/index.php?id=presse200904a. 29.02.2012
8. RheinfelsQuellen H. Hövelmann GmbH & Co. KG, Daten und Fakten. online. URL=www.rheinfelsquelle.de/content/unternehmen/datenundfakten.aspx. 29.02.2012
9. Hans-Jörg Bullinger, Dieter Spath, Hans-Jürgen Warnecke, and Engelbert Westkämper, editors. Handbuch Unternehmensorganisation: Strategien, Planung, Umsetzung. VDI-Buch. Springer, Berlin, 3 edn. (2009)
10. Scholz-Reiter, B., Sowade, S., Rippel, D.: Drivers for the configuration of autonomous logistic control systems' infrastructure. Int. J. Sys. Appl. Eng. Dev. **5**(3), 350–358 (2011)
11. Bernd Scholz-Reiter, Michael Freitag, Henning Rekersbrink, Bernd-Ludwig Wenning, Christian Gorldt, and Wolfgang Echelmeyer. Auf dem Weg zur Selbststeuerung in der Logistik—Grundlagenforschung und Praxisprojekte. In: Wäscher, G., Inderfurth, K., Neumann, G., Schenk, M., Ziems, D. (eds.) Intelligente Logistikprozesse—Konzepte, Lösungen, Erfahrungen. Begleitband zur 11. Magdeburger Logistiktagung, pp. 166–180, Magdeburg, Logisch-Verlag (2005)
12. Merkle, A., Terzis, A.: Digitale Funkkommunikation mit Bluetooth: Theorie und Praxis, Bluetooth-Simulator, konkurrierende Systeme. Telekommunikation, Franzis, Poing (2002)
13. Jörg F Wollert. Handbuch der Mess- und Automatisierungstechnik in der Produktion, chapter Drahtlose Netzwerke, pp. 625–654. VDI-Buch. Springer, Berlin, Heidelberg, 2 edn. (2006)
14. Klaus Pirklbauer, Peter Rechenberg, Gustav Pomberger (eds.) Informatik-Handbuch. Carl Hanser Verlag, München, 4 edn. (2006)
15. Bernd Scholz Reiter, Michael Freitag, Christoph de Beer, Thomas Jagalski. Analysing the Dynamics caused by Autonomously Controlled Logistic Objects. In: Proceedings of the 2nd Int. Conf. on Changeable, Agile, Reconfigurable and Virtual Production, Toronto, Canada (2007)
16. Erich Stein. Taschenbuch Rechnernetze und Internet. Carl Hanser Verlag, München, 3 edn. (2008)
17. Uwe Brinkschulte, Theo Ungerer. Mikrocontroller und Mikroprozessoren. eXamen.press. Springer, Berlin, Heidelberg, 2 edn. (2007)
18. Marco Block, Adrian Neumann. Haskell-Intensivkurs: ein kompakter Einstieg in die funktionale Programmierung. Xpert.press. Springer Verlag, Berlin, Heidelberg, 1 edn. (2011)
19. Vahrenkamp, R., Mattfeld, D.C.: Logistiknetzwerke: Modelle für Standortwahl und Tourenplanung. GWV Fachverlage GmbH, Wiesbaden (2007)
20. Verein Deutscher Ingenieure. Lastenheft/Pflichtenheft und Leistungsbeschreibung für die Simulationsstudie, VDI 3633–2 (1997)

21. Hans Peter Wiendahl. Logistikorientierte Kennzahlensysteme- und kennlinien. In Dieter Arnold, Heinz Isermannand Axel Kuhn, Horst Tempelmeier, and Kai Furmanns, editors, Handbuch Logistik, pages 228–248. Springer, Berlin, 3 edn. (2008)
22. Lotter, B., Wiendahl, H.-P.: Montage in der industriellen Produktion: Ein Handbuch für die Praxis. VDI-Buch-Springer, Berlin (2006)
23. Steffen Kinkel and Gunther Lay. Der Leistungsstand der deutschen Investitionsgüterindustrie, 1998. online. URL=http://hdl.handle.net/10419/29507. 29.02.2012
24. Mittwollen, M.: Technische Zuverlässigkeit und Verfügbarkeit. In: Arnold, D., Isermann, H., Kuhn, A., Tempelmeier, H., Furmanns, K. (eds.) Handbuch Logistik, pp. 865–872. VDI-BuchSpringer, Berlin Heidelberg (2008)

Robustness of Complex Adaptive Logistics Systems: Effects of Autonomously Controlled Heuristics in a Real-World Car Terminal

Christoph Illigen, Benjamin Korsmeier and Michael Hülsmann

Abstract Logistics systems have to cope with different challenges like unforeseeable machine failures leading to an increase of dynamics and complexity. Accordingly, a system's robustness (i.e. the ability to resist against a number of endangering environmental influences and the ability to restore its operational reliability after being damaged) might be decreased. Thus, this paper aims to answer the following research question: How do selected exemplarily heuristics (Minimum Queue-length Estimation, Minimum Cumulative Processing, Simple Rule-based, Holonic, Ant Pheromone, and Neural Net) contribute to a real world Hamburg Harbour Car Terminal's robustness? Thereby, the research focus in this investigation is on throughput time. As a main result it could be shown that all selected heuristics could contribute to a positive development of the system's robustness in case of machine failures. Thus, from a practical view potentials for the improvement of real-world scenarios might be assumed.

Keywords Complex adaptive logistics systems · Robustness of logistics systems · Autonomous control

C. Illigen (✉) · B. Korsmeier · M. Hülsmann
Jacobs University Bremen, Campus Ring 1, Bremen, 28759, Germany
e-mail: c.illigen@jacobs-university.de

B. Korsmeier
e-mail: b.korsmeier@jacobs-university.de

M. Hülsmann
e-mail: m.huelsmann@jacobs-university.de

K. Windt (ed.), *Robust Manufacturing Control*, Lecture Notes in Production Engineering, 161
DOI: 10.1007/978-3-642-30749-2_12,

1 Introduction

Robustness of logistics systems—i.e. the quality to remain effective for all plausible futures [1]—is widely accepted as one performance measure to value sustainable returns in a perturbed business environment [1–3]. Evidence from different industries emphasizes this need for robust systems that are exposed to many potential threats such as equipment failures or supplier discontinuities. Moreover, recent events have shown even higher risks arising from disruptions [4, 5]. Thus, it is desirable to maintain or increase a system's performance by increasing its robustness.

One approach to increase a logistics system's robustness is the concept of complex adaptive logistics systems (CALS), since [6] mention a high robustness as one key feature of CALS. Thereby, the higher robustness is achieved by the CALS concept through an autonomous reaction to complex and changing environments whereas a company strives for being successful in highly competitive and fast changing markets [6]. However, since the effects of the CALS concept on robustness has been investigated most prominently from a theoretical perspective, the corresponding research question of this paper is: How does the CALS approach contribute to the robustness of a real-world scenario? Hence, the paper implements one design option of a CALS according to [7] in a Hamburg harbour car terminal to investigate the effects on the system's robustness. In order to estimate the performance of the design option, five autonomously-controlled benchmark control heuristics are applied. The following goals are subject of the investigation:

The descriptive goals are to describe logistics systems as CALS, to introduce the relevant key performance indicator (KPI) throughput time, to describe the five heuristics and the CALS design option and finally to sketch the scenario. The analytical goals depict the timely effects of the five heuristics and the CALS design option towards robustness. This includes the effects on the average total throughput time (indicator 1) as well as the effects on the duration the system needs to restore the former system state after an incident (indicator 2). The pragmatic goal is to identify the most efficient approach related to the indicators mentioned above. In order to achieve these goals, the paper proceeds as follows: Sect. 1 starts with an introduction to the research field. In Sect. 2, system robustness as a challenge of CALS is introduced followed by a description of logistics systems as CALS. In Sect. 3, the underlying scenario including potential incidents is introduced comprehending the heuristics and the CALS design option. In Sect. 4, the results are presented, interpreted and critically discussed in a logistics context. Section 5 exhibits the conclusion and hints for further research.

2 Robustness in Complex Adaptive Logistics Systems

2.1 The Need for Robustness in Logistics Systems

The trend of logistics companies to become global players [8] leads to the progressive development of logistics systems from linear chains to international networks [9]. In consequence, they are part of global acting International Supply Networks, which are confronted with higher complexity (through number of actors) and dynamics (through the rate of change) [10]. Thus, logistics companies have to differentiate from competitors and simultaneously to cope with the increasing complexity and dynamics in a successful manner. In other words, they have to establish robust structures and processes. But what does robustness mean in the context of logistics systems?

The term robustness can be explained from different perspectives: According to Wycisk et al., robustness is the ability of the system to restore itself after being damaged [6]. Meepetchdee & Shah call a system robust if it is able to cope with complexity and dynamics in a way that an optimal target fulfilment can be achieved [11]. This paper focuses on the overall target achievement as well as on the restoring ability of the system as the research variables, since the robustness shall be understand as the optimization of logistics processes in terms of quality, time, and cost through optimal target fulfilment and restoring ability. However, the question occurs: How can this robustness be increased?

2.2 Logistics Systems as Complex Adaptive Logistics Systems

One approach being discussed for the achievement of higher robustness in logistics systems is the concept of Autonomous Control (AC) [12–14]. AC originates from the concept of self-organization and is enabled by the combination of software applications, sensor networks and new communication technologies (e.g. RFID). AC aims for an increased robustness of the overall logistics system through the improved ability to cope with complexity and dynamics [15]. However, how exactly can AC contribute to robustness?

Current research has already originated some findings regarding this question: Hülsmann et al. demonstrated that AC enlarges the robustness of logistics systems through a higher capacity for information processing [16]. Moreover, the contribution of AC to the robustness of logistics systems was researched through the analysis of AC's constitutional characteristics on the flexibility and stability and therewith on the adaptivity of the overall system [17]. While research was done on the effects of AC on the robustness of logistics systems, there is a lack of an empirical evaluation and findings. Thus, an approach to establish AC in a real-world scenario is required in order to research effects on robustness.

Wycisk et al. propose their CALS concept to realise a higher degree of AC in a logistics system [6]. They define logistics systems as Complex Adaptive Systems (CAS), since they identified various common properties between supply chains and CAS like heterogeneous agents or interaction. Thereby, they observed a higher adaptivity, positive emergence, and flexibility as CALS characteristics. In turn they also detected a vulnerability of CALS to all no-linear extreme dynamics respectively so called bullwhip-effects [6]. CAS in turn exhibit a certain degree of AC and thus can serve as an enabler of AC in a logistics system. As one major outcome of CALS is a higher system robustness [6], this concept shall be applied to the underlying scenario. However, based on the implications stated above the question occurs: How can the concept of CALS contribute to the robustness of a logistics system?

For the evaluation the throughput time will be the indicator 1, since it was shown that the throughput time rises with increasing the complexity level of logistics systems [13]. Thus, it constitutes an indicator of the logistics system's ability to cope with complexity and dynamics. Indicator 2 holds which AC heuristic features the best performance in restoring the system after a system breakdown. This constitutes an adequate indicator, since robustness beside others describes the system's ability to restore itself after a breakdown (see above). For answering these questions, control heuristics and a CALS design option for a real-world scenario are introduced next.

3 A Real-World Scenario for Investigating Effects of Autonomous Control Heuristics on Robustness

3.1 The Hamburg Harbour Car Terminal

In order to investigate the research question, an appropriate scenario is required that can adequately represent a CALS. The real-world car terminal is chosen, since it can be modelled as a CALS as it exhibits clearly distinguishable logistics objects (cars) [18] that can be equipped with CALS characteristics which are e.g. autonomy or interaction [6]. Hence, by implementing a CALS design option the model becomes a real CALS. Consequently, the car terminal can be utilized to estimate effects on the system's robustness through the CALS concept.

In addition to the suitability the terminal is readily described and has a flexible production sequence, a large amount of available real-world data and various changes and flexibilities in the car-flow process [19]. In a nutshell: the scenario exhibits a high degree of complexity and dynamics, system breakdowns can be modelled properly and it is adequately representable in a model. Hence, a subsequent simulation and thus concrete effects of different AC heuristics and the CALS design option on the scenario's robustness can be identified and investigated.

3.2 *The Simulation Environment*

The simulation model of this paper builds from a prior simulation developed by Windt et al. (2010b), since it is already operational and correctly represents the real world in the harbour context. The model contains information about e.g. number of cars per year, various routings of car flows in and out of parking, or various ways available to speed as many cars through the Terminal as possible. The information was decomposed into the 'boxes' shown in Fig. 1. Then, the means and standard deviations were calculated for the time each of 46,574 cars took to travel on the roads, to go through the required treatment queues and stations and then through the various subsequent stations, as required.

Cars arrive at the incoming parking area (I) (in Fig. 1). They are tagged with a parking order and then driven to a parking area (P1, P2, …, P9).[1] At DP1 one of several heuristics (we describe these below) is applied to determine which treatment-station queue to drive the car to. If incoming cars lack a parking tag they are immediately driven to the next decision point (DP1). Next, they are temporarily stored in the TPA parking area, or are driven directly to the queue of some treatment station (T1 to T7 in Fig. 1)—if a treatment order is shown. Otherwise, they are driven directly to the exit (O). Available treatment stations are (1) gasoline station; (2) diesel station; (3) de-waxing; (4) car-body repair; (5) car wash; (6) paint shop; and (7) final inspection. Each car entering the Terminal comes with pre-defined orders, which can be divided into treatment and parking orders. Treatment orders call for from one up to six treatment steps; some of the treatment steps have sequence constraints (e.g., removal of transport protection (de-waxing) needs to come before washing and painting). After all treatments are finished cars are driven to the exit O (and are out of the simulation).

3.3 *Implemented Incidents in the Car Terminal*

In order to investigate effects of the CALS concept on the robustness of the chosen scenario, some kind of disturbances need to be added. Therewith, effects on throughput time as well as information about restoring capability of the system can be created. There are numerous sources for supply chain (SC) deficiencies that have prominently been investigated in recent research [20–23]. Several categories of SC risks were elaborated [24, 25] and extended lists of risk drivers were identified [20, 26].

---

[1] For this baseline simulation accuracy (docking) test and application of the NN model, we ignore all parking waiting times except parking queues directly affecting the car-flows through specific treatment stations, since none of the heuristics involve any parking waits before cars get to DP_1.

This paper focuses on internal machine failures, since the model and the real world data offers most adequate information for these issues. In that context, the question of whether the system's robustness can be increased can be investigated: Firstly, the effects on logistics processes in terms of throughput time can be analysed (indicator 1). Secondly, the capability of the system to restore itself after an incident (i.e. a machine failure) can be estimated (indicator 2). Thus, the paper answers the research question by investigating how the different heuristics react to these incidents and keep the whole scenario effective as well as how the heuristics are able to restore the former system state after being exposed to an unforeseen incident.

If a machine breakdowns occurs, the respective machine cannot perform the required treatment for one time of its expected treatment time. Breakdowns are put into the scenario once at a particular date and for three machines in a way that each car is affected. Thus, a visible effect over the whole simulation time can be created and different effects of capacity bottlenecks on the scenario can be investigated for the applied heuristics as well as for the baseline scenario.

3.4 The Applied Car-flow Routing-Heuristics and a CALS Design Option

Ideally, each car-routing decision should attempt to minimise car-flow time. Thus it is possible that for any given car, the waiting time before entering into any particular station is constantly changing because of the routing effects of the queue-waits and station-processing speeds, say, of the 10 cars ahead of it in the various queues. To navigate the cars through the terminal, five AC heuristics—beside the Standard Method (i.e. the real-world situation in the terminal)—are applied in the simulation: The 'Minimum Queue-length Estimation Heuristic', 'Minimum Cumulative Processing Heuristic', 'Simple Rule-based Heuristic', 'Ant Pheromone Heuristic,' and the 'Holonic Heuristic'. These five heuristics are used as benchmarks in order to estimate the performance of the CALS design option 'Neural Net'[2] towards the system's robustness. This selection of the five benchmark heuristics is based on a literature search covering the current state of AC methods applied to logistics [18].

The Standard's Method is a pre-determined ordering of how cars are sent through the treatment stations, as follows: (1) incoming delivery; (2) gas or diesel refuelling; (3) de-waxing; (4) car-body repair; (5) car wash; (6) painting; (7) final inspection; (8) outgoing delivery [18].

[2] The Design Option 'Neural Net' is selected, since it is the first CALS design option of a bunch of consecutive design options, that require all the preceding design options [7].

The Minimum Queue Length Estimation Heuristic gives top priority to the station with the fewest number of cars in its queue [27]. If all queues have equal length, a random selection is made.

The Minimum Cumulative Processing Heuristic sends the car to the station that shows the lowest 'QP time', where QP = (station processing time × (car in station + number of cars in station queue)). This ignores driving times and possible overnight times [18].

The Simple Rule-based Heuristic compares estimated waiting times at station-queue points based on the processing time of the previous ten cars. This Heuristic proceeds as follows: (1) When a car leaves a station it transfers information about its processing time to a central location where it is collected and stored; (2) the information from the most recent ten cars is averaged; (3) the next car is sent to the treatment-station queue showing the quickest total treatment time (driving + queue waiting + treatment) based on the averaged times of the previous ten cars [28].

The Ant Pheromone Heuristic starts from how ants gather food: Cars following this heuristic leave virtual (pheromone) marks at each station when they leave. The more marks a station accumulates over a specific number of hours, the more likely it is that a subsequent car is sent to this station [29].

The Holonic Heuristic sends a car to the treatment station offering the shortest total time (based on driving + station-queue waiting + treatment times) gets the car. If several stations have the same queue length (delay), the car showing a quicker final release time is chosen. Whereas the Simple Rule Heuristic focuses on the average time of the previous ten cars, the Holonic Heuristic focuses on which next station is available the quickest (with soonest release time used to break ties) [30, 31].

The CALS design option[3] 'Neural Net' [7] sends the car to the next station based on six internal rules and according weightings. These rules define the optimal next station based on current information (e.g. utilization) as well as on information available from the last 50 cars that went through the station. Thus, the neural net not only considers the current state but also considers past developments and can thus identify potential problems in the system.

To benchmark the effects of the introduced heuristics with the CALS design option on the system's robustness, Sect. 4 presents the obtained results in terms of throughput times in the standard scenario and the scenario with machine failures as well as the effects on the time the system needs to restore after these incidents. These indicators were chosen for the evaluation since 'time' constitutes one main success factor for car terminals caused by the importance of customer orientation (delivery times are perceived as one major quality indicator) [32]. However, also other indicators (e.g. utilization) might be feasible for the evaluation of suchlike scenarios and remain still subject to further investigations in the underlying scenario.

---

3 For an overview of further design options of CALS beside the Neural Net see [7].

Table 1 Average throughput times in scenario without machine failures

Heuristic	Δ [%] Standard scenario	Δ [%] Machine failures	p-Value[a] (*t* test)
Standard method	Reference method		
Ant pheromone	+6.53	−15.64	P < 0.001
Holonic	−4.02	−19.85	P < 0.001
Minimum buffer	−8.71	−21.54	P < 0.001
Neural net	−12.73	−24.32	P < 0.001
Queue-length Estimation	−6.22	−18.39	P < 0.001
Simple rule-based	−8.32	−12.73	P < 0.001

[a] Since the sample size exceeds 46,000 cars, we assume normal distributions and thus the *t* test is applicable

4 Evaluation of the Performance Indicators

4.1 Results of the Simulation Runs

The simulation was implemented and executed in the simulation software Tecnomatix Plant Simulation (v.9) from the Siemens AG. Based on the simulations, the succeeding results in Table 1 for the total average throughput times in the standard scenario (column 2) and the total average throughput times in the scenario with machine breakdowns (column 3) were obtained. Additionally, in order to check for the fit between the model and the real world, the p-Values for every heuristic based on a *t* test between the Standard Method and each particular heuristic was accomplished for both scenarios.

Table 1 demonstrates that every heuristic except the Ant Pheromone Heuristic reduces the total average throughput time in the standard scenario. The Ant Pheromone Heuristic increases the total average throughput time by 6.53 % whereas the reductions are in the interval between −4.02 % (Holonic) and −12.73 % (Neural Net). In the scenario with machine failures (column 3), all heuristics reduce throughput time. Thereby, the reductions of the single heuristics are different. They range between the standard scenario and the scenario with machine failures from ~4 % (Simple Rule-Based: −8.32 to −12.73 %) up to ~22 % (Ant Pheromone: +6.53 to −15.64 %). The *t* test additionally presents a fit between the model and the real world (all p Values < 0.001).

Figure 1 contains the days during (Feb 29th to Mar 3rd) and right after (Mar 4th onwards) machine disturbances and the average buffer level of the heuristics in comparison to the buffer level of the Design Option 'Neural Net' in the standard scenario. This depicts the time the different heuristics require to restore the former system state (see Standard Method representing stable system state) (Fig. 2).

The figure displays relatively high buffer levels of the heuristics (between 25 and 30 % from Feb 29th to Mar 3rd) in comparison to the average utilization in a stable system state (~4 %). Contrariwise, the figure shows low fill levels after all machines are restored on March 4th.

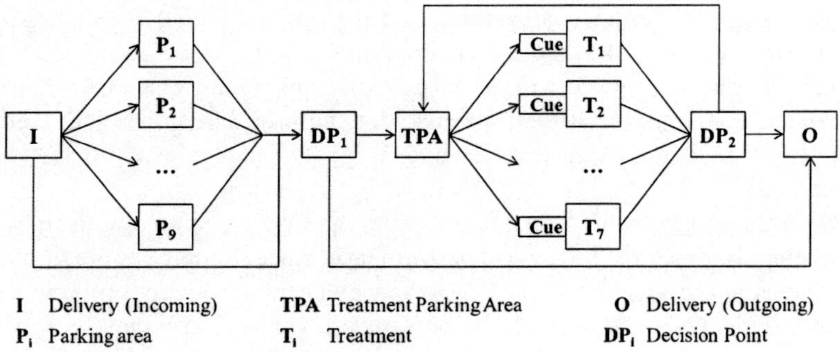

Fig. 1 Schematic representation of the model

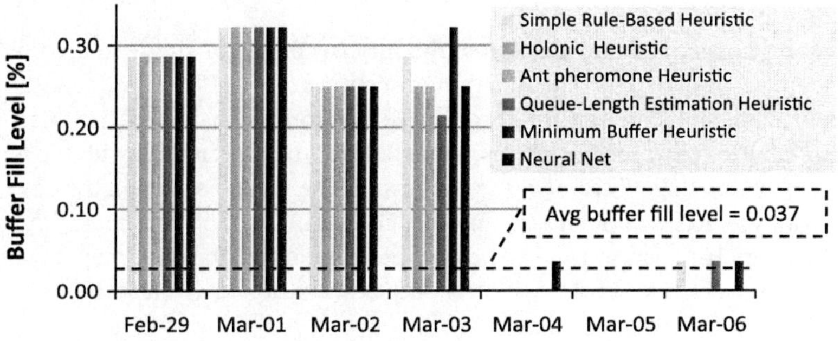

Fig. 2 System restore time after machine failures

4.2 Discussion of the Results

In order to evaluate the final results of the real world scenario simulation it has to be discussed how the heuristics effect the two indicators outlined in the research question. Thus, it has to be revealed how the average total throughput times (indicator 1) as well as the time for restoring the system after a machine failure (indicator 2) are influenced.

With regard to the average total throughput time **(indicator 1)** it can be stated that in the standard scenario all heuristics beside one (Ant Pheromone) could improve the throughput times. Thus, it is implied that through the implementation of suchlike heuristics cars can better navigate through the terminal. This finally leads to a better achievement of logistics goals like throughput time as well as due date reliability—as cars are routed faster through the terminal more due dates can be matched. This ultimately leads to a higher robustness in the standard scenario, since the whole system has more reserves due to a higher efficiency and is thus enabled to react more flexible to unexpected incidents finally increasing robustness.

For the machine failure scenario it can be observed that all heuristics generate a decreased throughput time in comparison to the reference heuristics. Thus, for all heuristics a logistics target achievement and therewith a higher robustness of the overall system through the heuristics' implementation can be assumed. From a practitioners point of view it can be stated that implementing suchlike heuristics can mitigate the negative effects of machine failures like jams and therewith potentially more missed due dates, which might happen in real-world working processes.

However, comparing the AC heuristics with the CALS design option shows that the design option features the highest performance, since it improves the throughput time in the standard scenario (−12.73 %) as well as the throughput time in case of machine failures (−24.32 %) with the best overall value. Consequently, following the assumption that an increased logistics target achievement leads to higher system robustness and having the result described above it can be stated that designing the underlying scenario as a CALS leads to the highest robustness for the overall system.

However, it has to be addressed that the improvements of single heuristics from the standard scenario to the machine failure scenario are highly volatile. Although a clear improvement trend can be observed, it has to be considered that also small scenario changes (e.g. more or less machine failures in combination with the number of cars) can create huge differences in the results like it can be seen in the case. This results from the AC characteristic emergence leading to non-predictability of the overall system [6]. Thus, it also has to be mentioned that the results are not assignable to every real world scenario—other scenarios might lead to completely different results for the heuristics.

According to the duration the system needs to restore itself after an incident **(indicator 2)** it could first be observed, that all heuristics were able to restore the system within one day after the machine failure. However, differences occurred regarding the time needed for the restoring. In comparison to the first indicator the CALS design option does not constitute the best solution approach. Thus, it can be mentioned that from a practical perspective all heuristics provide the restoring ability for the system after a breakdown. However, since the neural net does not deliver the best results here the question remains how to improve the neural net in order to improve also the restoring ability of this CALS design option. In this specific investigation a recommendation for the overall scenario considering both indicators cannot be given for one specific heuristic—the final decision has to be made based on the individual goal settings (which goals are the most important ones) or rather based on practical experiences (e.g. probabilities for machine failures in specific real world scenarios).

Finally, this leads to the conclusion that in sum all heuristics are able to increase the system's robustness in the scenario with machine failures, since they exhibit a lower total average throughput time than the Standard Method (keep system effective) and are able to restore the system after a failure. Thus, this case indicates that having the right circumstances in a real-world scenario (e.g. high complexity and dynamics, lots of options and entities) the selected AC heuristics can improve the system's robustness. Additionally, the CALS design option 'Neural Net' features in both scenarios the best throughput time. On the other hand, the time to restore is not as short as with the other heuristics. This shows that the CALS design option might feature additional potential to improve a system's robustness but in contrast to other heuristics the design option needs more time to restore the system. Thus, the total performance of the Neural Net in contrast to the other heuristics has to be evaluated in more detail and both indicators have to be weighed against each other depending on their economic impact as well as on the particular context.

5 Conclusion

This paper intended to give an idea about how the CALS approach contributes to the robustness of a real-world scenario in comparison to existing control heuristics. The main outcome is that in a scenario with high complexity and dynamics and in combination with machine failures all selected heuristics and the CALS design option can contribute to a positive development of the system's robustness. Additionally, from a practitioners perspective one can see that in real-world scenarios there is potential for improvements in system flows and AC heuristics can be one way to realize this potential.

However, this research features some limitations. First, it does not cover all relevant logistics KPIs—which are partially conflicting like low work in progress and high utilization [32]. Then, the introduced heuristics are complicated to apply in a real-world scenario. The reason is, that AC heuristics as well as the introduced design option 'Neural Net' demand for a complicated set of technological and computer components (e.g. sensor networks, communication interfaces), which also have to meet the organizational and structural requirements of the given institution [33]. Moreover, these components are in sum expensive, what can lead to financial risks resulting from the implementation [34]. Since a testing of the system's outcome is nearly impossible—due to the characteristics of AC like emergence [35]—it might be hard to convince investors of an acceptable return of their investment into the required technologies. Finally, people might avoid suchlike concepts due to their loss of control that comes along with the system's self-organization.

Further research should focus on the given limitations. Hence, additional logistics KPIs could be included in further research as well as a transfer of the gained insights to other simulation scenarios. Moreover, since the research also shows only very little differences between the particular heuristics the investigation horizon and the input parameters should be adapted in order to generate more detailed findings for the different approaches.

Acknowledgments This research was supported by the German Research Foundation (DFG) as part of the Collaborative Research Centre 637 "Autonomous Cooperating Logistic Processes—A Paradigm Shift and its Limitations".

References

1. Klibi, W., Martel, A., Guitouni, A.: The design of robust value-creating supply chain networks: A critical review. Eur. J. Oper. Res. **203**(2), 283–293 (2010)
2. Kouvelis, C., Yu, G.: Robust Discrete Optimization and its Applications. Kluwer Academic Publishers, Berlin (1997)
3. Snyder, L., Daskin, M.: A random-key genetic algorithm for the generalized traveling salesman problem. Eur. J. Oper. Res. **174**(1), 38–53 (2006)
4. Barrionuevo, A., Deutsch, C.: A distribution system brought to its knees. New York Times, Sept 1, (2005)
5. Mouawad, J.: Katrina's shock to the system. The New York Times, Sept 4, (2005)
6. Wycisk, C., McKelvey, B., Hülsmann, M.: Smart parts' supply networks as complex adaptive systems—analysis and implications. Int. J. Phys. Distribution Logis. Manage. **38**(2), 108–125 (2008)
7. McKelvey, B., Wycisk, C., Hülsmann, M.: Designing an electronic auction market for complex 'smart parts' logistics: Options based on LeBaron's computational stock market. Int. J. Prod. Econ. **120**(2), 476–494 (2009)
8. Klaus, P., Kille, C.: Die Top 100 der Logistik—Marktgrößen. Marktsegmente und Marktführer in der Logistik—Dienstleistungswirtschaft, Hamburg (2006)
9. Mason, R.B.: The external environment's effect on management and strategy- a complexity theory approach. Manag. Decis. **45**(1), 10–28 (2007)
10. Sydow, J.: Zum Management von Logistiknetzwerken. Logistik Manage. **4**(4), (2002)
11. Meepetchdee, Y., Shah, N.: Logistical network design with robustness and complexity considerations. Int. J. Phys. Distribution Logis. Manage. **37**(3), 201–222 (2007)
12. Hülsmann, M., Cordes, P.: Autonomous co-operation and control in complex adaptive logistic systems—contribution and limitations for the innovation capability of international supply networks. In: Zhou, J. (eds.): COMPLEX 2009, Part I, LNICST 4—Proceedings of the First International Conference on Complex Sciences: Theory and Application. Springer, Berlin, pp. 1023–1032 (2009)
13. Hülsmann, M., Scholz-Reiter, B., Austerschulte, L., de Beer, C., Grapp, J.: Autonomous cooperation—a capable way to cope with external risiks in international supply networks? In: Pawar, K.S., Lalwani, C.S., de Carvalho, J.C., Muffatto, M. (eds.) Proceedings of the 12th International Symposium on Logistics (12th ISL), Loughborough, pp. 172–178 (2007)
14. Windt, K., Hülsmann, M.: Perspectives on initial ideas and conceptual components of autonomous co-operation and control. In: Hülsmann, M., Windt, K. (eds.) Understanding Autonomous Cooperation and Control: The Impact of Autonomy on Management, Information, Communication, and Material Flow, Berlin (2007)
15. Hülsmann, M., Windt, K.: Development of a terminological system of autonomous cooperation, in: Hülsmann, M., Windt, K. (eds.) Understanding Autonomous Co-operation—The Impact of Autonomy on Management, Information, Communication and Material Flow, Bremen (2006)
16. Hülsmann, M., Scholz-Reiter, B., deBeer, C., Austerschulte, L.: Effects of autonomous cooperation on the robustness of international supply networks—contributions and limitations for the management of external dynamics in complex systems. In: Haasis, H.-D., Kreowski, H.-J., Scholz-Reiter, B. (eds.) Dynamics in Logistics—Proceedings of the 1st International Conference on Dynamics in Logistics. Springer, Berlin, pp. 241–250 (2008)
17. Hülsmann, M., Korsmeier, B., Illigen, C., Cordes, P.: autonomous co-operation of "smart-parts"–contributions and limitations to the robustness of complex adaptive logistics systems. In: Proceedings of the 2nd International Conference on Dynamics in Logistics (LDIC), Universität Bremen, Bremen (2009)
18. Windt, K., Becker, T., Kolev, I.: 2010. A comparison of the logistics performance of autonomous control methods in production logistics. In: Sihn, W., Kuhlang, P. (eds.)

Sustainable Production and Logistics in Global Networks. Proceedings of the 43rd CIRP International Conference on Manufacturing Systems, Vienna, pp. 576–583 (2010)
19. Windt, K., Becker, T., Jeken, O., Gelessus, A.: A Classification Pattern for Autonomous Control Methods in Logistics. CRC, University of Bremen, Bremen (2010)
20. Chopra, S., Sodhi, M.S.: Managing risk to avoid supply-chain breakdown. MIT Sloan Manage. Rev. **46**, 52–61 (2004)
21. Helferich, O.K., Cook, R.L.: Securing the Supply Chain. Council of Logistics Management (CLM) (2002)
22. Sheffi, Y.: The Resilient Enterprise: Overcoming Vulnerability for Competitive Advantage. MIT Press Books, Cambridge (2005)
23. Sweet, K.: China earthquake hits home for US companies. Fox Business, May 12 (2008)
24. Christopher, M., Peck, H.: 2004. Building the resilient supply chain. Int. J. Logistics Manage. **15**(2), 1–13 (2004)
25. Tang, C.: Robust strategies for mitigating supply chain disruptions. Int. J. Logis. Res. Appl. **9**, 33–45 (2006)
26. Kleindorfer, P.R., Saad, G.H.: Managing disruption risks in supply chains. Prod. Oper. Manage. **14**(1), 53–68 (2005)
27. Philipp, T., de Beer, C., Windt, K., Scholz-Reiter, B.: Evaluation of autonomous logistic processes—Analysis of the influence of structural complexity. In: Hülsmann, M., Windt, K. (eds.), Understanding Autonomous Cooperation and Control in Logistics—The Impact on Management, Information and Communication and Material Flow. Springer, Berlin, pp. 303–324 (2007)
28. Scholz-Reiter, B., Freitag, M., de Beer, C., Jagalski, T.: The influence of production networks' complexity on the performance of autonomous control methods. In: Teti, R. (ed.): Intelligent Computation in Manufacturing Engineering 5. Proceedings of the 5th CIRP International Seminar on Computation in Manufacturing Engineering (CIRP ICME '06), Naples, 317–320 (2006)
29. Armbruster, D., de Beer, C., Freitag, M., Jagalski, T., Ringhofer, C.: Autonomous control of production networks using a pheromone approach. Phys. A **363**(1), 104–114 (2006)
30. Markus, A., Kis Vancza, T., Monostori, L.: A market approach to holonic manufacturing. CIRP Ann. Manuf. Technol. **45**(1), 433–436 (1996)
31. Van Brussel, H., Wyns, J., Valckenaers, P., Bongaerts, L., Peeters, P.: Reference architecture for holonic manufacturing systems. PROSA. Comput. Ind. **37**(3), 255–274 (1998)
32. Nyhuis, P., Wiendahl, H.P.: Fundamentals of Production Logistics: Theory, Tools and Applications. Springer, Berlin (2008)
33. Hülsmann, M., Scholz-Reiter, B., Austerschulte, L., Wycisk, C., de Beer, C.: Autonomous cooperation—a way to cope with critical incidents in International Supply Networks (ISN)? An Analysis of Complex Adaptive Logistic Systems (CALS) and their Robustness. In: 24th EGOS Colloquium. Upsetting Organizations, web-publication, (2008)
34. Hülsmann, M., Illigen, C., Korsmeier, B., Cordes, P.: Risks resulting from autonomous co-operation technologies in logistics. In: Li, Y., Desheng, W. (eds.) Proceedings of 2010 IEEE International Conference on Advanced Management Science, China, vol. 1, pp 422–426, (2010)
35. Hülsmann, M., Windt, K.: Changing paradigms in logistics—understanding the shift from conventional to autonomous cooperation and control. In: Hülsmann, M., Windt, K. (eds.) Understanding Autonomous Co-operation—The Impact of Autonomy on Management, Information, Communication and Material Flow, Bremen, (2006)

A Pedestrian Dynamics Based Approach to Autonomous Movement Control of Automatic Guided Vehicles

Maik Bähr, Reik V. Donner and Thomas Seidel

Abstract Automatic guided vehicles (AGVs) are a prospective concept for optimizing transportation capacity and reducing the costs of material transport and handling in manufacturing systems. Besides the careful allocation of individual transportation tasks, single units have to be able to freely move in a given two-dimensional space possibly restricted by a set of fixed or variable obstacles in order to use their full potentials. One particular possibility for realizing an autonomous movement control is utilizing self-organization concepts from pedestrian dynamics like the social force model. Since this model itself does not explicitly prohibit possible collisions, this contribution discusses necessary modifications such as the implementation of braking strategies and approaches for anticipating deadlock situations, which need to be additionally considered for developing a generally applicable autonomous movement control. By means of numerical simulations, different operational situations are investigated in a generic scenario in order to identify the practical limitations of our approach. The presented work suggests considerable potentials of pedestrian dynamics-based self-organization principles

M. Bähr · R. V. Donner · T. Seidel
Institute for Transport and Economics, Dresden University of Technology,
Würzbuger Str. 35, 01187 Dresden, Germany

M. Bähr
Institute of Traffic Telematics, Dresden University of Technology,
Andreas-Schubert-Str. 23, 01069 Dresden, Germany

R. V. Donner (✉)
Research Domain IV-Transdisciplinary Concepts & Methods, Potsdam Institute for Climate Impact Research, Telegrafenberg A31, 14473 Potsdam, Germany
e-mail: reik.donner@pik-potsdam.de

T. Seidel
AMC Managing Complexity GmbH, An der Tongrube 1-3,
40789 Monheim am Rhein, Germany

K. Windt (ed.), *Robust Manufacturing Control*, Lecture Notes in Production Engineering,
DOI: 10.1007/978-3-642-30749-2_13,

for establishing a flexible and robust movement control for AGVs, which shall be further studied in future work.

Keywords Automatic guided vehicles · Autonomous movement control · Pedestrian dynamics · Social force model

1 Introduction

These days, production and logistics systems become more and more automated to achieve a higher cost-efficiency of the manufacturing and delivery processes. For a robust operation of individual production units, the absence of substantial perturbations and the simultaneous availability of all necessary resources such as raw material, tools, and energy are typical requirements. These idealized requirements are, however, often not completely met in reality because of the presence of unpredictable factors such as machine failures or human interference. In order to better cope with such problems, recent work has concentrated on elaborating the potentials of autonomous control approaches in various areas of application [1, 2] which promise higher flexibility and robustness in the presence of unexpected situations than traditional centralized control approaches.

Many real-world examples for material handling and storage systems (for example, container terminals) are characterized by a high number of individual transportation tasks each requiring a significant amount of time. In such cases the utilization of automated transfer cars can provide a feasible solution to keep the internal transportation costs as low as possible. Specifically, the installation and operation costs of modern, complex technical solutions need to be over-compensated by a reduction of labor costs and an overall increase of transportation efficiency in the mid- to long-term.

Automatic guided vehicles (AGVs) are already widely used in intralogistics or ports of transshipment. Their practical operation is challenged by two mutually interdependent problems for planning and control: task allocation to the entire set of available vehicles (scheduling) and routing of the individual AGVs. The latter aspect can be realized by either prescribing a fixed route the vehicle has to follow exactly, or utilizing autonomous control concepts allowing the vehicle to move freely in space in a cost-efficient and safe way. This contribution exclusively focuses on the latter aspect. Specifically, the potentials of pedestrian dynamics-based self-organization concepts are studied. Conversely, it is not the aim of this work to provide a thorough and complete review of existing alternative solutions or to discuss the closely related problem of job scheduling.

At present, AGVs are still often subject to a centralized movement control allowing for the anticipation and thus preventation of potential collisions or deadlock situations [3, 4]. One corresponding strategy is that the individual vehicles have to pass certain checkpoints allowing for a forecast of possible conflicts. In addition, gridding of the available space simplifies the distribution of priorities and permissions for entering specific areas: if the next checkpoint or the next grid cell cannot be reached (for example, due to an occupation by another transfer car) a vehicle remains at rest until the blocking situation is resolved. Alternatively, the network of possible transportation routes can be decomposed into paths that can be reserved for usage by individual vehicles. The latter strategy typically requires a central control unit with complete on-line coverage information of the entire network. Furthermore, the paths can be divided into sections allowing for a forecast of the future occupation and, hence, for possible preventive actions [5, 6].

As an alternative to such centralized approaches, efforts have been reported recently to design and practically realize a fully decentralized control strategy for AGVs in intralogistics applications,[1] where the availability of physical space is a crucial restriction to all transportation processes [7, 8]. As a basic condition, one has to require that such an approach must (i) adequately substitute the advantages of existing priority and permission assignment strategies by means of an intelligent control algorithm and (ii) avoid collisions and deadlock situations as far as possible. Specifically, individual vehicles shall freely move in space without any external actions and react autonomously and flexibly in the presence of possible conflicts. While the corresponding problem has already been widely studied in the field of robotics, this contribution discusses an alternative conceptual approach that is based on some fundamental self-organization principles realized in nature.

A paradigmatic example for successfully operating autonomous units are groups of human individuals. In the last decades, numerous efforts have been reported for mathematically modelling the dynamics of pedestrians and their mutual interactions on the microscopic (individual-based) level (i.e., as a multi-agent system [9]), including cellular automaton [10–12] and interaction force models [13–15]. The latter class of models postulates the existence of a repulsive short-range potential resulting in the avoidance of too close encounters between individuals. As a particularly useful and widely applicable approach, the social force model [14, 16, 17] has attracted great interest for appropriately describing the behavior of groups of human individuals in various situations. Meanwhile, it is being applied for simulating pedestrian dynamics in sophisticated state-of-the-art traffic simulators such as PTV VISWALK [18].

This contribution is intended to provide a first conceptual study of the potentials and practical limitations of a pedestrian dynamics-based strategy to establishing an autonomous movement control for AGVs. Specifically, the basic ingredients of the social force model are used for implementing corresponding control mechanisms that allow single AGVs to individually choose their trajectory according to a prescribed transportation task and gradually re-evaluate and adapt their motion to the presence of static as well as dynamic obstacles in real-time. By studying

[1] There are first successful implementations of autonomously controlled AGVs in container terminals were many problems that are expected to arise from a free movement are avoided by restricting the motion to designated one-way traffic lines with a Manhattan-type regular grid topology. However, in this contribution, the case of free motion is considered.

generic scenarios, possible limitations of this approach are identified, which call for modifications and extensions of the simple behavioral rules of the underlying pedestrian dynamics model to be applicable for the purposes of industrial logistics. In Sect. 2, the theoretical foundations of the social force model of pedestrian dynamics are briefly reviewed, with a particular emphasis on how to translate its basic ingredients for an autonomous movement control of AGVs. Section 3 describes some practical challenges for establishing such a control in the presence of real-world problems such as static and dynamic obstacles and high vehicle densities at specific sources or sinks in a logistics system. Possible solutions of these problems are discussed. Finally, the main findings of the presented work are briefly summarized and put into context.

2 Theoretical Background

2.1 The Social Force Model of Pedestrian Dynamics

Based on the fundamental Newtonian principle that each change in the state of motion of a physical body requires the action of a certain force, Lewin [19, 20] introduced the social field theory, postulating that social situations are influenced by factors (forces) that either drive (helping forces) or block (hindering forces) a movement towards a goal. Besides numerous applications in the social sciences, this theory has been used by Helbing and Molnár [14] as a basis for establishing a widely applicable mathematical model of pedestrian behavior. Specifically, they described the motion of human individuals as a superposition of a positive driving force towards a well-defined destination and a set of repulsive forces arising from non-physical ("social") interaction potentials caused by the presence of obstacles and other individuals. Specifically, according to their theory four different types of forces can be distinguished: a generic accelaration force due to the presence of a desired direction and speed of a human individual, repulsive forces due to static obstacles as well as other moving individuals (dynamic obstacles), and attractive forces due to the existence of potential points of interest. In the following, the corresponding mathematical formulation is briefly reviewed.

First, a pedestrian i aims to reach a certain destination $\boldsymbol{d}_i$ with a desired velocity v_i^0. The shortest path between this (static) destination and the (dynamically changing) position $\boldsymbol{r}_i(t)$ of the individual is described by a vector $\boldsymbol{d}_i - \boldsymbol{r}_i(t)$ $(k = 1, , K)$. This idea can be generalized by introducing a set of intermediate destinations $\boldsymbol{d}_i^k$ that make the pedestrian's trajectory (in the absence of other individuals or obstacles) a regular polygon. The desired direction of motion is then represented by the unit vector

$$\boldsymbol{e}_i(t) = \frac{\boldsymbol{d}_i^k - \boldsymbol{r}_i(t)}{\|\boldsymbol{d}_i^k - \boldsymbol{r}_i(t)\|}. \tag{1}$$

Any perturbation of the pedestrian's motion from the desired direction $\boldsymbol{e}_i(t)$ and speed v_i^0 implies a deviation of the actual velocity $\boldsymbol{v}_i(t)$ from the desired velocity $\boldsymbol{v}_i^0(t) = v_i^0 \boldsymbol{e}_i(t)$ due to deceleration or sidestepping to avoid collisions with other individuals. In such cases, the pedestrian has a tendency to approach $\boldsymbol{v}_i^0(t)$ again within a certain relaxation time τ_i after the corresponding conflict has been resolved. This gives rise to an acceleration force

$$\boldsymbol{F}_i^0(t) = \frac{1}{\tau_i}\left(v_i^0 \boldsymbol{e}_i(t) - \boldsymbol{v}_i(t)\right). \tag{2}$$

Second, a pedestrian intends to keep a certain distance from static obstacles (such as walls or streets with vehicular traffic). This fact can be modelled by means of a repulsive potential force

$$\boldsymbol{F}_{iB}(t) = -\nabla_{\boldsymbol{r}_{iB}(t)} U_{iB}(\|\boldsymbol{r}_{iB}(t)\|), \tag{3}$$

where $\boldsymbol{r}_{iB}(t)$ is the vector between the current position and the nearest point of the obstacle, and U_{iB} is a non-negative (repulsive) and monotonically decreasing potential.

In a similar spirit, human individuals consider the close proximity of other pedestrians as undesired and therefore aim to keep a certain distance with respect to their neighbors. In analogy to the treatment of static obstacles, the resulting repulsive force acting on an individual *i* due to the closeness of another individual *j* (acting as a dynamical, i.e., moving obstacle) can be described as

$$\boldsymbol{F}_{ij}(t) = -\nabla_{\boldsymbol{r}_{ij}(t)} U_{ij}(b(\|\boldsymbol{r}_{ij}(t)\|)), \tag{4}$$

where $\boldsymbol{r}_{ij}(t)$ denotes the shortest path between both individuals at time t, and b represents the semiminor axis of the ellipse around j,

$$b(\|\boldsymbol{r}_{ij}(t)\|) = \frac{1}{2}\left(\|\boldsymbol{r}_{ij}(t)\| + \|\boldsymbol{r}_{ij}(t) - v_j(t)\Delta t \boldsymbol{e}_j(t)\| - (v_j(t)\Delta t)^2\right)^{1/2}, \tag{5}$$

with $v_j(t) = |\boldsymbol{v}_j(t)|$. The latter accounts for the space required for j to take his/her next step that i anticipates in his/her own motion. Hence, the current velocity $\boldsymbol{v}_j(t)$ of the obstacle j (or, more specifically, its modulus) needs to be taken into account.

It should be noted that as a possible extension of this formulation, the velocity $v_i(t)$ of the considered vehicle could be considered as well when dealing with static as well as dynamic obstacles (i.e., a faster pedestrian will typically prefer a larger distance than a slower one) to account for safety considerations such as a velocity-dependent braking distance. This extension amplifies the effect obstacles on fast individuals, which leads to an enhancement of safety in terms of avoiding possible collisions. In the context of autonomous movement control of AGVs to be discussed here, the corresponding effect will, however, be considered negligible. Therefore, this possible extension shall not be further discussed here.

Finally, there can be points of attraction such as street artists or shop windows that potentially trigger detours or short stays at the respective places. Similar to the treatment of obstacles, such points of attraction can be modelled by an attractive interaction potential:

$$\boldsymbol{F}_{iA}(t) = -\nabla_{\boldsymbol{r}_{iA}(t)} U_{iA}(\|\boldsymbol{r}_{iA}(t)\|), \tag{6}$$

where $U_{iA}(\cdot)$ is (unlike $U_{iB}(\cdot)$ and $U_{ij}(\cdot)$) a negative (attractive) and monotonically increasing function.

The sum of the four aforementioned forces,

$$\boldsymbol{F}_i(t) = \boldsymbol{F}_i^0(t) + \sum_B \boldsymbol{F}_{iB}(t) + \sum_j \boldsymbol{F}_{ij}(t) + \sum_A \boldsymbol{F}_{iA}(t), \tag{7}$$

can be understood as determining temporal changes in the direction and speed of motion. A more detailed description can be found in [14]. One important aspect is the proper determination of the shape of the different interaction potentials, which are commonly assumed to be exponential [14] or power-laws. One recognizes that the latter choice would better account for a desired avoidance of collisions. In addition, more sophisticated aspects such as an anisotropy of all forces (due to the visibility of obstacles or points of interests only within a cone centered around the current direction of motion) can be easily included into the model. Detailed experimental studies on various aspects of real-world pedestrian dynamics are available [21, 22], but shall not further discussed here. Within the framework of the present study, the focus will mainly be on the consideration of the four discussed types of forces and their implementation in an efficient autonomous AGV control strategy.

2.2 Social Force Approach to Autonomous Movement Control of AGVs

Some of the basic ingredients of the social force model as described above can serve as foundations for establishing an autonomous movement control of AGVs. Specifically, there are existing approaches that already make (partial) use of several components inherent to this model:

- The use of AGVs implies a directed motion towards a predetermined final destination—or some prescribed intermediate checkpoints—that is typically realized with a designated optimum velocity. This is in complete analogy to the behavior of pedestrians, with the intermediate checkpoints playing the role of the points of attraction in the model of human behavior.
- Collisions with static as well as dynamic obstacles (e.g., stored material, other AGVs, etc.) need to be avoided. This can for example be reached by (e.g., reflection or RFID-based) sensors measuring the distance to possible obstacles

in real-time. The shorter the distance, the lower the velocity of the vehicle should be due to safety reasons, which can be mathematically described by the action of a non-physical, distance-dependent repulsive force.

- For the control of AGVs, obstacles behind the moving vehicle are not relevant for its further motion. This consideration has been realized in existing models for explaining the behavior of groups of animals (e.g., fishes, locusts, or bird flocks) [23]. In this spirit, it is sufficient to consider a certain range of angles around the instantaneous vector of motion within which static as well as dynamic obstacles need to be detected and considered for further planning of the AGV's individual trajectory. Thus, a non-isotropic version of the social force model is a promising candidate to account for the corresponding considerations.

In the following, it will be discussed in more detail which specifications and possible modifications of the basic social force model are necessary in order to achieve an efficient autonomous control strategy for AGVs in a generic setting.

3 Practical Challenges to Autonomous Movement Control

In order to better understand the dynamics of AGVs subject to a social force based autonomous control, different scenarios have been studied including such with either exclusively static or dynamic obstacles (i.e., only one AGV in a geometry with prescribed boundaries or several AGVs moving freely without external boundaries, respectively) as well as settings with fixed boundaries and a set of source and sink locations. In the following, some exemplary results taken from a more detailed simulation study [24] are discussed that highlight relevant points where clarifications or modifications of the basic social force model for pedestrian dynamics are necessary to achieve an industrially applicable solution for an autonomous movement control of AGVs. In order to keep the number of parameters in all simulations as low as possible, the following simplifying assumptions are made: First, the area usable for transportation processes has a flat profile. Second, all AGVs have the same physical properties, particularly the same size, mass and maximum acceleration. Third, the dynamics is considered as an ideal motion without friction losses. Finally, all accelerations and decelerations take place instantaneously without any reaction times.

3.1 Interaction Potentials and Avoidance of Collisions

According to Eq. (7) the total force "acting on" an individual AGV can be additively decomposed into different components. Due to this superposition of possibly conflicting force components, in a prescribed geometry collisions between individual vehicles can hardly be avoided completely. For example, there can be situations where some AGVs come very close to a single vehicle that is forced to move towards a wall. In the extreme case, such behavior can lead to a collision (if the resulting repulsive forces due to the static or the dynamic obstacles are different) either between different vehicles or with the wall.

One possibility to avoid this problem is using a sophisticated choice for the repulsive interaction potentials. For example, using power-laws or similar functions with a cutoff corresponding to a critical minimum distance in the direction of motion could be a feasible solution provided that a vehicle comes to rest as soon as its distance to the obstacle falls below this threshold. In the present study, situations leading to possible collisions are practically avoided by determining each force component separately to ensure that the AGVs do not get into physical contact with each other or with static obstacles.

A related problem is integrating information on both the distances to all relevant static as well as dynamic obstacles and their instantaneous velocities in the equations of motion, which is necessary in order to avoid abrupt breaking that is hardly possible in realistic scenarios. Here, the corresponding challenge is not treated explicitly, since the main goal is to achieve a minimization of the total transportation time and a reduction of the times during which conflicts between individual vehicles exist. Practically, everytime an AGV approaches the vicinity of an obstacle (e.g., a wall or another vehicle) it brakes and steers to the right. This simple rule mimics the behavior of pedestrians who often have a certain tendency of sidestepping by moving to a predetermined side. This preference can eventually be considered as originating from certain cultural conventions.

3.2 Avoidance of Deadlocks and Related Problems

Especially in case of a high vehicle density, the combined procedure of braking and eluding as described above can trigger the emergence of a deadlock situation. Specifically, due to the superposition of different mutually independent force components, units can reach a steady state or livelock (i.e., a situation where individual units alternate between different states in a periodic way) where no effective further motion is possible (see Fig. 1). Therefore, the affected units need to be released after undershooting a certain minimum speed (except for loading and unloading processes at the respective sources and sinks) to enable further movement. As a possible solution, deadlocks or livelocks can be detected by a simple logic comparing the instantaneous and preceding positions of each vehicle. If such a situation occurs, the unit has to be forced to calculate a new route, e.g., by making use of a prescribed set of checkpoints (see below).

The two-dimensional autonomous movement of an AGV can only be realized with a detailed knowledge of the available space. Figure 2 presents an example of a situation where a vehicle is forced by an oncoming unit to give way such that it fails to pass a static obstacle at the correct side. In order to avoid such problems and thus ensure an efficient operation, one possible solution is to specify distinct

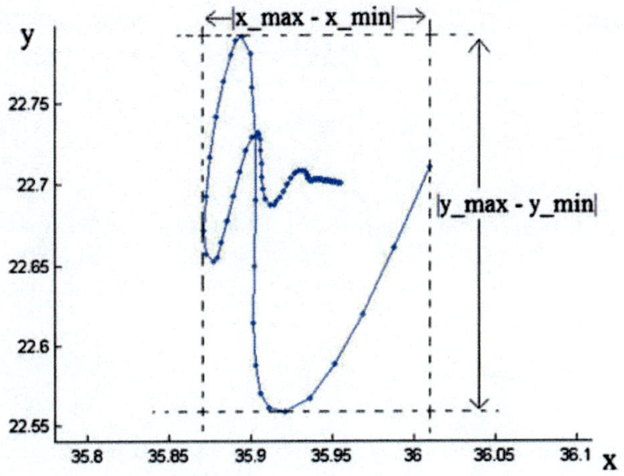

Fig. 1 Trajectory of an AGV that approaches a deadlock situation. The individual *dots* represent the respective positions of the vehicle at equally spaced sampling times

checkpoints that need to be passed on the way between the unit's origin and final destination. According to these prescribed checkpoints, a new route can be interactively recalculated onboard in case of an arising deadlock or livelock. For this calculation, it is necessary that each unit has access to full information on the positions of all static obstacles.

Another possible example for the emergence of deadlock situations are frequently used transportation relations where multiple vehicles have the same destination. For example, one could think of a machine which has to be continuously supplied with raw materials by AGVs (sink), or continuously produces semi-finished goods that need to be transported by AGVs to the next processing stage (source). If there is (intermittently or permanently) more transportation capacity (i.e., more AGVs) concentrated in the ultimate vicinity of this source or sink than necessary to serve the actual demand (which is determined by the production capacity and the time necessary for loading or unloading the transportation units) this can result in the mutual blocking of individual vehicles (see Fig. 3). Note that this is mainly a problem of properly designing the surroundings of sources and sinks in terms of available space in combination with a feasible job scheduling. However, given that a corresponding problem actually occurs due to whatever reasons, there are different possibilities to circumvent the resulting conflicts and avoid a deadlock. One simple approach is defining a certain area in the direct vicinity of the respective source or sink which all AGVs can only access after receiving an individual permission. In this case, the units have to communicate to receive this permission. This can be realized either centrally (by communication between the AGVs and some control entity representing the destination) or decentrally (by pairwise communication between the concerned AGVs). In the latter case, one could for example adopt the social goods concept [25, 26] to enforce individual AGVs to give way to others if they have a transportation job with lower priority (e.g., a later due date). A further possible improvement is the separate definition of incoming and outgoing routes by disjoint sets of checkpoints. As a

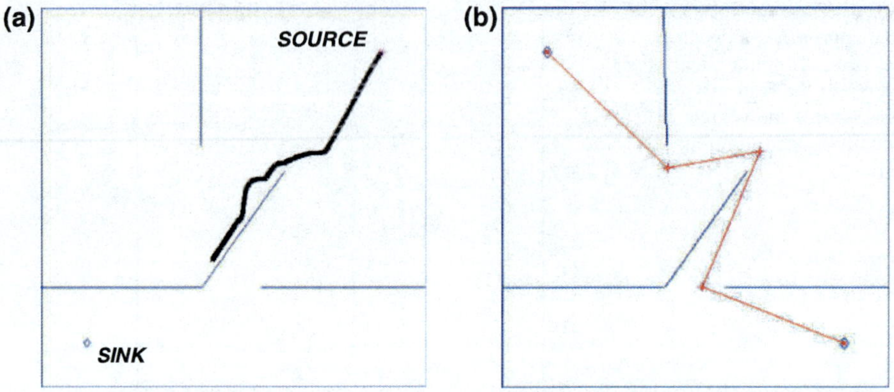

Fig. 2 **a** Trajectory of an AGV that is disturbed by an oncoming unit. Due to the policy of eluding to the right, the vehicle fails to pass the static obstacle (wall) at the correct side and, hence, to be able to reach its destination (sink). **b** Possible set of checkpoints to prescribe the desired path between origin and final destination that helps avoiding a possible deadlock situation at the inclined obstacle

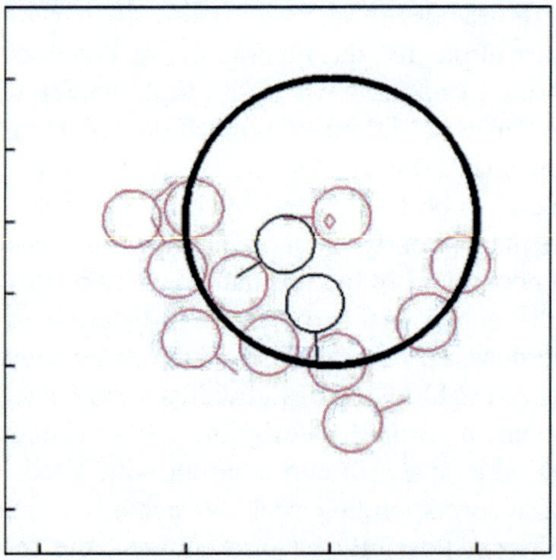

Fig. 3 Schematic illustration of a possible blocking situation at a sink (*small diamond*). For simplicity, all AGVs are assumed to have disk-like shape (*lines* indicate the instantaneous direction of motion). Some AGVs (*black*) that have already fulfilled their loading or unloading task are not able to leave the source, because the other AGVs (*magenta*) still want to reach their destination. The *thick black circle* determining a designated 'waiting area' avoids the emergence of the deadlock directly at the sink. In addition, splitting the vicinity of the sink (e.g., into two semi-circles) allows defining designated zones for entering and exiting vehicles (e.g., by means of distinct checkpoints). A combination of both mechanisms can help to widely reduce the risk of emerging deadlocks even in overcrowded situations

final (although typically less severe) problem, one has to examine the behavior of subsequent AGVs sharing the same route. Specifically, if two successive vehicles have the same trajectory, the repulsive force from the former would decelerate the latter. In order to avoid unnecessary braking and acceleration demands, adaptive methods should be used for adjusting the velocity of the second AGV according to the velocity and direction of motion of the first one.

4 Summary

This work presented a first conceptual study on the applicability of pedestrian dynamics based models for establishing an autonomous movement control of AGVs in intralogistics or container terminals. In particular, the case of free motion in a two-dimensional physical space has been considered which is more frequent and more flexibly applicable than the alternative of track-based vehicles.

The empirical, partially simulation-based considerations described in this contribution indicate that in combination with additional procedures ensuring the avoidance of collisions, deadlocks, and livelock situations, the social force concept provides a prospective foundation for the designated purpose. In particular, the proposed approach is applicable in the case of moderate to high vehicle densities [24]. However, without the discussed modifications the basic social force model is not directly applicable. Potential problems especially emerge in areas with particularly high vehicle density and/or few available space, such as bottlenecks in the given spatial geometry of the working area or around frequently visited sources and sinks. Most of these problems are practically avoided by introducing checkpoints at bottlenecks or diverges of different possible routes, as well as by establishing local permission strategies based on individual priorities. While the latter can be practically realized by means of a centralized control unit within a restricted spatial domain or sophisticated self-organisation concepts [25, 26], the major part of the transportation process is still subject to a fully autonomous movement control that is realized according to an on-line evaluation of the positions of static and dynamic obstacles.

In summary, the proposed pedestrian dynamics-based approach appears suitable and interesting for potential industrial applications. However, further research is necessary to fully explore its corresponding potentials and limitations in more detail.

Acknowledgments This work has been financially supported by the German Research Foundation (DFG project no. He 2789/8-1,8-2) and the Leibniz Society (project ECONS). Inspiring discussions with Stefan Lämmer and Dirk Helbing are gratefully acknowledged.

References

1. Hülsmann, M., Windt, K. (eds.): Understanding Autonomous Cooperation and Control in Logistics. Springer, Berlin (2010)
2. Hülsmann, M., Scholz-Reiter, B., Windt, K. (eds.): Autonomous Cooperation and Control in Logistics. Springer, Berlin (2011)
3. Zöbel, D.: The Deadlock problem: a classifying bibliography. ACM SIGOPS Operating Syst. Rev. **17**, 6–15 (1983)
4. Silberschatz, A., Galvin, P.B., Gagne, G.: Operating System Concepts. Wiley, Hoboken (2009)
5. Möhring, R.H., Köhler, E., Gawrilow, E., Stenzel, B.: Conflict-free real-time AGV routing. In: Operations Research Proceedings 2004, Part 1, pp. 18–24. Springer, Heidelberg (2005)
6. Hartwig, J.: Modellierung und Steuerung von Systemen kooperierender Automated Guided Vehicles. Diploma thesis, Dresden University of Technology (2006) (in German)
7. Berman, S., Edan, Y.: Decentralized autonomous AGV system for material handling. Int. J. Prod. Res. **40**, 3995–4006 (2002)
8. Srivastava, S.C., Choudhary, A.K., Kumar, S., Tiwari, M.K.: Development of an intelligent agent-based AGV controller for a flexible manufacturing system. Int. J. Adv. Manuf. Technol. **36**, 780–797 (2008)
9. Ferber, J.: Multi-Agent Systems: An Introduction to Distributed Artificial Intelligence. Addison Wesley, Harlow (1999)
10. Blue, V.J., Adler, J.L.: Emergent fundamental pedestrian flows from cellular automaton microsimulation. Transp. Res. Rec. **1644**, 29–36 (1998)
11. Muramatsu, M., Nagatani, T.: Jamming transition in two-dimensional pedestrian traffic. Physica A **275**, 281–291 (2000)
12. Burstedde, C., Klauck, K., Schadschneider, A., Zittartz, J.: Simulation of pedestrian dynamics using a two-dimensional cellular automaton. Physica A **295**, 507–525 (2001)
13. Okazaki, S.: A study of pedestrian movement in architectural space. Part 1: Pedestrian movement by the application of magnetic models. Trans. of AIJ **283**, 111–117 (1979) (in Japanese)
14. Helbing, D., Molnár, P.: The social force model for pedestrian dynamics. Phys. Rev. E **51**, 4282–4286 (1995)
15. Yu, W.J., Chen, R., Dong, L.Y., Dai, S.Q.: Centrifugal force model for pedestrian dynamics. Phys. Rev. E **72**, 026112 (2005)
16. Parisi, D.R., Dorso, C.O.: Microscopic dynamics of pedestrian evacuation. Physica A **354**, 606–618 (2005)
17. Parisi, D.R., Gilman, M., Moldovan, H.: A modification of the social force model can reproduce experimental data of pedestrian flows in normal conditions. Physica A **388**, 3600–3608 (2009)
18. PTV AG. http://www.ptv.de/software/verkehrsplanung-verkehrstechnik/software-und-system-solutions/viswalk/
19. Lewin, K.: The Conceptual representation and the measurement of psychological forces. Duke University Press, Durham (1938)
20. Lewin, K.: Defining the "Field at a Given Time". Psychol. Rev. **50**, 292–310 (1943)
21. Hoogendoorn, S., Daamen, W.: Pedestrian behavior at bottlenecks. Transp. Sci. **39**, 147–159 (2005)
22. Moussa, M., Helbing, D., Garnier, S., Johansson, A., Combe, M., Theraulaz, G.: Experimental study of the behavioural mechanisms underlying self-organization in human crowds. Proc. R. Soc. B **276**, 2755–2762 (2009)
23. Katz, Y., Ioannou, C.C., Tunstrom, K., Huepe, C., Couzin, I.D.: Inferring the structure and dynamics of interactions in schooling fish. Proc. Natl. Acad. Sci. USA **108**, 18720–18725 (2011)

24. Bähr, M.: Anwendbarkeit eines fußgängerdynamischen Modells für die autonome Steuerung fahrerloser Transportfahrzeuge. Technical report, Dresden University of Technology (unpublished, available from the authors upon request) (in German)
25. Seidel, T.: Modellierung von Produktionsnetzwerken aus der Perspektive interagierender Transportprozesse. Ph.D. thesis, Dresden University of Technology (2007) (in German)
26. Seidel, T., Hartwig, J., Sanders, R.L., Helbing, D.: An agent-based approach to self-organized production. In: Blum, C., Merkle, D. (eds.) Swarm Intelligence: Introduction and Applications, pp. 219–252. Springer, Berlin (2008)

Using a Clustering Approach with Evolutionary Optimized Attribute Weights to Form Product Families for Production Leveling

Fabian Bohnen, Marco Stolpe, Jochen Deuse and Katharina Morik

Abstract Production leveling aims at balancing production volume as well as production mix. Conventional leveling approaches require limited product diversity and stable, predictable customer demands. They are well-suited only for large scale production. This paper presents a methodology that enables the leveling of low volume and high mix production. It is based on two fundamental steps. In the first step, which is focused on in this paper, product types are grouped into families according to their manufacturing similarity. In the second step, a family-oriented leveling pattern is generated. This paper presents an innovative clustering approach for product family formation regarding leveling. It employs evolutionary strategies to optimize the weights of the attributes which are used for clustering according to their impact on the grouping result. The paper refers to an industrial application and also shows how product families can be utilized for leveling.

Keywords Production leveling · Clustering · Evolutionary strategies

F. Bohnen (✉) · J. Deuse
TU Dortmund University, Chair of Industrial Engineering, Dortmund, Germany
e-mail: fabian.bohnen@tu-dortmund.de

J. Deuse
e-mail: jochen.deuse@tu-dortmund.de

M. Stolpe · K. Morik
TU Dortmund University, Chair of Artificial Intelligence, Dortmund, Germany
e-mail: marco.stolpe@tu-dortmund.de

K. Morik
e-mail: katharina.morik@tu-dortmund.de

K. Windt (ed.), *Robust Manufacturing Control*, Lecture Notes in Production Engineering,
DOI: 10.1007/978-3-642-30749-2_14,

1 Introduction

Production leveling also referred to as production smoothing or heijunka, depicts a key element of the Toyota Production System and lean production, respectively [1]. The objective of production leveling is to balance production volume as well as production mix by decoupling production orders and customer demand [1, 2]. Thus, unevenness in form of variation in the production schedule is reduced. This comes along with a decrease in overburden and waste [1]. Leveling distributes production volume and mix to equable short periods. The sequence of these periods describes a kind of manufacturing frequency. According to this leveling pattern, every product type is manufactured within a periodic interval, which is reflected in the so called EPEI-value (Every Part Every Interval) [1, 3].

Conventional leveling approaches aim at manufacturing every product type within a periodic interval. This typically requires a limited product diversity combined with stable and predictable customer demands [4, 5]. Hence, the application of conventional approaches is limited to large scale production [6]. For leveling of low volume and high mix production, the Chair of Industrial Engineering at TU Dortmund University developed an adapted methodology, which is presented here.

The paper is organized as follows: After a brief state of the art review in Sect. 2, an overview of the methodology for leveling of low volume and high mix production is given in Sect. 3. The methodology is based on the formation of product families for leveling and the creation of a family-oriented leveling pattern. The problem of forming product families for leveling is described formally in Sect. 4. Afterwards, Sect. 5 presents an innovative clustering approach that has been developed at the Chair of Artificial Intelligence at TU Dortmund University. The multi-objective clustering uses evolutionary strategies in order to form product families for leveling. This approach outperforms conventional clustering techniques as is shown referring to a real life application in Sect. 6. Finally, a conclusion is given in Sect. 7.

2 Literature Review

Most of the literature dealing with production leveling focuses on large scale production, especially in form of mixed-model assembly lines in the automotive industry. Without reference to the application context, literature dealing with production leveling can be divided into two classes. The first class concentrates on procedure models, i.e. systematic procedures for leveling. The second class describes leveling as an optimization problem in context of production sequencing. Both, procedure models and optimization models for leveling are handled in the following subsections.

2.1 Procedure Models for Production Leveling

Concerning large scale production, several procedure models for leveling can be found in the literature. Most of them describe systematic procedures without focusing on the application in specific industry sectors. [7–9], for example, present procedure models which describe systematic procedures for leveling in a quite general way. These procedures are based on two fundamental steps. The first step aims at distributing production volume to equal-sized planning increments, such as, e.g., a day or a shift. The production mix is harmonized by determining a repetitive production sequence in the second step. In contrast, [3] and [10–13] also distribute production volume to planning increments in the first step, but use a so called pitch increment for sequencing in the second step. This pitch increment represents the product of takt time and packing unit.

While the procedure models just quoted focus on large scale production in general, [14] develops a specific procedure for leveling of electronic control units manufacturing in the automotive industry. Another procedure model is presented in [15] concentrating on leveling of low volume and high mix production. The latter procedure model is enhanced in this paper.

2.2 Optimization Models for Production Leveling

The majority of the optimization models for production leveling focus on synchronized mixed-model assembly lines. In this context, the objective of production leveling, i.e. harmonization of production volume and mix, is transferred to an analytical level considering the so called production smoothing problem (PSP). According to the PSP, objectives of leveling are mapped on a sequencing problem. In this context, approaches dealing with the description and solution of the PSP are also referred to as level scheduling approaches. In general the PSP represents the problem of finding a production sequence that minimizes variation concerning one or more objectives.

Literature surveys on level scheduling approaches and related work are given in [16–18]. According to these surveys, literature dealing with the PSP for mixed-model assembly lines can be classified concerning the considered production levels and their objective functions. A large amount of approaches only concentrates on one production level, generally the final assembly [19]. There are level scheduling approaches aiming at minimizing variation in production rates [20], material consumption rates [7], or workload at the stations of the line [21]. Additionally, some approaches aim at minimizing more than one of these objectives, simultaneously [22]. Furthermore, there are level scheduling approaches considering more than one production level. Here, the majority of the approaches in literature focuses on the objective of minimizing variation in material consumption rates [23]. Additionally, some approaches combine this objective with those mentioned before [19].

The majority of level scheduling approaches in literature focuses on mixed-model assembly lines. Nevertheless, there are few papers dealing with level scheduling for flow shop environments [24–26] or the single machine case, respectively [27–29]. Referring to the application context, it is important to note that these approaches also focus on large scale production. In contrast, [15] presents an approach for leveling of low volume and high mix production, which is also based on a single machine case. This approach does not contain an optimization model comparable with the ones quoted before. Nevertheless, it describes an analytical model which aims at creating a leveling pattern based on product families with minimal overall changeover time.

3 Leveling of Low Volume and High Mix Production

The approach that enables leveling in cases of high product diversity is presented in [15]. In its first step, product types are grouped into a manageable number of product families according to their manufacturing similarity. Hence, product types of the same family can be manufactured in an almost arbitrary sequence without or with minimal losses caused by changeover. Manufacturing similarity is reflected in different grouping criteria. These criteria represent attributes that describe characteristics of product types, which are relevant for leveling. Adequate grouping criteria for the formation of product families are operation sequences, required equipment and staff, process times, setup times for changeover, and share of identical components, parts, or raw material [30]. The step of forming product families will be described in detail in the next two sections. Before, an overview on the step of creating a family-based leveling pattern is given.

Conventional leveling approaches aim at creating a leveling pattern that is characterized by the EPEI-value (e.g. a day or a shift), the product type sequence, and the production volumes for each product type. Following such a pattern, every product type is manufactured within a periodic interval. In contrast, using a product family-oriented leveling pattern, not every product type but every product family is manufactured within a repetitive period. The latter is reflected in the so called EFEI-value (Every Family Every Interval). Creation of such a leveling pattern starts with a family-oriented ABC/XYZ-analysis, i.e. a Pareto analysis considering production volume combined with an analysis of variation in customer demand. Based on this analysis, leveling families, i.e. families, which have to be scheduled cyclically in the leveling pattern, are chosen. The remaining families are considered in an aggregated form, e.g. in form of a capacity time slot for strangers.

After selecting the leveling families, a family-oriented manufacturing sequence is determined that causes minimal overall changeover time. In the next step an EFEI-value for the chosen leveling families is calculated considering production capacities, especially available and required time for changeover. Based on this EFEI-value, capacity slots for every leveling family are determined. The sequence and the length of these slots describe the leveling pattern. More details on the step of creating a family-oriented leveling pattern are given in [15] and [31].

4 Problem Description (Family Formation Problem)

Formally, the problem of forming product families for leveling is to group a set $X = \{X_1, X_2,\ldots,X_n\}$ of n product types with p attributes representing the selected grouping criteria into the most adequate partition $C^{(k)} = \{C_1, C_2,\ldots, C_k\}$, $k \le n$. Here, k is the partition size, i.e. the number of formed product families. In Statistics and Data Mining, this problem is also known as the problem of cluster analysis [32]. In general, Data Mining techniques aim at discovering interesting patterns in large data sets [33]. In this context several algorithms have been developed that try to solve the problem of cluster analysis based on formal objectives, like minimizing the dissimilarity between objects in each group (cluster). In addition to these formal criteria, application specific objectives may be defined, that can be either incorporated directly into the optimization problem of a clustering algorithm or applied ex post for evaluating the quality of a resulting partition.

For evaluating the quality of a partition in the application context of low volume and high mix production, [34] defines a desirability index W according to [35] as the geometric mean of four objective functions.

$$W : \{w_1, w_2, \ldots, w_4\} \rightarrow [0,1], \quad w_i \in [0,1], \quad W = \sqrt[4]{\prod_{i=1}^{4} w_i} \tag{1}$$

The index is used successfully in [34] to evaluate the resulting partitions of different clustering algorithms, over a range of different partition sizes. Each objective function w_i delivers a value in the interval [0,1] where 0 denotes the worst and 1 the best achievement of the respective objective [35]. Since the clustering approach introduced in Sect. 5 is partly based on these objective functions, they are described in more detail in the following.

4.1 Homogeneity

Given a function $d : \mathbb{R}^p \times \mathbb{R}^p \rightarrow \mathbb{R}$ which measures the dissimilarity between two objects with p attributes, a general objective often stated is that objects belonging to the same group should be more similar to each other than objects of different groups. For example, the k-Means clustering algorithm [36] directly optimizes this criterion, which can be formally defined as objective (2) [32, 34].

$$w_1(C^{(k)}) = 1 - \frac{1}{k}\sum_{q=1}^{k} \frac{\sum_{X_i,X_j \in C_q} d(X_i,X_j)}{\left(|C_q| - 1\right)^2} \tag{2}$$

For dissimilarities $d(X_i,X_j)$ (normalized to the interval [0,1]) between objects X_i and X_j in each group C_q function w_1 represents the objective of minimizing the normalized total sum over all distances. In this context, $|C_q|$ represents the number of products types in group C_q.

4.2 Limited Partition Size

The exact number of product families usually is not known ex ante. Nevertheless, the partition size can be limited to a desired interval $[k_{min}, k_{max}]$ for a particular application using expert knowledge. Hence, objective function w_2 is defined as [34]:

$$w_2(C^{(k)}) = \begin{cases} 0, & k = 1 \\ \left(\frac{k-1}{k_{\min}-2}\right)^2, & k \in \{2, 3, \ldots, k_{\min}-1\} \\ 1, & k \in \{k_{\min}, k_{\min}+1, \ldots, k_{\max}\} \\ \left(\frac{100-k}{100-(k+1)}\right)^3, & k \in \{k_{\max}+1, \ldots, 99\} \\ 0, & k \in N : k > 99 \end{cases} \tag{3}$$

4.3 Balanced Family Sizes

The formed families are desired to be roughly equal-sized. Especially very small and very large product families have to be avoided. Otherwise, leveling pattern creation becomes a highly challenging task when very small product families, consisting of one product type with high variation in customer demand in the worst case, have to be integrated into the leveling pattern. Formally, the stated requirement can be expressed as objective (4).

$$w_3(C^{(k)}) = 1 - \frac{n_w - \min C^{(k)}(n_w)}{\max C^{(k)}(n_w) - \min C^{(k)}(n_w)} \tag{4}$$

In (4) n_w represents the total number of distances between the pairs of objects within each group over all groups in partition $C^{(k)}$ and min $C^{(k)}(n_w)$ and max $C^{(k)}(n_w)$ are the minimal and maximal number of distances over all possible partitions with k product families and n product types [34]. For a fixed k, n_w is smallest when all objects are equally spread over the k groups. The value n_w is largest when n-k+1 objects are in a single group and the remaining k-1 groups only contain a single object.

4.4 Reasonable Large Families

Very small product families with equal to or less than an amount of a product types allocated to the respective family have to be avoided. Since this objective is not fully covered by objective function w_3, in [34] a further objective (5) that punishes too small product families is introduced.

$$w_4(C^{(k)}) = 2^{-a} \tag{5}$$

In [34], the desirability index W serves to evaluate the quality of resulting partitions ex post, but is not used for optimization. In the same way, the results of the new clustering approach are evaluated by the geometric mean of all the criteria. The new clustering itself, however, moves beyond the desirability index. It uses some of the aforementioned criteria to find an optimal weighting of object attributes such that the quality of the resulting clustering is improved. It follows the multi-objective optimization approach to clustering as has been proposed in [36, 37].

5 Clustering with Evolutionary Optimized Attribute Weights

To solve the product family formation problem of Sect. 4 different approaches can be found in literature. [38] for example differs between part coding systems, manufacturing sequence-oriented methods (like production flow analysis), cluster analysis, and neural networks. Among these approaches cluster analysis represents the most flexible and therefore most reasonable method [39]. While [15] and [31] use conventional cluster analysis to form product families for leveling, an innovative clustering approach is presented subsequently that solves the product family formation problem.

It is well known that the partitions achieved for the same data set with the same clustering algorithm can vary largely with different weightings of the attributes [32, 34]. Usually, a natural weighting is already given, for example, when attribute values have a different scale or a different variance. If an equal weighting is required, the attribute values need to be normalized. In both cases, there is no guarantee that the weighting will lead to clustering results, which fit the particular application needs. Approaches like [40], which optimize attribute weights for clustering, usually rely on formal criteria like (2) and user-specified constraints describing which objects belong to the same cluster, at least for parts of a given data set. In a similar way, [41] successfully optimizes attribute weights while only probabilistic information about the group membership is given. [37] states the problem of attribute selection for clustering in terms of a multi-objective optimization problem, based on three different formal objectives. In the following, their set of objectives is adapted to the correct formation of product families for leveling. Moreover, the approach is modified to deliver attribute weights instead of a binary selection of attributes.

First of all, a dissimilarity measure is needed which respects the weighting of attributes. In the following, the assumed measure is the weighted Euclidean distance

$$d(X_i, X_j) = \sqrt{\sum_{l=1}^{p} a_l (X_{il} - X_{jl})^2} \tag{6}$$

where X_{il} and X_{jl} represent the values of attribute l for objects i and j respectively and vector $a = (a_1, a_2, \ldots, a_p)$ consists of the different weighting factors $a_i \in [0, 1]$ for each attribute.

The objective is to find a weighting a of the attributes such that the criteria w_1, w_2, w_3 and w_4 are optimized simultaneously. One possible approach would be to merge all four criteria into a single objective value, like the desirability index W, and state a single-objective optimization problem. However, since the relative importance of the objectives is unknown, good solutions cannot be guaranteed. In comparison, multi-objective optimization strives to find all solutions that are optimal.

A solution A dominates another solution B if all objective values of A are greater than or equal to the objective values of B and at least one objective value of A is truly greater. A solution A is non-dominated by a set of solutions if none of these solutions dominates A. A solution A is called *Pareto-optimal* if it is non-dominated by the whole solution space. The set of all non-dominated solutions is called a Pareto set. The solutions are optimal in the sense that no objective can be improved without worsening at least one of the other objectives.

A precondition for the successful use of multi-objective optimization is that at least two of the objectives w_i have to contradict each other. Otherwise it might happen that the Pareto set collapses to a single solution. A striking result from [37] is that in comparison to the selection of attributes for supervised learning, where the number of attributes should be minimized, the number of attributes needs to be maximized for clustering. The reason is that objective (2) decreases with a smaller number of attributes, since the pairwise distances between objects become smaller. The same is true for the weighting of attributes, since smaller weight values imply smaller differences between attribute values. In addition to the objectives w_1, w_2, w_3, and w_4, it is therefore necessary to maximize the components of weight vector a, leading to the additional objective (7).

$$w_5(C^{(k)}) = 1 - \sqrt{\sum_{l=1}^{p} \left(\frac{1 - a_l}{p} \right)^2} \tag{7}$$

(7) aims at minimizing the normalized Euclidean distance between the vector of maximum weights and the current attribute weights.

Another important observation from [37] is that objective (2) not only depends on the number of attributes p, but also increases with an increasing partition size k. Although the partition size k is already limited by the objective function w_2, following [37] the first objective w_1 should be replaced by the (in this case normalized) Davies Bouldin index [42]

$$w_1(C^{(k)}) = 1 - \frac{\frac{1}{k} \sum_{q=1}^{k} \max_{q,r \neq q} \left\{ \frac{s_q + s_r}{d(c_q, c_r)} \right\}}{p \cdot k}, \tag{8}$$

$$s_q = \frac{1}{|C_q|}\sum_{x_i \in C_q} d(x_i, c_q), \quad s_r = \frac{1}{|C_r|}\sum_{x_i \in C_r} d(x_i, c_r) \tag{9}$$

where s_q and s_r are the average within cluster distances for cluster C_q and C_r respectively and c_q and c_r are the corresponding cluster centroids. The index measures the relative separation of the two clusters which are worst separated and therefore is less sensitive to the number of clusters and attributes (for more details, see [42]).

Once all objectives are defined, an evolutionary strategy like the Non-dominated Sorting Genetic Algorithm II (NSGA-II) [43] can be used for optimizing the attribute weights and finding a set of Pareto-optimal solutions. First, the algorithm creates an initial population P_0 of N random weight vectors. For each weight vector, a clustering algorithm partitions the data set according to the weighted Euclidean distance. For each resulting partition, the objective functions $w_1, \ldots, w_4, w_5$ are evaluated. All solutions then receive a fitness value according to their non-domination. A population Q_0 of N children is generated by usual evolutionary operators: tournament selection, recombination of the weight vectors and a Gaussian mutation of the attribute weights. In subsequent steps t, the parent and children populations P_t and Q_t are combined to a single population R_t of size 2 N. Iteratively, the whole population is partitioned into Pareto sets of different levels according to non-domination, such that the highest level contains all best solutions, the next the second best, and so on. By decreasing level, these Pareto sets are added to the new population P_{t+1} until the total number of individuals N is reached. If not all points from the last added Pareto set fit into the population, those non-dominated points are preferred for inclusion that come from lesser crowded regions of the solution space (for the according distance measure, see [43]). An offspring generation Q_{t+1} is generated from P_{t+1} and the algorithm repeats until the maximum number of user-specified generations is reached or no better solutions can be found.

6 Industrial Application

The objective functions $w_1, \ldots, w_5$ presented above have been implemented as a performance measure for clustering and combined with an already existing operator for weighting attributes by NSGA-II in RapidMiner 5.1, which is a commonly used software tool for Data Mining and Machine Learning [44]. The approach was successfully validated using real data from an industrial application. Product families were formed to level an assembly line in the machine building industry with characteristics of low volume and high mix production. A set of about 250 product types had to be grouped into product families according to their share of identical components. These attributes were chosen for grouping because in the considered assembly system changeover times were primarily caused by

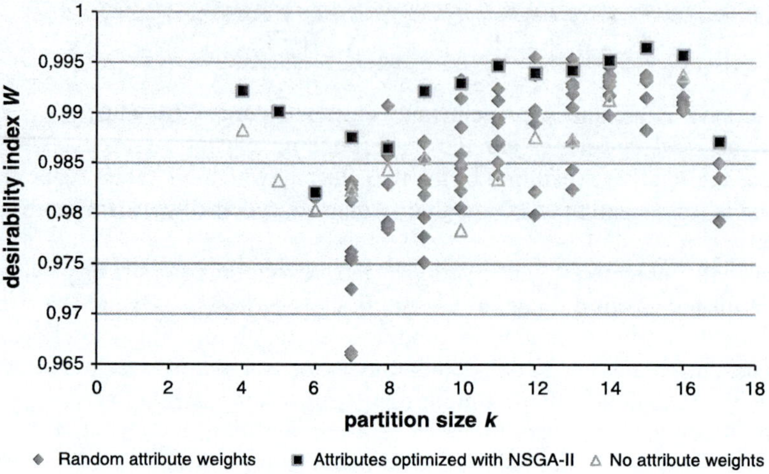

Fig. 1 Comparison of clustering with evolutionary optimized attribute weights, random weights and no weighting at all

material changeover. Hence, only minor changeover times were expected between product types that share a large amount of components. In this case, 700 binary attributes had to be considered. Regarding constraints resulting from the application context and additional expert knowledge, a minimal partition size of 4 and a maximal partition size of 16 product families was determined. For each k, a Pareto set of optimal solutions was determined by running the NSGA-II algorithm over 60 generations with population size 25, uniform crossover (recombination) with probability 0.75 and a mutation variance of 1.0. The k-Means algorithm was used for clustering in conjunction with NSGA-II, because k-Means directly tries to maximize the homogeneity within the formed groups for a given number of groups k. As mentioned before, k usually is not exactly known ex-ante. Due to that, the attribute optimization procedure is iterated over different k. More details about the k-Means can be found in [45].

Figure 1 compares the evolutionary weighting of attributes to 10 different random weightings for each k and to running k-Means with no weighting at all. Here, for each k only a single solution was chosen from each Pareto set, according to the highest geometric mean of the first four objectives, the desirability index (w_5 was not considered, because it was only introduced to avoid trivial solutions for the evolutionary weighting, see Sect. 5). As is clearly seen in most cases, the evolutionary weighting leads to partitions, which evaluate to the highest index value according to the objectives stated in Sects. 4 and 5. Since evolutionary algorithms find locally (not globally) optimal solutions, it can happen that random weighting finds a better solution. In any case, clustering without any attribute weighting always leads to worse solutions in comparison to the evolutionary weighting.

Table 1 Non-dominated solutions for clustering with $k = 15$

Solution	w_1	w_2	w_3	w_4	W	w_5
1	0,99840	1,00000	0,97210	1,00000	0,99255	0,43671
2	0,99860	1,00000	0,95509	1,00000	0,98823	0,42193
3	0,99862	1,00000	0,94943	0,50000	0,82977	0,44221
4	0,99845	1,00000	0,95825	1,00000	0,98901	0,44553
5	0,99838	1,00000	0,97933	1,00000	0,99439	0,43812
6	0,99855	1,00000	0,96540	0,50000	0,83322	0,43014
7	0,99838	1,00000	0,97355	1,00000	0,99292	0,43477
8	0,99846	1,00000	0,97842	1,00000	0,99418	0,42340
9	0,99855	1,00000	0,97006	1,00000	0,99207	0,44199
10	**0,99837**	**1,00000**	**0,98802**	**1,00000**	**0,99659**	**0,45467**
11	0,99850	1,00000	0,97293	1,00000	0,99279	0,42059
12	0,99848	1,00000	0,97151	1,00000	0,99242	0,43420
13	0,99842	1,00000	0,98095	1,00000	0,99481	0,41942

For the comparison, only a single solution was chosen from each Pareto set. However, the evolutionary multi-objective optimization finds many Pareto-optimal solutions instead of only a single one. This is especially important for unsupervised methods like clustering, where different partitions can have a similar quality and no unique solution exists.

For example, for the most promising value $k = 15$ (see Fig. 1), 13 different solutions were found. Table 1 shows the objective values and the geometric mean W over $w_1, \ldots, w_4$ for these solutions. The best solution according to the desirability index W (which is also plotted in Fig. 1) is marked in bold. A list of non-dominated solutions such as this gives the user insight into the structure of the solution space. For instance, it can be seen that a high value for the homogeneity sometimes comes at the expense of objective w_4, large enough family sizes. Moreover, in many cases the families are not as equally sized (objective w_3) as for the lower homogeneity values. Such insight would be missed by returning only a single solution. The evolutionary approach thus provides the user with more flexibility. Nevertheless, if this flexibility is not needed, a quite balanced single solution can also be found based on the geometric mean of the objective values.

Based on the formed product families, the family-oriented leveling pattern has been created. Therefore, 4 leveling families out of 15 product families were identified considering a family-oriented analysis on production volume and customer demand distribution. The other 11 families were classified as stranger families and included in the leveling pattern in an aggregated form. For the leveling families, a manufacturing sequence causing minimal overall changeover time was determined. Afterwards, an EFEI-value of 1 day was calculated taking into account required and available manufacturing capacities. Choosing an EFEI-value of 1 day means that the leveling pattern is repeated in the frequency of 1 day which corresponds to 2 shifts. Finally the size of the capacity slots was calculated. For a leveling period of 1 week, the resulting leveling pattern is shown in Fig. 2.

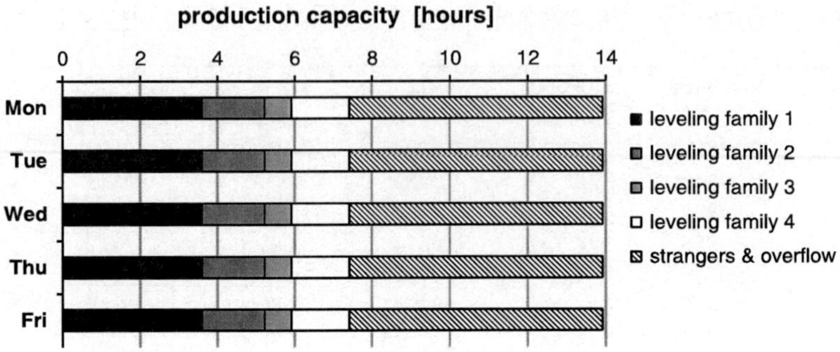

Fig. 2 Leveling pattern based on 4 leveling families

The pattern consists of 4 capacity slots for leveling families which correspond to the first shift and a capacity slot in the second shift which can be used for overflow and stranger production.

7 Conclusion

This paper presents a methodology for leveling of low volume and high mix production which utilizes an innovative clustering approach to form product families for leveling. It employs evolutionary strategies to optimize the weights of the grouping attributes according to their impact on the objective function. This clustering approach was tested with a data set from an industrial application. The experiments prove that the presented clustering approach outperforms conventional clustering techniques. To demonstrate how a family-oriented leveling pattern can be realized, leveling pattern creation is also described in theory with respect to the industrial application. Hence, this paper presents an advanced methodology that can be used to implement leveling in low volume and high mix production.

There still remains future research work to do. For example, the step of leveling pattern creation should consider level scheduling objectives and be lifted to a more analytical level. Additionally, leveling pattern creation should consider inventory levels and customer demand variation.

Acknowledgments Parts of the work on the proposed paper have been supported by Deutsche Forschungsgemeinschaft (DFG) within the Collaborative Research Centre SFB 876 "Providing Information by Resource-Constrained Analysis", project B3.

References

1. Liker, J.K.: The Toyota Way. McGraw-Hill, New York (2004)
2. Dennis, P.: Lean Production Simplified, 2nd edn. Productivity Press, New York (2007)
3. Rother, M., Harris, R.: Creating Continuous Flow. Lean Enterprise Institute, Brookline (2004)
4. Hüttmeir, A., de Treville, S., van Ackere, A., Monnier, L., Prenninger, J.: Trading off between heijunka and just-in-sequence. Int. J. Prod. Econ. **118**, 501–507 (2009)
5. de Smet, R., Gelders, L.: Using simulation to evaluate the introduction of a kanban subsystem within an MRP-controlled manufacturing environment. Int. J. Prod. Econ. **56–57**, 111–122 (1998)
6. Slomp, J., Bokhorst, J.A.C., Germs, R.A.: Lean production control system for high-variety/low-volume environments. Prod. Plann. Control **20**, 586–595 (2009)
7. Monden, Y.: Toyota Production System. Industrial Engineering and Management Press, Norcross (1983)
8. Shingo, S.: A Study of the Toyota Production System. Productivity Press, Cambridge (1989)
9. Takeda, H.: Das synchrone Produktionssystem, 4th edn. Redline Wirtschaft, Frankfurt/Main (2004)
10. Duggan, K.J.: Creating Mixed Model Value Streams. Productivity Press, New York (2002)
11. Erlach, K.: Wertstromdesign. Springer, Berlin (2007)
12. Rother, M., Shook, J.: Learning to See. Lean Enterprise Institute, Brookline (1999)
13. Smalley, A.: Creating level pull. Lean Enterprise Institute, Cambridge (2004)
14. Wuthnow, A.: Steuerung und Nivellierung von Wertströmen in der Automobilsteuergerätefertigung. Shaker, Aachen (2010)
15. Bohnen, F., Buhl, M., Deuse, J.: Systematic procedure for leveling of low volume and high mix production. In: 44th CIRP conference on manufacturing systems 2011
16. Boysen, N., Fliedner, M., Scholl, A.: Sequencing mixed-model assembly lines: Survey, classification and model critique. Eur. J. Oper. Res. **192**, 349–373 (2009)
17. Kubiak, W.: Minimizing variation of production rates in just-in-time systems. Eur. J. Oper. Res. **66**, 259–271 (1993)
18. Yavuz, M., Akcali, E.: Production smoothing in just-in-time manufacturing systems. Int. J. Prod. Res. **45**, 3579–3597 (2007)
19. Yavuz, M.: An iterated beam search algorithm for the multi-level production smoothing problem with workload smoothing goal. Int. J. Prod. Res. **48**, 6189–6202 (2010)
20. Miltenburg, J.: Level schedules for mixed-model assembly lines in just-in-time production systems. Manag. Sci. **35**, 192–207 (1989)
21. Merengo, C., Nava, F., Pozzetti, A.: Balancing and sequencing manual mixed-model assembly lines. Int. J. Prod. Res. **37**, 2835–2860 (1999)
22. Korkmazel, T., Meral, S.: Bicriteria sequencing methods for the mixed-model assembly line in just-in-time production systems. Eur. J. Oper. Res. **131**, 188–207 (2001)
23. Kubiak, W., Steiner, G., Yeomans, S.J.: Optimal level schedules for mixed-model, multi-level just-in-time assembly systems. Ann. Oper. Res. **69**, 241–259 (1997)
24. McMullen, P.R.: The permutation flow shop problem with just in time production considerations. Prod. Plann. Control **13**, 307–316 (2002)
25. Yavuz, M., Akcali, E., Tufekci, S.: A hybrid meta-heuristic for the batching problem in just-in-time flow shops. J. Math. Model. Algorithm **5**, 371–393 (2006)
26. Yavuz, M., Tüfekci, S.: Dynamic programming solution to the batching problem in just-in-time flow-shops. Comput. Ind. Eng. **51**, 416–432 (2006)
27. Kubiak, W., Yavuz, M.: Just-in-time smoothing through batching. Manuf. Serv. Oper. Manag. **10**, 506–518 (2008)
28. McMullen, P.R.: JIT mixed-model sequencing with batching and setup considerations via search heuristics. Int. J. Prod. Res. **48**, 6559–6582 (2010)

29. Yavuz, M., Tufekci, S.: Analysis and solution to the single-level batch production smoothing problem. Int. J. Prod. Res. **45**, 3893–3916 (2007)
30. Bohnen, F., Deuse, J.: Leveling of low volume and high mix production based on a group technology approach. In: Sihn, W., Kuhlang, P. (eds.) Proceedings of the 43rd CIRP International Conference on Manufacturing Systems, pp. 949–956 (2010)
31. Bohnen, F., Maschek, T., Deuse, J.: Leveling of low volume and high mix production based on a Group Technology approach. CIRP-JMST **4**, 247–251 (2011)
32. Hastie, T., Tibshirani, R., Friedman, J.: The Elements of Statistical Learning, 2nd edn. Springer, New York (2009)
33. Han, J., Kamber, M.: Data Mining—Concepts and Techniques. 2nd edn., Morgan Kaufmann, Los Altos (2006)
34. Weihs, C., Szepannek, G.: Distances in classification. In: Goebel, R., Siekmann, J., Wahlster, W. (eds.) ICDM 2009. LNCS, vol. 5633, pp. 1–12. Springer, Heidelberg (2009)
35. Harrington, E.C.: The desirability function. Ind. Qual. Control **21**, 494–498 (1965)
36. Morik, K., Kaspari, A., Wurst, M., Skirzynski, M.: Multi-objective frequent termset clustering. Knowl. Inf. Syst. **30**, 715–738 (2012)
37. Mierswa, I., Wurst, M.: Information preserving multi-objective feature selection for unsupervised learning. In: Proceedings of the 8th Annual Conference on Genetic and Evolutionary Computation (GECCO), pp. 1545–1552 (2006)
38. Deuse, J.: Fertigungsfamilienbildung mit feature-basierten Produktmodelldaten. Shaker, Aachen (1998)
39. Yin, Y., Kaku, I., Tang, J., Zhu, J.: Data Mining. Springer, London (2011)
40. Halkidi, M., Gunopulos, D., Vazirgiannis, M., Kumar, N., Domeniconi, C.: A Clustering Framework Based on Subjective and Objective Validity Criteria. ACM Trans Knowl Discov Data 1, pp. 18:1–18:25 (2008)
41. Stolpe, M., Morik, K.: Learning from label proportions by optimizing cluster model selection. In: Gunopulos, D., Hofmann, T., Malerba, D. et al. (eds.) ECML PKDD 2011, LNAI, vol. 6913, pp. 349–364. Springer, Berlin (2011)
42. Davies, D.D., Bouldin, D.W.: A Cluster separation measure. IEEE Trans. Pattern Anal. Mach. Intell. **1**, 224–227 (1979)
43. Deb, K., Pratap, A., Agarwal, S., Meyarivan, T.: A fast and elitist multi-objective genetic algorithm: NSGA-II. IEEE Trans. Evol. Comput. **6**, 182–197 (2002)
44. Mierswa, I., Wurst, M., Klinkenberg, R., Scholz, M., Euler, T.: Rapid prototyping for complex data mining tasks. In: Proceedings of the 12th ACM SIGKDD 2006
45. MacQueen, J.: Some methods for classification and analysis of multivariate observations. Symp. Math. stat. prob., 281–297 (1967)

Data Mining as Technique to Generate Planning Rules for Manufacturing Control in a Complex Production System

A Case Study from a Manufacturer of Aluminum Products

Christian Rainer

Abstract This paper presents a case study from a manufacturer of aluminum products characterized by a complex and flexible production system with a broad product variety. We examine the application of the data mining process for generating planning rules. The resulting planning rules can be implemented in a manufacturing execution system to support the decision process in decentralized manufacturing control. The aim is to discover patterns and drivers for high manufacturing lead time from ERP data in order to define planning rules with the objective to reduce lead time.

Keywords Data mining · Manufacturing control · Planning rules

1 Introduction

Rising product variety and complex material flows in the metal processing industry sector increase not only the need for highly skilled technologists but also the need for adaption of the production planning and control system. Steel, which constitutes around 95 % by volume of the total output of metals, and aluminum, which is second in volume to steel, are both mainly fabricated as rolled sheet and foils produced by hot and cold rolling of cast slabs followed by numerous further diverse annealing and finishing steps [1]. Manufacturing control in this process industry sector thus involves a variety of complex technological processes, which

C. Rainer (✉)
Department of Economics and Business Management, University of Leoben,
Franz-Josef-Straße 18, 8700 Leoben, Austria
e-mail: christian.rainer@wbw.unileoben.ac.at

K. Windt (ed.), *Robust Manufacturing Control*, Lecture Notes in Production Engineering,
DOI: 10.1007/978-3-642-30749-2_15,

implicate numerous production constraints. Machine scheduling in a rolling mill has therefore been recognized as a difficult industrial scheduling problem [2].

This paper describes the application of the data mining process for generating planning rules and is based on a case study in an integrated aluminum rolling mill. The goal is to discover patterns and drivers for high manufacturing lead time from ERP data that deliver useful information to derive planning rules. The global aim is to cope with the complexity of a broad production variety and to reduce overall lead time and thus also inventory. The created planning rules support the scheduling decision process in decentralized manufacturing control.

2 Production Planning in Aluminum Industry

The basic manufacturing process begins with the transformation of molten metal into solid metal blocks through the ingot casting process. The metal ingots are then processed through a rolling mill to yield rolled metals such as sheets, plates and foils. They are mainly made by rolling thick aluminum between rolls that reduce the thickness and lengthen, but the rolling process is only one step in a wide-ranging sequence. Each flat-rolled product has its customer specific manufacturing procedure [3]: the preparatory steps of alloying, casting, scalping and pre-heating; hot and/or cold rolling; intermediate annealing; and such later finishing steps as solution heat treatment or final annealing, stretching, leveling, slitting, edge trimming and aging.

Hot rolling scheduling is usually a multi-objective and multi-constraint optimization problem, because of its logistical complexity and plenty of technological constraints. A hot strip mill production scheduling problem is a NP-hard problem [4]. It can hardly be modeled with classical mathematical methods, mainly due to the variety of products that have their own rolling requirements. The hot rolling process has been described and examined in detail with proposed algorithms for an approximate optimal schedule [4, 5]. Less research work considers planning in an integrated aluminum rolling mill, most papers deal with scheduling hot or cold rolling only [5, 6].

Due to practical constraints, most of the established models are limited to specific issues and cannot easily be applied for other producers. A number of features cause problems in production planning in the aluminum conversion industry [7]:

- The convergent and divergent nature of the material flow.
- The flexibility in the choice of raw materials used in the casting unit.
- The need to synchronize the products at different steps of the production process.
- A variety of processing times.

Complexity of the material flow is further increased as it is determined by the number of predecessors and successors of an operation, the number of reflows and

Table 1 Characteristics of a complex and flexible integrated aluminum rolling mill

Casting process	Casting output (input for rolling)
Make-to-order, small lot sizes	Product variety: over 450 cast products (ingots)
Over 100 different alloys (alloys 1xxx to 8xxx)	(Vary in dimension, alloy composition)
Process manufacturing, batch processing	Over 20.000 rolling slabs/year
Rolling process	**Rolling output**
Make-to-order, small lot sizes	Customized products, high flexibility
Profound technological know-how	Product variety: around 4.500 rolled products
Flexible adaption of rolling and heating processes	(Plates, sheets and foils)
Complex material flow	Over 500 customers
(Linear, cyclic, converging and divergent)	Just-in-time (automotive, ...)
Sequence dependent setup times	High quality requirements (aerospace, ...)

the product variety [8]. An integrated aluminum rolling mill such as the one in the case study is thus characterized as a complex production system as outlined in Table 1.

3 Data Mining in Manufacturing

3.1 Data Mining and Knowledge Discovery in Databases

As a result of recent technical advances in computers and manufacturing, vast amounts of data are collected in database management systems and data warehouses in today's manufacturing enterprises. A lot of data related to bill of materials, product design, manufacturing process planning and scheduling, production process and systems, monitoring and diagnosis, and market forecasting are collected and stored as valuable knowledge for enterprises at various stages and levels [9].

Organizations have therefore undertaken many data warehousing projects in the last decade in anticipation of potential benefits. After data mining components are added, organizations have experienced payback of 10–70 times their data warehouse investment [10]. Data mining has appeared as an important tool for knowledge acquisition from the manufacturing databases and recent research addresses knowledge discovery together with data mining applications in manufacturing [11].

The term data mining is often used as a synonym for knowledge discovery in databases (KDD). KDD refers to the overall process of discovering useful knowledge from data, with the goal of mapping low-level data into other forms that might be more compact, more abstract, or more useful [12]. Data mining is a particular step in this process and applies specific algorithms from machine learning, pattern recognition and statistics to extract understandable patterns from large sets of data. The further KDD steps of data selection, data preparation, data cleaning and appropriate interpretation of the data mining results, make sure that valid, novel, and useful knowledge is derived from the data.

Useful patterns, which were previously not known, may be revealed in manufacturing data. But compared to other domain areas, there has been less research interest in the manufacturing domain inter alia for the following reasons [9]:

- Most of the researchers in the manufacturing domain area are not familiar with data mining algorithms and tools.
- Most of the theoretical data mining researchers are not familiar with the complex manufacturing domain area.
- Researches who are skilled in both, data mining and manufacturing, are not able to access manufacturing enterprise data, which are often proprietary and sensitive.

3.2 Literature Review

From the years 1987–2005 there was a significant growth in the number of publications about data mining in manufacturing in some areas of manufacturing, such as fault detection, quality improvement, manufacturing systems, and engineering design [13]. On the other hand, the data mining community has paid comparatively less attention to shop floor control. Over the last decade, a lot of studies have been carried out to examine how enterprise data could be mined to generate useful models and knowledge for running the business more efficiently and effectively. Liao [14] provides a comprehensive overview of previous studies on enterprise data mining.

Choudhary et al. [11] have shown that there is a rapid growth in the application of data mining in the context of manufacturing processes and enterprise. Recently Windt et al. [15] used data mining methods adapted from gene expression analysis to identify causes of lateness in a multistage production system. The potential of data mining to improve due date reliability has been investigated in a case study [16]. Harding et al. [13] expected that future research would focus on analyzing data that are related to shop floor control, scheduling and ERP.

4 Generating Planning Rules Using the KDD-Process

4.1 The Procedure Model

The procedure model consists of seven steps derived from the original KDD process as illustrated in Fig. 1. An initial assessment in the business understanding phase helps to determine the requirements and objectives. The five original KDD steps are used to explore implicit knowledge from ERP data. We refer to the

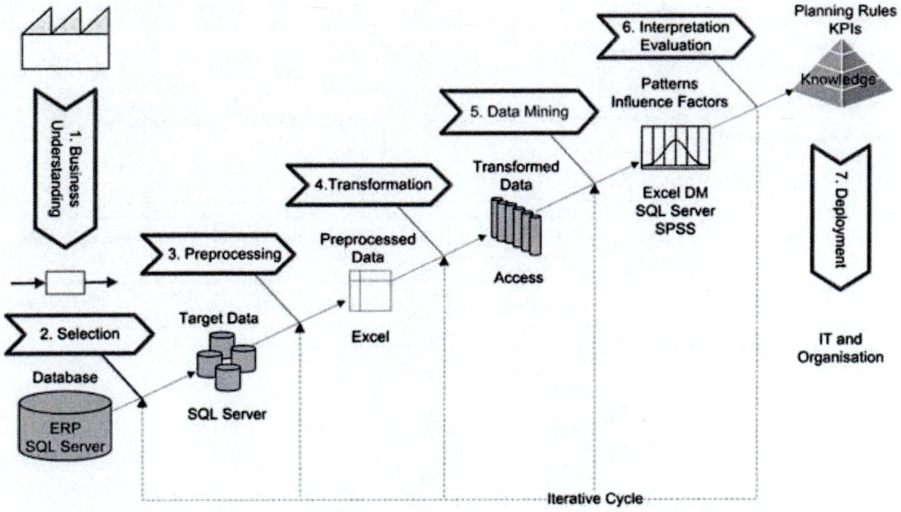

Fig. 1 Data mining process for generating planning rules

original KDD-process as data mining process, as data mining is often used as a synonym for knowledge discovery in databases. The critical phase is the interpretation and evaluation of the results from the data mining algorithm. The goal is to identify patterns and influence factors. They should support the human experts in deriving useful planning rules. The concluding deployment phase assures the conversion of tacit knowledge into explicit knowledge and creating organizational knowledge as a competitive resource [17].

4.2 Business Understanding

At the beginning of a data mining project the business understanding phase focuses on the comprehension of objectives and requirements from a business perspective and involves several steps such as assessing the current situation, determining the business objectives, establishing data mining goals, and developing a project plan [18].

In the context of production logistics the fundamental goal can be formulated as the pursuance of greater delivery capability and reliability with the lowest possible logistic and production costs [19]. Due to the conflicting objectives known as dilemma of operations planning it is necessary to define a measurable objective for the data mining project out of the main logistic key performance indicators inventory (work in progress), delivery reliability, lead time and capacity utilization.

The focus in the case study was on reducing manufacturing lead time and thus reducing inventory and increasing delivery reliability and flexibility while trying to keep high capacity utilization. To get an understanding about the business a

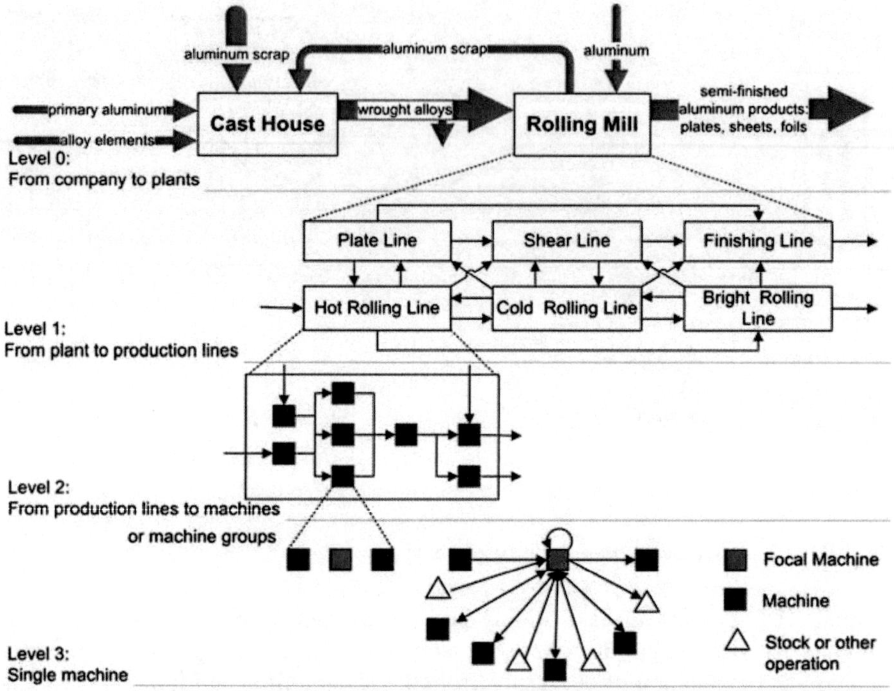

Fig. 2 Levels of a production system exemplified for an aluminum rolling mill

material flow analysis gives an overview of the different levels of the production system as illustrated in Fig. 2. Level 0 shows the big picture of an integrated aluminum rolling mill with the relation to the casting house. Level 1 shows the production lines that correspond with the autonomous planning units. Each production unit is specified with its machine or machine groups at level 2.

Level 3 displays the lowest level of the single machine. This is the target level for data selection to reveal knowledge for scheduling machines on shop floor level. Therefore, the examination considers a focal machine respectively with all inflows that can come from other machines, stock or other operations like quality inspections or external processing.

4.3 Selection and Preprocessing

The main source of data is the ERP system (SAP) that contains all relevant data beginning from the sales order up to the collected data from the operations and machines on the shop floor. The main task is to identify the relevant data objects and right relationships between them.

The goal of preprocessing is to improve the quality of the selected data by data cleaning and preprocessing. Usually there exists no data warehouse that contains all needed valid and consistent data. Therefore, data have to be integrated from different sources. For example, in an ERP system planning is performed on aggregated level on machine. The real short term scheduling is done by the planners of a planning unit on the shop floor on a single concrete machine. These data from different planning levels as illustrated in Fig. 2 have to be consolidated according to certain rules.

To get valid data for further processing only lots that started and finished in the observed year have been included. Production orders that where marked for research and development have been excluded as they are not relevant for due date reliability. The number of data sets was thus reduced by round about 5–10 %.

The phases for collecting and cleaning of datasets are known as the most time consuming activities that make around 80 % of the data mining process and are critical to success of data mining [9]. Usually there are some iterative steps necessary to get the required data, e.g. after the determination of the relevant production orders again a new selection was necessary to collect operating records of the predecessor and successor of the focal machine at level 3.

4.4 Transformation

This step involves the preparation of data in a form of a data matrix as shown in Table 2 that can be processed by a data mining tool. The main task is to define and calculate the additionally needed data.

The requirements for our data mining model are as follows:

- A single key column: unique identifier of the lead time objects for each record
- Input columns: discrete or discretized characteristics of the object
- At least one predictable column: e.g. lead time

A line in the matrix represents a lead time object on a certain level with all relevant characteristics in the rows. Depending on the level a lead time object can be a business order, a lot, a partial lot or an operation. The characteristics are related to:

- Orders: customer related data, customer specification, strategic business unit, …
- Products: raw material, alloy category, temper, quality grade, surface, …
- Operations: setup group, predecessor, successor, process parameters, …

The third element is the predictable column that is the manufacturing lead time. In process industry an important part of lead time such as process time and cooling time is given by technological specifications and restrictions and can thus not be affected by scheduling. The remaining parts of the overall lead time that can be influenced by scheduling are waiting time and setup time. The calculated manufacturing lead time for data mining included negligible transport times, because

Table 2 Structure of a data matrix for the application of a data mining tool

Characteristics	Order		Product		Operation		Lead time
Object key	X_1	X_2	X_3	…	X_j	…	X_k
o_1	X_{11}	X_{12}	X_{13}	…	X_{1j}	…	X_{1k}
:	:	:	:	…	:	…	:
o_v	X_{v1}	X_{v2}	X_{v3}	…	X_{vj}	…	X_{vk}
:	:	:	:	…	:	…	:
o_n	X_{n1}	X_{n2}	X_{n3}	…	X_{nj}	…	X_{nk}

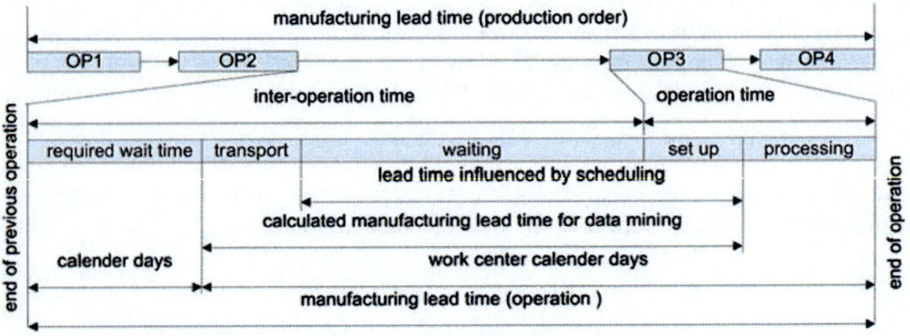

Fig. 3 Calculated manufacturing lead time for data mining

there were no data available to subtract time for transport between work centers. As shown in Fig. 3, on the order level there are separate operations OP and each operation is split into five more components on the operational level [20]: required waiting time after processing (cooling time), transportation, waiting before processing, setup and processing.

4.5 Data Mining

After accomplishing the preparation of the data an appropriate data mining method for the specific problem situation has to be chosen. To explore critical product characteristics cluster analysis has been shown to be an effective method [16]. The classical k-means algorithm from the group of deterministic clustering methods has been applied to explore the causes of due date reliability. To classify product orders and explore critical characteristics that can explain lead time we use the naïve Bayes algorithm. The algorithm is based on Bayes theorem and generally used in predictive modeling as a classification algorithm delivering quick results and easy to use [21].

In classical clustering, also known as crisp or deterministic clustering, objects are assigned to exactly one cluster. In contrast in fuzzy clustering, objects are not only assigned to a particular cluster, but they also possess a membership function indicating the strength of membership in all or some of the clusters. The strength of membership can also be interpreted as the probability of belonging to a cluster [22]. Fuzzy clustering techniques are therefore also known as probabilistic cluster analyzing methods [23] and can be seen as generalization of naïve Bayes classifiers [24].

The Microsoft Naïve Bayes algorithm can be applied for clustering as well as for analyzing and visualizing the key influencers for a specific target [25]. Figure 4 illustrates the exemplarily results for the exploration of key influencers of the target lead time on operational level for a focal machine. It shows the generated clusters and an excerpt of calculated probabilities for the particular characteristics as relative impact factor calculated as conditional probabilities using Bayes theorem.

4.6 Interpretation and Evaluation

The results from the data mining analysis have to be validated, evaluated and interpreted by experts from the particular planning unit. Additional charts such as a frequency charts or box-plots support the detection of key influence factors. The overall data mining process is an iterative process and usually there is the need of a second or third data mining cycle with new additional characteristics. As shown in Fig. 4 there is e.g. one particular setup family and alloy category that is strong related with a short lead time. On the other hand middle, high and very high lead time is mainly caused by the predecessor AA. Hence a second data mining analysis with the predecessor AA as new focal machine should deliver more information.

For example, quality grade was one of the key influencing factors for lead time for such a focal machine. After further evaluation by the planners and analyzing of past data, planning rules as exemplarily illustrated in Table 3 were derived. The planning rules define maximum values for an average weekly work load for different product characteristics that are mainly due to technologically caused setup times. These planning rules correspond approximately to Ho:No-sequencing rules, used for car sequencing in the automotive industry. There they restrict the maximum occurrence of a work-intensive option o to at most Ho out of No successive car models launched down the production line [26]. Instead of work-intensive options we have certain characteristics, which lead to a maximum weekly workload for certain values due to technological conditions.

As every planning unit is trying to minimize setup times and optimize capacity utilization, they now get real-time information about the implication of their short term schedule outcome. Such planning rules can also be used to establish WIP control loops between the manufacturing units similar to a decentralized WIP-oriented manufacturing control [27]. The planning rules serve as preventive indicators and can reduce the risk of setup optimizing sequencing that usually delays orders with high setup times. They support the decision process but the final decision and responsibility is still up to the planner who has to find the right

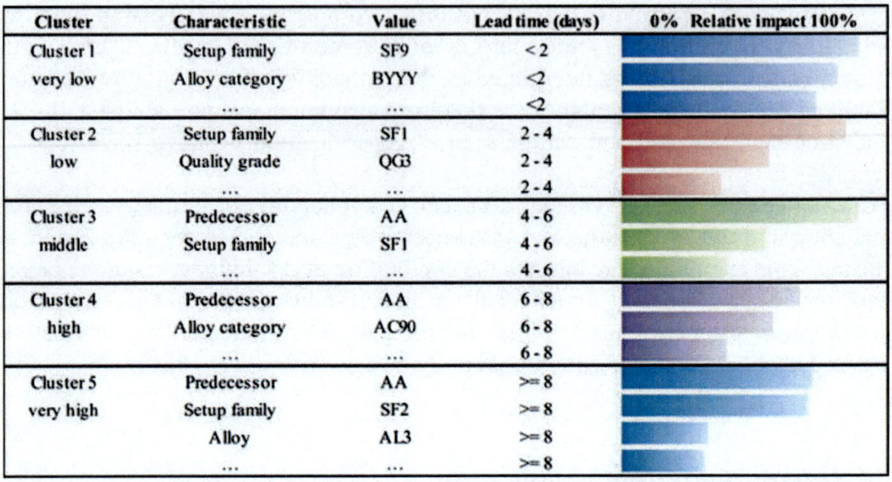

Fig. 4 Exemplarily detected clusters and influence factors from naïve Bayes algorithm

Table 3 Conceptual design for planning rules

Machine: AA			Generated by: planning unit PU01		
Planning rule P01: weekly work load for order release					
Planning rule number		P01–01	P01–02	P01–03	P01–04
Name		Optimal number of coils	Maximum number of coils	Maximum number of coils	Relation successor
Characteristic 1	Characteristic 2				
QG1	SF1	12		120	2
QG2	SF2	36			3
QG4	SF3	24	30		
…	…	…	…		

balance. Further coordination between the planning units is improved as dependencies are getting more transparent.

4.7 Deployment

The gained knowledge from interpreting and evaluating data mining results in form of planning rules must be organized and presented in a way that the end user can apply it within an organization's decision-making process. The planning rules are thus implemented in the manufacturing control system in a form that the human planner gets an alert when a planning rule is violated while he is scheduling the production orders. According to the type of the planning rule the planner is allowed to save his schedule if the rule is defined as warning only or he has to reschedule the orders if the rule is defined as must. In the case study the planning rules have turned out to be an appropriate support for the concerned planning unit in releasing and scheduling the local queue of production orders. The effect on lead time and inventory is positive but due to the changing product mix difficult to compare with the initial situation.

5 Conclusion

This paper analyzed a production system in the aluminum processing industry, which due to its complexity is predestinated for the application of data mining. The data mining process turned out to be an appropriate method to generate planning rules by obtaining data from the ERP system and implementing the results in a manufacturing control system. Such planning rules are especially useful under autonomous manufacturing control where planning decisions are based on the knowledge of the human planners. Even though the application of a naïve Bayes algorithm delivers fast and useful information about clusters and factors influencing lead time, the generation of planning rules cannot only be performed automatically but with the interpretation of the human experts and further additional information. The implemented planning rules can support the human planners by showing an alert when certain rules are violated while performing a real-time scheduling. Planners can compare different planning scenarios and evaluate them through the number of rule violation as key performance indicator. Further research has to deal with the quantification of a rule violation to allow a more accurate comparison of different planning scenarios.

References

1. Samarasekera, I.V.: Hot rolling. In: Cahn, R.W., Flemings, M.C., Ilschner, B., Kramer, E.J., Mahajan, S., Veyssière, P., and Buschow (eds.) Encyclopedia of Materials: Science and Technology, pp. 3836–3843. Elsevier, Oxford (2001)
2. Tang, L., Liu, J., Rong, A., Yang, Z.: A review of planning and scheduling systems and methods for integrated steel production. Eur. J. Oper. Res. **133**, 1–20 (2001)
3. Association, A.: Rolling Aluminum: From the Mine Through the Mill. Aluminum Association, Arlington (2007)
4. Lopez, L., Carter, M.W., Gendreau, M.: The hot strip mill production scheduling problem: a tabu search approach. Eur. J. Oper. Res. **106**, 317–335 (1998)
5. Stauffer, L., Liebling, T.M.: Rolling horizon scheduling in a rolling-mill. Ann. Oper. Res. **69**, 323–349 (1997)
6. Nishi, T., Konishi, M., Ago, M.: A distributed decision making system for integrated optimization of production scheduling and distribution for aluminum production line. Comput. Chem. Eng. **31**, 1205–1221 (2007)

7. David, F., Pierreval, H., Caux, C.: Advanced planning and scheduling systems in aluminium conversion industry. Int. J. Comput. Integr. Manuf. **19**, 705–715 (2006)
8. Lödding, H.: Verfahren der Fertigungssteuerung Grundlagen, Beschreibung, Konfiguration. Springer, Berlin u.a (2008)
9. Wang, K.: Applying data mining to manufacturing: the nature and implications. J. Intell. Manuf. **18**, 487–495 (2007)
10. Chen, L.-D., Sakaguchi, T., Frolick, M.N.: Data mining methods, applications, and tools. Inf. Syst. Manage. **17**, 1–6 (2000)
11. Choudhary, A.K., Harding, J.A., Tiwari, M.K.: Data mining in manufacturing: a review based on the kind of knowledge. J. Intell. Manuf. **20**, 501–521 (2008)
12. Fayyad, U., Piatetsky-Shapiro, G., Smyth, P.: From data mining to knowledge discovery in databases. AI Magazine **17**, 37–54 (1996)
13. Harding, J.A., Shahbaz, M., Srinivas, S., Kusiak, A.: Data mining in manufacturing: a review. J. Manuf. Sci. Eng. **128**, 969–976 (2006)
14. Liao, T.W.: Enterprise data mining: a review and research directions. In: Liao, T.W., Triantaphyllou, E. (eds.) Recent Advances in Data Mining of Enterprise Data: Algorithms and Applications. World Scientific, Singapore (2008)
15. Windt, K., Hütt, M.-T.: Exploring due date reliability in production systems using data mining methods adapted from gene expression analysis. CIRP Ann. Manuf. Technol. **60**, 473–476 (2011)
16. Windt, K., Knollmann, M., Meyer, M.: Anwendung von Data Mining Methoden zur Wissensgenerierung in der Logistik - Kritische Reflexion der Analysefähigkeit zur Termintreueverbesserung. In: Spath, D. (ed.) Wissensarbeit - Zwischen strengen Prozessen und kreativem Spielraum, pp. 223–249. GITO, Berlin (2011)
17. Nonaka, I., Takeuchi, H.: The knowledge-creating company: how Japanese companies create the dynamics of innovation. Oxford University Press, New York (1995)
18. Olson, D.L., Delen, D.: Advanced Data Mining Techniques. Springer, Berlin (2008)
19. Nyhuis, P., Wiendahl, H.-P.: Fundamentals of Production Logistics: Theory, Tools and Applications. Springer, Berlin (2009)
20. Wiendahl, H.-P.: Load-Oriented Manufacturing Control. Springer, Berlin (1995)
21. Mitchell, T.M.: Machine Learning. McGraw Hill, New York (1997)
22. Everitt, B.S., Landau, S., Leese, M., Stahl, D.D.: Cluster Analysis. Wiley, Chichester (2011)
23. Bacher, J., Pöge, A., Wenzig, K.: Clusteranalyse: Anwendungsorientierte Einführung in Klassifikationsverfahren. Oldenbourg Wissenschaftsverlag, München (2010)
24. Borgelt, C., Timm, H., Kruse, R.: Probalistic networks and fuzzy clustering as generalizations of Naive Bayes classifiers. Comput. Intell. Theory Pract. **8**, 121–138 (2001)
25. MacLennan, J., Tang, Z., Crivat, B.: Data Mining with Microsoft SQL Server 2008. Wiley, Indianapolis (2009)
26. Boysen, N., Golle, U., Rothlauf, F.: The Car Resequencing Problem with Pull-Off Tables. BuR - Business Research. 4 (2011)
27. Lödding, H., Yu, K.-W., Wiendahl, H.-P.: Decentralized WIP-oriented manufacturing control (DEWIP). Prod. Planning Control **14**, 42–54 (2003)

Striving for Zero Defect Production: Intelligent Manufacturing Control Through Data Mining in Continuous Rolling Mill Processes

Benedikt Konrad, Daniel Lieber and Jochen Deuse

Abstract Steel production processes are renowned for being energy and material demanding. Moreover, due to organizational and technological restrictions in flow production processes, the intermediate product's internal quality features cannot be assessed within the process chain. This lack of knowledge causes waste of energy and material resources, unnecessary machine wear as well as reworking and rejection costs, when defective products are passed through the entire process chain without being labeled defective. The process control approach presented in this paper provides the opportunity of gaining transparency on quality properties of intermediate products. This aim is achieved by predicting intermediate product's quality by means of data mining techniques. This approach can be applied in a wide field of production environments, ranging from steel and rolling mills to automated assembly operations. Concerning this concept, the authors derive a methodology for representing different quality properties in a way that it can be applied in the process control. Beyond that, first results of statistical analyses on the quality-related significance of process parameters are disclosed.

Keywords Process data mining · Real-time quality prediction · Intelligent manufacturing process control

B. Konrad (✉) · D. Lieber · J. Deuse
Chair of Industrial Engineering, TU Dortmund University, Dortmund, Germany
e-mail: benedikt.konrad@tu-dortmund.de

D. Lieber
e-mail: daniel.lieber@tu-dortmund.de

J. Deuse
e-mail: jochen.deuse@tu-dortmund.de

K. Windt (ed.), *Robust Manufacturing Control*, Lecture Notes in Production Engineering,
DOI: 10.1007/978-3-642-30749-2_16,

1 Introduction

Resource and energy efficiency of interlinked manufacturing processes can be increased in many ways: process engineering solutions or design improvements, higher quality of raw materials as well as optimized operating parameters [1]. In this context, based on a rolling mill case study, real-time optimization and adaptation of predefined operating modes on the basis of continuous inline quality control by means of data mining techniques poses a promising approach to sustainably increase the efficiency of production processes [2, 3].

Continuous quality control across all stations of the process chain is intended to ensure processing conditions within certain tolerance limits in order to guarantee the final product's quality standards defined by the customer. Nevertheless, up to now, processing by rigid and inflexible program guidelines is still common practice in continuous flow production systems. In case of missing quality gates due to organizational and technological restrictions, processing steps are basically targeted to pre-fixed operation settings instead of orienting and adjusting on the basis of actual product quality levels or states of processing.

Based on the Toyota Production System, the development of a new production control approach is motivated by the Jidoka principle, which focuses on avoiding waste, such as rejection and rework, as early as possible within production processes [4]. Although these process-immanent quality checks are state of the art in the automotive industry, they cannot be transferred to high complex and interlinked production systems in the steel industry one-to-one [5].

The rigidly linked sequence of production steps as well as rough environmental conditions during processing and technological restrictions within the value chain obstruct assessing the physical quality of intermediate products [6]. For that reason, steel production is still characterized by final quality diagnostics measurements that are done at the finishing stands at the end of the process chain [7]. This is a major drawback as failures during production lead to high internal costs and waste of resources when only the final product's quality properties can be tested [6].

Therefore, this paper is focused on avoiding inferior product quality by real-time Inline Quality Prediction (IQP) tools based on data mining and artificial intelligence and its integration into a comprehensive Intelligent Manufacturing Process Control (IMPC) approach for industrial application.

The considered case study is provided by a leading German steel producer and is representative for the production control approach discussed in this paper. The process chain consists of five major processing steps starting with the heating process at the rotary hearth furnace where steel bars are heated to forming temperatures. In the following facilities the bar's profile is reduced to customer specifications before the entire bar is separated into rods of customer defined length (see Fig. 1). A more detailed description of the production steps as well as preliminary work on processing and storing the collected static, process and quality data can be found in [6, 8].

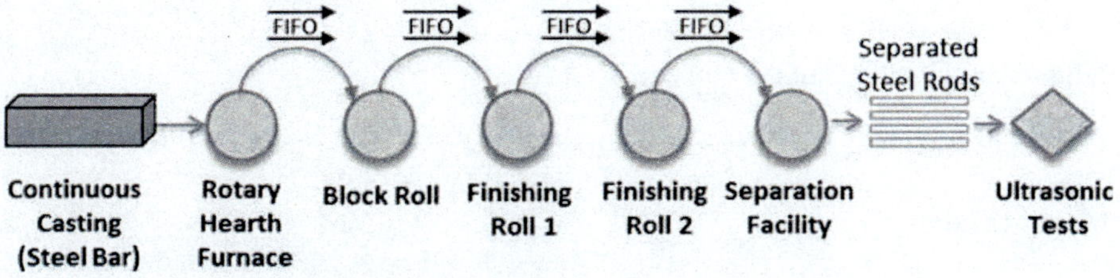

Fig. 1 Process chain of rolling mill case study [6]

The remainder of this paper is arranged as follows: Sect. 2 summarizes the IT-infrastructure and schemes applied for production control in the case study. Section 3 introduces the Intelligent Manufacturing Process Control concept as an answer to the previously identified drawbacks. Sections 4 and 5 focus on deriving quality criteria and statistically analyzing the impact of various process parameters on quality properties on the basis of data collected at the first step of the rolling mill process chain, the rotary hearth furnace. Finally, in Sect. 6 a conclusion is given and the next steps in developing the process control concept are lined out.

2 Manufacturing Process Control Tools in Flow Production

2.1 Manufacturing Process Control IT-Infrastructure in a Rolling Mill Case Study

IT infrastructures—in the case study considered—are characterized by a hierarchical organization of all required process planning, control and execution entities. Figure 2 depicts information flows and communication interfaces of the process control tools within the IT-structure of the case study.

On the top level, the production planning and control level, customer order specifications such as material, order volume and final dimensions are transformed to unique internal production order specifications including a predetermined work flow sequence as well as a specific assignment of predefined rolling schedules dependent on input and output dimensions of the material. This basic production guideline is transmitted to the master level, in which a centralized master computer coordinates and traces all individual intermediate products of the corresponding production order within the value chain by their unique identification code. The master level spreads all relevant processing information to the machine level, so that the operator is informed of current and upcoming tasks. Machine-to-Machine communication is not provided. After confirming the incoming production telegram, the corresponding rolling program is loaded and automatically passed to the automation level, where each processing step of the predefined rolling schedule is executed, supervised and controlled by a Programmable Logic Controller (PLC).

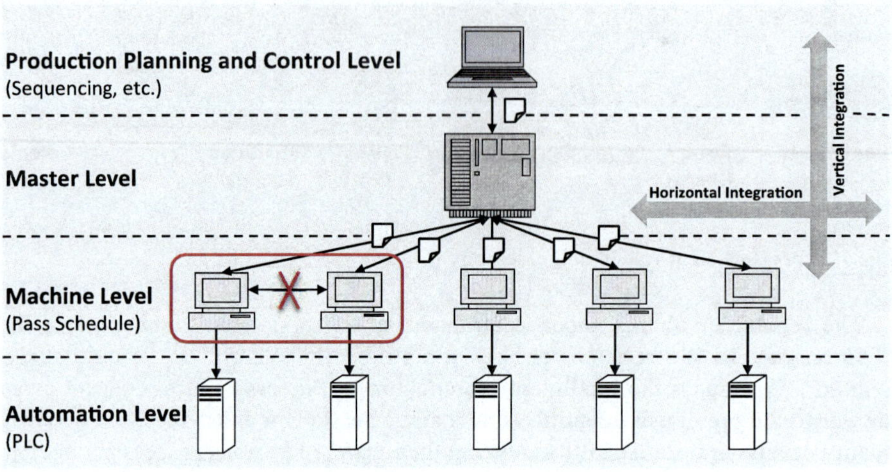

Fig. 2 Information flows across function levels

Nevertheless, the operator is able to intervene and adapt rolling parameters manually at any time according to unforeseen interruptions or failures during processing. After finishing the intermediate product's processing steps, the operator confirms the finished production telegram which is then automatically fed back to the master level and transferred with new input parameters to the next production step.

2.2 *Structural Weaknesses of Process Control Systems in Flow Production*

This type of manufacturing process control strategy can be described as a non-adaptive automated production control, which is based exclusively on predefined pass schedules in a one-way feed-forward approach. Despite of manual interventions by the operator due to obvious abnormalities from regular production, there is no additional supporting entity integrated that can monitor and verify the status of processing in real-time or even derives adequate counteractions in case of deviations to guarantee quality specifications of the final product. In fact, no feed-back loop which, for instance, compares target and actual processing parameters regarding the intermediate product's quality properties is provided.

To react on quality deviations of intermediate products in real-time, the automated process control has to be enhanced with intelligent process control modules based on artificial intelligence which allow real-time optimization and adaptation of pass schedule parameters to face challenges such as resource and energy efficient production processes [2, 9]. Therefore, an Intelligent Manufacturing Process Control applicable in hot-rolling production processes in the steel industry has to be developed.

3 Intelligent Manufacturing Process Control in Flow Production Systems

As lined out in the previous chapters, a new control scheme has to be developed. With the intention of implementing the Jidoka principle in steel industry, this approach has to incorporate features such as process data acquisition and interpretation besides the fundamental aspects of order control and tracking. The general idea of continuously monitoring and adjusting process parameters is known as Advanced Process Control in the process industry [10, 11]. The Intelligent Manufacturing Process Control (IMPC) introduced here transfers this basic idea from process industry to other industry branches by introducing data mining techniques to predict the quality of an intermediate product and adjust further processing steps according to the prediction.

The data mining process on which this approach is based is the knowledge discovery in databases (KDD) process [12]. In this context data mining is defined as the "application of data analysis and discovery algorithms that produce a particular number of patterns (or models) over the data" [13]. The KDD process is divided in nine steps [12]:

1. Developing an understanding of the application domain
2. Creating a target data set, selecting a data set
3. Data Cleaning and preprocessing
4. Data reduction and transformation
5. Choosing the data mining task
6. Choosing the data mining algorithms
7. Data mining
8. Evaluating output of step 7
9. Consolidating discovered knowledge.

The IMPC concept presented in this section follows these steps. Steps 1 and 2 were executed while developing the IMPC, whereas the remaining steps 3 to 9 are automated and can be found in the modules described below.

The developed IMPC approach consists of different functional modules for distinct purposes. The core modules are (1) Data Acquisition and Storage (DAS), (2) Data Monitoring (DM), (3) Inline Quality Prediction (IQP) and (4) Process Parameter Optimization (PPO). Modules (2–4) represent the different process control stages and form the IMPC. While DAS is a prerequisite for implementing the IMPC, the IMPC itself can be realized to two different extents. Firstly, including the PPO-module, which allows an online optimization of process parameters depending on the current state of the production process in order to compensate previous process deviations. And secondly, just including the DM and IQP-module in terms of a decision support system. This type of process control relies on generating real-time quality estimates for the intermediate product. If inferior quality is predicted, the production process is halted. In contrast to the previous option, the decision on further actions to be taken to improve the products quality is not made by the IMPC, but relies on expert knowledge. This modular design of the entire process control approach allows an on demand integration in existing process control infrastructures. Due to the early stage of the research project, the paper's scope is limited to this approach and therefore to the first three modules.

3.1 Data Acquisition and Storage

The extent to which the DAS module has to be introduced into a production system is determined by the complexity of production that has to be supervised. Obviously, complexity in this context reflects the number of distinct influencing parameters. These can range from environmental factors, e.g., temperatures, humidity, noise or vibrations caused by other machinery, to machine states like applied torque, engine speed or execution time for a predefined task. This short overview illustrates that a general concept for DAS cannot be given, as it rather depends on the actual process the entire IMPC concept is applied to.

In case of the rolling mill case study, processing parameters as well as machine states of all major processing stations are relevant. Examples for these are zone temperatures, material temperatures and time spent in each zone at the rotary hearth furnace, or rolling speed and forces, material temperatures at each forming step and grooves used at each of the rolls. A comprehensive description of relevant factors as well as information on the actual implementation of the DAS is given in [5, 6].

3.2 Intelligent Manufacturing Process Control

The IMPC concept as implemented in the case study can be interpreted as a separate building block in a company's process control landscape (see Fig. 3). The IMPC does not affect the process of production planning, i.e., it generates no advice on optimized production sequences. Instead, the focus is on analyzing processing states in real-time aiming at forecasting the intermediate product's quality properties. This knowledge is used for either deriving recommendations on whether the product should be processed any further or for optimizing the next processing step's parameters so that the required product quality can be obtained.

In both cases the IMPC requires information on the current processing parameters which are gained from the DAS module. Moreover, information on historic processing data, merged in a quality prediction model has to be available for assessing the intermediate product's quality correctly. When it comes to optimizing processing parameters, even more knowledge is demanded. In this case it is mandatory to have a material forming simulation, i.e., a finite element method (FEM) simulation, at hand which verifies the impact of altered processing parameters on the product. Figure 4 summarizes the required knowledge and infoData Monitoring and Inline Quality Prediction. Based on the real-time

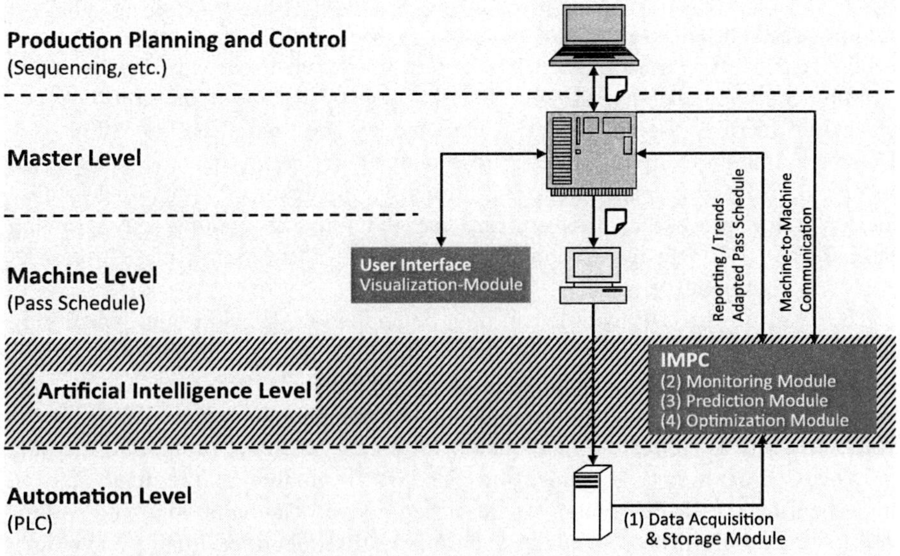

Fig. 3 Integration of artificial intelligence in operational process control

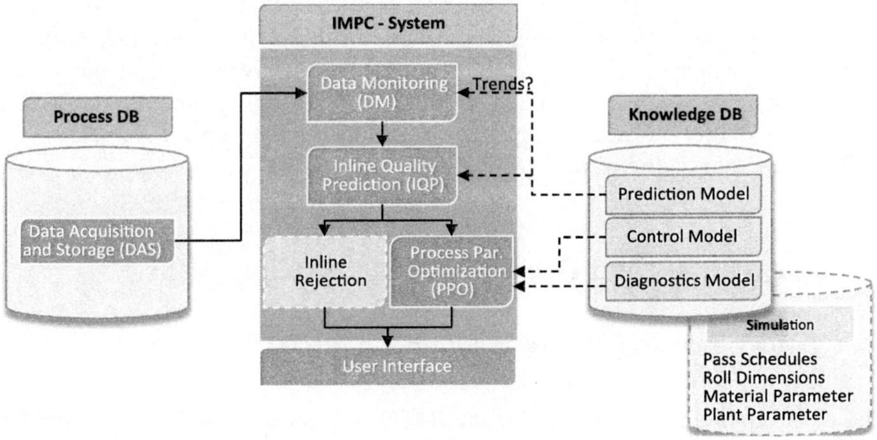

Fig. 4 Intelligent manufacturing process control model (*IMPC*)

processing of data available from the DAS module, the DM module generates all statistics and information demanded by the IQP module. These tasks also include all necessary pre-processing steps on the data gathered: e.g., depending on the type of data missing values are replaced either by a default value or average values in the case of time-series data. Additionally, for time series data statistics such as maximum and minimum values, standard deviation and the maximum gradient are computed. These characteristics are used for a first evaluation of the current processing step before an in-depth analysis is conducted. All results in the DM module are then fed into the IQP module.

The IQP module translates all available information irrespective of its actual meaning into a quality assessment. This is done by means of data mining, i.e., supervised learning models [5, 14]. These models are trained on labeled historic datasets so that these algorithms can judge on the currently processed quality based on actual processing parameters. Additionally, the results of the final quality checks at the process chain's end are merged into the continuously growing database. By this, the prediction accuracy of the data mining models can be increased continuously.

The result of the IQP module can then be applied to decision rules, specifying whether an intermediate product is ejected from the process or processed further depending on the quality prediction and its position in the process chain. In a second step, the IMPC can be extended by the implementation of the PPO module, which also makes recourse on the quality estimates generated in the IQP module.

Process Parameter Optimization. The PPO module is the final step in implementing the IMPC model. As described above, it aims at adjusting parameters of upcoming process steps in such a way that identified quality deviations caused by previous processing steps are compensated in the remainder of the process chain. To accomplish this, the PPO module needs comprehensive access to information on process parameters and especially on FEM simulation results to validate the impact of altered parameters on product quality properties. This module is obviously the most challenging one in the IMPC concept. It relies on completely implemented DM and IQP modules. Due to the early state of the research project, the PPO module is still a theoretical concept and not jet defined on a detailed level. Consequently, the remainder of the paper focusses on the IMPC model consisting of the DM and IQP modules.

4 Quality-Levels and Decision-Rules in Rolling Mills

4.1 Quality-Levels for Intermediate Products

The most important prerequisite for the IMPC concept is the definition of quality in the context of intermediate products. Final products, i.e., steel rods in the case of the rolling mill case-study, are assessed according to various distinct quality parameters. The intermediate product's quality, i.e., the quality of the steel bar, which is predicted by the IMPC is not assessed physically at any point in the process chain. In this special case the intermediate product's quality cannot be related to the final product's quality one by one as each intermediate product (each steel bar) results in a variable number of steel rods depending on the profile specified in the customer's order. Hence, the intermediate product's quality has to be composed of the quality of the different final products.

Due to these facts, intermediate product quality is determined according to following sum function to aggregate the final products' quality properties to one embracing quality label for the training of the quality prediction models:

$$Q_b = 1 - \frac{1}{R_b}\sum_{r=1}^{R_b}\sum_{p=1}^{P}\left(w_p \cdot \frac{\lambda_{p,r,b}}{\lambda_{p,max}}\right); \forall b \in B, where \quad (1)$$

$Q_b \in [0, 1]$, quality level of bar b

$w_p \in [0, 1], \sum_{p=1}^{P} w_p = 1$, weight of quality property p

$\lambda_{p,r,b} \in [1, \lambda_{p,max}]$, value of quality property p of rod r in bar b

$b \in \{1, \ldots, B\}$, steel bar b in set of allbars B

$r \in \{1, \ldots, R_b\}$, steel rod r in set of all rods resulting from bar b

$p \in \{1, \ldots, P\}$, quality property p in set of all properties

$B, R_b, P \in \mathbb{N}$

Q_b represents the quality level of a certain steel bar (see Formula 1). It does not represent the actual physical quality according to some specified criterion such as "type of error", but allows an aggregated estimate on the overall quality of each bar. The quality level is composed of various different quality properties ($\lambda_{p,r,b}$) that are normalized on a 0–1 interval by dividing each value by its maximum value ($\lambda_{p,max}$). The quality level represents the weighted average of these distinct quality properties.

Thus, a quality level of 1 implies that the current bar does not show any quality deviations at all. On the other hand, a quality level of 0 emphasizes that all rods from this bar obtained the maximum error value. Hence, all values between zero and one have to be interpreted more carefully. The aggregation of various quality properties from different products to one single label leads to an information loss so that it cannot be judged whether one single rod has a quality level of almost zero or all rods show a slightly minor quality.

When it comes to predicting the intermediate product's quality, this loss of information can be neglected, because the resulting decision in the IMPC on whether the steel bar should be processed further or ejected is the same in both cases: assuming that one rod out of a bar is defective it is advantageous to completely process the bar and to discard the defective bar. On the other hand, if all rods have a slightly minor quality the customer might accept the rods despite of the quality deviations or the rods can be assigned to a different customer order which requires lower quality properties.

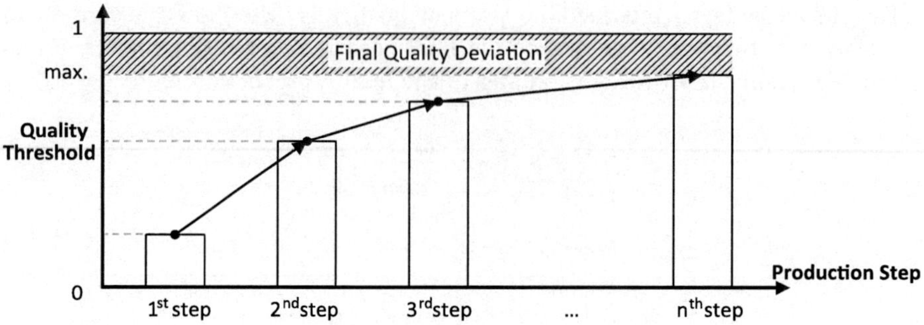

Fig. 5 Required quality thresholds at different production steps

4.2 Application of Quality-Levels in the IMPC Concept

The quality level of a certain intermediate product is assessed by the IMPC at each processing step. In order to meet customer requirements, the quality level has to meet the defined target at the end of the process chain. Moreover, quality assessments are becoming more precise, with increasing processing information on the intermediate product. Consequently, the tolerated quality deviations at the end of the process chain will be the smallest.

The IMPC accounts for these characteristics, as it applies a threshold method for deriving recommendations for the decision on ejecting the intermediate product or not. The threshold at the beginning of the process chain is rather low as on the one hand quality level predictions might be somewhat fuzzy and on the other hand the quality level can be influenced positively to a significant degree in the remainder of the processes. Thus, the threshold increases with the number of processing steps completed. At the last process step the difference between threshold and $Q_b = 1$ corresponds to quality deviations accepted by the customer and the remaining fuzziness of the quality prediction model (see Fig. 5).

The quality level thresholds at all processes have to be determined by means of expert knowledge as well as data mining and multivariate statistics. Analyzing the quality related significance of each process's parameters reveals its impact on the quality level of the final product. The following chapter presents the analysis conducted on the rotary hearth furnace's process parameters.

5 Fundamental Process Parameter Studies and First Results

This chapter focusses on statistical analyses conducted on the effects of process parameters of the rotary hearth furnace. The analyses are performed on a sample of about 800 steel bars of one certain material and size.

Table 1 Correlation coefficients of quality level and characteristic parameters

	Min.	Max.	Max.	Std. Dev.	Max. Grad.
Furnace temperature					
Qual. level	−0.045	0.065	0.032	−0.099	−0.093
Significance	0.106	0.037	0.189	0.003	0.005
Temperature at top of steel bar					
Qual. Level	−0.094	0.025	0.081	−0.146	−0.115
Significance	0.005	0.247	0.012	0.000	0.001
Temperature at top in core bar					
Qual. Level	−0.048	0.069	0.103	−0.124	−0.119
Significance	0.094	0.027	0.002	0.000	0.000
Temperature at bottom of steel bar					
Qual. Level	−0.093	0.067	0.101	−0.145	−0.119
Significance	0.005	0.031	0.003	0.000	0.000

5.1 *Correlation Analyses on Process Parameters of Rotary Hearth Furnace*

The DAS module at the furnace gathers information on four parameters: the temperature in the furnace and the temperature at the bottom, top and core of each steel bar. Each of these parameters represents time series data, as they are collected with a frequency of up to 100 Hz [8], while the bar is in the furnace. For analyzing the effect and significance of these factors regarding the quality level, the time series data is preprocessed in order to determine characteristics that describe each of the four value series. For each time series of each steel bar passing the furnace the maximum, minimum and average temperatures are computed as well as standard deviation and largest gradient.

Analyzing the correlation coefficients (see Table 1) shows that the standard deviation and the maximum gradient have the biggest impact on the quality level and influence it negatively in all analyses. This result shows that not the temperatures themselves but their variance and the speed of temperature changes during heating are the most important determinants of quality levels. Process parameters having a significance level ≤ 0.05 have a statistically significant effect on the quality level.

5.2 *Regression Analyses on Process Parameters of Rotary Hearth Furnace*

On the sample set regression analyses are conducted in order to judge on the size of each parameter's impact and its significance for the prediction of quality levels. As the residuals of the linear regression are not distributed normally a logistic regression is carried out on the sample set [15, 16]. The required binning of the dependent variable, i.e., the quality level, is performed according to expert knowledge leading to two classes

Table 2 Regression coefficients and significances from logistic regression

	Regression coefficients	Significance
Furnace temperature		
Min.	−0.006	0.052
Max. Grad.	0.039	0.367
Avg.	−0.005	0.173
Std. Dev.	−0.023	0.008
Max. Grad.	0.291	0.365
Constant	−34.972	0.536
Temperature at top of steel bar		
Min.	−0.012	0.000
Max. Grad.	−0.090	0.154
Avg.	0.002	0.438
Std. Dev.	−0.030	0.000
Max. Grad.	2.324	0.018
Constant	118.763	0.116
Temperature in core of steel bar		
Min.	−0.007	0.104
Max. Grad.	−0.053	0.039
Avg.	0.011	0.000
Std. Dev.	−0.012	0.183
Max. Grad.	−0.076	0.940
Constant	60.616	0.025
Temperature at bottom of steel bar		
Min.	−0.012	0.000
Max. Grad.	−0.049	0.195
Avg.	0.009	0.001
Std. Dev.	−0.028	0.004
Max Grad.	1.872	0.151
Constant	61.134	0.149

of quality levels, those in [0; 0.75] labeled 0 for insufficient quality and those in [0.75; 1] labeled 1 for good quality. Instead of a comprehensive regression on all parameters, four separate analyses are performed as the process parameters analyzed are interdependent. The results of the regression models are shown in Table 2. The significances state, that the standard deviation is the major determinant of the quality level in most cases. Moreover, it has the strongest influence on quality levels of all significant parameters. That the steel bar's core is the only case in which the standard deviation of temperatures has no significant effect on the quality level can be explained with the heating process itself. The temperatures at the core are increasing markedly slower and more steadily, as the surrounding material attenuates the temperature differences. The same reasoning can be applied for the maximum gradient's values. Instead of the standard deviation the average temperature influences the quality most in this sce-nario.

The results for the minimum temperatures are the most striking ones in this analysis. The regression coefficients show that lower minimum temperatures, i.e., starting temperatures in the heating process, add to final quality. This effect is marginal yet mostly significant. This result can be explained by the sample for minimum temperatures which is not representative as it is skewed towards lower minimal temperatures. Actually it is expected, that higher minimum temperatures lead to higher quality. This can also be reasoned using standard deviation, as an increased minimum temperature will lead to a lower standard deviation in a continuous heating process as the difference between minimum and maximum temperatures decreases.

5.3 Conclusion on Statistical Tests

The correlation analysis shows that especially the standard deviation of temperatures is critical for producing good quality as it is highly significant and besides this has the highest negative correlation coefficients. The same is true for the maximum gradient. The regression analyses conducted confirm the relevance of the standard deviation, whereas the maximum gradient is found to be insignificant. This can be explained by the correlation of these two parameters. The relevance of average bar temperatures at all three locations in the regression model can be proven by the same facts. Developing the learning algorithm for quality predictions these factors with high significances have to be considered in the model.

The results presented of the logistic regression are in fact quite similar to those results generated by the linear regression analysis regarding input parameter significances and relative impact of each parameter. Even though these models include twenty parameters computed from only four separate time series gathered at the very first step in the process chain, the coefficient of determination (Nagelkerkes-R^2) indicates that the bar's models account for 7–9 % and the furnace's model for 4 % of the overall variability. The remaining 92 % have to be accounted for by models of the remaining processes.

This result directly influences the decision on the quality threshold of the rotary hearth furnace. Knowing all prediction models' coefficients of determination sheds a light on the distribution of quality thresholds. The higher the percentage of variation accounted for at a certain point in the process, the higher this process's quality level threshold has got to be. Thus, the quality threshold at the furnace will be rather low, compared to the remaining ones, which is a result that has been expected by the process experts, who assume the heating process not to be the most critical one. Nevertheless, a smooth heating-up process, avoiding high temperature changes, will have a positive effect on the product's quality.

6 Summary and Future Work

This paper proposes a production control concept based on data mining techniques. This concept offers the chance to reduce waste of energy and material resources as well as reworking and rejection costs resulting from producing, if inferior quality is not identified in the process. The proposed IMPC concept identifies those products by assessing their quality based on a prediction model. At each processing step a certain quality threshold has to be attained in order to permit further processing. To select quality relevant parameters for prediction, first statistical analyses were conducted on the data gathered at the first process step which show that the standard deviation of the temperature value series data recorded in the furnace and at three points of the steel bar has the most significant effect on quality levels. Moreover, the statistical models account for 7–9 % of variability, which is a satisfactory result due to the models' complexity and coinciding with experts' expectations.

The statistical analyses have to be conducted for all remaining steps in the process chain. Based on the results of these, the relevant parameters for the IQP module of the IMPC concept have to be identified. Thresholds for each process will be derived from both models' coefficients of determination and expert knowledge. Given this information local quality prediction model will be trained for each process. Once this is done, the IMPC can be applied in practice and helps to reduce waste due to processing products of inferior quality.

Acknowledgments This work has been supported by the DFG, Collaborative Research Center 876 "Providing Information by Resource-Constrained Data Analysis", project B3 "Data Mining in Sensor Data of Automated Processes". http://sfb876.tu-dortmund.de

References

1. Otte, R., Otte, V., Kaiser, V.: Data Mining für die industrielle praxis (Data Mining for Industrial Application). Hanser, Munich (2004)
2. Alvarez, E.G.: Advanced process control to meet the needs of the metallurgical industry. World Metall. ERZMETALL **58**(3), 123–128 (2005)
3. Morik, K., Bhaduri, K., Kargupta, H.: Introduction in data mining for sustainability. Data Mining and Knowledge Discovery, vol. 24, No. 2, pp. 311–324, Springer (2012)
4. Ohno, T.: Toyota Productions System, pp. 6–8. Productivity Press, Portland (1982)
5. Morik, K., Deuse, J., Faber, V., Bohnen, F.: Data mining in sensordaten verketteter prozesse (data mining in sensor data of interlinked processes). ZWF **105**(1–2), 106–110 (2010)
6. Lieber, D., Konrad, B., Deuse, J., Stolpe, M., Morik, K.: Sustainable interlinked manufacturing processes through real-time quality prediction. In: Leveraging Technology for a Sustainable World: Proceedings of the 19th CIRP Conference on Life Cycle Engineering. Springer, Berkeley (2012) (accepted for publication)
7. Haapamäki, J., Tamminen, S., Röning, J.: Data mining methods in hot steel rolling for scale defect prediction. In: International Conference on Artificial Intelligence and Applications, Innsbruck, Austria, pp. 90–94 (2005)
8. Stolpe, M., Morik, K., Konrad, B., Lieber, D., Deuse, J.: Challenges for data mining on sensor data of interlinked processes. In: Next Generation Data Mining Summit: Ubiquitous Knowledge Discovery for Energy Management in Smart Grids and Intelligent Machine-to-Machine (M2 M) Telematics, Athens, Greece (2011). Available at: http://www.kd2u.org/NGDM11
9. Oh, S., Han, J., Cho, H.: Intelligent process control system for quality improvement by data mining in the process industry. In: Braha, D.: Data Mining for Design and Manufacturing, Academic Publishers, Norwell pp. 289–309 (2001)

10. Dittmar, R., Pfeiffer, B.-M.: Modellbasierte prädiktive Regelung (Modell-based Predictive Control), pp. 1–4. Oldenbourg, Munich (2011)
11. Seborg, D.E., Edgar, T.F., Mellichamp, D.A.: Process Dynamics and Control, 2nd edn, pp. 411–414. Wiley, Hoboken (2004)
12. Fayyad, U.M.: Data mining and knowledge discovery: making sense out of data. IEEE Expert **11**(5), 20–25 (1996)
13. Fayyad, U.M., Piatetsky-Shapiro, G., Smyth, P.: From data mining to knowledge discovery in databases. AI Magazine **17**(3), 37–54 (1996)
14. Stolpe, M., Morik, K.: Learning from label proportions by optimizing cluster model selection. In: Gunopulos, D., Hofmann, T., Malerba, D., Vazirgiannis, M. (eds.) ECML PKDD 2011, Part III, vol. 6913, pp. 349–364, Springer, Berlin, Heidelberg (2011)
15. Menard, S.: Applied logistic regression analysis. 2nd edn. Sage University Papers Series on Quantitative Applications in Social Sciences 07–106, Sage, Thousand Oaks (2001)
16. Sethi, I.: Data mining: An introduction. In: Braha, D.: Data Mining for Design and Manufacturing, Kluwer Academic Publishers, Norwell pp. 1–40 (2001)

Part III
Robustness in Manufacturing Networks and Adaptable Logistics Chains

Role and Novel Trends of Production Network Simulation

Giacomo Liotta

Abstract Production networks are complex systems consisting of distributed entities that cooperate in manufacturing scenarios in a long term/stable collaboration time horizon. In order to govern the network complexity, manage the risks and dynamically predict the impacts of decisions before their implementation, simulation can be applied. The paper presents an overview on the state of the art of production network simulation, including also supply networks, and outlines evolution trends and challenges for further developments in this field. Trends and proposed challenges mainly deal with interdisciplinary research approaches, inclusion of sustainable development dimensions in the application scopes of simulation, and enabling simulation technologies and architectures.

Keywords Global production · Simulation · Complexity · Sustainability

1 Introduction

Current industrial scenarios are characterized by a raising complexity due to the number of variable factors, interdependencies, uncertainties and data lack affecting the decisions related to manufacturing and logistics network design, planning and control. At global scale, a Production Network (PN) can be considered as a complex system consisting of distributed entities, i.e., manufacturers, suppliers, linked by material, information as well as financial flows. Members cooperate with a long term/stable collaboration time horizon and the network can be heterarchic

G. Liotta (✉)
National Research Council of Italy, Institute of Industrial Technologies and Automation (CNR—ITIA) Strada Della Neve, via Salaria Km 29,300, 00010 Montelibretti, RM, Italy
e-mail: giacomo.liotta@itia.cnr.it

K. Windt (ed.), *Robust Manufacturing Control*, Lecture Notes in Production Engineering, 233
DOI: 10.1007/978-3-642-30749-2_17, © Springer-Verlag Berlin Heidelberg 2013

or focused on one focal company [1]. On a lower scale, i.e., within a single industrial plant, PNs consist of work systems (the network nodes) that exchange material and information flows through links to jointly realize products. In Economics and Finance-oriented models, PNs are usually described by vertices representing firms on different levels and customers; several types of edges represent product, information and financial flows.

In order to govern the network complexity, manage the risks and dynamically predict the impacts of decisions before their implementation, simulation can be applied. Simulation allows to simultaneously consider several parameters that can influence the network performance and evolution patterns. It can be used as an experimental arena where other methods, modeling paradigms and interdisciplinary approaches can be tested. However, simulation of PNs requires heavy modeling, implementation, verification and validation efforts. This paper presents a literature review on supply and PN simulation on different scales. It then investigates evolution trends and proposes challenges for this method by highlighting interdisciplinary research contributions, sustainable development issues and potential areas for technological advances. The remainder of the paper is organized as follows. Section 2 presents the role and the state of the art of simulation related to networked logistics-production systems. Evolution trends and challenges are discussed in Sect. 3. Conclusions follow.

2 Simulation of Networked Logistics-Production Systems

The traditional role of simulation is to provide decision support from the strategic, tactical or operational standpoints via the execution of *what-if* analyses based on experiments purposively designed. Nonetheless, the role of simulation in business games for learning, teaching, training and research purposes is also remarkable. Simulation in production and logistics can be devoted to different system levels: from the evaluation of single nodes of a network (e.g., production cell, factory, warehouse, container terminal, etc.) to an entire network of interacting nodes (e.g., a PN with the related different tiers and logistic flows). Simulation in Supply Chain Management (SCM) revealed a more relevant research effort than simulation focused on PNs, probably due to the wide interpretation of the Supply Chain Network (SCN) concept commonly accepted and introduced by seminal studies in the past. Kleijnen (2005) [2] discusses the application of four different simulation types in SCM, more specifically, spreadsheet simulation, system dynamics, discrete event dynamic systems simulation, business games. Terzi and Cavalieri [3] discuss the important role of simulation techniques in SCM while developing a survey based on more than 80 papers. Surveys explicitly dedicated to PN simulation have not been found in the literature. The purpose of the following literature review is to present an overview of SCN and PN simulation due to the similarities, links and overlaps existing between these two general concepts. The review is based on the following categories (see Table 1):

Table 1 Literature review focused on specific simulation applications

	Papers													
	Persson [4]	Toshniwal et al. [15]	Duffie et al. [16]	Donner et al. [20]	Renna and Aragoneto [17]	Armbruster et al. [30]	Hirsch et al. [40]	Kara et al. [41]	Confessore et al. [39]	Özbayrak et al. [10]	Pierreval et al. [18]	Scholz-Reiter et al. [31]	Sun et al. [37]	Deleris et al. [35]
Simulation														
DES	x	x	x	x	x	x	x	x	x					
SD										x	x	x		
MCS													x	x
ABS														
SG														
H/IA				x	x									x
OTH														
SP	L	L	L	L	L	L	L	L	L	L	L	L	L	L
Network														
SCN	x									x			x	
PN		x	x	x	x	x	x	x	x		x	x		x
Scale														
Single site		x	x			x			x			x		
Multi-site	x			x	x		x	x		x	x		x	x
Interdisc.						x						x		x

	Papers																			
	Chan and Chan [7]	Kaihara and Fujii [8]	Li et al. [24]	Li et al. [25]	Mizgier et al. [27]	Schwesig et al. [22]	Holweg and Bicheno [13]	Meijer [14]	Tunali et al. [5]	Shukla et al. [26]	Almeder et al. [6]	Lanza and Ude [19]	Herrmann et al. [38]	Rodriguez et al. [9]	Battiston et al. [34]	Weisbuch and Battiston [33]	Ueda et al. [29]	Becker et al. [32]	Uygun et al. [21]	Iannone et al. [12]
Simulation																				
DES									x	x	x	x								
SD																				
MCS												x								
ABS	x	x	x	x	x															
SG						x	x	x												
H/IA		x	x	x					x	x	x	x	x							
OTH														x	x	x	x	x	x	x
SP	L	L	L	L	L	L	L	L	L	L	L	L	L	L	L	L	L	L	D	D
Network																				
SCN	x	x	x	x	x		x	x	x	x	x			x	x	x				x
PN						x						x	x				x	x	x	
Scale																				
Single site													x				x	x		
Multi-site	x	x	x	x	x	x	x	x	x	x	x	x		x	x	x			x	x
Interdisc.			x	x	x					x					x	x	x	x		

- *Simulation method and paradigm*: Discrete Event Simulation (DES), System Dynamics (SD), Monte Carlo Simulation (MCS), Agent Based Simulation (ABS), business Simulation Games (SG), Hybrid/Integrated Approaches (H/IA) intended as a combination of different simulation techniques or as an integration of optimization/mathematical modeling and simulation, other computer simulations not specified in a paper (OTH), Simulation Paradigm (SP): Local (L) vs. Distributed (D).
- *Network type*: SCN, PN.
- *Scale*: single site (i.e., single plant/facility) or multiple site (i.e., a network of plants, multiple tiers or actors involved in a network).
- *Interdisciplinary approach* (Interdisc.): presence of modeling approaches simultaneously involving several disciplines, e.g., Engineering, Biology, Economics, etc.

Several analyzed papers rely on approaches concurrently including different simulation methods and mathematical models (e.g., in the H/IA simulation category).

Moreover, since the purpose of the paper is to provide an overview on the most relevant, recent papers on the subject while identifying research contributions of different science disciplines, the literature review presentation consists of two parts: the former is devoted to single-discipline approaches (e.g., Engineering, Computer Science) while the latter on interdisciplinary approaches involving also, e.g., Biology, Physics. In each part, a further distinction in terms of network type is provided.

2.1 Single-Discipline Approaches

SCN Simulation. SCN simulation typically entails the modeling of several tiers/ actors and sites over the network. The following references present structured scenarios consisting of suppliers, manufacturers, logistics facilities, distributors or customers. DES is applied in [4] with the aim of comparing different scenarios in terms of different routings between production sites. Tunali et al. [5] study a supply chain by using a hybrid approach that integrates mathematical modeling and (discrete) simulation modeling to set realistic order due dates. Almeder et al. [6] present a robust solution approach based on the combination of optimization techniques and DES in a general SCN context. Chan and Chan [7] investigate a Make-To-Order (MTO), 2-echelon, multi-product manufacturing SCN by using ABS. In [8], a simulation model to form a virtual enterprise environment with multiple negotiations among several potential partners (agents) in the supply chain is developed. Rodriguez-Rodriguez et al. [9] present a simulation study for a business scenario with the aim of studying negotiation processes in a supply network while addressing manufacturing capacity issues. In [10], the case of a supply chain with 4 echelons in a MTO system is faced by means of SD from an operational viewpoint. In [11], the author discusses the applicability of Distributed Simulation (DS) technology for across-echelon and within-echelon SCM with the latter viewed as likely the most promising application scenario. Iannone et al. [12] propose an architecture (SYNCHRO) able to synchronize simulation models located in different geographical areas such as in supply chain contexts. The relevance of game-based simulation approaches is essentially related to the involvement of humans (players) and to the observation and analysis of their behaviors in interaction processes related to particular business decision scenarios. In [13], a logistics game related to the case of UK automotive steel supply chain is presented. The concept and purposes of gaming simulation are deepened in [14] where two gaming simulations concerning supply networks are presented and discussed.

PN Simulation. PN simulation can refer to different scales: a network of production resources within a single plant (site) or a network of distributed plants (multi-site).

Single Site Simulation. Toshniwal et al. [15] use DES models and an industrial dataset in order to assess the fidelity of control-theoretic models in a PN through comparisons with DES. Duffie et al. [16] present a dynamic model of PN consisting of a large number of autonomous work systems with local capacity control. The results of a control-theoretic model are compared with a DES model.

Multi-site Simulation. Renna and Aragoneto [17] present a methodology based on game theory to support co-opetition networks for capacity management in response to unexpected events: a Multi Agent System (MAS)-based architecture and DES are exploited. Pierreval et al. [18] make use of SD for reproducing various types of production units for complex products in a supply chain of the automotive industry for, e.g., investigating the global dynamic trends of the set of collaborating units. Lanza and Ude (2011) [19] present an integrated approach based on multi-criteria analysis, DES and MCS for evaluating different configurations of value added networks. Donner et al. [20] investigate a manufacturing network consisting of 4 cooperating companies by using DES and the analysis of time series of logistic quantities. Uygun et al. [21] present an overview of distributed manufacturing simulation and the related information representation by High Level Architecture (HLA) and its Object Model Template: a simple scenario is discussed. Schwesig et al. [22] present a web based group simulation game in which a second layer of the game reproduces a PN environment: the players representing different companies participate in the inter-organizational product development process.

2.2 Interdisciplinary Research Approaches

In order to tackle the complexity of SCNs and PNs, interdisciplinary research approaches have been investigated in the literature. This Section distinguishes between SCN and PN simulation while highlighting, in the third part, simulation approaches potentially very relevant for addressing sustainability issues that may require multidisciplinary contributions.

SCN Simulation.

Complex Adaptive System (CAS)-Based Simulation Approaches. Pathak et al. [23] discuss the CAS perspective in the study of supply networks, thus introducing the concept of Complex Adaptive Supply Network (CASN). In [24] and [25], the evolution of, respectively, supply networks and CASNs is investigated while exploiting MAS simulation, CAS paradigm and fitness landscape theory for modeling the dynamic behavior of the networks and the interactions among the firms and the environment (demand, laws, etc.).

Psychology and Biology Inspired Simulation Approaches. Shukla et al. [26] present a hybrid approach consisting of DES, Taguchi method, robust multiple non-linear regression analysis and a psychoclonal algorithm with the aim of studying the complexity of interactions in a supply chain.

Economics and Finance-Oriented Simulation Approaches. Economics and Finance-oriented research has addressed SCN and PN issues through simulation. Artificial Economics, for instance, encompasses also these issues among other social, behavioral and macro-economic aspects. Mizgier et al. [27] exploit agent-based modeling for simulating the dynamics of a SCN in an uncertain environment for analyzing the propagation modes of local defaults of supply chain actors over the network entailing bankruptcy effects. Their approach is based on a model of dynamics for investigating particular phenomena in Physics [27].

PN Simulation.

Biology Inspired Simulation Approaches. The analogies between biological entities and manufacturing facilities as complex systems operating in dynamically changing environments are discussed in [28].

The following references concern single-plant simulation. Ueda et al. [29] present a Biological Manufacturing System (BMS) model based on self-organization while introducing bounded-rational agents: discrete-time ant system simulation (with bounded-rationality) and BMS simulation models (including also a self-organization method) are developed and compared. Armbruster et al. [30] propose an approach for autonomously controlled PNs by comparing the results of DES and fluid models and testing a pheromones-based control strategy. Scholz-Reiter et al. [31] present two bio-inspired control strategies for a multi work system PN based, respectively, on an ant-like pheromone strategy as well as on bee's foraging behavior. The strategies are tested through SD simulation while highlighting a better performance in terms of WIP control with respect to real data. Becker et al. [32] propose a framework consisting of network descriptions, commonalities and differences in order to model metabolic, traffic, and PNs in a unified manner: network simulation experiments are then conducted.

Economics and Finance-Oriented Approaches. The two following research works entail the Economics and Finance perspectives merged with models for the production dynamics deriving from Chemical Physics. In [33], the authors investigate via simulation the consequences of simple local processes of orders/production/delivery/profit/investment on the evolution of PNs and wealth while addressing local failures conditions and spatial economics issues. Battiston et al. [34] make use of simulation for investigating credit chains and bankruptcy propagation in PNs.

Deleris et al. [35] reproduce a General Semi Markov Process via MCS for analyzing the risks of disruptions generated by hazard events for a PN consisting of 80 plants. Their approach links risk modeling based on insurance data with the structure of the SCNs while supporting network design decisions.

Sustainability-Oriented Simulation Approaches. Manring and Moore [36] discuss interdisciplinary perspectives of networks, virtual webs, and learning organizations while addressing sustainable industrial development and production issues in terms of economic, ecological, and social dimensions: the authors discuss the impacts on dynamics of business simulation exercises. Sun et al. [37] present a study based on MCS concerning an industrial symbiosis network where exchanges of energy and material take place by analyzing interdependencies among the industrial chains and including financial data.

Under the economic and environmental sustainability standpoints, it is important to underline the increased importance of energy efficiency issues in manufacturing. Simulation-based research approaches bounded to the evaluation of single factories have been developed in the literature (see for instance [38] and [39]). Nonetheless, these research approaches can represent basic building blocks the simulation of energy consumption at multi-site (global) network scale could be based on.

The importance of reverse logistics networks for managing the End-of-Life (EoL) phases of the product lifecycle recalled the attention of research scientists who developed simulation-based approaches. For instance, in Hirsch et al. [40] a discrete simulator is presented for the environmental assessment of logistic operations (i.e., transport, stock, transshipment) in terms of types of emissions, total energy consumed and noise penetration in a globally distributed PN. Kara et al. [41] present a DES model for analyzing a reverse logistics network dedicated to EoL management of discarded white goods.

3 Trends and Challenges for PN Simulation

3.1 Major Findings and Novel Trends Observed

The cited papers exhibit the use of the various simulation techniques for facing strategic, tactical and operational problems in production and logistics mainly at multi-site scale. From this perspective, research works related to specific applications did not include aspects concerning environmental, economic and social sustainability with comprehensive research approaches (i.e., including, in parallel, transport, emissions, land use and social impacts). On the other hand, sustainability and economics issues have been addressed only in focused researches that make use of simulation.

On the basis of the literature review, simulation of SCNs and PNs through interdisciplinary research approaches has been used in about 32 % of the cases.

These approaches include relevant and innovative aspects such as, e.g.: (i) the CAS-based investigation of the dynamicity and self-adaptation of network structures and collaboration mechanisms in SCN evolution, (ii) the effectiveness and good performance of control strategies in PNs at single plant scale using biology inspired approaches, (iii) the consideration of bankruptcy and failures in PNs from an Economics and Finance viewpoint using study approaches deriving from Physics. Other studies addressed sustainability in terms of reverse logistics simulation for product reuse, remanufacturing and recycling purposes but also basic elements for energy oriented simulation that could be useful for PN simulation at higher scales (i.e., network of plants). From the viewpoint of simulated network type, about 50 % of the cited papers related to specific applications explicitly addressed the simulation of PNs while the others dealt with the wider concepts of SCN that include manufacturers, suppliers and customers. The model scale is predominantly focused on multi-site simulation (76 % of the cases) while in 24 % of the cases on the simulation of PN resources within a single plant.

Simulation of SCNs and PNs is performed by means of various techniques. The most applied technique is DES that is used in about 38 % of the cited papers focusing on specific applications. ABS revealed a significant usage rate (about 15 %) as well as other computer simulation techniques not explicitly mentioned in a paper. MCS, business SG and SD account for about 9 % each. H/IA approaches combining simulation techniques and/or analytical models have been observed in about 32 % of the cases. Generally speaking about simulation in manufacturing and business, it is important to underline that, according to Jahangirian et al. [42], DES, although being a popular technique, does not reveal the same level of stakeholder engagement with respect to, e.g., SD and simulation gaming, likely because it can entail difficulties and time for data gathering. SD is based on standardized conceptual modeling techniques and it is less reliant on hard data with respect to DES while simulation gaming is widely applied for education and training purposes [42]. However, the cases examined in this paper have shown the use of DES enriched by combining simulation and analytical models, e.g., optimization models, as well as discrete simulation and continuous models. This aspect represents a very promising evolution trend in order to (i) optimize the number of experiments to be performed for achieving the desired system performance in the virtual environment, (ii) embody intelligence in a simulation control logic, (iii) perform comparisons between solutions statically and dynamically obtained for verification and validation purposes, and (iv) include into discrete systems the effects of several phenomena described by continuous functions in the reality. Indeed, in [42], it is highlighted that hybrid simulation based on the combination of DES and SD is becoming popular. Business/logistics games have been mainly used for educational and training purposes or for research purposes in gaming simulation (see for instance [14]).

Regarding the simulation paradigms, i.e., local simulation executed on a single computer vs. DS or Parallel Simulation (PS), recent studies revealed a growing interest. Its application can effectively reproduce in virtual environments distributed SCN and PN structures consisting of different, autonomous nodes. It offers the possibility to adopt architectures such as the HLA which enables the integration of separate simulation models (see for instance [21]). Although these technologies still need further developments for being more widely and economically deployable in real industrial cases (see [43, 44]), they can represent a novel evolution trend of SCN and PN simulation if industry utilization requirements will be fully met.

Finally, the raising interest on complexity governance in SCNs and PNs triggered relevant studies based on: simulation approaches including the CAS concept in combination with MAS modeling paradigm, and modeling approaches for combining the interactions in the design of production systems, product and environment (see [45]). This trend could be very promising for simulating the behavior of complex socio-technical-environmental systems in production and logistics.

3.2 *Potential Challenges*

PN simulation has to face new challenges beyond the traditional industrial use oriented to network design, production planning and control for the evaluation of production and logistic performance indicators. According to Váncza et al. [46], in production engineering and management, operations of enterprises are performed while interacting with consumers, market competitors, suppliers, technology and service providers, authorities and agencies that define the business environment. The study of complex PNs has to include further elements related to the simultaneous consideration of economic, environmental and social sustainability. Public bodies and industry need to use in decision practices valid models as well as user-friendly, cost-reasonable tools for the analyses related to the sustainable development of territories and businesses. Therefore, future challenges for PN simulation should deal with:

1. The inclusion of sustainable development issues on a multi-site scale.
2. Suitable, enabling simulation technologies and architectures for interoperability.

Concerning sustainable development, manufacturing industry daily copes with scenario changes such as, e.g., energy, material and labor cost variability, financial turbulence, demand fluctuations, new regulations, new social and environmental concerns. Hence, interdisciplinary approaches in logistics-production problem setting and solving need a wider spread. The aims are to successfully transfer methods to different application fields and to face problems simultaneously addressing several analysis dimensions such as the logistics-production performance as well as the economic, financial, social and environmental ones. There are still lacks in the concurrent inclusion of transport, emissions, land use and social issues in SCN and PN simulation. PN simulation should then include the harmonization of development aspects (e.g., infrastructures, socio-economic activities' development, transportation modes and logistics platforms, etc.) of regional/local areas where production facilities (will) operate. In these cases, the validation of complex models could represent a hard challenge especially when the simulation time horizon is very extensive (i.e., years). In terms of sustainability, from the standpoint of tool requirements, PN simulation should enable more easily the inclusion of energy procurement and consumption/emissions elements in parallel with the traditional logistic/production processes. These features could be included by enriching the parameter and variable settings within simulation tools. From the standpoint of usage scopes, simulation should address with heavier efforts the modeling of reverse logistics systems possibly in parallel with the PN simulation in order to mimic forward and reverse material flows. This aspect can be interesting in order to pave production scenario settings simultaneously including de-manufacturing.

Concerning the simulation technologies, PN simulation seems to need further advances for the combined use of optimization and simulation techniques, thus facilitating the integration and interoperability of optimization and simulation engines. The current elaboration speed of commercial computers often does not allow to rapidly test complex optimization models in combination with simulation, although notable improvements have been made for the combination of optimization and simulation, e.g., *Simulation Optimization* (see for instance [47]). From the viewpoint of simulation techniques, DES is demonstrating to be extensively used and exploited in hybrid approaches including, e.g., optimization, SD. ABS and business SG are other promising techniques for reproducing multi-objective, multi-decision maker PN scenarios. In terms of simulation paradigm, DS and PS can represent a challenging development line. They can be suitable for simulating networked logistics and production systems with several actors/modelers involved in distributed scenarios controlled by a single company, or in scenarios with the engagement of multiple autonomous companies using different simulation tools. However, there is still the need of overcoming some lacks in technologies that hinder its wide and cost-effective adoption in the industrial practice (see [43] and [44]).

4 Conclusions

SCN and PN simulation is currently performed by means of different research approaches (single-discipline and interdisciplinary) and techniques (mainly DES). It is used for carrying out the estimation of logistics-production performance but also for addressing economics, finance as well as sustainability issues in some focused cases. This work points out the challenge of concurrently including in PN simulation applications sustainable development concerns in addition to the logistics-production performance estimates at global scale. For this purpose, interdisciplinary approaches should be fostered. From the technology viewpoint, future challenges deal with the continuous improvement of enabling simulation technologies and architectures for model interoperability as well as for the combined use of simulation and optimization.

References

1. Wiendahl, H.P., Lutz, S.: Production in Networks. CIRP Ann. Manuf. Technol. **51**, 573–586 (2002)
2. Kleijnen, J.P.C.: Supply chain simulation tools and techniques: a survey. Int. J. Simul. Process Model. **1**, 82–89 (2005)
3. Terzi, S., Cavalieri, S.: Simulation in the supply chain context: a survey. Comput. Ind. **53**, 3–16 (2004)
4. Persson, F.: SCOR template—A simulation based dynamic supply chain analysis tool. Int. J. Prod. Econ. **131**, 288–294 (2011)
5. Tunali, S., Ozfirat, P.M., Ay, G.: Setting order promising times in a supply chain network using hybrid simulation-analytical approach: an industrial case study. Simul. Model. Pract. Theory **19**, 1967–1982 (2011)
6. Almeder, C., Preusser, M., Hartl, R.-F.: Simulation and optimization of supply chains: alternative or complementary approaches? OR Spectrum **31**, 95–119 (2009)
7. Chan, H.K., Chan, F.T.S.: Comparative study of adaptability and flexibility in distributed manufacturing supply chains. Decis. Support Syst. **48**, 331–341 (2010)
8. Kaihara, T., Fujii, S.: Virtual enterprise coalition strategy with game theoretic multi-agent paradigm. CIRP Ann. Manuf. Technol. **55**, 513–516 (2006)
9. Rodriguez-Rodriguez, R., Gonzalez, P.P., Leisten, R.: From competitive to collaborative supply networks: a simulation study. App. Math. Model. **35**, 1054–1064 (2011)
10. Özbayrak, M., Papadopoulou, T.C., Akgun, M.: Systems dynamics modelling of a manufacturing supply chain system. Simul. Model. Pract. Theory **15**, 1338–1355 (2007)
11. Lendermann, P.: About the need for distributed simulation technology for the resolution of real-world manufacturing and logistics problems. In: Perrone, L.F., Wieland, F.P., Liu, J., Lawson, B.G., Nicol, D.M., Fujimoto, R.M. (eds.) Proceedings of the 2006 Winter Simulation Conference, pp. 1119–1128. IEEE, Piscataway (2006)
12. Iannone, R., Miranda, S., Riemma, S.: Supply chain distributed simulation: an efficient architecture for multi-model synchronization. Simul. Model. Pract. Theory **15**, 221–236 (2007)
13. Holweg, M., Bicheno, J.: Supply chain simulation—A tool for education, enhancement and endeavour. Int. J. Prod. Econ. **78**, 163–175 (2002)
14. Meijer, S.A.: The Organisation of Transactions: Studying Supply Networks Using Gaming Simulation. PhD Thesis. Wageningen University, The Netherlands (2009)
15. Toshniwal, V., Duffie, N., Jagalski, T., Rekersbrink, H., Scholz-Reiter, B.: Assessment of fidelity of control-theoretic models of wip regulation in networks of autonomous work systems. CIRP Ann. Manuf. Technol. **60**, 485–488 (2011)
16. Duffie, N.A., Roy, D., Shi, L.: Dynamic modeling of production networks of autonomous work systems with local capacity control. CIRP Ann. Manuf. Technol. **57**, 463–466 (2008)
17. Renna, P., Argoneto, P.: A game theoretic coordination for trading capacity in multisite factory environment. Int. J. Adv. Manuf. Technol. **47**, 1241–1252 (2010)
18. Pierreval, H., Bruniaux, R., Caux, C.: A continuous simulation approach for supply chains in the automotive industry. Simul. Model. Pract. Theory **15**, 185–198 (2007)
19. Lanza, G., Ude, J.: Multidimensional evaluation of value added networks. CIRP Ann. Manuf. Technol **59**, 489–492 (2010)
20. Donner, R., Scholz-Reiter, B., Hinrichs, U.: Nonlinear characterization of the performance of production and logistics networks. J. Manuf. Syst. **27**, 84–99 (2008)
21. Uygun, Ö., Öztemel, E., Kubat, C.: Scenario based distributed manufacturing simulation using HLA technologies. Inform. Sciences **179**, 1533–1541 (2009)
22. Schwesig, M., Thoben, K.D., Eschenbacher, J.: A simulation game approach to support learning and collaboration in virtual organizations. In: Camarinha-Matos, L.M., Afsarmanesh, H., Ortiz, A. (eds.) IFIP TC5 WG 5.5 Sixth IFIP Working Conference on Virtual Enterprises, Collaborative Networks and their Breeding Environments, IFIP AICT, vol. 186, pp. 547–556 (2005)

23. Pathak, S.D., Day, J.M., Nair, A., Sawaya, W.J., Kristal, M.M.: Complexity and adaptivity in supply networks: building supply network theory using a complex adaptive systems perspective. Decision Sci. **38**, 547–580 (2007)
24. Li, G., Ji, P., Sun, L.Y., Lee, W.B.: Modeling and simulation of supply network evolution based on complex adaptive system and fitness landscape. Comput. Ind. Eng. **56**, 839–853 (2009)
25. Li, G., Yang, H., Sun, L., Ji, P., Feng, P.: The evolutionary complexity of complex adaptive supply networks: a simulation and case study. Int. J. Prod. Econ. **124**, 310–330 (2010)
26. Shukla, S.K., Tiwari, M.K., Wana, H.-D., Shankar, R.: Optimization of the supply chain network: simulation, taguchi, and psychoclonal algorithm embedded approach. Comput. Ind. Eng. **58**, 29–39 (2010)
27. Mizgier, K.J., Wagner, S.M., Holyst, J.A.: Modeling defaults of companies in multi-stage supply chain networks. Int. J. Prod. Econ. **135**, 14–23 (2012)
28. Mill, F., Sherlock, A.: Biological analogies in manufacturing. Comput. Ind. **43**, 153–160 (2000)
29. Ueda, K., Kito, T., Fujii, N.: Modeling biological manufacturing systems with bounded-rational agents. CIRP Ann. Manuf. Technol. **55**, 469–472 (2006)
30. Armbruster, D., de Beer, C., Freitag, M., Jagalski, T., Ringhofer, C.: Autonomous control of production networks using a pheromone approach. Physica A **363**, 104–114 (2006)
31. Scholz-Reiter, B.; Karimi, H.R.; Duffie, N.A.; Jagalski, T.: Bio-inspired capacity control for Production Networks with autonomous work systems. In: 44th CIRP international conference on manufacturing systems, Madison, USA (2011)
32. Becker, T., Beber, M.E., Windt, K., Hütt, M.T., Helbing, D.: Flow control by periodic devices: a unifying language for the description of traffic, production, and metabolic systems. J. Stat. Mech. Theory E., pp. 1–27 (2011). doi:10.1088/1742-5468/2011/05/P05004
33. Weisbuch, G., Battiston, S.: From production networks to geographical economics. J. Econ. Behav. Organ. **64**, 448–469 (2007)
34. Battiston, S., Delli Gatti, D., Gallegati, M., Greenwald, B., Stiglitz, J.E.: Credit chains and bankruptcy propagation in production networks. J. Econ. Dyn. Control **31**, 2061–2084 (2007)
35. Deleris, L.A., Elkins, D., Paté-Cornell, M.E.: Analyzing losses from hazard exposure: a conservative probabilistic estimate using supply chain risk simulation. In: Ingalls, R.G., Rossetti, M.D., Smith, J.S., Peters, B.A. (eds.) Proceedings of the 2004 Winter Simulation Conference, pp. 1384–1391. IEEE, Piscataway (2004)
36. Manring, S.L., Moore, S.B.: Creating and managing a virtual inter-organizational learning network for greener production: a conceptual model and case study. J. Clean Prod. **14**, 891–899 (2006)
37. Sun, J.C., Wang, N.M., Cheng, S.C.: Measuring interdependency in industrial symbiosis network with financial data. Energy Procedia **5**, 1957–1967 (2011)
38. Herrmann, C., Thiede, S., Kara, S., Hesselbach, J.: Energy oriented simulation of manufacturing systems—concept and application. CIRP Ann. Manuf. Technol. **60**, 45–48 (2011)
39. Confessore, G., Liotta, G., De Luca, P.: A simulation model for estimating energy and plant utilities consumption in a pharmaceutical production system. In: Teti, R. (ed.) Proceedings of 6th CIRP International Conference on Intelligent Computation in Manufacturing Engineering, pp.121–126 (2008)
40. Hirsch, B.E., Kuhlmann, T., Schumacher, J.: Logistics simulation of recycling networks. Comput. Ind. **36**, 31–38 (1998)
41. Kara, S., Rugrungruang, F., Kaebernick, H.: Simulation modelling of reverse logistics networks. Int. J. Prod. Econ. **106**, 61–69 (2007)
42. Jahangirian, M., Eldabi, T., Naseer, A., Stergioulas, L.K., Young, T.: Simulation in manufacturing and business: a review. Eur. J. Oper. Res. **203**, 1–13 (2010)
43. Straßburger, S., Schulze, T., Fujimoto, R.: Future trends in distributed simulation and distributed virtual environments. In: Alexopoulos, C., Goldsman, D., Wilson, J.R. (eds.) Advancing the Frontiers of Simulation: A Festschrift in Honor of George Samuel Fishman, International Series in Operations Research and Management Science, vol. 133, pp. 231–261. Springer, Heidelberg (2009)

44. Lendermann, P., Heinicke, M.U., McGinnis, L.F., McLean, C., Straßburger, S., Taylor, S.J.E.: Panel: distributed simulation in industry—A real-world necessity or ivory tower fancy? In: Henderson, S.G., Biller, B., Hsieh, M.H., Shortle, J., Tew, J.D., Barton, R.R. (eds.) Proceedings of the 2007 Winter Simulation Conference, pp. 1053–1062. IEEE, Piscataway (2007)
45. Schuh, G., Monostori, L., Csáji, B.Cs., Döring, S.: Complexity-based modeling of reconfigurable collaborations in production industry. CIRP Ann. Manuf. Technol. **57**, 445–450 (2008)
46. Váncza, J., Monostori, L., Lutters, D., Kumara, S.R., Tseng, M., Valckenaers, P., Van Brussel, H.: Cooperative and responsive manufacturing enterprises. CIRP Ann. Manuf. Technol. **60**, 797–820 (2011)
47. April, J., Glover, F., Kelly, P., Laguna, M.: Practical introduction to simulation optimization. In: Chick, S., Sánchez, P.J., Ferrin, D., Morrice, D.J. (eds.) Proceedings of the 2003 Winter Simulation Conference, pp. 71–78. IEEE, Piscataway (2003)

On the Configuration and Planning of Dynamic Manufacturing Networks

Nikolaos Papakostas, Konstantinos Efthymiou, Konstantinos Georgoulias and George Chryssolouris

Abstract Manufacturing organizations have been attempting to improve the operation of supply networks through efficient supply chain management. Dynamic Manufacturing Networks (DMNs) constitute chains of diverse partners, whose operation and interaction may change in a rapid and often not predictable way. While the existing supply chain models are quite static, and examine transportation modes, product changeover and production facility options with fixed suppliers and over a long period of time, the DMNs address operations and risks on a daily basis. In this paper, a novel decision-making approach is proposed for supporting the process of configuring a DMN from a holistic perspective, taking into account production, transportation and time constraints as well as multiple criteria, such as time and cost.

1 Introduction

In a volatile market environment, today's manufacturing organizations strive to improve their performance, whilst providing customers with more customization options [1]. The main classes of attributes to be considered when making manufacturing decisions, i.e. cost, time, quality and flexibility, are closely interrelated and have been investigated towards optimization, in an attempt to improve product quality, to confront market competition, to shorten lead times, as well as to reduce

This contribution was previously published in Logistics Research (2012) pp. 105–111. DOI:10.1007/s12159-012-0086-9.

N. Papakostas · K. Efthymiou · K. Georgoulias · G. Chryssolouris (✉)
Laboratory for Manufacturing Systems and Automation, Department of Mechanical Engineering and Aeronautics, University of Patras, Patras, Greece
e-mail: xrisol@lms.mech.upatras.gr

K. Windt (ed.), *Robust Manufacturing Control*, Lecture Notes in Production Engineering, DOI: 10.1007/978-3-642-30749-2_18, © Springer-Verlag Berlin Heidelberg 2013 costs. These aspects constitute the main reason for the increasing complexity met in modern manufacturing systems. Controlling this complexity with conventional methods, such as the approaches based on Manufacturing Resource Planning (MRP II) principles and concepts, require more and more data and is becoming extremely difficult to manage. One of the top business pressures, dealt by enterprises, is the need to react to demand changes in a timelier manner. Further to having to address the increase in year-over-year fulfillment and transportation costs per unit, companies have been attempting to improve the cross-channel supply chain flexibility in order to achieve a faster reaction to demand changes and to improve supply chain responsiveness [2]. Manufacturing companies should be able to quickly restructure or transform the supply chain execution (source–deliver processes) in response to an evolving global, multi-channel supply chain scenario. However, a lot of companies still do not have the ability to respond to dynamic demand cycles, while, at the same time, the increased globalization pushes the demand uncertainty at even higher levels [2]. In the retail domain, for instance, the demand has been so uncertain in the time span between mid 2010 and end of 2011 that the volume of inventory has either been too high or too low [2]. The recent events, concerning the volcano's eruption in Iceland and the nuclear disaster in Fukushima, have reaffirmed the need for greater flexibility in order for manufacturing organizations to cope with the dynamic nature of the market and its fluctuations.

At the same time, the existing, off-the-shelf Supply Chain Management software platforms and tools are too expensive to be implemented and deployed at a broader networked enterprise scale, including smaller companies with limited Information and lower Communication capacity, and are unable to:

- Cover all actual phases of a manufacturing network lifecycle and
- Cope with the highly dynamic and uncertain nature of demand.

It is not enough for today's manufacturing enterprises to be networked: they have to be able to change and adapt to a continuously evolving environment and to form dynamic alliances with other companies and organizations in a fast and cost-efficient manner.

2 Current Approaches for Manufacturing Network Management

The variations at trade barriers level and the worldwide evolution of the transportation and communication means have led to the globalization of manufacturing activities [3]. New global strategies have pushed forward the internationalization of manufacturing systems [4]. The manufacturing landscape has become more competitive, dynamic and complex.

A large number of studies have addressed various aspects of the supply chain management problem. The initial configuration of supply chains and the selection of partners constitute one of the most critical phases in the lifecycle of a supply network.

A few research efforts have proposed the use of mixed-integer mathematical models with the objective to maximize profits or minimize the overall supply chain operation costs [5]. Others have focused on the identification of the optimum transportation modes for minimizing the total transportation and inventory costs, including those addressing multi-product cases for identifying optimal shipping times and loading policies [6]. Production planning and transportation problems have also been addressed jointly [7]. Another stream of research work has dealt with the problem of having the supply chain flexibility increased, whilst retaining the capability to produce towards satisfying demand, by leveraging the alternative supply chain options and the routing flexibility within a pre-defined planning horizon [8]. The problem of locating or relocating production facilities for satisfying the varying local demand has also been modeled by a few researchers. In some cases, transportation mode and product switching decisions have been addressed jointly [9, 10].

Collaborative planning of fixed supply networks is another issue that has attracted the interest of many research teams. The objective is to align the plans of the individual supply chain partners and coordinate the production of the supply chain towards achieving a series of common, or in some cases partner-specific, objectives [11]. Hierarchical approaches, initiated by the Original Equipment Manufacturer (OEM) have also been proposed, where each partner's tier performs all production planning activities and then provides these plans to the next tier for carrying out its own process of production planning, until all tiers have completed their production planning activities [12]. Merging the planning activities of several partners into one planning domain may improve the results of the upstream collaboration [13]. Negotiation-based collaborative planning approaches have been reported, focusing on the use of upstream planning at the beginning and then on the employment of a negotiation process in order for the overall performance to be improved [14].

The vast majority of the research work reported, dealing with the supply chain management and optimization, dealt with very specific parts of the phases of a supply chain lifecycle. A few recent studies have dealt with the challenges related to each phase of the supply chain lifecycle in a more integrated manner. The combined problem involving multiple transportation modes, diverse supply chain flexibility options and dynamic facility locations has been tackled in [8], experimenting with different adaptability schemes of a supply chain.

In [4], the integrated planning and transportation problem is addressed, proposing a mathematical model with production and transportation capacity constraints.

In general, so far, the approaches towards managing supply chains have dealt with static instances of their operation: parts or the entirety of the supply chain model are fixed and only a few alternative options are available. A few attempts deal with different transportation modes, some others take into account alternative facility locations and product changeover options and very few, in principle the recent ones, propose a more sophisticated methodology in order for more facets of the problem to be addressed simultaneously.

Our modeling approach allows for the formation of alternative dynamic production network configurations as well as for their validation via simulation in a series of network and demand settings, ensuring that the network be adaptive and capable of addressing the demand requirements. It may take into consideration partners who have not been part of the network in the past, requiring minimal information from their part regarding the initial configuration and planning of the manufacturing network. This way, a significant number of suppliers may be considered initially and therefore the chances towards achieving an adaptive network configuration are significantly increased. At the same time, the uncertainty related to the demand, the production process and the transportation of products, subassemblies and parts may also be considered, so that the risks regarding the operation of the network be taken into account.

The development of highly adaptive manufacturing networks is a very important objective in today's volatile environment. The proposed approach employs an integrated holistic view of the network and attempts to evaluate the performance of the network against multiple criteria, such as time and cost. At the same time, it offers a mechanism for generating, evaluating and ranking a set of alternatives, so that the stakeholders involved be provided with more options, when having to decide about the configuration of a manufacturing network.

3 Dynamic Manufacturing Networks Modeling

The manufacturing networks have to be more adaptive to the fluctuating demand in order for a more responsive and efficient operation to be achieved. Towards this direction, a new modeling approach, employing a holistic view of the overall network performance, is proposed. The major steps are depicted in Fig. 1.

The principle objective is to use minimal information, so that potential partners with minimal Information and Communication capacity may take part in a Dynamic Manufacturing Network (DMN).

3.1 Information Requirements

This approach requires that some minimal information regarding the production orders and the partners' capacity and network be available, in order for different alternative DMN configurations to be generated and evaluated.

Assuming that:

- S: The overall number of partners (including manufacturers, suppliers and customers),
- P: The number of products, subassemblies and parts,
- O: The overall number of orders,
- M: The number of different modes of transportation (e.g. ground, air, etc.),
- t: The time unit (e.g. day, shift, hour, etc.), $t = 1 \ldots T$,

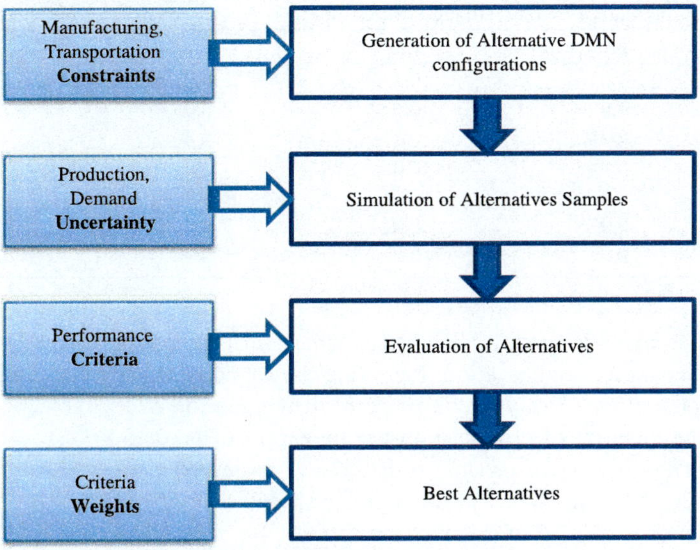

Fig. 1 Overview of the proposed approach

- T: The scheduling horizon,
- A: The number of alternative DMN configurations to be generated,
- N: The number of samples (simulation runs) for each alternative,

the following information is required:

- PP_{ij}: This variable represents the bill of materials (BOM) of all products, subassemblies and parts that may be produced or are available; when $PP_{ij} = 1$, with $i = j$, product i does not require other parts for being produced.
- SPC_{sp}: The cost of manufacturing one unit of product p in partner s.
- SPI_{sp}: The inventory cost per unit of product p in the facilities of partner s.
- $SSR_{ss'pm}$: The cost of transferring one unit of product p from partner s to partner s' using transportation mode m.
- $SST_{ss'pm}$: The time required for transferring one unit of product p from partner s to partner s', using transportation mode m.
- $SSTV_{ss'pm}$: The stochastic variation of the time required for transferring one unit of product p from partner s to partner s', using transportation mode m, following a uniform distribution $[-SSTV_{ss'pm}, SSTV_{ss'pm}]$.
- SP_{sp}: The capacity per time unit required for producing product p in the facilities of partner s, with $0 \leq SP_{sp} \leq 1$, $s = 1 \ldots S$, $p = 1 \ldots P$.
- SPV_{sp}: The stochastic variation of capacity per time unit required for producing product p in the facilities of partner s, following a uniform distribution $[-SPV_{sp}, SPV_{sp}]$.
- S_{max}: the maximum number of partners that may produce the same part within the DMN.

Table 1 An alternative DMN configuration example

Partner	Product P_1	Product P_2	Product P_3	Product P_4
S_1	0.0	0.4	0.0	0.0
S_2	1.0	0.6	0.0	0.0
S_3	0.0	0.0	0.8	0.0
S_4	0.0	0.0	0.0	0.7
S_5	0.0	0.0	0.2	0.3
S_6	0.0	0.0	0.0	0.0
S_7	0.0	0.0	0.0	0.0

- ST_{st}: The capacity already allocated in time unit t for partner s.
- STV_{st}: The stochastic variation, regarding the capacity already allocated in time unit t for partner s, following a uniform distribution $[-STV_{st}, STV_{st}]$.
- SY_{sp}: The quantity of product p in the inventory of partner s.
- PO_{ops}: The quantity of product p of order o, issued by partner s.
- DD_o, ED_o: The due date and the simulation end date of order o.
- AD_o: The arrival date of order o.

The above is the minimal information required for generating alternative DMN configurations, without having to take into account the process plans and the specific details of each partner's production equipment.

3.2 Generation of Alternative DMN Configurations

We define as an alternative DMN configuration the SxP matrix A_{sp}, where each element of this matrix a_{sp} represents the probability that partner s produces product p.

This probability actually defines which partner will be producing which product, part or subassembly, when an order (either for an end product or for a subassembly or a part required for manufacturing the end product) arrives or is issued within the DMN.

An example of an alternative DMN configuration (matrix A_{sp}) is shown in Table 1: with reference to the case scenario described in Sect. 4 (alternative #4 of Table 4), where 5 suppliers (S_1 to S_5) and 2 customers (S_6 and S_7) have to collaborate for the dispatch of a number of orders, product P_1 will entirely be produced by S_2, whilst partner S_1 will produce 40 % of the quantity ordered of P_2 and S_2 will produce the remaining 60 % of the quantity ordered of P_2. We consider as a DMN the set of all potential partners that could take part in the dispatching of an order. Contrary to the existing hierarchical approaches, the cooperation among the DMN members is considered being loose, without having to identify which partners have a leading role or not. Orders may actually be received by all partners. In this paper, however, it is assumed that the partners who can manufacture and deliver a specific product are the ones who usually receive an order for this product and therefore initiate the DMN configuration process.

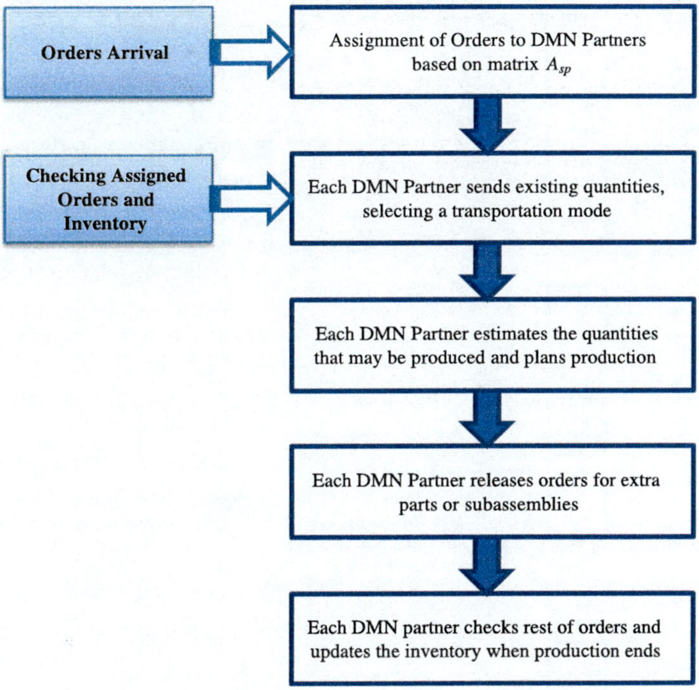

Fig. 2 Overview of the simulation process

3.3 Simulation of Alternatives Samples

For each alternative DMN configuration a number of samples is simulated (Fig. 2). For each sample, in each time unit, the orders received are randomly assigned to the partners available, the ones who can produce the products ordered, as per the matrix A_{sp}. Each partner checks the assigned orders and in case a part of an order may be fulfilled, a transfer order is released towards the partner who has released the original order. In order to take into account different transportation options in all samples, thus considering how adaptive the DMN configurations, in terms of transportation efficiency, are, a random transportation mode m from the ones available is selected for each sample. The associated transportation cost and time $SSR_{ss'pm}$, $SST_{ss'pm}$, $SSTV_{ss'pm}$ are used in the process of calculating the corresponding transportation cost and time of order o for sample n (TC_{on}). The remaining product quantities of the assigned orders are then checked against their requirements of subassemblies and parts. If the production for a part of the order may be initiated, a production order is released and planned, having taken into account the production capacity already allocated (ST_{st}, STV_{st}) as well as the capacity requirements of the products to be produced (SP_{sp}, SPV_{sp}). In case extra subassemblies or parts are required for the fulfillment of an order, new ones are

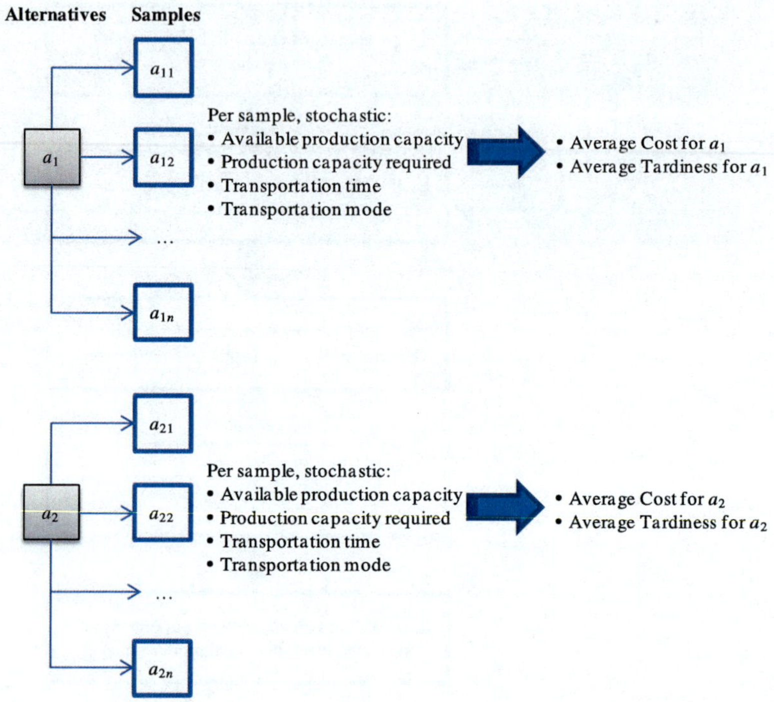

Fig. 3 An example with alternatives and samples

released towards the DMN partners. When all orders have been dispatched, the simulation of the samples is completed and other ones are then simulated until all *N* samples of all *A* alternatives are evaluated (Fig. 3).

3.4 Evaluation of Alternatives

All the samples of alternatives are evaluated against the criteria of average tardiness and cost. In particular:

$$Tard_a = \frac{\sum_{n=1}^{N} \sum_{o=1}^{O} \{max(ED_{on} - DD_{on}), 0\}}{n} \quad (1)$$

$$Cost_a = \frac{\sum_{n=1}^{N} \sum_{o=1}^{O} TC_{on}}{n} \quad (2)$$

Using the simple additive weight method and having already identified the criteria weights for defining their relative importance, the overall utility of each alternative may be calculated with the aid of a software application. This way, all alternatives may then be ranked and presented to the user. The average cost and

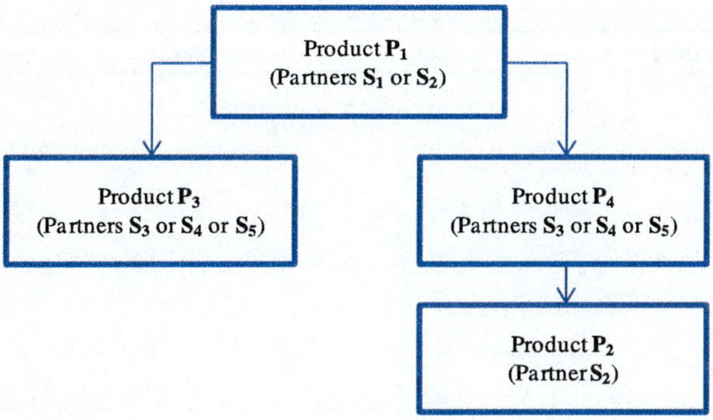

Fig. 4 Case scenario: bill of materials and suitable partners

Table 2 Description of the case scenario

DMN properties	Value
Number of partners	7
Number of products	4
Number of tiers	3
Transportation modes	2
Evaluation criteria and weights	Cost: 50 %, tardiness: 50 %

tardiness values of the alternative DMN configurations are considered as a measure of the DMN's adaptability towards demand requirements. Apparently, future demand scenarios may also be taken into consideration for each alternative.

4 Implementation and Experiments

For the purpose of testing and validating this proposed approach, a software application with a simulation engine has been implemented and a series of experiments has been carried out. A 3-tier case scenario is demonstrated with 7 partners (including 2 customers) and 4 products. Part P_1 may be produced by partner S_1 and S_2, whereas, P_2 is produced by S_2 only and P_3 and P_4 may be produced by partners S_3, S_4, S_5 (Fig. 4).

The properties of the DMN are shown in Table 2.

The information regarding the orders is depicted in Table 3.

Four experiments have been carried out with a different number of alternatives (A) and a maximum number of partner (S_{max}) who could take part in the manufacturing of the same product or part. For the first two experiments only one partner may produce each part, while in experiments 3 and 4, up to 2 partners may produce each part. The results of the best alternative generated in each experiment are shown in Table 4.

Twenty samples were generated per alterative for all four experiments. The performance of the best alternatives suggested in these experiments is compared

Table 3 Orders information

Order#	Product	Customer	Quantity	Due date (days)
1	P_1	S_6	1	2
2	P_1	S_7	2	4
3	P_1	S_6	2	7
4	P_2	S_7	1	2
5	P_3	S_7	2	3

Table 4 Experiments and performance of best alternatives

#	A	S_{max}	$Cost_a$ (€)	$Tard_a$ (days)	Util	P_1 partners	P_2 partners	P_3 partners	P_4 partners
1	5	1	55400	5.24	0.00	S_2 (100 %)	S_1 (100 %)	S_3 (100 %	S_4 (100 %)
2	50	1	36765	4.20	0.88	S_1 (100 %)	S_2 (100 %)	S_3 (100 %)	S_4 (100 %)
3	5	2	41747	4.53	0.63	S_1 (40 %)	S_1 (70 %)	S_3 (40 %)	S_3 (50 %)
						S_2 (60 %)	S_2 (30 %)	S_5 (60 %)	S_4 (50 %)
4	50	2	38275	3.87	0.96	S_2 (100 %)	S_1 (40 %)	S_3 (80 %)	S_4 (70 %)
							S_2 (60 %)	S_5 (20 %)	S_5 (30 %)

and their utility is estimated, taking into account the criteria weights. It is obvious that the more alternatives are generated, simulated and evaluated the more promising the best alternative DMN configuration looks. It is also interesting to note that the performance of the DMN is better when more options are available, in terms of the maximum number of partners that can produce the same part.

5 Conclusions

A novel approach for modeling Dynamic Manufacturing Networks as well as for generating and evaluating alternative configurations has been proposed. This method requires minimal information regarding the status of the manufacturing systems belonging to the network partners. This information is in principle limited to the capacity available per partner over the scheduling horizon, their production capabilities, the status of their inventory and the existing modes of transportation.

The dynamic nature of the manufacturing network is addressed in the following ways:

- The uncertainty associated with the production and transportation times, as well as with the demand profile is also considered via the sampling mechanism of the proposed approach: many different scenarios are therefore simulated beforehand, in order to ensure that the manufacturing network may operate efficiently under different conditions.
- This method enables collaboration schemes of specific products, subassemblies and parts, i.e. their production may be distributed to many partners. The uncertainty related to the partners' production capacity is taken into consideration and

therefore collaborative schemes with more partners are proposed in case it is likely that a partner cannot deliver.
- The different transportation modes provided are also taken into account, along with the corresponding costs and times for each alternative via the sampling mechanism. This way, the adaptability of the proposed DMN configurations in terms of how well they behave in terms of transportation efficiency is considered; in case any transportation problems emerge, the proposed DMN configurations are expected to cope well with these problems.
- Whenever a disruption in the operation of a DMN occurs, the proposed approach may be executed again, towards modifying the initial DMN configuration.

Nevertheless, a series of assumptions were made for testing, validating and presenting the proposed approach:

- Production capacities have been assumed to be evenly distributed,
- A randomly generated demand profile was used including the orders' due dates.

However, without loss of generality, the proposed methodology may easily be used with other statistical distributions and demand profiles.

Through the simple case scenario given and the experiments carried out, it has been shown that the proposed approach could be used for determining adaptive DMNs in a volatile and highly uncertain global market environment. The problem of integrating complex products/parts and suppliers' interrelationships, the finite production capacity of the potential partners, different transportation modes and the uncertainty pertaining to available and required production capacities and process times cannot be handled by conventional Mathematical Programming and Operations Research approaches.

Going beyond the configuration and planning phases, further features would include options for lot sizing within the DMN as well as options for expanding the use of the proposed approach in the domain of the manufacturing scheduling, where detailed process plans and configurations have to be considered at each partner's level. Integrating data from the shop floor and the logistics network for monitoring the operation of a DMN is also another idea that is worth experimenting with. More sophisticated scenarios may also be tested, involving the transportation activities and organizations as part of the DMN.

DMNs are expected to be in charge of an increasing part of the global manufacturing activity and therefore, providing new methods and tools for improving their operation and overall efficiency is of paramount importance.

Acknowledgments The work in this paper has been partially supported by the FP7 Integrated Project "IMAGINE—Innovative End-to-end Management of Dynamic Manufacturing Networks", funded by the CEU.

References

1. Chryssolouris, G.: Manufacturing Systems—Theory and Practice. Springer, Berlin (2006)
2. Permenter, K., Anand, S.: State of Cross Channel Retail Supply Chain Execution. Aberdeen Group, Boston (2011)
3. Chryssolouris, G., Papakostas, N., Mavrikios, D.: A perspective on manufacturing strategy: Produce more with less. CIRP J. Manuf. Sci. Technol. **1**, 45–52 (2008)
4. Scholz-Reiter, B., Morosini Frazzon, E., Makuschewitz, T.: Integrating manufacturing and logistic systems along global supply chains. CIRP J. Manuf. Sci. Technol. **2**, 216–223 (2010)
5. Viswanadham, N., Gaonkar, R.S.: Partners selection and synchronized planning in dynamic manufacturing networks. IEEE Trans. Robotics Autom. **19**, 117–130 (2003)
6. Speranza, M.G., Ukovich, W.: Minimizing transportation and inventory costs for several products on a single link. Oper. Res. **42**, 879–894 (1994)
7. Theattre, K., Graves, S.: Tactical shipping and scheduling at Polaroid with dual lead-times. In: Innovation in Manufacturing Systems and Technology (IMST) Report, http://web.mit.edu/sgraves/www/SMA%20paper%20Oct%2001.pdf
8. Seifert, R.W., Langenberg, K.U.: Managing business dynamics with adaptive supply chain portfolios. Eur. J. Oper. Res. **215**, 551–562 (2011)
9. Vidal, C.J., Goetschalckx, M.: A global supply chain model with transfer pricing and transportation cost allocation. Eur. J. Oper. Res. **129**, 134–158 (2001)
10. Klincewicz, J.G., Luss, H., Yu, C.-S.: A large-scale multilocation capacity planning model. Eur. J. Oper. Res. **34**, 178–190 (1988)
11. Stadtler, H.: A framework for collaborative planning and state-of-the-art. OR Spectr. **31**, 5–30 (2009)
12. Bhatnagar, R., Chandra, P., Goyal, S.K.: Models for multi-plant coordination. Eur. J. Oper. Res. **67**, 141–160 (2004)
13. Pibernik, R., Sucky, R.: An approach to inter domain master planning in supply chains. Int. J. Prod. Econ. **108**, 200–221 (2007)
14. Dudek, G., Stadtler, H.: Negotiation-based collaborative planning between supply chain partners. Eur. J. Oper. Res. **163**, 668–687 (2005)

What Can Quality Management Methodology and Experience Contribute to Make Global Supply Networks More Robust?

Werner Bergholz

Abstract The microelectronics industry is characterized by a worldwide supply network and a complex production process with up to 1000 consecutive production steps which interact and take 1–3 months to complete the product. Under such circumstances, robust manufacturing and a stable supply chain can only be maintained if stringent QM System is implemented. Essential QM tools to ensure stable processes are SPC and FMEA. Although QM systems which satisfy the standard ISO TS 16949 or ISO 9001 are, in the first place, examples of clock-work—type organizational structures with centralized control, there is a built-in local control aspect which is needed if the QM system is to function efficiently.

Keywords Microelectronics · Quality management system · Statistical process control · FMEA · ISO TS16949 · ISO 9001

1 Introduction

The microelectronics industry started in the 1960s as a fledgling technology for space and military applications, and has experienced a strong growth to a 300 billion dollar industry with a highly specialized truly global supply chain.

Last year's tsunami in Japan, which caused the destruction of the Fukushima nuclear power plants was just one example of the vulnerability of the supply chain for many industries, last but not least the automotive industry which suffered from a shortage of electronic components some of which are only manufactured in

W. Bergholz (✉)
Jacobs University Bremen, Campus Ring 1, 28759 Bremen, Germany
e-mail: w.bergholz@jacobs-university.de

K. Windt (ed.), *Robust Manufacturing Control*, Lecture Notes in Production Engineering, 259
DOI: 10.1007/978-3-642-30749-2_19, © Springer-Verlag Berlin Heidelberg 2013

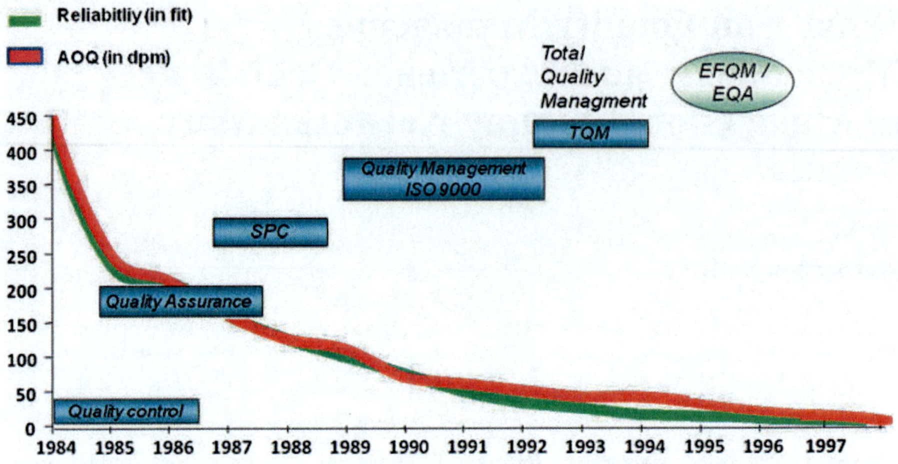

Fig. 1 Improvements in the Reliability (measured in fit and the Average Outgoing Quality (AOQ, measured in ppm defective) of microelectronic products. The source of the figure is a Siemens Semiconductor Division document which had been distributed to customers (and which the author contributed to). Similar numbers have been published by Intel, no more up to date data have been published since (due to the confidential nature of such data). However, it is safe to assume that the trend has continued and that e.g. AOQ is now below 1 ppm

Japan. Given the importance of the microelectronics industry, the question to be examined is: What are the important factors that determine robustness of the supply chain for the microelectronics industry itself, since many critical materials and equipment to manufacture microelectronic circuits are manufactured in parts of the world where disaster is almost certain to strike at some point in time.

- Japan is the most important location for the manufacturing of silicon raw wafers, the same is true for some ultraclean chemicals and other materials
- The largest equipment manufacturer for that industry is located in California

Additional intrinsic risks for the robustness of the manufacturing process and the stability of the microelectronics supply chain arise from the following additional factors:

1. The level of quality, as measured by the early fail rate and the average outgoing quality (AOQ) (Fig. 1) has improved by almost a factor of 1000, to extraordinary low single digit ppm values for AOQ) (1 ppm (parts per million) means only one of 1 million delivered parts are defective) or single digit fit-values (1 fit = one failure in 1 billion accumulated hours of operation).
2. The cost degression curve of the industry has been about 40 % per year, for more than 4 decades, i.e. there is an immense pressure for continuous improvement [1].
3. The manufacturing process is long and complex: Based on personal experience in the industry, to manufacture e.g., a CPU, about 1000 consecutive manufacturing steps are necessary, due to the confidential nature of these issues,

there is practically no published material on this. Since some of these steps take hours to complete, the cycle time can be as long as 2–3 months.

4. The drive to smaller and smaller minimum feature sizes is the strongest driver for productivity: Reducing the dimensions of the circuit by a factor of 2 results in a productivity increase in a factor of 4, since the area of the circuit is reduced by a factor of 4, so the number of microchips on a wafer can be 4 times as high, this state of affairs has been captured by Moore's law [2] and in the International Roadmap for Semiconductors [3]. As a result of the relentless drive to miniaturization (minimum feature size at present around 30 nm) means that technologies limits are constantly driven to new limits of what is technically feasible.
5. The manufacturing process requires an ultraclean environment and materials, as due to factor 4 there is a constant pressure to push out the technical limits.

As a result of 2, there is a strong pressure for continuous improvement of productivity and cost reduction for materials, maintenance, infrastructure etc., while the high level of quality must not only be maintained but has to be continuously improved. Obviously, any change in materials, reduction of material consumption, change in process and, last but not least pushing out the limits of technology (4) and purity (5) comes with risks attached. This inherent problem is amplified by the fact that the problem of a process change may only become apparent months later because of factor 3. The simultaneous achievement of high quality and continuous improvements in productivity in this highly competitive industry is, to a large part, due to the implementation of a stringent Quality Management (QM) System. Without it, production of microelectronic devices nowadays would simply be impossible!

It is the purpose of this paper to elucidate how a QM System and the QM toolbox makes the manufacturing process and the supply chain in this industry relatively robust, and in what way this centralized system at the same time enables and effectively implements decentralized control. Most of the discussed issues and concrete figures are based on the author's personal experience in various functions in a major microelectronics manufacturing company over 17 years, and work in various electronics industry standard development organizations. The paper is organized as follows:

- Section 2 gives a brief introduction to QM systems and QM tools
- Section 3 deals with the process stability and robustness in a microelectronics manufacturing process with the help of QM tools and a QM system
- Section 4 covers the microelectronics supply chain, in particular how to ensure the quality of materials and therefore a robust and steady flow of materials
- Section 5 examines the role of QM systems in the context of stabilizing supply chains and improving the resilience to process variations and supply chain excursions and other unforeseen events

2 QM Systems and QM Tools

2.1 *Quality Management Systems and Standards*

Quality Management has started as an engineering activity in the 1920s by Deming and Shewhart [4], which initially focused on the statistical analysis of production data. Quality engineering has emerged as a systematic activity to improve and control not only manufacturing, but also the service industries in the 1980s, and has been gradually extended to cover not only technical but also administrative and management processes. In that decade, the first ISO 9001 standard for a QM system was published by the ISO organization in Geneva [5], which has been adopted by more than 1 million companies worldwide by now, the numbers of ISO 9001 certified organizations keeps growing continuously. For many industries, certification is mandatory for business-to-business supplier–customer relationships.

In such a situation it is unavoidable that poor implementation of QM Systems leads to a state that the cost of ownership of a QM system can exceed its benefits. In several studies [6, 7] involving thousands of companies, it has been shown that

- ISO 9001 certification can result in significant economic benefits. A critical factor appears to be that top management has understood the potentials of a functional QM system and has succeeded to communicate this to the ranks and file so that staff has "bought into the idea".
- If, on the other hand, implementation and certification of the QM system is only done because the ISO 9001 label is needed, then economic benefits do not materialize, as demonstrated by several studies.

Due to the apparent success of QM systems based on ISO 9001, which is designed in a very general way to be applicable to any production or service organization, the automotive industry introduced another ISO standard which has been "tailor-made" for the industry ISO TS 16949 [5]. Remarkably, this standard contains the ISO 9001 standard as a verbatim "copy", and for almost all sections, additional requirements have been added. Since nowadays, more than 30 % of the cost to build a car is for electronic components and software, it is obvious that the microelectronics industry has to fulfill the more demanding ISO TS 16949. As the logistics industry is a significant service provider for the automotive industry, it follows that a logistics company must also comply with ISO TS 16949.

2.2 *Quality Management Tools*

To implement a QM system, a number of tools have to be used. There are more than 10 QM tools [8, 9], only the 2 most important ones will be discussed here.

2.2.1 FMEA

To realize ambitious goals for cost reduction and performance improvement by technical and logistic innovation, it is obviously necessary to "chart the unknown territory" for potential problems *without* having to learn from mistakes. The Failure Modes and Effects Analysis tool (FMEA) [9] is one of the main techniques to achieve this goal. The main idea is to identify a representative number of potential failure modes, and then to assign to each of the failure modes a Risk Priority Number (RPN). The RPN is calculated from the 3 dimensions, each of which is assigned a number between 1 and 10, namely

- *S*everity (1 = marginal effects, 10 = extreme effects)
- Probability of *O*ccurrence (1 = extremely unlikely, 10 = almost certain)
- *D*etectability (1 = easy and fast to detect, 10 = difficult to detect)

The RPN is defined as

$$\mathrm{RPN} = \mathrm{S} \times \mathrm{O} \times \mathrm{D} \tag{1}$$

The failure modes are then ranked according to the RPN, and subject to pre-defined limit for the RPN, the risk is found either acceptable or, if the RPN number is larger than the limit, an action must be identified and implemented to mitigate the risk to result in an acceptable RPN.

An FMEA is not a strict mathematical tool, but a pragmatic, efficient and structured approach for risk evaluation. The "art" of an FMEA is to define technology (and sometimes company-specific) standardized scales for the assignment of the S, O and D values. It is also noteworthy that standardized scales and FMEA documents based on them are a valuable knowledge management tool.

The authors experience from 15 years of practical experience is that in the microelectronics industry an FMEA can typically prevent at least 70 % of all failure modes by identification and implementation of suitable countermeasures. No significant published data exist on the success rates for FMEAs, though. It also appears likely that the success rate to some degrees depends on the industry type.

The FMEA technique has been criticized from the perspective of financial risk management [10] in which only the dimensions Severity (=financial damage) and Occurrence (probability) are evaluated. The assignment of the scales 1···10 is deemed to be not exact, as an alternative the calculation of expectation values for damage based on the 2 dimensions is proposed. While this approach seems reasonable under certain circumstances, from a practical point of view it is too time consuming, and not as efficient as the FMEA to prevent risks in operations.

2.2.2 Statistical Process Control

While the FMEA is used to predict failure modes and prevent them from happening, statistical control is a tool to assess the stability of a running process [8, 9] by statistical criteria and to derive "rules" how to "manage" the process stability so that the desired process result is achieved with a predefined and *predictable* probability. A process is considered stable, if the probability for failure is smaller than 3.4 ppm (part per million, see [8, 9]).

An essential feature of Statistical Process Control (SPC) is that there are well-defined criteria ("Western Electric Rules") to assess

- whether the process is still stable = "under control" (and it is FORBIDDEN to adjust process parameters, since that would decrease stability!) or
- whether the process is out of control and needs attention to identify and eliminate the root cause of the process excursion.

The essential principle of SPC (according to standard textbooks) is, that the process stability is determined by comparing

- the variability of the process result (expressed as the standard deviation sigma) (e.g. the diameter of a tube) with
- the technical specification interval (=acceptable process result)

The stability parameter cpk for a process is defined as

$$\text{cpk} = \{\min(\text{USL} - \text{AVG}; \text{AVG} - \text{LSL})\}/3\ \text{sigma} \qquad (2)$$

where: USL = upper specification limit, LSL = lower specification limit, AVG = mean value of process result determined in a pre-run, Sigma = Standard deviation of process results, from a pre-run.

A cpk value of 1.67 (or alternatively 1.5) is the generally accepted minimum value for sufficient process stability (which for a maximum of 1.5 sigma decentering of the process average AVG corresponds to the failure rate of 3.4 ppm mentioned previously). Every process in a production process must either fulfill this stability criterion, or must be improved towards this goal.

FMEA and SPC are the main tools to make production and/or service processes resilient against external perturbations. In addition there are many more tools (see [8, 9] that help to improve the robustness of processes and to promote continuous improvement of processes and products.

In the following two sections a few representative examples will be described to demonstrate how QM is deployed in microelectronics production and supply chain management.

3 Microelectronics Production Process

As mentioned previously, microelectronic production processes have an unusual long cycle time and up to 1000 process steps. To keep a process stable, a stringent process management is implemented, in particular with regards to improvement activities and trouble shooting.

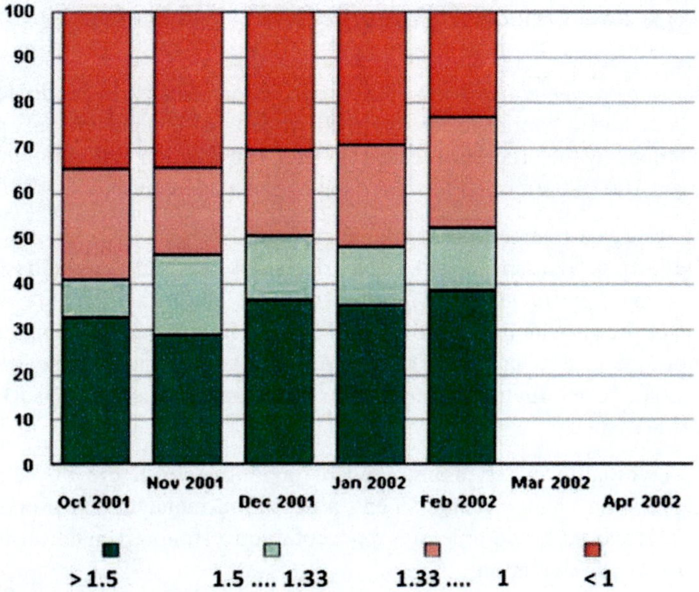

Fig. 2 Stacked bar graph of the percentages of cpk values of a microelectronics production process after startup of the process. The increase of the fraction of larger cpk values reflects the improvement of process stability (figure from own training material)

3.1 Process Robustness Ensured by SPC and FMEAs

In a typical microelectronics manufacturing process there are more than 5000 process parameters to be kept under control. To assess the overall trend in improvements of process stability, graphical aggregation of data is used in the form of process control charts and aggregated views of the distribution and time evolution of cpk values for all processes, as shown in Fig. 2.

Thus, the SPC method is an effective tool to make a complex and long production process stable to ensure a reasonable yield (80–90 %) and, most important, reliable products which can *only* result from a stable process, quality cannot be tested into a product, it has to be made.

When a new production technology is developed (or a product design), there is a stringent qualification process after the "process freeze", which does not only involve the process stability study, but consists of a number of reliability and life tests. Only once these tests are all passed, mass production and customer delivery can start on the basis of the "frozen" process of record (POR).

3.2 Change and Deviation Management

Any process improvement means that the "frozen" process has to be modified. So how can this be done and deliveries to customers still maintained, which are only permitted for material made by the process of record (POR)? The solution to this conundrum is a structured and 100 % controlled *technical* change management process:

- The first step in a planned process change is an FMEA
- If the FMEA risk assessment is positive, decisions are made on the basis of data from process experiments for a limited number of production lots, including evaluation criteria which have to be met if the process can be changed.
- Usually before a full changeover of production, a partial trial change 20–50 % is implemented, to evaluate whether or not new unidentified risks surface with improved statistics.

In the unavoidable event of *unintended* deviations from the POR, a similar methodology as the change management, a deviation management process has to be activated. It would be completely unacceptable to scrap all material for which there was a process deviation:

- Once necessary risk assessment has been carried out, often rework or special quality checks can still lead to salable product.
- Unintended process deviations are a valuable source of knowledge which are transferred into the FMEA documents.

Both Change Management and Deviation Management documentation and well-structured methodologies are a solid basis to respond to unforeseen events, this point will be detailed in Sect. 5.

As a result of applying stringent quality management the production processes have been rendered more and more robust. Due to FMEAs and SPC already at the development stage, the ramp up times from pilot production with a total process yield of initially 40–60 % have decreased from 2–3 years in the 1980s to less than 6 months!

3.3 SPC, FMEA and the Automotive Standard IS TS 16949

As mentioned before, that standard TS 16949 is significantly more demanding than the ISO 9001 standard. The most significant add-on is the requirement of a complete documentation of all cpk values at regular intervals (e.g. monthly) for all process/product parameters significant for quality and performance (termed "special characteristics" or "key control characteristics" (KCCs) and a full documentation of FMEAs for design, development, and a process FMEA for EACH of the up to 1000 process steps, the latter being a document of several hundred pages! This requires significant engineering resources for the creation of such documentation, but experience shows that it pays in terms of

- Much smoother and more efficient work in change and deviation management
- A high flexibility in terms of response to unforeseen events either in the process or in the supply chain this kind of seamless process documentation and knowledge base is a valuable asset, more details in Sect. 5

4 Supply Chain Management in Microelectronics

In most industries there is a strong trend to outsource processes; this is not only true for materials, components and certain logistics services, but even for some development activities. It is clear that it is not sufficient to limit QM activities to the own organization but that the QM system has to be extended to suppliers if a stable production is to be ensured.

4.1 QM Extended to Suppliers to Make the Supply Chain More Robust too

Like most other industries, microelectronics depends critically on stable quality of materials and equipment purchased from suppliers, therefore to improve the own production processes to improve process stability and the quality of the products is obviously insufficient.

A traditional way to ensure the consistent quality of supplied products has been incoming inspection. However, this is neither efficient nor effective: Products manufactured by an unstable process will contain a significant fraction of marginal products which will fulfill the technical specifications but will be prone to early failure. Moreover, incoming inspection is expensive and wasted time compared to “ship to line”.

The alternative is to verify by regular QM audits and by a continuous monitoring of the suppliers cpk values (via online access to the suppliers SPC data) that the production and logistic processes are under the same stringent QM control at the various suppliers sites as the own production.

4.2 Technical Cooperation With Suppliers

Experience in microelectronics has shown that in addition to this, an intensive and trustful technical cooperation is another way to increase the stability of the value adding chain and to accelerate technical development. Depending on the level of trust that has developed over time, suppliers give early warning in potential problems, irrespective whether they materialize later or not.

4.3 Multiple Sourcing and More

The ISO TS 16949 standard *explicitly* requires that production should be safeguarded against interruption by unforeseen events by planning or implementing contingency measures. In terms of supplies of essential resources the best solution is dual or multiple sourcing, the increased overhead is offset by the possibility of benchmarking the suppliers and increasing the share of the better supplier.

If there is only one source, then a combination of buffer stocks and the application of SPC *to the arrival rate* of deliveries and other relevant performance indicators for suppliers have proven to be instrumental for the early detection of supplier problems. A most impressive example is the early detection of significant supply risks by Nokia after a fire in the Philips Albuquerque chip factory in 2000. As a consequence, Nokia could secure additional supplies before the limited capacity was sold out, whereas Ericsson could not, the long interruption in production of Ericsson eventually resulted in the decision to terminate production of mobile phones at Ericsson [11].

5 QM, Stability of Manufacturing and Resilience Against Unforeseen Events

From what has been described up to now, a QM system appears to be close to "machine organization" with strong centralized control, which could imply that response to unforeseen events is slow. There is strict control in a QM system in terms of coordination activities in an organization and making sure that risks for quality are minimized. However, in a QM system, at the same time, there is a certain level of autonomy and local control:

- The backbone of a QM system is the system of key processes and sub processes, with well-defined interfaces and input/output requirements.
- *Local* control is enabled by the authority (and duty of the process "owner") to continuously improve, i.e. change the process, autonomously, of course subject to the change and deviation management rules. No decisions from top management are needed.

5.1 Stability and Fast Response in Production/Service Processes

A QM system for reasons explained in Sect. 2 leads to a significantly more robust and stable production and supply chain due to the benefits of the system and the QM tools, the most important ones being SPC and FMEA and internal auditing of the QM system (not mentioned before).

A *fast response to perturbations* is facilitated by the following inherent characteristics of a QM system:

- All necessary performance and quality-relevant data are readily available in the data warehouse of the QM system
- Clear processes how to proceed in case permanent or temporary changes are defined and can be executed without top management attention
- Risk Evaluation for containment and corrective actions can be done on the basis of the QM documentation, in particular if the QM system satisfies ISO TS 16949
- Last but not least: Documentation of all actions and effects happens almost automatically and can be retrieved years later, if needed

5.2 *Robustness of the Supply Chain*

For a truly global industry with major suppliers in the US, Japan, Korea, Taiwan, and last but not least Europe it is a challenge to maintain a robust and stable supply chain.

The control of suppliers via QM audits and QM tools, and their performance evaluation jointly between QM, Purchasing, Production and Logistics is a solid baseline to judge potential risks, even in terms of cultural differences in how to do business, since these become elucidated during QM audits, this has been personally experienced by the author in his role as lead auditor. It has not been mentioned yet that a change of supplier needs qualification of EVERY supplier critical to quality before delivery of product made with that material in it can be delivered to customers. Such strict qualification rules certainly reduce the ability to react quickly to supplier problems or sudden increases in demand for that material, but there is no alternative, and another good reason for a dual supplier strategy.

In combination with Change and Deviation management a reasonable quick reaction to supplier problems is possible.

6 Conclusions

Quality Management Systems which satisfy the standards ISO 9001 or ISO TS 16949 are a mainstream management tool to ensure robust manufacturing and a stable supply chain. Industries like the microelectronics industry with a worldwide supply chain and up to 1000 consecutive production steps which take several months would simply not be viable without stringent process control and QM.

The QM methodology has been summarized and visualized in Fig. 3 [12]: Ideally, there is seamless process control throughout the entire supply chain.

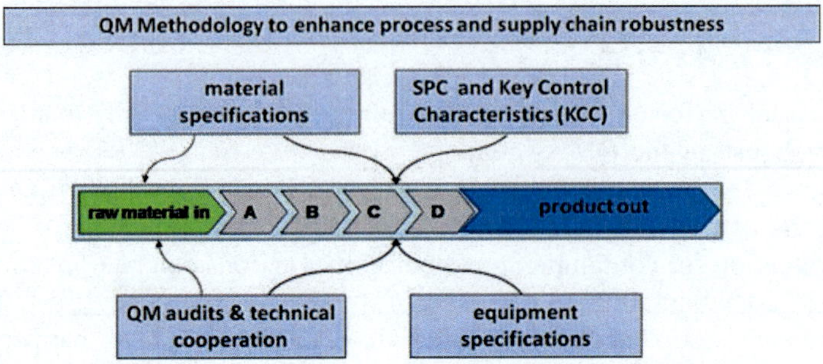

Fig. 3 Visualization of the QM methodology to enhance process robustness

As pointed out in Sect. 5, a QM system can be viewed as a large clock-work organization, but with "embedded" local control. A more systematic development and emphasis on local control in the current versions of the QM standards would certainly worth to be considered for the next updates of ISO 9001 and ISO TS 16949.

References

1. Bergholz, W., Weiss, B., Lee, C.: Benefits of standardization in the microelectronics industry and its implications for nanotechnology and other innovative industries. In: International Standardization as a Strategic Tool, Commended Papers from the IEC Centenary Challenge, pp. 35–50 (2006)
2. Moore, G.E.: Some personal perspectives on research in the semiconductor industry. In: Rosenbloom, R.S., Spencer W.J. (eds.) Engines of Innovation, pp. 165–174. Harvard Business School Press, Boston (1996)
3. http://www.itrs.net/
4. Deming, W.E.: Some principles of the Shewhart methods of quality control. Mech. Eng. **66**, 173–177 (1944)
5. http://www.iso.org/iso/home.html
6. Corbett, C.J., Montes-Sancho, M.J., Kirsch, D.A.: The financial impact of ISO 9000 certification in the US: an empirical analysis. Manage. Sci. **51**, 1046–1059 (2005)
7. Terziovski, M., Power, D., Sohal, A.S.: The longitudinal effects of the ISO 9000 certification process on business performance. Eur. J. Oper. Res. **146**(3), 580–595(16) (2003)
8. Pfeifer, T.: Qualitätsmanagement, Strategien, Methoden, Techniken. Carl Hanser, München (2001)
9. Wadsworth, H.M. Jr, Stephens, K.S., Godfrey, A.B.: Modern Methods for Quality Control and Improvement, 2nd edn. John Wiley, New York (2002)
10. Blum, U., Gleißner, W.: Unternehmensbewertung, Rating und Risikobewältigung, Wissenschaftliche Zeitschrift der Technischen Universität Dresden **55**, pp. 113–122 (2006)
11. http://www.ftpress.com/articles/article.aspx?p=1244469[02.03.201212:02:50]
12. Bergholz, W., Fabricius, N.: Performance standards. In: Murashov, V., Howard, J. (eds.) Nanotechnology Standards, pp. 89–116. Springer, New York Dordrecht Heidelberg London (ISBN 978-1-4419-7852-3) (2011)

Innovative Quality Strategies for Global Value-Added-Networks

Gisela Lanza, Johannes Book, Kyle Kippenbrock and Anamika Saxena

Abstract Many companies no longer act locally within their domestic markets, but have established a global network of worldwide production sites. Due to the long and diversified structures of supply chains and differences in the maturity levels of suppliers, distributed networks develop various fluctuations, in terms of varying product quality and delivery times, which can result in image loss and financial losses to the companies of the network. Moreover, an improperly implemented quality strategy in a network will result in higher costs. The preliminary idea is to make these networks insensitive to such fluctuations by identifying and evaluating suitable quality strategies. Due to the absence of site-specific optimization and the complex structures of networks, it is difficult to find suitable quality strategies for production networks. The complexity in the networks includes unknown defect propagation, limited influence due to decentralized structures and conflicting objectives and unknown inter-relationships amongst the various supply chain members. The research project IQ.net deals with these problems by developing innovative methods, models and practical tools for planning, optimization and control of quality strategies for globally distributed production networks, thus obtaining zero-defect production networks. This chapter aims to discuss various aspects of IQ.net including, the definition of quality in networks, the analysis and evaluation of various systems for managing network-wide quality data considering local versus global data, as well as, three core methods to identify robust quality strategies for specific network configurations.

G. Lanza · J. Book (✉) · K. Kippenbrock
Karlsruhe Institute of Technology, Kaiserstaße 12,
76131 Karlsruhe, Germany
e-mail: Johannes.Book@kit.edu

A. Saxena
Indian Institute of Technology, Kharagpur,
West Bengal, India

K. Windt (ed.), *Robust Manufacturing Control*, Lecture Notes in Production Engineering,
DOI: 10.1007/978-3-642-30749-2_20,

Keywords Defect propagation · Maturity levels · Quality strategy · Value-added-networks

1 Introduction

Rapid technological advancements have allowed the increasing specialization and distribution of activities in all stages of a production value chain. Some segmented activities can be performed in different locations worldwide and reintegrated into global value chains and global production networks. Many transnational companies are playing a key role in organizing and controlling these production systems, benefiting from location differences in costs, infrastructure, capabilities in manufacturing, marketing and logistics, and in trade and investment regimes. This is having far-reaching effects on competitiveness, cross-national transfer of new technology, ideas, skills, knowledge and learning, and potentially offers greater opportunities for reaching welfare gains, but it also brings new challenges [1].

Due to challenges in global production networks (as discussed in the next section), these networks develop various fluctuations, in terms of product quality and delivery times, which can result in image losses and financial losses to the companies of the network. The preliminary idea is to make these networks insensitive to the fluctuations by identifying and evaluating suitable quality strategies. The proper and systematic choice of suitable quality management (QM) measures for managing globally distributed networks has become crucial for a company's success [2].

This chapter discusses various aspects of the research project IQ.net, which focuses to solve the challenges in global production networks by developing suitable quality strategies for networks. The various aspects of IQ.net include, definition of quality in networks, development of a quality-based product model, determination of site-specific maturity levels, analysis and evaluation of various systems for managing network-wide quality data, as well as, three core methods to identify robust quality strategies for specific network configurations viz. optimal inspection strategy based on the locational factors of a site, dynamic tolerance strategy, and agent-based modeling for value-added-networks.

The next section illustrates the problem statement stating the challenges in global production networks. In Sect. 3, the state of the art for various aspects of IQ.net is discussed. Section 4 explains various aspects of IQ.net by discussing in detail suitable quality strategies at site as well as network level. Both Sects. 4.1 and 4.2 are comprised of three sub-sections: definition of targets and influencing factors, analysis and evaluation, and modeling, evaluation and optimization of quality assurance (QA) strategies. Finally, the summary of the chapter and an outlook of future work are presented.

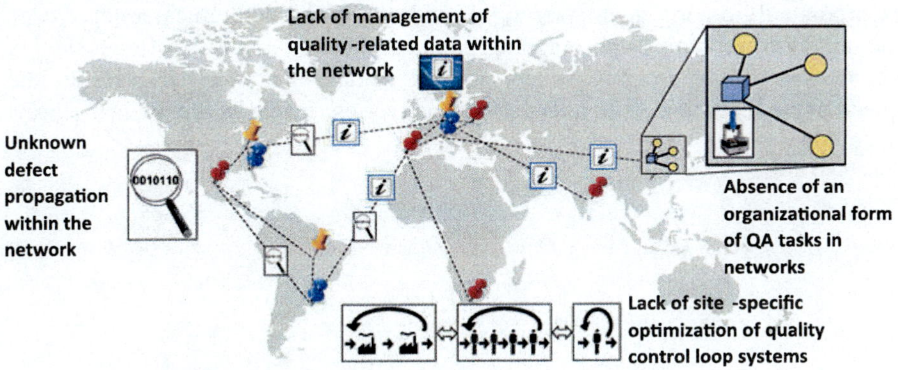

Fig. 1 Challenges in global production networks

2 Problem Statement

The field of global production has been addressed from several perspectives in research and practice in last years. The focus has mainly been put on network configuration, local sourcing or adaptation to locational factors [3]. In practice, companies face problems in operating these kinds of global networks in terms of managing the product quality. High efforts are needed for assuring quality in global production networks, because QA aspects are not sufficiently integrated in the modeling and design of these networks.

Due to long and diversified structures of supply chains and differences in cultural backgrounds, maturity levels and objectives of supply chain members, it is difficult to find suitable quality strategies for networks [4, 5]. An improperly implemented quality strategy in a network will result in higher total costs.

Suppliers from low cost countries are often characterized by a low awareness of quality and a high variation of product characteristics (e.g. dimensional tolerances). These suppliers often have conflicting objectives and in many cases are more strongly focused on monetary objectives. Moreover, customer–supplier relations between different companies from different industry sectors can result in a mismatch in product characteristics due to different requirements from their main customers (e.g. integration of electronic components in the automotive sector). Furthermore, improper coordination of various suppliers in a network may lead to a difficulty in assuring the quality of the products during the production ramp-up. Figure 1 shows the major challenges in globally distributed networks.

An unidentified defect in manufactured products, which originates at a point in the network, propagates through the whole network. As a result, all the costs and efforts invested in producing defective items in the network are wasted. A lack of site-specific optimization of quality control loops systems leads to unnecessary quality costs and varying quality of products. Furthermore if quality assurance strategies are not coordinated at a network level, extra costs can accrue due to redundant functions and capabilities at various sites. It may also be necessary to incorporate third party quality service providers, which may implement checks on the quality of incoming and outgoing products. Moreover, a lack of data-sharing among the supply chain members and its management may also result into uncertainties caused due to the delayed deliveries, machine breakdowns, order fluctuations etc.

3 State of the Art

3.1 Product Model

A product's quality depends on the quality of its components and materials. The quality requirements of a product cannot be met if components and materials of insufficient quality are used.

To describe the creation of product quality throughout a production network, an abstraction in a product model can offer a suitable way [6]. In the literature there are several perspectives or models for regarding a product, like i.e. product structure models. However, an approach to describe the interrelations of characteristics, parts, modules and functions regarding its overall quality is not existent yet.

3.2 Maturity Model

Maturity models help companies to identify how efficient and effective they are, in terms of their QM systems. Some approaches exist in the literature for establishing a framework for evaluating the QM system of a company with the help of quality awards. Among these, the European Quality Award of European Foundation for Quality Management (EFQM) is well known among Total Quality Management (TQM) practitioners in Europe. On the basis of the EFQM's Excellence Model, initial TQM maturity models have already been developed. However, one main deficit of quality awards is that their underlying models do not give management concrete suggestions on how progress in the development of QM systems can be achieved [7]. Although, the EFQM model aims at improving the overall quality of a company, it does not lay a specific focus on the organizational requirements for identifying and implementing customer needs and producing goods of high quality efficiently. Another maturity model for QM systems can be found in DIN EN ISO 9004. Like the EFQM-Model, this approach also lacks concrete characteristics that describe the maturity levels from an operational point of view but rather addresses strategic concerns of a company regarding quality management in a very general manner. Concrete hints to follow-up actions for managers and measures for improving the maturity level are not part of this model but are consciously left open. Furthermore existing maturity models do not attempt to quantify qualitative locational factors for determining the maturity level of a company [8]. The locational factors help to evaluate the quality maturity level of a site with respect to the potential that can be achieved in fulfilling its quality management aspects.

3.3 Site Level Quality Control Loops

Manufacturing companies often employ different quality control measures to obtain the desired quality of their final product. This can be achieved by finding the suitable quality control loops in the company. Many models exist in the literature to identify the site level quality control loops by finding the optimal inspection strategies for multi-stage manufacturing processes [9]. However, current approaches to find the optimal inspection strategy for multistage manufacturing processes do not include various site-level locational factors into account such as working environment and its impact on the worker, workers' qualification, skills and problem solving competencies, tracing of quality costs, and machinery and maintenance.

3.4 Definition of Quality in Global Production Networks

Manufacturing of quality products in value-added networks is the result of successfully meeting the customers' needs and demands. Quality is achieved by delivering a product with certain functionality, which arises from a compilation of quality characteristics, to the customer at the right time to the right place and in the right quantity.

The two aspects of quality in value-added-networks are functional quality and logistical quality. The production in a value-added-network comprises of intertwined manufacturing and logistics processes. As discussed earlier, a product's manufacturing quality depends on the quality of its components and materials. The quality requirements of a product cannot be met if components and materials of insufficient quality are used [10]. Due to the inter-linkage of the enterprises in a value-added-network and the existing transport delays, duration and impact of manufacturing quality induced production stops add up in the direction of the material flow. Additionally the logistic processes themselves contribute to the punctual delivery of quality products in several ways. In general the relevant logistic processes are external and internal transport, handling of cargo, order picking, storage and packaging [11, 12].

The logistic performance is described by the delivery time, the delivery reliability, the delivery quality and the delivery service. The delivery time is defined as the time between order and delivery to the customer. The compliance with defined delivery times and its variations are measured by the delivery reliability. Quality of delivery describes the conformity of the order in product type and quantity as well as the condition of the product in regard of transport damages or soiling. A more qualitative performance indicator is the delivery service which summarizes the link to the customer and the ability to react to the customers' needs [11, 12].

There is a connection between the two aspects of manufacturing and logistics though, as manufacturing of products and parts with good functional quality is a precondition to the punctual delivery of the finished products or assemblies.

3.5 Production Ramp Ups in Networks

The quality of a product often results from the combination of components from different suppliers. Therefore in order to assure the functionality of the assembled product, component suppliers are required to deliver parts, which fulfill the dimensional and form tolerances defined during the development of the product. However, during production ramp-ups, frequent process and tool changes are usually necessary before all specified tolerances can be fulfilled. Currently the measures for individual components are derived independently of each other, which can lead to unnecessary process and tool changes. Until now there is no method to identify the optimal technical change strategy during a manufacturing ramp-up in a distributed network with respect to time, cost and risk.

3.6 Agent Based Modeling for Networks

One suitable approach to map a network model with several target systems is the application of software agents [13]. In general, a software agent denotes a self-contained autonomous system that interacts with its environment and pursues its own goals [14]. These characteristics of software agents correlate highly with value-added-network's peculiarities, thus, software agents have already been employed in several ways in planning and operation support tools for value-added-networks.

Some practitioners used the concept of agent-based system for the operative cross-company management. Giannakis and Louis developed a theoretic framework for an agent-based risk management approach in demand-driven supply chains. The focus of this work was on ordering and inventory politics only. The applications of QM measures were not integrated into the model [15].

It can be said that agent-based modeling and optimization of value-added-networks allows the decision-making structures and individual target systems to be considered. Current approaches on agent-based simulation do not consider the quality aspects in operative management of networks.

To eliminate the above mentioned deficits, attempts have been made in the scope of IQ.net to develop new models, methods and phenotypes for assuring quality at site as well as at network level.

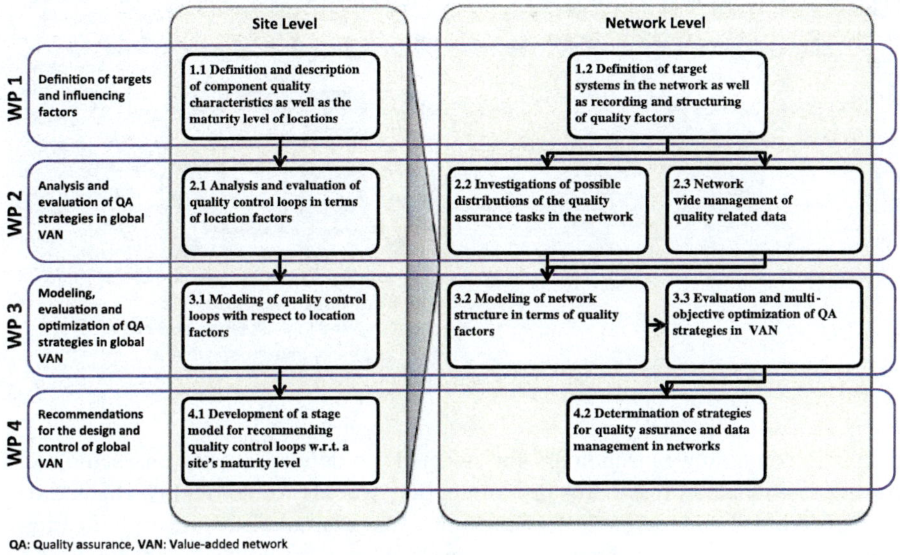

Fig. 2 Solution approach to achieve the research objectives of IQ.net

4 Quality Strategies in Global Production Networks

The objective of the research project IQ.net is to develop innovative methods, models and practical tools for planning, optimization and control of quality strategies for globally distributed production networks, thus obtaining zero-defect production networks. Figure 2 shows the overall solution approach of the project to achieve the research objectives.

4.1 Quality Strategies at Site Level

4.1.1 Definition of Targets and Influencing Factors

Product Model

The product model describes the modeling of product and part characteristics and their creation during the production process. Therefore, a suitable solution to a simple mapping of the complex and company-individual interrelations of a product's cross-company creation process, with its composition from self-contained parts, that each have individual quality characteristics and their respective significance for the products functionality, has to be found.

The selected model (as shown simplified in Fig. 3) is set apart from the actual composition of a product [6]. In the product model the product is regarded as a

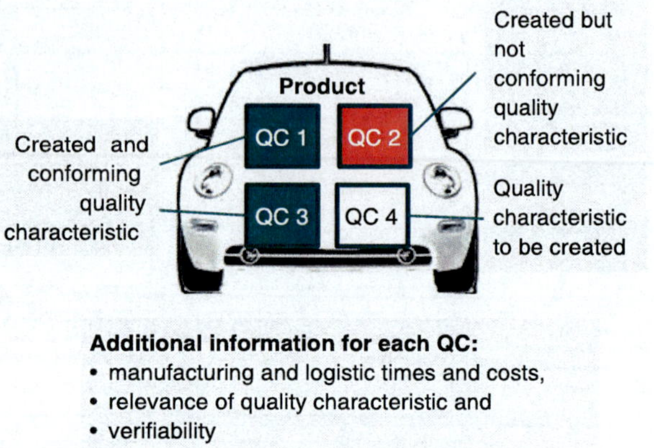

Fig. 3 Model of the final product with selected quality characteristics (QC)

template, containing placeholders for selected, predefined quality characteristics. During a simulated cross-company production process, these quality characteristics are created and defined. Assigned to each characteristic are specifications about the degree of requirement fulfillment of this characteristic (in the figure: QC 1 and 3—conforming, QC 2—not conforming, QC 4—not created), the significance of this characteristic concerning the product's functionality, and information about the verifiability and resulting costs.

Although a cross-company sharing of quality-related product data is promising high benefits, tools or methods to communicate the quality-related product data throughout the network are still not existent. As a prerequisite to this a standardized description of the product quality for electronic data exchange is required. In a next step the product model should be extended into this direction.

Maturity Model

The target at the site level is to determine the maturity level of the sites based on a maturity grid by taking QM dimensions, locational factors (factors that affect QM of a particular location or site) and associated key figures into account. The QM system of a production site can be regarded as dependent on mainly three factors: company-specific factors, regional context factors [16] and Supply Chain factors. At site-level, the main focus is on company-specific factors. Although being important but hard to standardize, regional context factors are not considered here while examining a company's quality performance on site-level. Company-specific factors can be broadly categorized under the following QM dimensions which influence the overall quality at site-level: Product Development, Information Management (Quality Data and Reporting), Process Management and Employees. The classification into these categories follows an approach to empirically examine influence factors on TQM Systems. Furthermore, to account for supply chain factors, the quality dimension named as "Network Capabilities" is included.

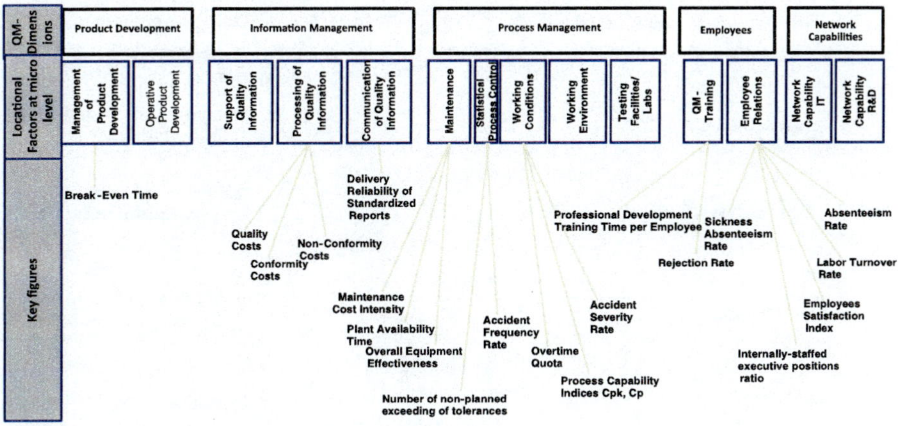

Fig. 4 Relationships between QM dimensions, locational factors and key figures

As a first step to build the maturity model, several locational factors are assigned to each of the QM dimensions. Then, within a refinement step, wherever suitable, several key figures are assigned to the locational factors (See Fig. 4). QM dimensions and locational factors are purely qualitative in nature. The evaluation of such qualitative factors can be done with the help of the associated key figures, which are quantitative in nature.

Having established the framework consisting of locational factors and key figures, a maturity grid is derived by assigning maturity level-dependent manifestations of the characteristics describing each locational factor. Wherever possible, these manifestations can refer to key figure value ranges. The maturity grid can be used to determine the maturity level of a company. For a particular site, each locational factor is evaluated on a scale from 1 to 4. To evaluate an assessment of the maturity level graphically, spider charts can be used. Figure 5 illustrates an example of using the spider charts to evaluate the maturity level of a company (labels 1.1–5.2 in the figure show the related locational factors at micro level).

4.1.2 Analysis and Evaluation

Site Level Quality Control Loops

Consider a multi-stage manufacturing system, in which the items pass through several stages and the items are inspected immediately after each process stage (see Fig. 6). At each inspection station, n items are randomly sampled from a lot of size N ($n < N$) and the number of defective items d is determined. If d is less than or equal to a predetermined number, c, ($d \leq c$) defective items in the sample are replaced with good items and the lot is accepted and released to the next

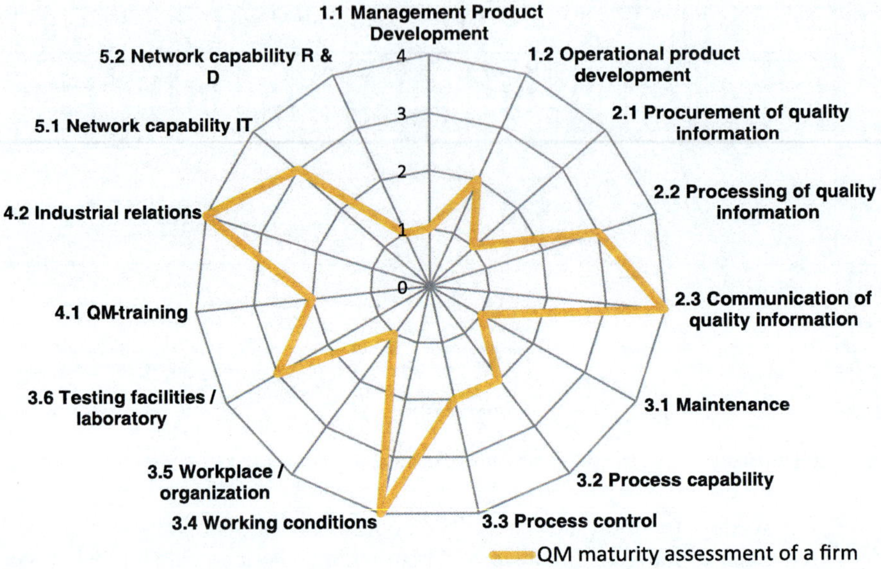

Fig. 5 An assessment of the maturity level of a site

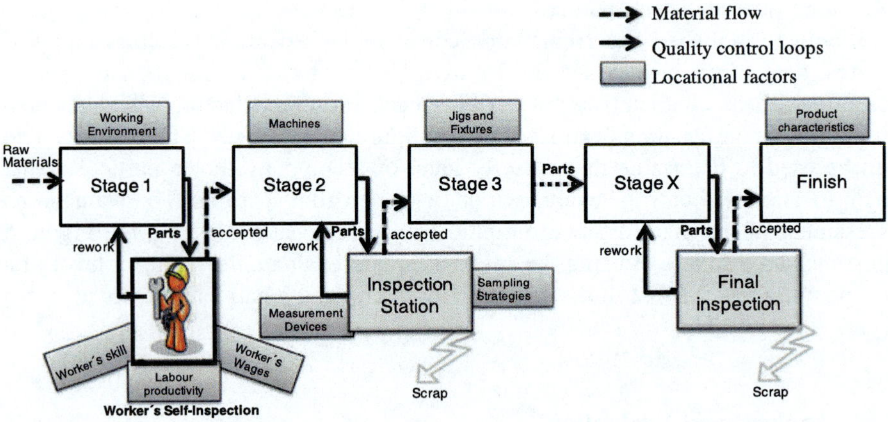

Fig. 6 A multi-stage manufacturing system

manufacturing stage. However, if d > c, the entire lot is screened and all defective items are either reworked or replaced with acceptable ones [17]. Such inspections are often called rectifying inspections [18]. Inspection of an entire lot can also be done by worker's self-inspection. An index called the "worker's capability index" (WCI) is proposed here to obtain the possibility of a worker doing 100 % inspection after a particular stage instead of inspectors doing the full inspection at that stage. WCI explains how good a worker is, to inspect the items of a lot after a

particular stage. It can be determined with the help of associated key-figures such as workers' skills, wages, and their productivity.
The analysis and evaluation of various site level quality control loops (Fig. 6) is done based on the four alternatives below, for inspecting the items after each stage:

1. Sampling inspection by inspectors at the inspection station.
2. 100 % inspection (full inspection) by inspectors at the inspection station.
3. Full inspection by worker himself (Workers' self-inspection).
4. No inspection.

The evaluation of each quality control loop is done on the basis of the total costs associated with them. The total costs are classified as: manufacturing, quality and penalty costs. Quality costs are further sub-divided into inspection costs (comprised of, inspection costs from inspectors and worker's self-inspection costs) and costs of poor quality (comprised of, cost of rework or repair and cost of replacing a defective item with a new one). Penalty costs are further sub-divided into internal penalty costs, if defective items enter into the next stage and external penalty costs, if defective products go to the customers.

4.1.3 Modeling, Evaluation and Optimization of QA Strategies

Optimal Inspection Strategy Based on the Locational Factors of a Site

A simulation based approach to model multi-stage manufacturing systems is proposed here. The aim is to identify the optimal inspection strategy out of the previously analyzed and evaluated site level quality control loops. Thus, the objective of the optimization problem is to minimize the overall costs that exist in multi-stage manufacturing processes subject to the quality constraints that have been put on Average Outgoing Quality (AOQ) and the probability of acceptance of any lot (Pa). If for any stage, AOQ is less than or equal to the target quality level AOQ* and Pa is greater than or equal to target quality level Pa*, the lot is accepted [17].
Furthermore, if the value of the WCI for any stage is greater than or equal to a certain value (say 60 %) and the overall cost of involving a worker in doing self-inspection is less than the cost of carrying out a 100 % inspection at the inspection station, then the worker's self-inspection is performed in place of a 100 % full inspection done by inspectors.

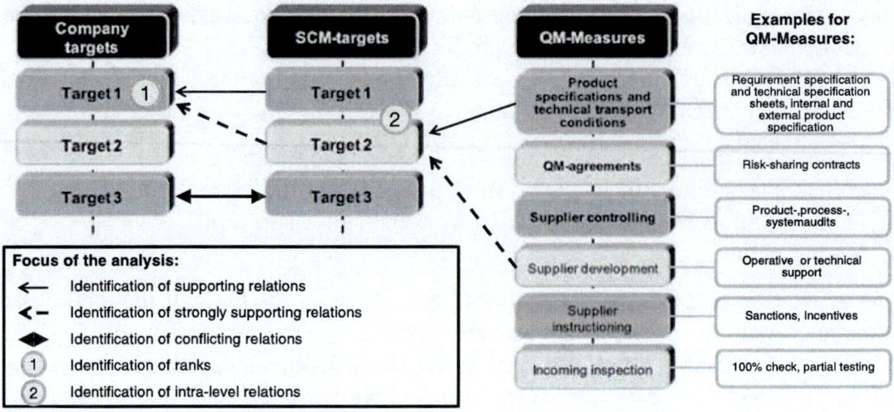

Fig. 7 Focus of target system analysis

4.2 *Quality Strategies at Network Level*

4.2.1 Definition of Targets and Influencing Factors

Enterprise Target Systems

To identify company internal relations between target systems on company level, supply chain management targets as well as the qualitative relations between inter-company QM measures and supply chain management targets, an expert consultation was conducted. The participating experts came from three German companies, two from the machinery and equipment industry and the other as a service provider for quality management in the electronics industry with an external view on several other companies. In Fig. 7, the focus of this consultation is shown [2]. The presented results of the expert consultation show, that on all levels relevant targets and measures can be identified and that there are substantial influences and conflicts between the levels. Figure 8 depicts the summarized results. The interdependencies are expressed by arrows. The weighted importance of the company targets identified in this consultation is shown through the ranking with the most important targets on top. Supply Chain Management targets and QM measures are ranked by their influence on the next upper level.

4.2.2 Analysis and Evaluation

The analysis and evaluation of the quality strategies at network level is done by identifying the following network phenotypes for quality management and its boundaries for application.

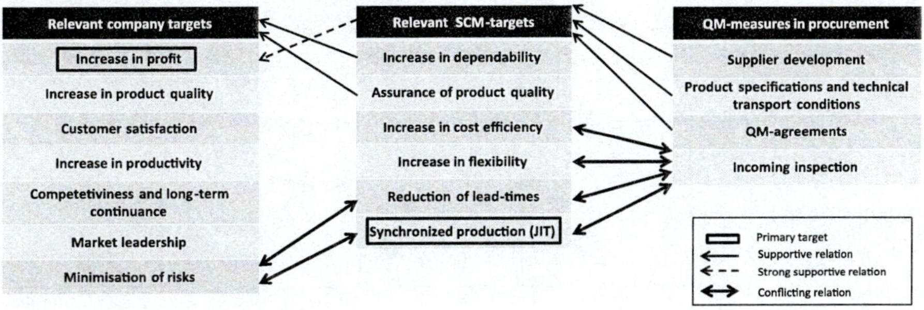

Fig. 8 Target system hierarchy

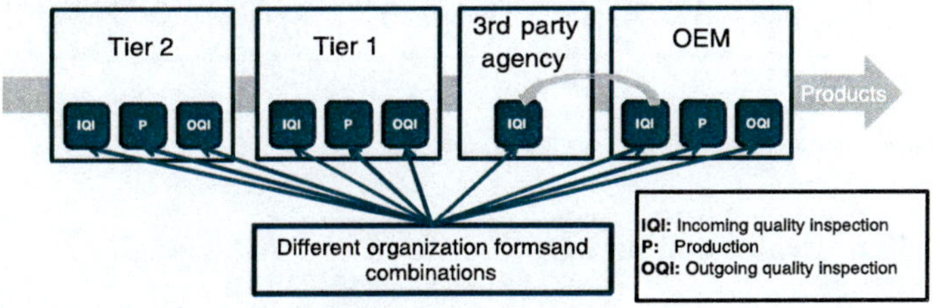

Fig. 9 Introduction of third party agency

Distribution of QA Tasks in the Networks

QA tasks can be distributed amongst all the supply chain members in various ways. In theory each company should maintain a high quality of its own products by incoming inspections for supply parts and process control or outgoing inspections for its own products. However, to reduce quality related-costs there are several other forms to manage the product quality. One possibility, which is especially common in low cost countries and for long transport distances, is an introduction of a third party agency that inspects quality of the incoming products from the suppliers to the customers (see Fig. 9).

Phenotypes of Network-Wide Management of Quality Related Data

To coordinate quality assurance strategies, supply chain members should be open to share quality-related data with their network partners. Storing, processing and managing quality related data amongst all network members can only be achieved with global standards for quality-related data. Therefore an initial catalogue of quality characteristics, which could serve as a basis for such a standard, is being developed within IQ.net.

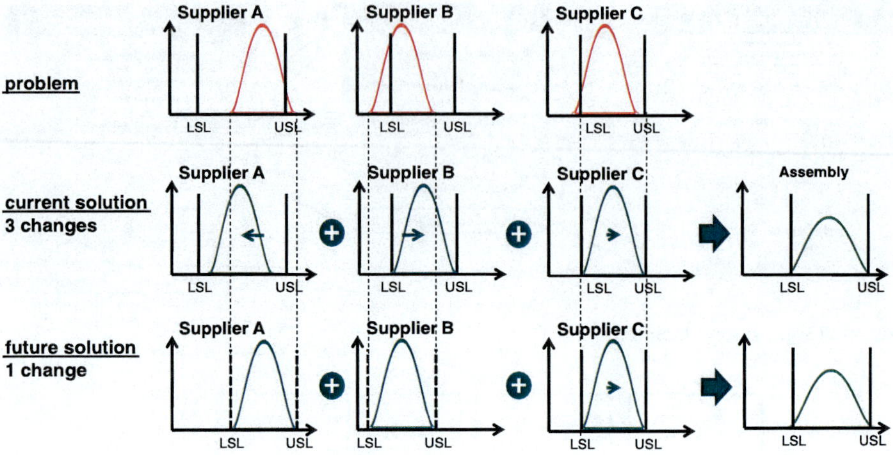

Fig. 10 Quality assurance in production ramp ups

4.2.3 Modeling, Evaluation and Optimization of QA Strategies

Dynamic Tolerance Strategy

Since the quality of the resulting product depends on various components from different suppliers, the focus of this method is to identify the optimal technical change strategy for obtaining the desired quality of an assembly during a production ramp-up in a distributed network. To implement the approach a validated, functional tolerance model of the product is used. By replacing the assumed production variations with actual measurement data of pre-series components, the effects of process and tool changes can be simulated with a statistical tolerance analysis tool. Possible combinations of process and tool changes can then be evaluated based on the resulting quality as well as the difficulty, risk and necessary time for the changes to be implemented. By dynamically adjusting the tolerances and nominal values during the production ramp-up phase, the total number of tool changes as well as the duration of the ramp-up phase could potentially be reduced, see Fig. 10.

Agent Based Modeling

A simulation model was developed that can be used for agent-based simulation of the QM behavior of supply chains. The functions and processes in companies were modeled with respect to certain aspects in real production networks like autonomous decisions of each company, realized with messages based on the agent communication language to maintain autonomous decisions of the agents, and a cooperation of different functions inside the company, depicted by multiple and parallel working reasoning engines. The universal modeling concept of the agents, with the possibility of parameter adaptation according to the structure and properties of company-specific value-added-networks, allows an easy reconfiguration of the model. It could be showed that the behavior of the model and its company agents resembles the conditions in real supply chains [2].

5 Summary and Outlook

Many companies have now-a-days established a global network of worldwide production sites. Due to the long and diversified structures of supply chains and differences in the maturity levels of suppliers, distributed networks develop various fluctuations, in terms of varying product quality and delayed delivery times, which can result in customer dissatisfaction, image loss and financial losses to the companies of the network. Moreover, an improperly implemented quality strategy in a network will result in higher costs. The preliminary idea is to make these networks insensitive to such fluctuations by identifying and evaluating suitable quality strategies. So, there is a need to identify and evaluate suitable quality strategies for making distributed networks insensitive to the fluctuations.

The research project IQ.net aims to develop innovative methods, models and practical tools for planning, optimization and control of quality strategies for globally distributed production networks. In this chapter, various aspects of the research project IQ.net have been presented for assuring quality at site as well as at network level. Various recommendation models for the design and control of global value-added-networks will be developed as the part of IQ.net in the near future. These include a stage model for recommending quality control loops with respect to a site's maturity level and an another model for determining quality strategies and data management in networks.

Acknowledgments The IGF promotion plan 17219N of the Research Community for Quality (FQS), August-Schanz-Str. 21A, 60433 Frankfurt/Main has been funded by the AiF within the programme for sponsorship by Industrial Joint Research and Development (IGF) of the German Federal Ministry of Economic Affairs and Technologies based on an enactment of the German Parliament.

References

1. Memedovic, O.: Inserting local industries into global value chains and global production networks. SME Technical Working Paper Series, Working Paper, UNIDO, Vienna (2004)
2. Book, J., Lanza, G.: Quality strategies for global supply chains. Working paper for GCOM conference, Lyon, France (2011)
3. Creazza, A., Dallari, F., Melacini, M.: Evaluating logistics network configurations for a global supply chain. Supply Chain Manag. Int. J. **15**(2), 154–164 (2010)
4. Dreyer, H., Alfnes, E., Strandhagen, J.O., Thomassen, M.K.: Global supply chain control systems: a conceptual framework for the global control centre. Prod. Planning Control **20**(2), 147–157 (2009)

5. Reefke, H., Sundaram, D., Ahmed, M.D.: Maturity progression model for sustainable supply chains. Lecture Notes in Business Information Processing 46 LNBI, pp. 308–319 (2010)
6. Book, J., Lanza, G.: Modeling and simulation of quality control strategies in value-added-networks under consideration of individual target systems and product characteristics using software agents. In: Proceedings of 44th CIRP conference (2011)
7. Hanselmann, M., Roman, S.: Qualitätsfähigkeit : eine praxisorientierte Anleitung zum Aufbau von Total-quality-Management in mittelgrossen Unternehmen (German). Haupt, Bern. ISBN: 3-258-05298-0 (1996)
8. Eickelpasch, A., Lejpras, A., Stephan, A.: Hard and soft locational factors, innovativeness and firm performance: an empirical test of porter's diamond model at the micro-level, Discussion papers of DIW 723, Berlin, pp. 1–26 (2007)
9. Sarhangian, V., Vaghefi A., Eskandari, H., Ardakani, M.K.: Optimizing inspection strategies for multi-stage manufacturing processes using simulation optimization. Winter Simulation Conference, vol. 15, pp. 1974–1980 (2008)
10. Bowersox, D.J., Closs, D.J., Cooper, M.B.: Supply Chain Logistics Management. McGraw-Hill/Irwin, Boston (2002)
11. Heid, T.: Begriff der Logistik, logistische Systeme und Prozesse (German). Handbuch Logistik, Springer, Berlin, pp. 3–34 (2008)
12. Rushton, A.: The Handbook of Logistics and Distribution Management. Kogan Page, London (2009)
13. Macal, C.M., North, M.J.: Tutorial on agent-based modeling and simulation part 2: how to model with agents. In: Proceedings of the 2005 winter simulation conference, pp. 73–83, Association for Computing Machinery, Piscataway (NY) (2005)
14. Wooldridge, M.: Intelligent Agents: The Key Concepts. Springer, Berlin (2002)
15. Giannakis, M., Louis, M.: A multi-agent based framework for supply chain risk management. J. Purch. Supply Manag. **17**(1), 23–31. Elsevier (2011)
16. Kull, T.J., Wacker, J.G.: Quality management effectiveness in Asia: the influence of culture. J. Oper. Manag. **Band 28**(3), 223–239 (2010)
17. Volsem, S.V., Dullaert, W., Landeghem, H.V.: An evolutionary algorithm and discrete event simulation for optimizing inspection strategies for multistage processes. Eur. J. Oper. Res. **179**, 621–633 (2007)
18. Choi, T.Y.; Rungtusanatham, M.: Comparison of quality management practices: across the supply chain and industries. J. Supply Chain Manag. **Band 35**(1), 20–27 (1999)

From Collaborative Development to Manufacturing in Production Networks: The SmartNets Approach

Armin Lau, Manuel Hirsch and Heiko Matheis

Abstract Innovative, knowledge-intensive products are essential for companies to stay competitive in a globalised market. For the development of such products, small and medium-sized enterprises combine their core competences and resources in dynamic, loosely coupled networks. The basis for efficient and robust production of the newly developed subject can already be provided in the early phases of such collaboration. However, a continuous manufacturing process can only be achieved by a conscious and guided transformation from development network to production chain. This paper presents an approach from the European research project SmartNets to control and support this transformation methodologically and technologically, considering not only the organizational perspective, but also information and communication technologies, and knowledge aspects. A case study will illustrate how the transformation could be implemented in one of the project's industrial networks.

Keywords Collaborative innovation · Smart network · Production network · Knowledge orientation

A. Lau (✉) · M. Hirsch · H. Matheis
DITF Denkendorf—Centre for Management Research, Denkendorf, Germany
e-mail: armin.lau@ditf-mr-denkendorf.de

M. Hirsch
e-mail: manuel.hirsch@ditf-mr-denkendorf.de

H. Matheis
e-mail: heiko.matheis@ditf-mr-denkendorf.de

K. Windt (ed.), *Robust Manufacturing Control*, Lecture Notes in Production Engineering, 287
DOI: 10.1007/978-3-642-30749-2_21,

1 Introduction

For complex and knowledge-intensive products and processes, cooperation of companies with partners along the value chain and with complementary service providers is one of the most important factors both for successful product and process development as well as for robust manufacturing,[1] especially for small and medium-sized enterprises (SME) [1]. Frequently changing customer demands, shorter lead times and condensed product life cycles increase the need for intensified collaboration [2].

To respond to the dynamics of the global market, flexible organizational forms have emerged to complement the strengths and weaknesses of more stable, but also rigid networks [3]. The concept of dynamically networked enterprises has lead to a paradigm shift both in product and process development and in manufacturing [4].

Still, innovation and production networks are mostly regarded separately both in research and in practice, even though considering and enabling robust manufacturing should be an intrinsic target in any development process. Methods and tools to support a conscious transformation from a development network into a production network are still lacking.

The European research project SmartNets is focusing on the elaboration of such methods and tools to support this transformation [5]. This paper presents the overall approach and describes preliminary results from the project addressing three different perspectives of collaboration—organizational issues, hyperlinking through information and communication technologies (ICT) and conjoint knowledge exploration, application and exploitation—that support:

- Combining development-related activities in ad-hoc workflows towards a sophisticated manufacturing process architecture.
- Documenting tacit knowledge and evolving structures from explicit, but unstructured, development-related knowledge.
- Integrating IT services (value creation services and utility services) into interoperable, harmonized service parks with complementing functionalities to support production processes.

As a practical example, the overall approach and its implications regarding the three perspectives will be demonstrated in a case study from one of the three industrial networks in the project.

[1] Robust manufacturing describes the proper functioning, repeatable and stable behavior of manufacturing systems in terms of e.g. quality, cost effectiveness and sustainability even under the influence of uncertain parameters and disturbances.

2 SmartNets Transformation

The SmartNets concept of describing collaboration by three complementary perspectives is based on the idea of 'Smart Organizations' [6]. Such organizations are 'knowledge-driven, internetworked, dynamically adaptive to organizational forms and practices, learning as well as agile in their ability to create and exploit the opportunities offered in the digital age' [7]. Organizations exhibiting these abilities are well-prepared to collaborate with others in dynamic networks in order to explore and to exploit opportunities in changing markets. Such networks—often referred to as 'Smart Networks' [8]—are an outstanding basis for the development of innovative, knowledge-intensive products and processes. Furthermore, they are showing exactly the characteristics, which are required for a conscious and systematic transformation into a production network, taking into account not only the organizational view on collaboration, but knowledge and ICT aspects as well.

By characterizing starting and end point of such a transformation as well as its (implicit) complexity along a generic innovation process (see Fig. 1), the need for appropriate method and tool support becomes obvious. From an organizational point of view, the development phase is focused towards individual, ad-hoc activities while in the production phase repeatable processes are carefully designed. While the emphasis from a knowledge point of view is on implicit and evolving knowledge during the development, manufacturing depends a lot on explicit and reliable knowledge available in documents, manuals, and also in ICT systems. A third observation can be made in the field of ICT, as in the production phase, whole application systems will be used, while for the support of the development phase, flexible stand-alone IT services are not only sufficient but crucial for its success.

Principles from control theory are applied to support the transformation in all three perspectives. For each phase of development and production, targeted (or desired) states will be generically defined based on literature and industrial practice. Methods and tools will be provided to the networks to analyze their respective current status. To support the transformation from the current to the desired state, adaptive control techniques [9] will be implemented, which are capable of handling uncertainties and under-determined systems by providing dedicated knowledge services in the *SmartNets Transformation Manager*:

- It will provide functionalities of a *state observer*, gathering knowledge about the current status of the system from various sources and identifying lacking knowledge.
- It will serve as a *classical controller*, offering proposals to the user how to transform the system based on the desired state and a rule set for control.
- It will serve as an *adaptive controller*, adjusting its rule set by documenting actual user decisions and thus learning rules not explicitly modeled.

In the following, implications of this approach will be discussed under the three Smart Networking perspectives of organizational, knowledge, and ICT networking.

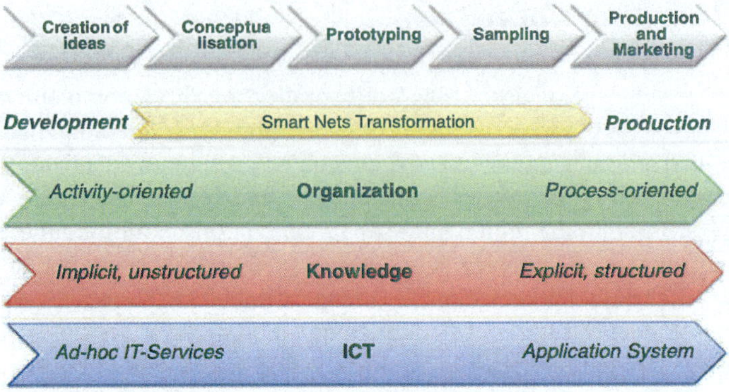

Fig. 1 Transformation from development to production

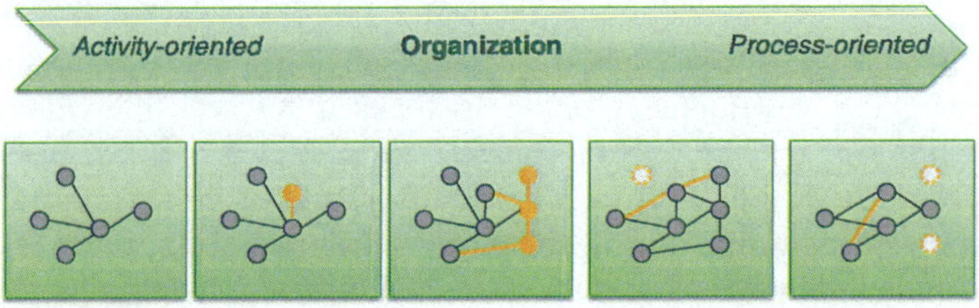

Fig. 2 Changing partnership structures from development to production

2.1 *Organizational Transformation*

Networking from an organizational perspective can be best described by the term "organizational teaming" [7]. People from different organizations with complementary competences are closely working together on specific tasks and activities and are facing social challenges of collaboration like decision making and trust. To solve that, each contributor has to be provided with the appropriate infrastructure and with the respective resources to enable a timely delivery of the input required by others. When considering the development and production process, the importance of particular partners may grow or shrink along the way. And while earlier phases of development are characterized by a high degree of uncertainty [10], stability and robustness of processes are indispensable in production. In order to provide methods that support a conscious and guided transformation from an organizational point of view, both of these aspects have to be carefully considered.

Changes in the network partnership will be necessary with respect to the contributions that are expected of partners in the different phases of development and production. Figure 2 shows exemplarily possible changes in a network partnership along the development and production process.

In the following, the Co-opetition model [11], which distinguishes suppliers, customers, complementors, and competitors as potential external partners for innovation, will be used to describe some of the changes of roles that might occur. Within the first phases of the development process, knowledge is gathered and explored, options for realization are discussed and evaluated, and first tangible results will be developed and subjected to testing. In these phases, service providers and complementors will contribute their competences, e.g. by providing basic research on a recent material development, by helping to investigate various design options or by adapting machinery to new handling requirements. The role of suppliers and customers in this phase is mainly to give valuable input on available materials and about market expectations.

Turning towards production, vertical collaboration along the value-added chain becomes even more important. First prototypes and samples will be developed and tested, the production process has to become stable and robust, material supply needs to be secured, and continuous production will be prepared. While for the development, one supplier of material might have been enough, now several suppliers could be required to scale up the production. Customers can test first samples in their application system and their feedback on the quality of the product may allow conclusions regarding the reliability of the process. Some service providers still have an important role in preparing the production phase, e.g. machine providers taking care of the process implementation or complementors offering ancillary products or services to the customer. Contributions of other service providers, like basic researchers or IPR consultants will scarcely be required when approaching production.

Trust and confidentiality considerations are crucial for the success of collaboration [12] and thus play an important role for the transformation as well. While confidentiality of information can be addressed with adequately designed ICT systems, e.g. by applying appropriate IPR means and bi-lateral (explotation) contracts, trust has to be constantly built up and strengthened in the network all along the life-cycle.

But not only partnership itself is altered along the process, but also the *way of collaboration from an organizational point of view* changes drastically. Work within the early phases of development and production has the characteristics of a project, and is governed by a high level of uncertainty. With each innovation having its own particularities, there will be only few repeatable activities that can be followed in each process. Thus, collaboration will be event-triggered and task-oriented. For example, based on the results of a specific test, next steps will be determined to reach a new decision point. This process is understood as ad-hoc workflow [13], which might be adapted based on incidents.

The preparation and execution of continuous production does not follow project characteristics. Well-established processes have to be followed to ensure reliable and robust manufacturing. In this phase, processes from different organizations must be connected to enable reliable material and information flows all along the value-added chain. This linkage can be prepared step by step along the development process by subsequently stabilizing ad-hoc workflows to resilient processes, which can then be interlaced with company-internal processes to create a flexible network process architecture.

In order to apply the concept of the above-mentioned *SmartNet Transformation Manager* to the organizational perspective on collaboration, first of all, generic descriptions of appropriate network partnerships and process structures for each phase of the development and production process are required. Furthermore, specific partnerships and actual ad-hoc workflows as well as process infrastructures have to be analyzed in the networks and have to be related with the current status of development. Based on the gap between "as-is" and "should-be", generic suggestions can be offered, like the proposal to include additional actors or to implement specific kinds of processes. Through feedback about the decisions made, and by analyzing the particularities of the development project, the *SmartNet Transformation Manager* will be able to adapt its underlying rule set to the specific characteristics of the development and will be able to give suitable advice in further phases. Regarding the social aspects of this transformation, guidelines and best practices for the creation and strengthening of trust between partners will be provided all along the life-cycle.

The organizational perspective of networking is obviously closely connected both with the knowledge perspective and the ICT perspective. All along the development process, partners collaborate closely to conjointly explore, apply and exploit knowledge. ICT systems enable the information exchange for locally dispersed teams and thus facilitate intensive cross-organizational collaboration. In both perspectives, there is a significant change from development to production, which will be described in the upcoming sections.

2.2 Knowledge Transformation

During each development and manufacturing phase of a product, knowledge and competences from various actors in the network are combined to elaborate new knowledge about the targeted innovation and to carry out knowledge-intensive tasks. The knowledge representations used and formed in these tasks range from fairly unstructured (e.g. textual, tabular, or graphical collection of product ideas) to highly structured (e.g. semi-formal machine settings and execution models or algorithms for the final product). Figure 3 illustrates how the evolution of knowledge structures along the life cycle of products and/or services can be captured.

At the beginning of development activities, ideas are usually drafted in textual descriptions and sketches. Accordingly, most details are stored implicitly or kept tacitly in human minds. During the transformation from development to manufacturing, more and more details are elaborated and documented in rudimentary knowledge structures that are shared between partners. This transformation is crucial for the success of a development initiative: on the one hand, knowledge can only be stored, when adequate and sufficiently detailed knowledge structures are

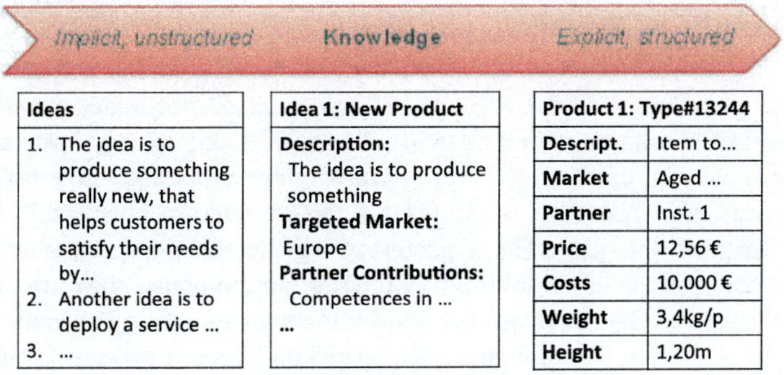

Fig. 3 Evolution of knowledge structures in innovation networks

provided, on the other hand, to store all the knowledge that is created or even talked about in innovation-related discussions might lead to information overload that has to be avoided [14]. In the final phase of manufacturing, all data for an efficient and effective production and/or deployment of an innovative product and/or service has to be presented in well-defined (ICT-compatible) knowledge structures.

Not only the heterogeneity of knowledge representation and knowledge structures, but also the evolution (e.g. product characteristics will be more detailed from one phase to another) and transformation (e.g. transformation of market survey to product features) of knowledge from the development to the manufacturing phase is part of the SmartNets challenge. As a consequence, the management of knowledge in networks has to deal explicitly with the availability of required knowledge at the right time (e.g. availability of a service provider for materials research during material testing) but also to conserve newly generated knowledge for later phases of development and manufacturing (e.g. experiences gathered during material testing could be helpful for anticipating problems in production processes). To sum up: It is essential to continuously manage the knowledge as well as knowledge structures used and created in course of concurrent idea generation activities, prototyping, and later on in collaborative manufacturing as well as marketing of valuable products and services along their whole life cycle.

Therefore, the *SmartNet Transformation Manager* comprises semantic tools for capturing the knowledge structures applied in a network as well as appropriate means for their harmonization, integration, and promotion among partners. Furthermore, it helps to ensure that already existing knowledge structures are used correctly and that adaptations of the structures can be merged coherently. Rule-based similarity checks provide context-sensitive suggestions about how to further elaborate the knowledge base of a network.

The transformation of knowledge and knowledge structures can additionally be supported by organizational and technological means. This will be outlined in the next chapter about ICT as well as in the use case provided.

2.3 ICT Transformation

ICT tools applied in industrial networks should provide adequate means for supporting the transformation of organizational as well as knowledge structures from development to manufacturing. While knowledge structures have to be well defined and a priori provided in context of the production phase, they are continuously adapted and provided a posteriori in the development phase. While knowledge contents are essential input in the production phase, they are object and output of the knowledge work in the development phase. As a consequence, ICT tools need to be able to handle both unstructured and structured knowledge as well as ad-hoc workflows in virtual development teams and robust production processes in well-defined supply chains. Furthermore, knowledge about product ideas, concepts, prototypes, sampling activities, and finally about manufacturing of products is the object of discourse in cross-sectoral development networks. Accordingly, ICT tools to support collaborative development and manufacturing activities have to guarantee instant, asynchronous, guided, and context sensitive access to the continuously evolving knowledge base of the network.

To be able to successfully manage these transitions, a conceptual, technological and methodological framework is necessary to adapt and transform the ICT itself along the transformation from development to production in networks. In SmartNets, this is realized by applying a flexible, need-driven (middle-out), and service-oriented approach. This approach incorporates ad-hoc generation of knowledge-based services in the development phase, aggregation of services in the transformation phase, and setup of a comprehensive service-based application system in the production phase.

Figure 4 shows the transformation of respective ICT components. In the idea creation and concept validation phase at the beginning of development-related activities within networks, ICT tools have to provide ad-hoc solutions for spontaneous knowledge needs of actors in the network. Information has to be gathered, complementary competences have to be aligned, and ad-hoc workflows, for instance for the application of dedicated methods for innovation management, have to be set up. Therefore knowledge workers in the development phase are supported by knowledge-based IT services that are flexible, modularized, and created by the end-user and domain experts themselves, e.g. web-services, Smart Services [15], macros, scripts, and others.

Accordingly, services for collaborative product development are successively deployed in response to concrete needs of knowledge workers by means of agile ICT development strategies like rapid IT-service development or Ontology-driven Service Development principles [15]. This leads to a heterogeneous set of loosely coupled services that have to be orchestrated, merged, and eventually deactivated in a respective Service Park [16]. Services Parks, comprising a set of well integrated high-quality IT services, can be used as flexible, easily extendible, and low-cost means to build up and fill a knowledge base in the development phase of

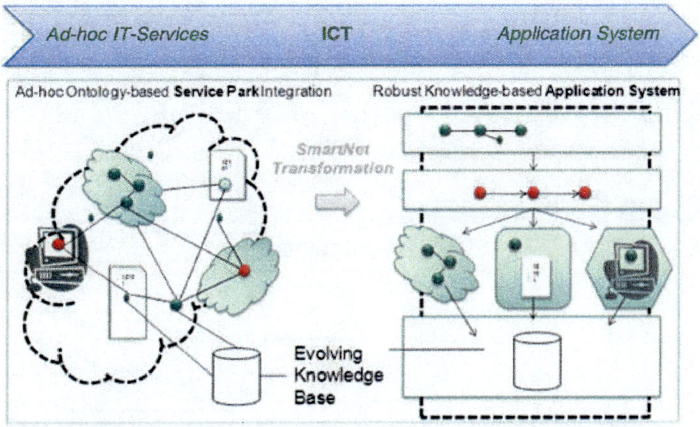

Fig. 4 Transformation of an ad-hoc generated knowledge-based Service Park (*left*) into a robust Application System (*right*)

production networks that can directly be used for robust manufacturing in the production phase.

As soon as knowledge structures are materialized, innovation-related knowledge is created and finalized, and organizational structures are set up in preparation of collaborative production of a newly designed product, a comprehensive Application System can be implemented top–down e.g. on basis of already existing services from the network's Service Park. While Service Park components might easily be re-used in context of other development projects, monolithic application systems for manufacturing are highly optimized and specialized for the purpose of supporting the effective and efficient production of a well-defined innovative product.

The *SmartNet Transformation Manager* provides knowledge-based services that can be applied both in the development as well as manufacturing phase of a network in order to analyze, optimize, and extend its Service Park—and finally to set up a comprehensive Application Systems.

3 Case Study

Within the SmartNets project, several already established networks will implement and evaluate the presented approach in their own collaborative development and production processes. This chapter describes implications of the concept for a project network in which six partners develop and produce an innovative medical device. It will demonstrate the current status of the network, and will show how its transformation along the development and production process could be supported by the outlined *SmartNet Transformation Manager*.

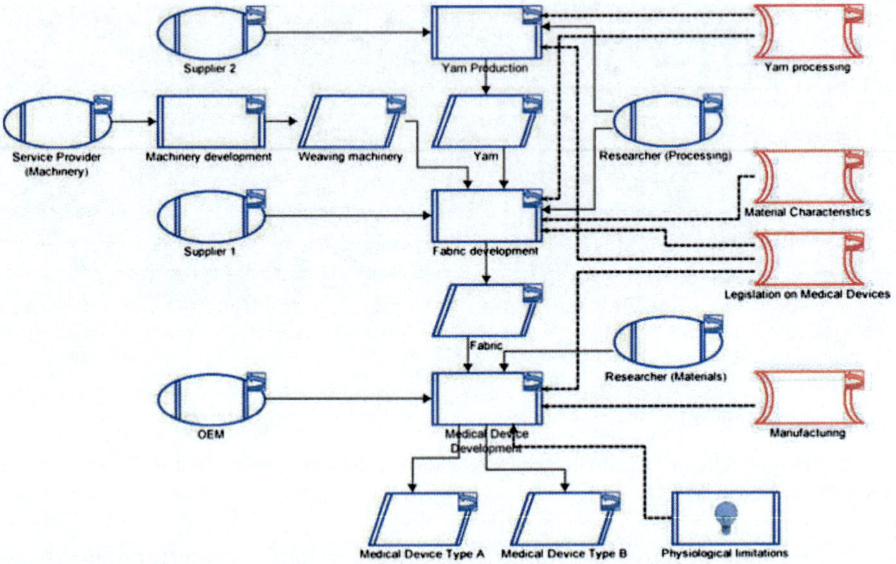

Fig. 5 Exemplary description of a product development process

The network partnership and its way of collaboration are documented in network topology models and process models based on a modeling framework, which has been developed to describe organizational, knowledge and ICT aspects of a network [17].

One example of such a model is shown in Fig. 5. Oval symbols resemble the actors in the networks, rectangles illustrate major processes, rhomboids depict main products obtained and symbols in red show crucial knowledge required for the identified process steps. These models are adapted whenever changes in the network structure or way of collaboration occur. Thus, the different versions of the models describe not only the current status of the network, but can be combined to analyze the complete history along each phase of the process.

The specific description of the network in such a model is related to a status in the development and production process and particular actors are connected to generic actor types. This allows a gap analysis and suggestions for adaptations regarding the project partnership. To give an example, *Service Provider* (*Machinery*) is currently supporting the development of *Supplier* 1 by implementing new machinery for the prototyping process. Thus, *Service Provider* (*Machinery*) is the core partner for the *Fabric development*. However, as soon as this development reaches production phase with validated machinery, the role of the service provider will dwindle and *Supplier* 2 and *OEM* will become the main partners along the value-adding chain. Another probable change in the network is the withdrawal of the research partners after the successful sampling of the product. Careful transformation has to ensure that these partners will no longer be needed for material and information flows in production.

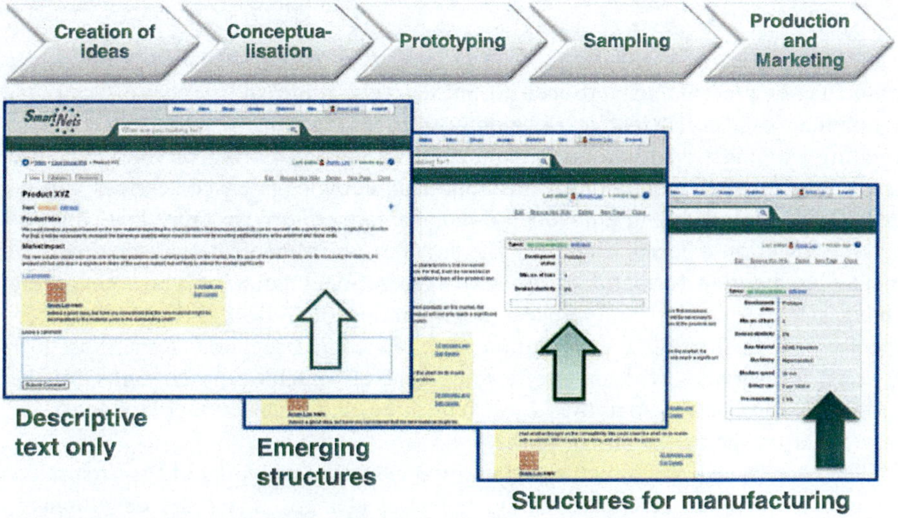

Fig. 6 Development of knowledge structures along life-cycle in a hybrid wiki

Available and missing knowledge is documented in the network models as well. In the medical device network, partners started with descriptive text and rough sketches when first describing and discussing the ideas of the medical device.

During conjoint knowledge exploration, the body of knowledge grew significantly. For documentation and distribution of this knowledge as well as to derive well-founded decisions, text was written, extended, shared, and discussed among SmartNets stakeholders on the project's collaboration platform *Tricia,* a hybrid wiki for enterprise collaboration and information management [18].

Along the process, initial structures for engineering tasks like product conceptualization, prototyping and testing have emerged which are now documented as attributes of specific types of pages. Figure 6 shows how this could lead to highly detailed tabular and fixed structures supporting robust manufacturing of newly developed products in highly innovative processes in an optimized organizational environment.

Neither organizational nor knowledge transformation could be achieved without the support of ICT systems. The example in Fig. 6 shows how a suitable collaboration platform can support the development of knowledge structures in the network. Specific services support the distribution of tasks and the definition of ad-hoc workflows which are critical for the development of the medical device. Services on the platform can be interlinked to create a SmartNets Service Park, which can be accessed by all project partners. Of course, also interfaces have to be provided to information on the platform with data already existing within monolithic application systems to enable network partners to follow their company-internal, well-proven process architectures.

4 Conclusions

This chapter presents an approach to pro-actively guide and facilitate the transformation from collaborative development to production in networks, thereby creating a foundation for robust manufacturing already during product development. For the implementation of this approach, knowledge-based services will be developed, deployed, and embedded in an adaptive control structure that can pro-actively react on external changes. The services will support network partners in managing the transformation from an organizational point of view, regarding changes in partnership and process configuration, and from a knowledge perspective by facilitating the evolution of knowledge structures. Regarding ICT aspects, the services will aid the development of scalable service parks for the support of production activities by establishing and connecting individual ICT services from the development phase.

The concept will be implemented in three industrial networks of the SmartNets project. In the presented case study, potential changes along the development process for a medical device are identified to demonstrate the tasks and challenges the concept needs to appropriately support in practice. The realization of the *SmartNets Transformation Manager* will be done in compliance with the presented concept and careful evaluation in all three project partnerships will help to identify the practical value of the tool.

Acknowledgments The work presented in this paper has been partly carried out in the context of the SmartNets project (http://www.smart-nets.eu) co-funded by the European Commission under the 7th Framework Program (NMP, project no. 262806). The authors would like to acknowledge the support of the European Commission and the comments and contributions of the SmartNets project partners.

References

1. Mazzarol, T., Reboud, S.: The role of complementary actors in the development of innovation in small firms. Int. J. Innov. Manag. **12**(2), 223–253 (2008)
2. Christopher, M., Lowson, R., Peck, H.: Creating agile supply chains in the fashion industry. Int. J. Retail Distribution Manag. **32**(8), 276–367 (2004)
3. Camarinha-Matos, L.M., Afsarmanesh, H.: Collaborative networks: a new scientific discipline. J. Intell. Manuf. **16**(4), 439–452 (2005)
4. Estruch, A., Vila, C., Siller, H.R., Romero, F., Abellán, J.V.: Enabling innovative concurrent engineering and collaborative manufacturing in extended enterprises. In: Thoben, K.-D., Kulwant, S. P., Goncalves, R. (eds.) A New Wave of Innovation in Collaborative Networks. Proceedings of the 14th International Conference on Concurrent Enterprising: ICE 2008. Centre for Concurrent Enterprise, Nottingham, pp. 759–766 (2008)
5. SmartNets: Welcome to the SmartNets EU Project! http://www.smart-nets.eu
6. Filos, E., Banahan, E.P.: Towards the smart organization. An emerging organizational paradigm and the contribution of the European RTD programmes. J. Intell. Manuf. **12**(2), 101–119 (2001)

7. Filos, E.: Smart organizations in the digital age, In: Mezgár, I. (ed.): Integration of ICT in Smart Organizations, pp. 1–37, Idea Group Publishing, Hershey (2006)
8. Lau, A., Fischer, T.: Cross-sectoral innovation networks for knowledge-intensive products and services. In: Spath, D., Ilg R., Krause, T. (eds.) Innovation in Product and Production. Proceedings of the 21st International Conference on Production Research (ICPR21). Fraunhofer-Verlag, Stuttgart (2011)
9. Åström, K.J.: Theory and applications of adaptive control: a survey. Automatica **19**(5), 471–486 (1983)
10. Kim, J., Wilemon, D.: Focusing the fuzzy front-end in new product development. R&D Manag. **32**(4), 269–279 (2002)
11. Nalebuff, B., Brandenburger, A.M.: Co-opetition. Harper Collins Business, London (1996)
12. Dooley, L., O'Sullivan, D.: Managing within distributed innovation networks. Int. J. Innov. Manag. **11**(3), 397–416 (2007)
13. Voorhoeve, M., van der Aalst, W.: Ad-hoc workflow: problems and solutions. In: Proceedings of Eighth International Workshop on Database and Expert Systems Applications, pp. 36–40 (1997)
14. Hinz, K.: Information Overload: Die Informationsüberlastung der Konsumenten im Zeitalter der Digitalisierung und Technisierung. GRIN Verlag (2007)
15. Hirsch, M.: Smart Services for Knowledge Integration—Ontologiebasierte Dienste zur Unterstützung der kollaborativen Wissensarbeit in Innovationsnetzwerken. Bd. 202. Fortschritt-Berichte VDI 16. Düsseldorf: VDI-Verlag (2012)
16. Petrie, C., Bussler, C.: The myth of open web services—the rise of the service parks. IEEE Internet Comput. **96**, 80–82 (2008)
17. SmartNets: SmartNet Modeller and application guidelines. Public Deliverable. https://www.smart-nets.eu/file/Downloads(Public)/PublicDeliverables/D1.2SmartNetsModellerandapplicationguidelines.pdf
18. InfoAsset: Tricia—Hybrid Wikis for Enterprise Collaboration and Information Management. http://www.infoasset.de/wikis/infoasset/tricia-hybrid-wikis

Service-Oriented Integration of Intercompany Coordination into the Tactical Production Planning Process

Christoph Besenfelder, Yilmaz Uygun and Sandra Kaczmarek

Abstract The processes of coordination and planning within networks are subject to constant change. A tool is necessary which supports production planning within networks in a lean way and which makes the on-demand and inexpensive institutionalization of coordination and planning processes among partners possible. The project Supply Chain Planning (SCP) of the EffizienzCluster LogistikRuhr deals with planning on the tactical level. The case of application 'production' (Technische Universität Dortmund and ABH Stromschienen GmbH) develops a service-based system for supporting the planning process. This paper is about the integration of intercompany coordination into the tactical production planning process flow of enterprises within supply nets. This includes the creation of a data structure which is independent of the respective IT infrastructure of the partners. It is called 'capacity corridors' and the processes of coordination are run with it. Especially small and medium-sized enterprises benefit from this independent service. They can profit from a higher planning reliability without having considerably to invest into their IT infrastructure.

Keywords Logistics as a service · Supply chain planning · Supply net management · Production planning · Robust coordination · Intercompany planning

C. Besenfelder (✉) · Y. Uygun · S. Kaczmarek
Chair of Factory Organization, Technische Universität Dortmund,
Leonard-Euler-Street 5, 44227 Dortmund, Germany
e-mail: besenfelder@lfo.tu-dortmund.de

Y. Uygun
e-mail: uygun@lfo.tu-dortmund.de

S. Kaczmarek
e-mail: kaczmarek@lfo.tu-dortmund.de

K. Windt (ed.), *Robust Manufacturing Control*, Lecture Notes in Production Engineering,
DOI: 10.1007/978-3-642-30749-2_22,

1 Initial Situation

Cooperation is getting more and more important within production. A low vertical range of manufacture and the specialization in core competencies are only a few reactions to increasing globalization and individualization. Cooperation in the field of production is the way of coping with these external factors and the pressure to expand. But this cooperation is often marked by insecurities, problems of planning, and information asymmetry. Especially small and medium-sized enterprises (SME) experience difficulties in entering successful cooperation without committing themselves on a long-term basis. An IT support is necessary which supports the initialization and execution of cooperative planning processes between companies on demand.

The guiding topic "Logistics as a Service" of the Cluster of Excellence called 'EffizienzCluster LogistikRuhr', which is founded and granted by the German Federal Ministry of Education and Research, has the development of a toolbox as its aim which is composed out of IT modules. This toolbox individually supports the logistical tasks within enterprises by combining the single IT services on demand. Within this context, there are four projects of the guiding topic (Supply Chain Design, Supply Chain Planning, Supply Chain Execution, and Service Design Studio) which deal with planning within different time horizons and abstraction levels in order to define and put exemplary logistical assistance systems to practice. The project Supply Chain Planning concentrates on the planning on the tactical level. The terms of reference of the tactical planning within the framework of this project correspond to the task model of Supply Chain Management according to Kuhn and Hellingrath [1]. In order to make an application-oriented and extensive examination of the planning tasks possible, the project is subdivided into three cases of application. Each of them works in cooperation with an industry partner: 'Distribution' (Fraunhofer Institute for Material Flow and Logistics IML and Continental Reifen GmbH), 'After Sales' (University Duisburg-Essen and SDZ GmbH) and 'Production' (Technische Universität Dortmund and ABH Stromschienen GmbH). In the following, the development in the case of application 'production' is described, since in this case the robust production planning within networks plays an important part. Therefore a service-based system for supporting the planning process is developed.

The requirements towards such planning supporting systems are diverse and often inhomogeneous, but the following common aspects can be named according to [2]:

Cooperation within the field of production expands the problem of production planning and control beyond company boundaries. But in today's IT systems, other enterprises are merely considered as suppliers or customers. They are only involved in planning processes through orders and purchases. This limited point of view allows a capacity planning and scheduling by means of the traditional, sequential Material Requirement Planning (MRP) methods [3]. These prerequisites lead to an inadequate solution with regard to a production network. Furthermore, real capacity limitations and other basic conditions are simply forgotten, so that a robust schedule generation is not possible. The consequences of this abstraction are known as Bullwhip, Forrester, and Burbidge Effect [4]. These result in avoidable stocks, ordering costs, and fluctuations in demand [1]. The actual approaches of Supply Chain Management deal with these problems. But in general they focus on the IT standardisation, the partners' integration, and inventory strategies. This often includes considerable capital expenditures [1]. In the case of enterprises which take simultaneously action in cooperative production processes within several networks and which often also target for inconstant or short-term relationships with changing partners, such methods are only suitable to a certain degree. Advanced Planning Systems (APS) try to solve tactical supply chain planning problems with deterministic optimization models, but two major problems arise. On the one hand, the model design, planning data, planning organization and the implementation of such an APS causes high expenses [5, 6]. On the other hand the APS systems need transparency of the whole network and centralized planning data [7]. These methods also reach their limits or their costs exceed the promised benefits in short-term relationships with changing partners and high complex multi-network cooperation [7].Further restrictions result from the technical requirements of small and medium-sized enterprises. They have got the most intense restrictions and at the same time these constitute the typical form of enterprise within short-term cooperation. Small and medium-sized enterprises often do not have a large or even an own IT department, are only seldom willing to invest in projects which are characterised by a long amortization period, and they work with most diverse partners [8]. In addition, the IT infrastructure is heterogeneous and the penetration level of electronic support differs considerably [9]. From these facts following requirements can be inferred: independence from existing IT and processes, low investment and implementation costs, low operating and maintenance costs, and uncomplicated integrability of different partners [8]. Today's rigid, monolithic IT systems, which often even have to be individualized by external software houses, are thus clearly not suitable for this kind of cooperation. Furthermore customized solutions are not either feasible due to the investment intensity and the high number of different partners. The cooperative planning activities have to be supported by means of IT systems which maintain the planning sovereignty at the enterprise and which are fast and individually available. Recapitulatory the main requirements at a production planning instrument for SME in short to medium-term networks are:

- Support of intercompany production planning process
- Possibility of multi-network cooperation
- Suitable for short-term cooperation
- Maintain the planning sovereignty at the enterprise
- Independence from existing IT and processes
- Low investment and implementation costs
- Low operating and maintenance costs
- Uncomplicated integratability of different partners

- Ad-hoc availability and easy customization

These requirements can be satisfied by decentralized interoperable systems [10]. The designing as an independent software service, which is the aim of the project, allows small and medium-sized enterprises to benefit from a higher planning reliability without having to invest a considerable amount of money in the IT infrastructure. In the following, the problem is located within the process model of production planning and control in order to identify existing weak points. After this, prevailing methods are introduced. Finally the principle of the capacity corridors as service-oriented solution in the field of cooperative, tactical production planning is explained.

2 Localization within the PPC Process Model

Production Planning and Control (PPC) is regarded as the core task of each producing enterprise and it is the central element of a production system. Cooperative production planning on a tactical level is especially relevant in the case of short to medium-term cooperation. In order to create a relevant demand for coordination, a repetitive service provision in an alliance is at least to be assumed. Especially the case studies of small-scale production on a short to medium-term level and repetitive partnerships in the field of components of contract manufactures are focused here. In order to localize the necessary supporting methods and processes, at first a generic process of PPC is examined and the potentials and weak points in the cooperative tactical planning are identified. The Aachen PPC Model according to Schuh et al. constitutes due to its detailed process representation the terms of Ref. [11, 12]. In order to judge the complexity of a holistic consideration fairly, the Aachen PPC Model subdivides the subject into specific related sections. These are also examined by different points of view [11]. Thus, Schuh et al. define the processes of PPC according to the respective company-specific production type. Here, it is differentiated between the typical PPC processes of a contract or make-to-order manufacturer, a blanket order manufacturer, a manufacturer of products with variants, and a make-to-stock manufacturer [12]. In addition, the reference model examines and explains PPC from different views: with regard to tasks, data, function, and processes. Tasks are divided in network and core tasks. Whereas the task-view is independent from the type of order processing as well as from the overall planning approach, the subtasks of production planning in the process-view are arranged into a type-specific temporally logical sequence and are combined with regard to the content [12]. At this point, it is most important to have a closer look at the processes view referring to the production type of the contract or make-to-order manufacturer. The research project mentioned here specifically starts and intervenes at this point.

Furthermore, only the core tasks of PPC can be used for describing the processes in SME, since the network tasks create a strategic level of planning, which requires a central entity in the production network [13]. A central entity like this is also needed for APS systems, as already mentioned above.

This condition cannot be fulfilled in the case of short to medium-term production networks composed of SMEs, the planning sovereignty of the single enterprises has to be maintained at the planning enterprise.

The internal planning process of a contract manufacturer starts with the production program planning in which the market trend is included. Subsequently, the order-anonymous product requirements are determined. For the next step, information is gathered from the inventory management and thus net requirements of order-anonymous components are determined. Both gross and net requirements of the identified components only refer to the in-house production planning. Subsequently, the basic resource planning is carried out and it is specifically checked if the present sales planning is feasible. But again, only the in-house production is considered. Components which have to be bought are not yet included in the planning.

On the basis of the compiled production program, the first process step of the production requirements planning is the determining of the gross dependent requirements, as it is depicted in Fig. 1. This process step involves for the first time purchased components. These requirements are compared with the relevant key figures from inventory maintenance and thus the net dependent requirements are identified.

Only at this point, after coordinating the capacities, the suggestion with regard to the procurement program is examined. This is inadequate, because it ignores real capacity restrictions and other external conditions of the cooperation partners. The procurement program includes both in-house production program and external procurement program. At this point, the order is presented to the corresponding network partner and he presents his offer in return. A trade-off between the network partners takes place for the first time during the described planning process. Subsequently the feasibility of the suggestion with regard to the order program is examined. If the external procurement program is not feasible, a loopback to order coordination of the order management is necessary and thus the complete planning is invalid [12].

In the above presented process map (Fig. 1) it is evident that this generic process procedure with its general procedural structure localizes the process of coordination between the network or cooperation partners only within the operational horizon of the overall process of PPC. But the processes of coordination have a considerable influence on the previous planning stages and process steps. If necessary, the steps have to be carried out once again at a later point in time and thus with a smaller operating horizon due to the downstream agreement between the network partners as well as the consequently resulting information gap. The recursion of certain planning steps which result from this problem can be avoided by a previous coordination process on the tactical level to balance the prevalent information gap and to guarantee a more robust general layout of the PPC.

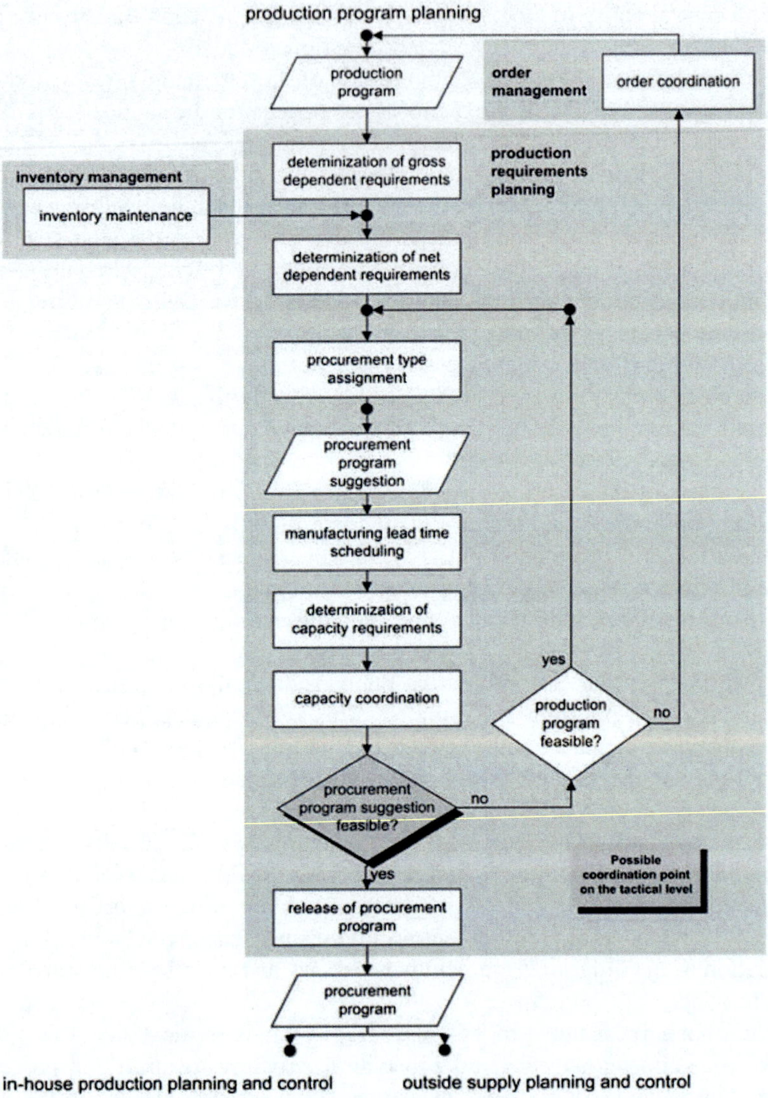

Fig. 1 Localization of the planning focus within the process model of PPC according to [12]

3 State of Research

The process of coordination between partners has already been identified as a central element of planning by myOpenFactory, concept of Quick Response, and Demand Capacity Planning.

In the case of myOpenFactory, a quasi-standard for the corporate data exchange of ERP systems via internet has been developed. It reduces the complexity of order processing within temporary production networks and facilitates the operational processes of coordination between network partners [14]. The challenges of corporate order processing are frequently underestimated in practice at the moment. Moreover a lack of necessary organizational and IT prerequisites for an efficient exchange of the central messages for order processing with regard to communication processes and means has already been identified [14]. myOpenFactory starts with the order processing, in which the mentioned data standard takes especially account of the challenge of data inconsistency within corporate processes of coordination. Furthermore, the standard has been exemplarily developed for machinery and plant construction and has been cut out for its characteristics. Due to this specific focus, myOpenFactory is not completely applicable to the model developed here. A specific industry reference is not used in the project SCP. Moreover the diversity and positioning of interfaces on the organizational, tactical process level for corporate capacity coordination is considered. This transcends the IT and system technical prerequisites from data view.

The concept of Quick Response starts at the point of communication and data transfer as well. In this present case time is focused as critical competitive factor [15]. Quick Response is a concept which reacts to changes in the market demand by means of a fast and flexible ability to react. Thus, the coordination of value-adding processes is supposed to be facilitated. The American textile industry is regarded as the starting point of the Quick Response concept [16]. It is especially made for use in outbound logistics [15]. In the whole value creation process the specific customer preferences are in the center of attention. At the point of sale product consumption is directly conveyed to the supplier, by means of bar code, transponder, and laser technology. Thereby, it is possible to track changes in consumption directly. Correspondingly, the concept makes use of push strategy of data transfer. Data do not have to be called in, but they are moreover automatically transferred from the end user to the upstream value creation levels [15]. This concept tries to improve the availability of information by transferring requirement information in a proactive and immediate way. But also the concept of Quick Response is due to its strong focus on distribution. The consideration of the end user as central issue is not directly comparable to the model developed here. The present case concentrates on the processes of coordination during the production requirements planning within a production network.

The methodology "Demand Capacity Planning" of Odette [17], which has been developed for collaboration with SME within automotive networks, also starts at this point of the planning process. Due to the high pressure of integration by the OEM within the automotive field, the reference process relies on a capacity model as basic information. This information automatically combines the pattern of demand of the customer with the pattern of capacity of the supplier. This model comprises relatively sensitive data for the supplier. Therefore, the methodology requires a high measure of interoperation of the used software systems. According to that, the process of Odette is not applicable to the present case.

There are several other instruments and methodologies which address coordination of demand in supply chains. The majority of those concentrate on the management of inventories. Inventory management can be used for means of control and flexibility in supply chain [18]. Strategies like Vendor Managed Inventory, Rapid Replenishment and Centralized Inventory Management try to optimize demand coordination in networks. This kind of coordination methods try to change the performance of networks without changing the production planning process at its own. It implements decision models into the inventory management processes that lead to beneficial behaviour in terms of overall network performance [19]. The costs and benefits arising are required to be shared among the network, so all partners benefit from the improved overall network performance [20]. To implement such inventory strategies together with cost-benefit-sharing models, there are non-negligible expenditures and transparency about the target collaboration necessary. In conclusion, these models are not suitable for short-term networks of SME with no or small collaboration experience.

4 Capacity Corridors for Coordination of Production Planning

Coordination with cooperation partners in the field of production requirements planning has to satisfy requirements in order to be used for the field of tactical planning in short to medium-term production networks. Especially the retention of planning sovereignty has to be taken into consideration when choosing a methodology.

The developed process is not supposed to centralize planning, but it is supposed to make a more efficient decentralized planning possible by means of an additional process of coordination.

The basic idea of cross-company coordination is an inquiry to corresponding cooperation partner, which has been set off in the framework of an availability planning on the level of production requirements planning. Furthermore, the message enquires about the availability of sufficient capacities for meeting the dependent requirements. It guarantees the successful ordering within the following external procurement planning and thus it secures the planning. Due to this simple methodology, the planning run (MRP run) would however be slowed down by a heteronomous factor. Responses, especially qualified information concerning capacities, can easily draw on several days.

The delay of the response can only be avoided by means of already present data or automatic implementation of inquiry processes. But since due to the requirements a continuous IT planning support of the partners or a uniform IT infrastructure cannot be assumed, a completely automated response cannot be sensibly realised. In order to avoid a delay during the planning run provided or already available data is used. A response which is based on necessarily calculated or due to the order processing already existing data would be suitable.

The tendering, which is done in the initial stage of a cooperation and even in the case of order request and uncomplicated customer–supplier-relations, deals with the issues “technical feasibility”, “delivery date”, “price”, and “terms of business" [21]. In this phase, in case of a positive technical feasibility a delivery date with price and terms of business is generated on the one hand. On the other hand a distinctness of a capacity load profile which is compiled for the enquired product at the enquired enterprise and which is compared with the available capacities is developed [22]. Exactly this data can be used for proactively improving the coordination between partners in case of further future relations and avoiding unnecessary planning recursions. The exact form of planning does not have to be determined with this method, since only a capacity comparison is necessary. The combination of planned delivery time and the linked possible quantity delivered, which refers from the use of the capacity load profile to the available capacities, is called capacity corridor. Within this corridor, the product can be planned as dependent requirement and the cooperation partner can deliver the product according to his planning. The capacity corridor is expatiated by comparing the remained capacity of an offer or contract to the capacity profile of the service (Fig. 2). This can occur regularly during each planning run or irregular with each scheduled offer. The principle is similar to the offer scheduling according to Bramkamp [23] and it is based upon exposure profiles which are compared with available capacity profiles [22]. The determined capacity corridor is reported to the IT service—which has to be developed—and thus it can be used for the cooperation partners planning runs at any time. The actuality of the data which are reported to the service can be improved by indicating the validity periods of the capacity corridors. If possible, these validity periods should match the planning periodicity of the cooperation partners. Consequently, a capacity corridor is composed out of the following three elements: feasible amount (PQ_{cc}), corresponding delivery period (DT_p), and a validity period (VP_{cc}) of these details. When the validity of the current capacity corridor expires, the service notifies automatically the partner and asks for updated data.

Thus, there are four different cases in which a capacity corridor is reported to the service (Fig. 3):

- During the initialization of the cooperation between partners a request is sent to the supplier after the specification of the partner within the service. The supplier responses to this request with the initial capacity corridor and thus he confirms the relation.
- A reactive notification of the capacity corridor is asked for by the service when the validity of the given capacity corridor is expired.
- A proactive notification of the capacity corridor should occur with every relevant change of the capacity situation of the cooperation partner.
- An updated notification is necessary in the case of a capacity reservation, because of the remaining capacity changes of the partner.

The planning is still decentralized, but thanks to the closer linkage of the individual planning processes a clearly higher robustness is achieved. Furthermore, a number of cooperation partners can be managed by means of the same service. Consequently, alternative performance allocation can be used if capacity corridors do not suffice or are not sufficiently used to capacity (Fig. 4).

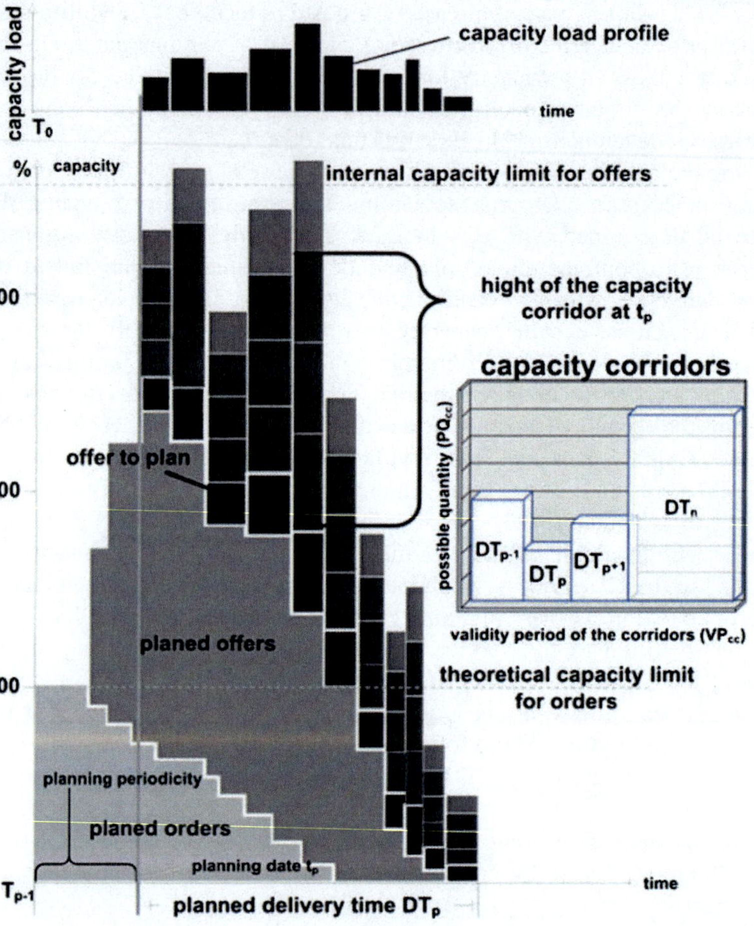

Fig. 2 The conception of capacity corridors according to [21]

Additionally, a control of planning with a rudimentary supplier assessment can be implemented in which the activity of the cooperation partners is assessed during the process of coordination. Frequent current notifications of the capacity corridor enhance the attractiveness of the cooperation, since more robust planning processes are made possible without reducing flexibility. In the case of expired and out-of-date capacity corridors the functionality of the method cannot be guaranteed anymore. This case has a negative effect on the assessment of the cooperation partner.

Further potentials could be achieved by means of coordination between cooperation partners which is induced by the service in case of inadequate capacity corridors. The partners could use different short and medium-term measures of capacity balancing jointly and thus they can avoid medium-term bottleneck situations in collaboration.

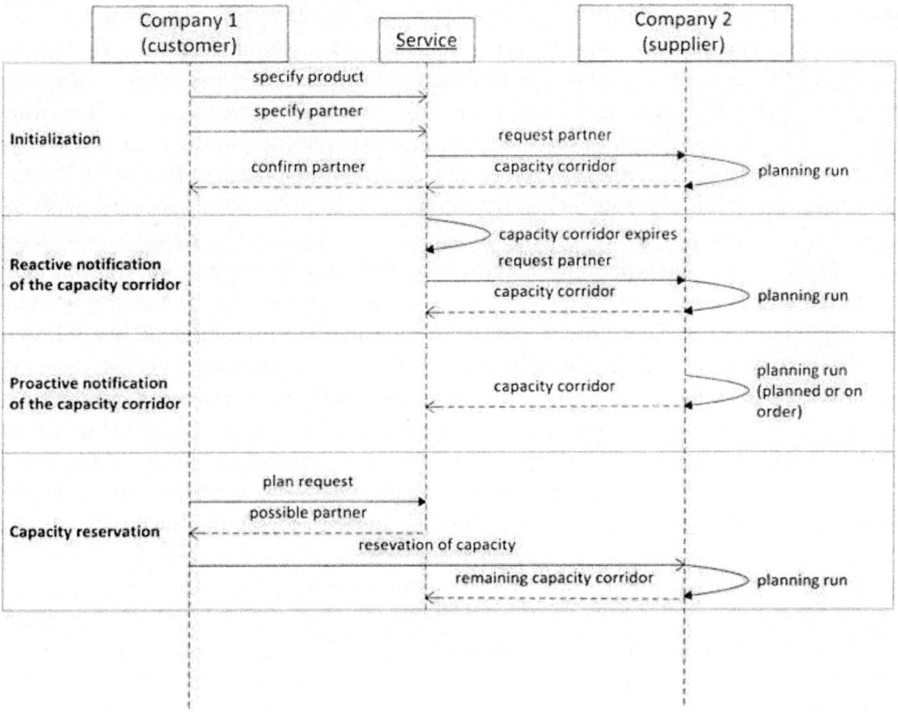

Fig. 3 Differentiation of cases when reporting capacity corridors

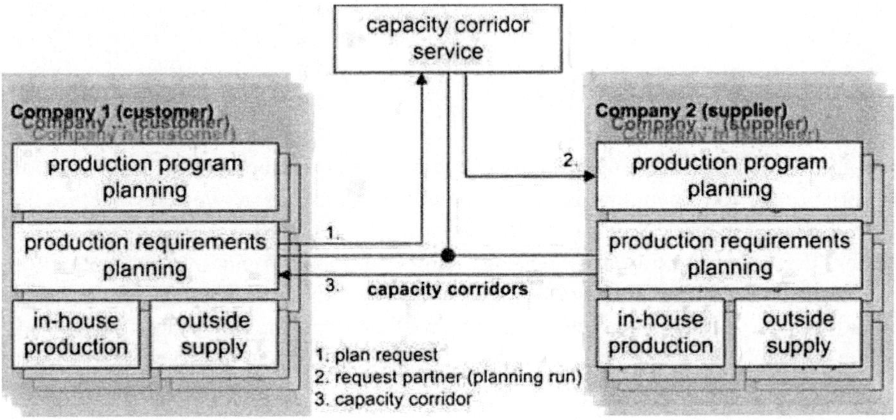

Fig. 4 Capacity corridor service

As already mentioned, the service is working without being connected directly to IT solutions, since no IT infrastructure is required. The organization as a stand-alone web service makes possible a well-directed guiding of user through the process. This takes place from the initialization to the operational capacity corridor notification. The partners are informed about changes in status and necessary input via email. Thus a data transfer between the enterprises and the service is made possible which is completely independent from the IT infrastructure. Therefore standardized data objects are created which make interoperation with other services possible. This 'Business Objects' are defined on basis of OAGIS [24, 25] in collaboration with other projects of the ECLR to allow the development of interface services to ERP or PPC systems.

The capacity data which is managed by the service is tailored for field of application and they are not directly extracted from the operational planning system of the manufacturer. Therefore, the transferred data is not understandable without the context of the cooperation, which improves the confidence into the system. Simultaneously, only in the case of planning inquiries the capacity corridors are compared, notifications generated, and passed on. The misuse by means of automated inquiries can be avoided by state-of-the-art systems (e.g. CAPTCHAs).

The optimization of the overall network performance with the described coordination methodology has to be proved by case studies and simulation experiments. Nonetheless the availability of capacity data of suppliers at production requirements planning makes planning more realistic and helps to avoid deterministic impossible plans.

Thus, the system could offer an improvement of robustness of production planning for the case of application of small-scale production in a short to medium-term horizon. Moreover repetitive cooperation in the field of components of contract manufacturers and similar cooperation in the short to medium-term interval can be offered by this system.

5 Conclusions and Outlook

The introduced methodology for the improvement of processes of coordination within production planning is prototypically realized as a web service in the project Supply Chain Planning in the framework of the EffizienzCluster LogistikRuhr. The coordination methodology starts in the domain of the production requirements planning and it carries out direct capacity coordination with partners already at this point. The development of a realizable production plan is facilitated, which recognizes predictable capacity bottlenecks early and avoids operational adjustments. The main features of the capacity corridor concept can be summarized as:

- Implementation of an additional coordination process on the tactical level of the production planning process.

- A web-based IT service is developed to organize the transaction of capacity data.
- Request at the service whether the planned demand is realizable by the supplier; answer on basis of previously provided data.
- The used data is generated during the common capacity planning of the supplier.
- Transmitted data: feasible amount, corresponding delivery time, and a validity period of these details.
- Reporting triggers: initial at start of the partnership, proactive at planning run, reactive at expiring corridor, proactive at capacity reservations.
- Cooperation partner assessment by means of actuality and frequency of capacity reports to the service.

The solution is cut out for small and medium-sized enterprises, since no uniform IT infrastructure is necessary and the expenditure for integration of the systems is low. This is on the one hand due to a form-based capacity notification and on the other hand due to an implementation and initialization support which is directly realized in the service. The proof of an enhanced performance of the whole network is intended in a case study with an industry partner, supported by simulation experiments at the end of the service development. Further developments allow the connection to other services or IT systems through standardized messages and data types.

References

1. Kuhn, A., Hellingrath, H.: Supply Chain Management: Optimierte Zu-sammenarbeit in der Wertschöpfungskette. Springer, Berlin (2002)
2. Besenfelder, C., Liesebach, T., Uygun, Y.: Supply Chain Planning. In: Zeitschrift für wirtschaftlichen Fabrikbetrieb, 12/2011; pp. 939–943 (2011)
3. Schuh, G.: Produktionsplanung und Steuerung: Grundlagen, Gestaltung und Konzepte. Springer, Berlin (2006)
4. Geary, S., Disney, S.M., Towill, D.R.: On bullwhip in supply chains—historical review, present practice and expected future impact. Int. J. Prod. Econ. **101**, 2–18 (2006)
5. Jonsson, P., Kjellsdotter, L., Rudberg, M.: Applying advanced planning systems for supply chain planning: three case studies. Int J Phys Distrib Logistics Manag **37**(10), 816–834 (2007)
6. Kjellsdotter Ivert, L., Jonsson, P.: Problems in the onward and upward phase of APS implementation: Why do they occur? In: Proceedings of the 22nd NOFOMA conference 451 (2010)
7. Stadtler, H., Kilger, C.: Supply Chain Management and Advanced Planning: Concepts, Models, Software and Case Studies. Springer, Berlin (2005)
8. Völker, R., Koehler, M.: Web-Applikationen als KMU-adäquate IT-Lösung. PPS Management. 2, pp. 44–47 (2007)
9. Meyer, M., Walber, B., Schmidt, C.: Produktionsplanung und -steuerung (PPS) in temporären Produktionsnetzwerken des Maschinen- und Anlagenbaus. In: Schuh, G: Produktionsplanung und Steuerung: Grundlagen, Gestaltung und Konzepte. Springer, Berlin (2006)
10. Hegmanns, T., Klingebiel, K., Winkler, T., Bruns, Ch.: Service-basierte Assistenz-systembausteine für die taktische Optimierung der Bestandsstrategie. In: Wolff-Kluthhausen, H. (eds.) Jahrbuch Logistik (2012)

11. Schuh, G., Giert, A.: 2.1. Aachener PPS-Modell. In: Schuh, G: Produktionsplanung und Steuerung: Grundlagen, Gestaltung und Konzepte. Springer, Berlin (2006)
12. Schuh, G., Schmidt, C.: 2.4. Prozesse. In: Schuh, G. Produktionsplanung und Steuerung: Grundlagen, Gestaltung und Konzepte. Springer, Berlin (2006)
13. Schuh, G., Stich, V., Schmidt, C.: Produktionsplanung und -steuerung in Logistiknetzwerken. In: Nyhuis, p. (eds.) Beiträge zu einer Theorie der Logistik. Springer, Berlin, pp. 249–273 (2008)
14. Schuh, G.: Effiziente Auftragsabwicklung mit myOpenFactory. Carl Hanser Verlag, München (2008)
15. Kimil, T.: Zeit als Faktor in der Distributionslogistik—Quick Response. Grin Verlag für akademische Texte (2009)
16. Stocker, S., Radtke, P.: Supply Chain Quality. Hanser Verlag, München (2000)
17. Scholz-Reither, B., Wolf, H.: B.8.4. Information und Kommunikation. In: Arnold, D., Isermann, H., Kuhn, A., Tempelmeyer, H (eds.) Handbuch Logistik. 2., aktualisierte und korrigierte Auflage. Springer, Berlin, pp. B8-14–B8-22 (2004)
18. Chikán, A.: The new role of inventories in business: Real world changes and research consequences. Int. J. Prod. Econ. **108**(1–2), 54–62 (2007)
19. Gunasekarana, A., Patelb, C., McGaugheyc, R.E.: A framework for supply chain performance measurement. Int. J. Prod. Econ. **87**(2004), 333–347 (2004)
20. Bookbinder, J.H., Gümüs, M., Jewkes, E.M.: Calculating the benefits of vendor managed inventory in a manufacturer-retailer system. Int. J. Prod. Res. **48**(19), 5549–5571 (2010)
21. Luczak, H., Eversheim, W (eds.).: Produktionsplanung und—steuerung. Grundlagen, Gestaltung und Konzepte. 2. Aufl. Springer, Berlin (1999)
22. Wiendahl, H.P.: Betriebsorganisation für Ingenieure. 6. Aufl. Hanser Verlag, München (2008)
23. Brankamp, K.: Ein Terminplanungssystem für Unternehmen der Einzel- und Serienfertigung. Physica-Verlag, Heidelberg (1968)
24. http://www.oagi.org, OAGi Open Application Group. Open Standards that open Markets, from 09th of May (2012)
25. Lampathaki, F., Mouzakitis, S., Gionis, G., Charalabidis, Y., Askounis, D.: Business to business interoperability: a current view of XML data integration standards. Comp. Stand. Interfaces **31**(2009), 1045–1055 (2009)

Description of a Configuration Model for Establishing Adaptable Logistics Chains

Markus Florian, Henrik Gommel and Wilfried Sihn

Abstract Logistics chains are mostly influenced by changes in their business environment. A system's adaptability is seen as one potential to effectively counteract these environmental changes. To consider the effects of adaptability on the whole supply chain, a framework for configuring adaptable logistics chain was developed within the research project "KoWaLo". This paper deduces an approach how to identify adaptable logistics configurations. Furthermore a case study shows the potential of the configuration model.

Keywords Logistics · Supply chain · Adaptability

1 Introduction

Nowadays manufacturing companies are increasingly exposed to changes in their business environment. Ongoing globalization does not only lead to new competitors, but also to new markets and new demand potential [1]. Furthermore a change in the customer market can be recognized. The shift from a seller's market to a buyer's market is reflected besides higher service levels in shorter reaction times, increasing individuality of products along with declining prices. The increasing individuality of products leads to a high number of different product variants [2–4]. A further indicator of intensified competition and the increasing impact of external influences

M. Florian (✉) · H. Gommel · W. Sihn
Fraunhofer Austria Research GmbH, Vienna, Austria
e-mail: florian@imw.tuwien.ac.at

M. Florian · H. Gommel · W. Sihn
Vienna University of Technology, Vienna, Austria

K. Windt (ed.), *Robust Manufacturing Control*, Lecture Notes in Production Engineering,
DOI: 10.1007/978-3-642-30749-2_23,

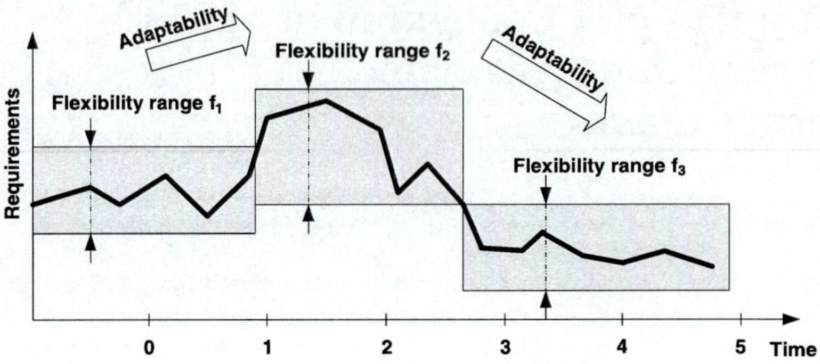

Fig. 1 Adaptability as the capability to shift flexibility ranges [10]

are declining innovation cycles and product life cycles, caused by the rapid development in information and communication technologies [5, 6].

Such events entail more and more turbulent business environment. In order to stay competitive, new strategies have to be applied to face the ongoing changes. Therefore the topic "adaptability", which can be described as the ability of a system to perform both reactive and proactive adaptations by specifically varying processes, is an important approach to deal with turbulence and retain competitiveness [8–11]. In comparison to flexible systems, which only can deal with changes within a certain range, adaptable systems allow to shift the range of flexibility to a higher or lower level by specific arrangements as shown in Fig. 1, e.g. through investments and/or organizational arrangements [11].

The main focus of previous research activities on adaptability has been on the factory level. Supply Chains as a whole have been taken into account to a lesser extent [12]. First research activities in this matter were carried out by Christopher on a conceptual level without discussing defined constitutive characteristics (i.e. number of warehouses or transport concepts) precisely [13] and Dürrschmidt by [14] developing a concept for planning adaptable logistics systems for serial production without disclosing approaches for configuring logistics chains. More recent research activities focusing on adaptability in logistics chains were carried out by Nyhius et al. by [15] evaluating intra-enterprise logistics chains based on the requirement and the economic value added of an adaptable configuration.

Within the research project "KoWaLo" a framework for configuring adaptable logistics chains based on concretely defined constitutive characteristics will be developed in order to consider the effects of adaptability on the efficiency of the whole supply chain. The Austrian Research Promotion Agency funded this project with the partners Knorr-Bremse GmbH Division IFE Austria, Seisenbacher GmbH and the Vienna University of Technology. The focus of this paper is to describe the procedural method to identify the main constitutive characteristics in order to set up the configuration framework.

2 Shifting Flexibility Ranges in Logistics Chains

As described in the introduction adaptability offers great potential to cope with turbulences. In this respect adaptability considers structural changes in three basic principles: rapidness, flexibility and costs. Given that until now adaptability was primarily discussed with focus on production systems, factory structures, organizational matters or order processing systems, thus primarily focusing on intra-enterprise issues, and the fact that in many industries 50–70 % value added is contributed within a supplier network and therefore the adaptable positioning of an individual company is not sufficient it is inevitable to identify options that allow shifts of flexibility ranges in logistics chains. From the supply chain management point of view, the stability of the supply chain needs to be preserved at its best. While meeting delivery times or coping with an increased demand, companies face the problem of increased logistics costs. This leads to extra or emergency transports with for example low capacity utilization and/or the usage of expensive carriers like planes instead of trucks or trains. Along with these financial issues there are issues like increased emissions and their ecological effects. Longer-term supply shortfalls due to production breakdowns or quality problems may be considered when choosing sourcing strategies whereas changes in demand may be considered when planning distribution networks. These examples show the importance of developing a framework helping companies to empower adaptability in their logistics chains.

In order to identify and asses the main constitutive characteristics and their respective specifications with regard to their ability to enable a logistics system to be (re-)configured continually, rapidly and in a cost effective manner as the major basis for configuring adaptable logistics chains, it is necessary to analysis different environmental dynamics scenarios and their effects on the logistics system [17]. By analyzing these effects together with the ability of the general constitutive characteristics of logistics systems to handle environmental dynamic, the main characteristics can be identified (Sect. 3.2). By modifying the respective specifications of constitutive characteristics different logistics chain configurations can be developed.

As to secure cost effectiveness the configuration of adaptable systems has to be carried out in consideration of the systems cost effectiveness during its life cycle or a given time horizon [18]. The total costs of adaptability can be divided in system-costs (initial investments) and process-costs. Process-costs can further be divided in direct costs comprising costs for operating the system and costs for flexibility shifts, whereas the indirect costs comprise inefficiencies of the system caused by over- or under-designed systems.

As there might be possibilities on how to set up an adaptable system, companies need to consider these with subject to the degree of adaptability and related total cost in order to be able to choose the most favorable configuration, i.e. the one with the best adaptability-cost-ratio. Therefore the different scenarios need to be evaluated by appraising the different types of costs for each scenario, as shown in Fig. 2.

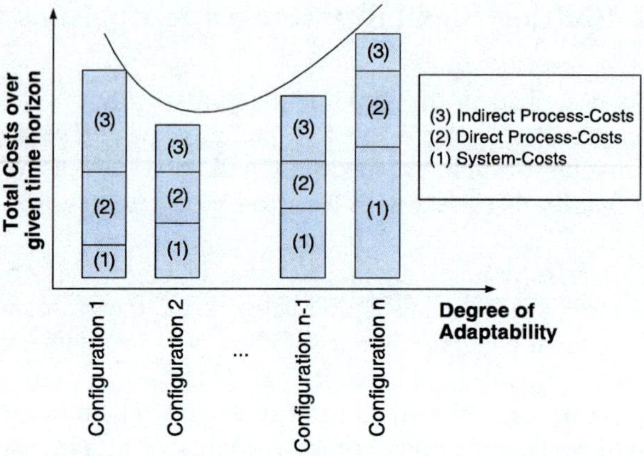

Fig. 2 Total costs consideration of different scenarios [5]

Chapter 3 presents the configuration model for establishing adaptable logistics chains. The configuration model is a process model which helps the operator to find and assess new logistics chain scenarios. Chapter 3 shows a case study where the configuration model has been applied.

3 Configuration Model for Establishing Adaptable Logistics Chains

The configuration model for establishing adaptable logistics chains is composed of six steps shown in Fig. 3. The focus of this chapter is to describe the approach of an operator to identify optimized logistics configuration in order to counteract external influences (load scenarios, i.e. major shifts in demand).

3.1 Definition of Load Scenarios

Logistics chains have to provide a robust configuration to handle different external influences. Influences which cannot be handled by the current flexibility ranges result in special charges like express or extra transports. Due to different impacts by the external influences on the logistics chain, a classification of these influences is necessary. To provide a thorough and generic approach, the classification was structured in terms of possible influence locations and influence factors. The influence location describes the place in the logistics chain, where the influence can have an impact on. Along the logistics chain the influences can affect the demand side, the supply side or the surrounding of the logistics chain. With regard to the influence factors the external influences can have an impact on the factors described by the 6 R's of logistics (goods, time, location, quantity, quality and

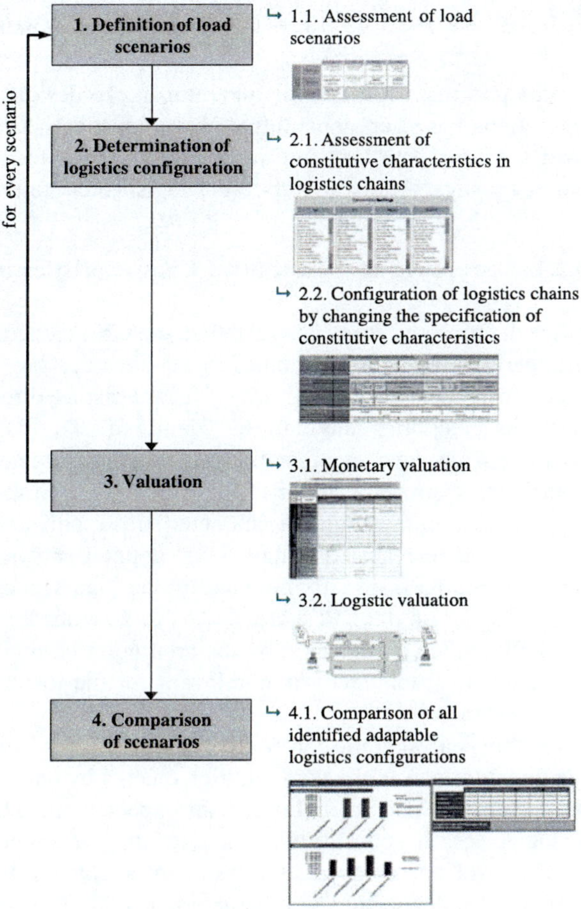

Fig. 3 Overview of the configuration model for establishing adaptable logistics chains

price). Within the configuration model only the factors time, location, quantity and quality are considered. The factor price will be considered within the process step monetary valuation and the factor product will not be considered because this model acts on the assumption that the right product is already available [19].

The aim of this process is to identify these external influences that can have an impact on the operator's supply chain. Therefore a logistical scope has to be identified (e.g. a region, a supplier) where the operator identifies the relevant load scenarios by the use of cost analysis, KPI analysis, environmental analysis or other analyses which are required to specify inferences from the external influences on the logistics chain. After the assessment the operator chooses one load scenario for the further approach.

3.2 Determination of Logistics Configuration

In this part of the process the operator has to design new adaptable logistics configurations based on constitutive characteristics which can counteract the chosen load scenario. The first step in this approach is to identify the right constitutive characteristics. Afterwards different logistics configurations can be generated.

3.2.1 Assessment of Constitutive Characteristics in Logistics Chains

After defining the scenarios or rather scenario categories it is essential to identify the relevant regulating variables in logistics dealing with the impacts of the load scenarios. Therefore constitutive characteristics directly influenced by the load scenario categories have to be identified. To identify adaptable constitutive characteristics, two separate analysis where conducted and merged by a multidimensional approach. Herein KPIs are used as linkage between load scenarios (analysis 1) and constitutive characteristics (analysis 2) [20].

After the relevant KPIs have been defined analysis 1 can be initiated. Within this analysis KPIs directly affected by the load scenarios (has the load scenario a direct impact on the KPI?) are chosen to be considered in the following. Because the KPI's value is regulated by the configuration and performance of constitutive characteristics, this analysis is relevant for the identification of adaptable constitutive characteristics.

Within analysis 2 the constitutive characteristics have to be evaluated considering the direct influence of a configuration change by changing the specifications within the constitutive characteristics and the expected impact on the value of the KPI (does a change within a constitutive characteristic has direct impact on the KPI?).

The third step combines the results of analysis 1 and analysis 2 by linking load scenarios and constitutive characteristics: if a specific KPI is expected to be influenced by a load scenario and at the same time is expected to be influenced by the change of a constitutive characteristic it is most likely that varying this characteristic allows counteracting the load scenario. In terms of adaptability those constitutive characteristics that allow counteracting the defined load scenarios form the basis for configuring an adaptable logistics chain. This task has only to be done once because every identified scenario can be supported by a specific load scenario group and the linkage between constitutive characteristics and KPIs are defined by the respective KPI definition [20].

3.2.2 Configuration of Logistics Chains by Changing the Specification of Constitutive Characteristics

Using the assessment of constitutive characteristics for every load scenario the operator can now identify the right levers to reconfigure the logistics chain. In addition to the identified constitutive characteristics the operators has to estimate which of these characteristics is relevant for his logistics chain. For example the constitutive characteristic supplier strategy is not a right lever when there is only one supplier for the whole industry.

3.3 Valuation

After selecting the new logistics configurations, the operator of the configuration model (i.e. logistics manager) has to evaluate each configuration concerning the factors costs and logistics performance.

3.3.1 Monetary Valuation

Within this approach process the different logistics scenarios are evaluated by the factors costs. Therefore the total cost of ownership approach has to be applied to provide a complete view of all investments needed to change the current logistics configuration to the new logistics configuration. The approach considers different categories. On one hand, costs concerning warehousing, transport and administration have to be identified. On the other hand, these costs have to be attributed to adaptable object costs, direct adaptable process costs, indirect adaptable process costs or operating costs, which are not considered in the logistic valuation step. Adaptable object costs bundle all investments needed to realize the new logistics configuration. These costs only occur once (e.g. purchasing of extra bins). Direct adaptable process costs bundle all costs needed to change the current logistics configuration to the new configuration (e.g. costs for adjustment or dismounting). The indirect adaptable process costs bundle all costs which accrue while reconfiguring the logistics chain and affect the logistics or production performance (e.g. loss of production output because of the non-availability of the equipment to be changed). The operating costs are considered, because of the high impact of the external influences on these operating costs. To measure the operating costs in detail, a simulation model has been deployed, which will be described in the next subsection. Underlying cost rates for transport, stock, delay penalties and raw materials allow the evaluation of the systems total cost alongside with time effects like delivery capacity and reliability [19].

3.3.2 Logistic Valuation

To assure a complete logistics valuation two models—the simulation model and the evaluation model—were developed (Fig. 4—for higher resolution see Appendix A). The simulation model represents a standard logistic chain, including inbound and outbound logistics as well as aggregated main production processes and their associated behavioral pattern concerning lead-times and deviation of

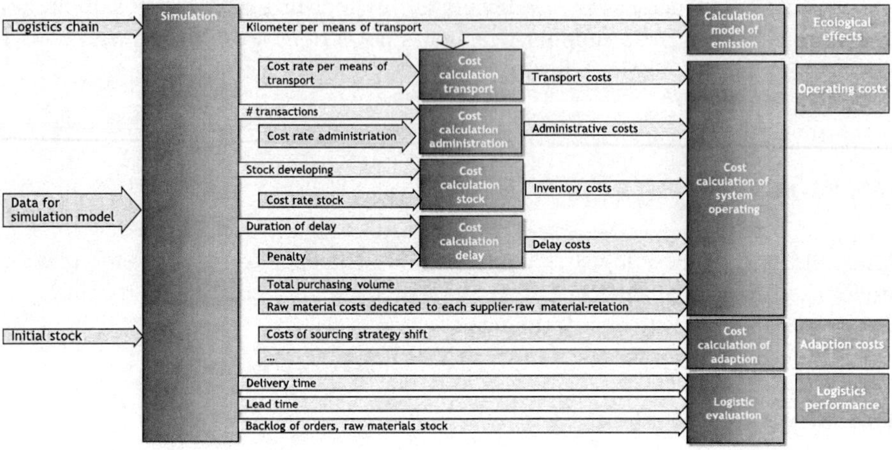

Fig. 4 Structure of the simulation and evaluation model

lead-times. In order to assess the effectiveness of one or a set of constitutive criteria and their respective configurations, the simulation model is based on a discrete event simulation algorithm. Furthermore it represents an exemplary order fulfillment process to simulate the entire information and material flow over the logistics configurations. In addition to the simulation model, which only simulates the information and material flow over the defined processes, an evaluation model is required to translate the result data into comparable KPIs. Beside the measurement of the logistics performance (e.g. delivery time, lead time, backlog of orders, stock) within the evaluation model, further results are generated. As mentioned within the subsection "monetary valuation" operation costs are also generated by the evaluation model. Furthermore ecological effects like the CO_2 emission can be identified. The identification of these factors within the evaluation model is done in order to generate a realistic image of the actual logistics chain and the processes represented by the simulation model. The identification of the adaption costs (described in subsection "monetary evaluation") is only a part of the evaluation model and not of the simulation model.

With the integrated application of the simulation and evaluation model dynamic effects of certain scenarios on the logistics configurations can be simulated (within simulation model) and bottlenecks can be identified (within evaluation model). By changing the specification of adaptable constitutive criteria and their theoretical effect on the systems behavior by the operator, a statement can be made concerning the probability of an adaptable logistics chains aptitude to counteract these dynamic effects.

3.4 Comparison of Scenarios

Using the results of the two valuation steps, a comparison of all logistic configurations can be provided in consideration of costs, the logistics performance (as depicted in Fig. 3) and ecological effects due to transport system changes. This step provides an overview over all logistic configurations and provides the operator a base to make decisions regarding the new logistics configuration.

4 Potential of a Configuration Model for Establishing Adaptable Logistics Chains

Companies in the railway vehicle manufacturing industry have to cope with different external influences. One of these influences is the high individuality of the products. Therefore, customers have to consider long delivery times because of some parts' long production and/or replenishment times. Another reason for these long delivery times are short-termed changes of orders, which can cause delayed delivery. As this long delivery times are not acceptable nowadays, some customers claim shorter delivery times.

To achieve this requirement the configuration model was applied. The first step was to identify the right load scenario. Because of the influence location demand and the influence factor time the right load scenario could be defined. Within the next process step new logistic configurations have to be developed. To shorten the delivery time and handle short-termed shifts of orders different configurations have been considered in the case study:

1. Customer-oriented consignment warehouse (LC1): Storage of three scheduled deliveries (24 items). The replenishment of the consignment stock is based on the scheduled deliveries and not on the stock withdrawal.
2. Production-oriented semi-finished product warehouse (LC2): Storage of three scheduled deliveries in terms of semi-finished products at the manufacturer. The replenishment process is similar to the case study 1.

4.1 Results

By reconfiguring logistic chains based on external influences improvements can be realized. The operation of adaptable logistics chains monetary improvements as well as improvements in the logistics performance can be achieved. Within in the case study, the configuration LC1 and LC2 could provide a payback period of nearly one month. Furthermore the logistics performance (delivery reliability, mean delay in delivery) could be improved by about 50 %. CO_2 emissions could be shortened by about 65 %. The immense reduction of CO_2 emissions traces back to long logistic chains where extra transports use air-transport and the standard transports uses see-transports.

Overall, an application of the described configuration model allows an operator to handle external influences in an optimal way. In addition to reduced costs, the logistics performance can be improved.

5 Summary

Adaptability constitutes high relevance for systems facing volatilities or continuously changing markets. According to the factors flexibility and reactivity as well as economic factors, structures can be (re-)configured to reach a high performance and low total cost. Adaptable concepts within production systems approve this statement. Nevertheless production systems only constitute one part of the value added chain and therefore adaptable concepts regarding the whole logistics chain have to be developed.

The presented configuration model for establishing adaptable logistics chains shows a structured approach to identify new logistics configurations based on load scenarios which can have an influence on logistic chains. Furthermore the results of the case study underline the need to reconfigure the logistic chains.

A.1 6 Appendix

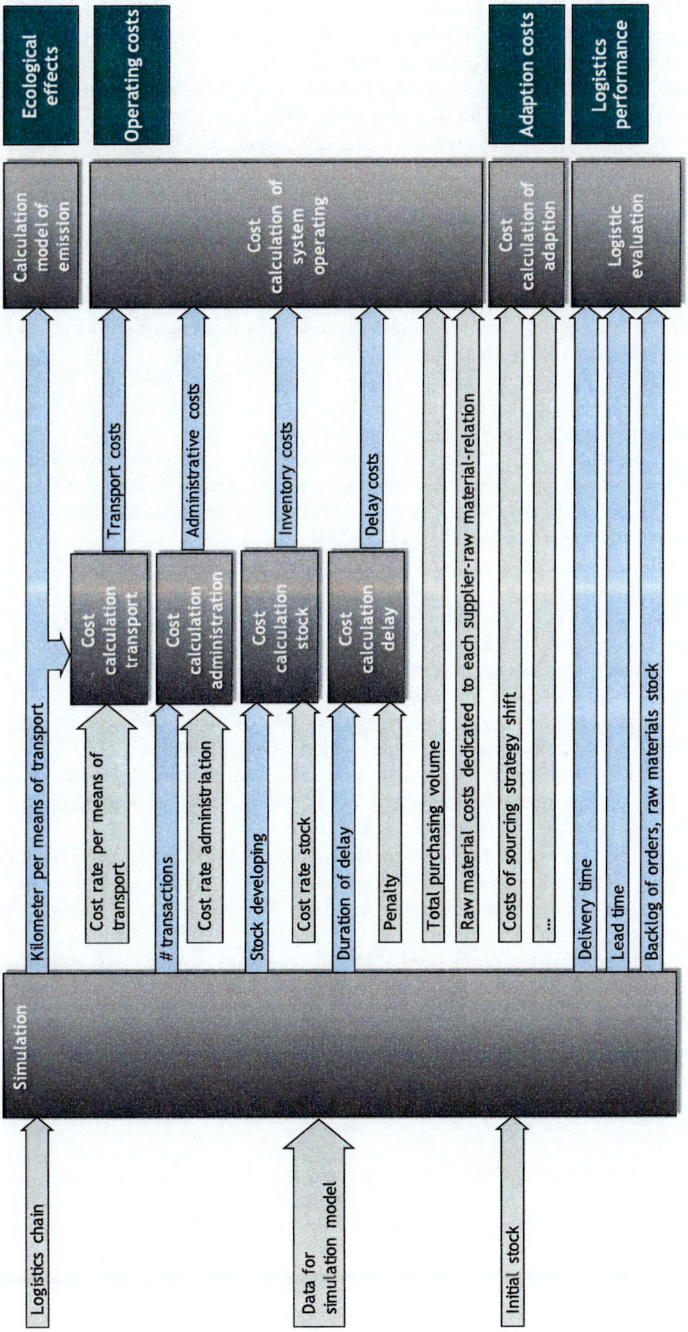

References

1. Kaluza, B., Blecker, T.: Erfolgsfaktor Flexibilität—Strategien und Konzepte für wandlungsfähige Unternehmen, p. 2. Erich Schmidt Verlag, Berlin (2005)
2. Keijzer, W.C.: Wandlungsfähigkeit von Entwicklungsnetzwerken—ein Modell am Beispiel der Automobilindustrie, p. 1. Munich University of Technology, Munich (2007)
3. Heinecker, M.: Methodik zur Gestaltung und Bewertung wandelbarer Materialflusssysteme, p. 1. Munich University of Technology, Munich (2006)
4. Schenk, M., Wirth, S.: Fabrikplanung und Fabrikbetrieb—Methoden für die wandlungsfähige und vernetzte Fabrik, p. 12. Springer, Berlin (2004)
5. Heger, C.L.: Bewertung der Wandlungsfähigkeit von Fabrikobjekten, pp. 1–25. PZH Produktionstechnisches Zentrum GmbH, Leibnitz (2007)
6. Eversheim, W., Berholz, M., Zohm, F.: Die Fabrik im 21. Jahrhundert—Werkzeuge für die Planung und Gestaltung der Fabrik von Morgen,Nr. 1 2002, p. 16, RWTH Themen, Hrsg. Aachen (2002)
7. Balve, P., Wiendahl, H.H., Westkämper, E.: Order management in transformable business structures—basics and concepts. Robot. Comput. Integr. Manuf **17**(2), 461–468 (2001)
8. Westkämper, E.: Strategic development of factories under the influence of emergent technologies. CIRP Ann. Manuf. Technol. **56**(1), 419–422 (2007)
9. Wiendahl, H.-P., ElMaraghy, H.A., Nyhius, P., Zäh, M.F., Wiendahl, H.-H., Duffie, N., Brieke, M.: Changeable manufacturing—Classification, design and operation. CIRP Ann. Manuf. Technol. **56**(2), 783–809 (2007)
10. Nyhuis, P., Reinhart, G., Abele, E.: Wandlungsfähige Produktionssysteme—Heute die Industrie von morgen gestalten. PZH Produktionstechnisches Zentrum, pp. 23–24 (2008)
11. Nyhuis, P., Fronia, P., Pachow-Frauenhofer, J., Wulf, S.: Wandlungsfähige Produktionssysteme. wt Werkstatttechnik online **99**(4), 205–210 (2009)
12. Christopher, M.: The agile supply chain—Competing in volatile markets. Ind. Mark. Manage. **29**(1), 37–44 (2000)
13. Dürrschmidt, S: Planung und Betrieb wandlungsfähiger Logistiksysteme in der variantenreichen Serienproduktion, Thesis Ph.D., Technical University ofMunich p. 4 (2001)
14. Nyhius, P., Nickel, R., Horváth, P., Seiter M., Urban G., Logistische Wandlungsfähigkeit von Lieferketten. Available from: http://logistics.de. 6 Oct 2009 (2008)
15. Spath, D.: Grundlagen der Organisationsgestaltung. In: Bullinger, H.-J. et al., eds. Handbuch Unternehmensorganisation, pp. 3–24. Berlin, Heidelberg, Springer (2009)
16. Zäh M. F., Möller, N., Vogl W.: Symbiosis of changeable and virtual production. In: Zäh, M. F. et al., eds. 1st International Conference on Changeable, Agile, Reconfigural and Virtual Production (CARV), pp. 3–10. Munich, Utz (2005)
17. Gommel, H.; Florian, M.; März, L.; Palm, D.: Identification of constitutive characteristics for adaptable logistics chains as basis for an assessment by simulation, International Conference on Harbor Maritime and Multimodal Logistics Modeling & Simulation, pp. 38–45. Rome (2011)
18. Sihn, W.; Florian, M.; Gommel, H.: Identification of constitutive characteristics for configuring adaptable logistics chains. In: 44th CIRP Conference on Manufacturing Systems in Madison, Wisconsin (2011)

Real-Time Logistics and Virtual Experiment Fields for Adaptive Supply Networks

Michael Toth and Klaus M. Liebler

Abstract Deciding quickly and reliably are key factors for the successful management of adaptive supply networks. This requires real-time information about the current situation and anticipated future behavior in the supply chain. Furthermore, the actors in distribution networks need fast and reliable decision support. This paper presents a process model and implementation guidelines for the concept of Virtual Experiment Fields. This approach combines the knowledge of experienced human planners with a powerful simulation tool and reasoning engines. Through synergetic interaction, verified decisions in complex supply network will be available near real-time.

Keywords Supply Chain management · Decision support · Real world awareness · Adaptive supply networks · Virtual experiment fields · Simulation · Logistics assistance systems

1 Challenges in Current Supply Networks

Analysts predict new challenges due to increasing global supply, changing markets, economic and ecological risks, complex products with short live-cycles, information overload, scarcity of raw materials and sustainability. All these megatrends have a high impact on logistics and supply chain management.

M. Toth (✉) · K. M. Liebler
Supply Chain Engineering, Fraunhofer-Institute for Material Flow and Logistics, Dortmund, Germany
e-mail: Michael.toth@iml.fraunhofer.de

K. M. Liebler
e-mail: Klaus.Liebler@iml.fraunhofer.de

K. Windt (ed.), *Robust Manufacturing Control*, Lecture Notes in Production Engineering, DOI: 10.1007/978-3-642-30749-2_24,

Companies are forced to plan their supply chains with more flexibility, effectiveness, cost reduction opportunities and ecological sustainability.

Today, a high percentage of value creation is done by global supply networks with long order lead times, dynamic and risky transport relations and high transfer stocks. Demand variation induces either supply shortfalls or high inventory levels, both causing additional costs [1]. Global sourcing and factory relocation can only be efficient as long as logistic costs do not exceed the expected savings through lower production costs. In addition to the mentioned attributes external risks, e.g. supplier insolvencies, transport damage or transport delays due to customs checks, strikes, bad weather conditions or accidents are also responsible for an increased complexity in the management of global supply networks [2]. These risks conflict with logistic targets and important competitive factors, e.g. short service times and high delivery reliability.

During the next years, higher supply chain flexibility will be the main target for logistics in supporting efficient and dynamic order processes. In order to reach high delivery reliability, avoid supply shortfalls, utilize the remaining flexibility and enable short service times, methods and processes for an effective joint planning and control are required [3]. Furthermore, supply chain partners shall act in a collaborative decision processes (e. g. [4, 5]).

This paper will introduce how these new challenges influence future supply chains and show which characteristics will be most important for IT-Systems dealing with these challenges. A holistic global management taking into account dynamic supply network behavior, collaborative partnerships and highly volatile markets is not covered by the given approaches [6]. The number of global supply chains will increase in the near future, yet major problems are not solved:

- Global supply chains have high in-transit stocks (inventory on transport and in warehouses), but reduced flexibility due to the availability of goods in-transit and high lead times.
- Global supply chains have low production costs, but high logistic costs due to additional and expensive transports in bottleneck situations or high capital commitment costs due to high inventory.
- Global supply chains produce a lot of data, but there is often lack of transparency due to inconsistencies and data gaps.
- New technology like RFID offer high potential, but there is no integrated approach to enable effective data-handling, data exchange and at least a resilient cost-benefit-sharing.

There are good and bad news for all those operating a global supply chain. The complexity of future networks and IT-solutions will increase even more, but new organizational concepts and technologies are able to help finding new ways to effectively, reliably and flexibly manage supply chains. The attributes of future supply networks are “real-time awareness” and “adaptive risk management”. During the next chapters new technologies for real-time awareness and an approach on how to apply this information for fast decision support in critical situations shall be introduced.

2 Real-Time Awareness for a Transparent Supply Chain

Currently, lack of transparency is a major problem in supply networks, especially concerning in-transit stocks and material flow. Often simple questions like "Where is container no. 36412 now?", "How long will it take for shipment 67890 to be available in my warehouse?" or "How many parts of type A will be available next week? "Cannot be answered. The main reason behind this is a very heterogeneous IT-infrastructure and insufficient communication between the supply chain partners. Furthermore, simple basic information regarding the current status and geographic position of an object in a supply network is often nonexistent or unavailable for partners or logistic service providers. The consequence is that shortage situations or delays in supply are often not detected in time. Therefore, either costly extra transports or expedited freight have to be initiated to avoid possible bottleneck situations, or high inventory buffers are held. Both result in increased costs.

Given identification technology (e. g. RFID) and high-performance IT-Infrastructures is the key for enabling real-time material flow and information event collection. Current projects intend to standardize the technological infrastructure, the event data and informational systems, so that supply chain information can be shared across company borders in real-time. Concepts like intercompany information brokerage or local infrastructures with ad-hoc connection via services (e. g. agent theory) are able to provide real-time data exchange and, based on these information; a more effective supply chain management shall be possible.

Today, a significant problem in gathering all relevant information for supply chain transparency is the heterogeneous IT-Infrastructure along the entire supply chain. Both the integration of IT-systems in one company and the synchronization of information from other supply chain partners cause plenty of problems. Data exchange with external partners often works on a daily basis without standardized formats. The synchronization of data along the supply chain is the major problem. Lack of information caused by different time horizons (e.g. daily), no data actualization, data aggregation and export times (e.g. 6 a.m. or in the evening) lead to visualizing a wrong picture of the supply chain. For instance, if a ship arrives in the destination harbor the transport management system will delete this ship from the list of active transports. In given examples it took up to 3 weeks until a container was registered at the container yard after customs. During this period it was not possible to find the container in given supply chain IT-Systems.

For an effective global supply chain management, the concepts of RFID (radio frequency identification) and company-wide information brokerage offer promising technological background. RFID enables capturing a real-time picture of the supply chain operations. Each movement of goods and each operational step are documented by a RFID read point event. Thus, the current as-is state of the system is accessible at any time and the system state is updated as each object passes through a read point.

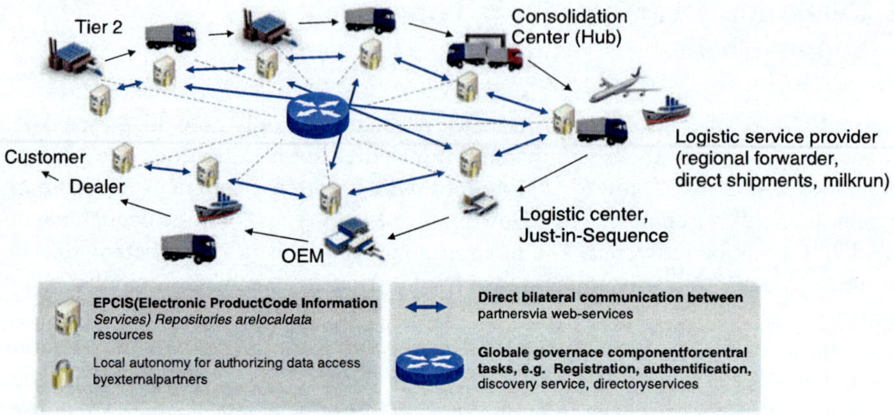

Fig. 1 The RAN Information-Broker Concept

However, up to now companies apply RFID solutions mainly in closed-loop systems. A standardized industry-wide exchange of RFID/Auto-ID information between manufacturers, logistic service providers and suppliers has not been accomplished so far ([7]). Some individual solutions exist, but the technology has not yet been exploited to its full potential for the management of supply chains due to missing standards and collaborative solutions. The automotive industry is a prominent example for this situation. Particularly first tier suppliers suffer from this situation, for they have to deal with multiple OEM-specific solutions.

The research project "RAN—RFID-based Automotive Network" funded by the Federal Ministry for Economics and Technology (BMWi) aims at increasing the transparency of information exchange in production and logistic networks of the automotive industry [8] through the application of RFID. The RAN project intends to standardize the technological infrastructure, the event data and informational systems so that event information can be shared across company boarders easily. The basis for cross-company data exchange is an identification standard for logistic or production objects. RAN utilizes the Electronic Product Code EPC as well as the ISO-standard. Both of these standards ensure that information sharing between suppliers, customers and logistic service providers works reliably and that the exclusive reference of a part or object to its manufacturer or owner is possible [9].

Building on the standardized processes and data structures a so called InfoBroker infrastructure serves as a basis for the effective exchange of object-related event data between companies. Distributed event repositories constitute the core of the InfoBroker (see Fig. 1). Each repository is assigned to a defined organizational domain and is able to store the event information generated within this domain in standardized form. Using query and subscription services companies can exchange event information between repositories and other peripheral systems. Information exchange is controlled by rules for security and privacy of data. Each domain owner must authorize any external partner to receive or retrieve information from his repository. Figure 1 shows the RAN InfoBroker concept.

The next step in real-time awareness is to use all kinds of additional information available through mobile phone data, GPS information, web-service based container tracking and social network data. With this regard, the major challenge that emerges is to enable an adequate visual logistics management based on data aggregation, critical event identification and suitable preparation of the huge amount of data available for management decisions. Furthermore, organizational concepts, which would allow the establishment of short-dates collaboration between supply chain partners with stable information exchange and inter-organizational planning and execution processes, have to be developed.

3 Real-Time Decisions through Virtual Experiment Fields

As previously shown, information on material flows in a distribution network is made available through an InfoBroker implementation nearly in real-time across borders of companies and even continents. This chapter presents a new methodology on how planning decisions can be derived from this information.

The challenges of decision making in complex supply networks are manifold. On the one hand, the future impact of decisions is not definite. On the other hand, the targets as well as the impacts are multidimensional and a decision maker can hardly evaluate all of them without systematic decision support [10]. Moreover, there is a need for fast decisions. Kuhn et al. have systematized a planning process [11], as described in Fig. 2.

The four phases of preparation shall be characterized briefly. The initial event is a signal from the real world indicating a change at time t_V. During the **identification phase**, the event is detected and distributed among systems. In the subsequent **analysis** phase it has to be assessed whether the shift in demands and resources may be balanced using the flexibility of the system. Only shifts beyond the allowed capacity corridors imply a need for change [12].

If a need for change has been identified at time t_B, certain measures have to be developed and evaluated in the **planning phase**. They shall be able to shift the flexibility corridor of the system into a region that covers the new demands of the distribution system. Of major significance is an exhaustive and holistic view on the system, i.e. all effects and side effects of new demands and all change options are considered. At the end of the planning phase, a list of several evaluated change scenarios, each defined with fully described change options, shall be provided.

The final **decision** on any of the evaluated change scenarios takes place in the last phase of the preparation sequence. Generally, the list of evaluated scenarios contains performance indicators from technical, ecological and economic domains. A decision maker has to select the experiment that yields results that match his preferences.

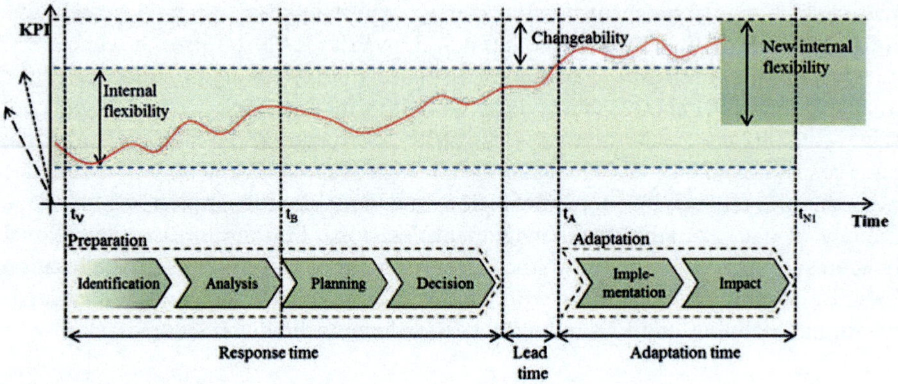

Fig. 2 The planning process model ([1])

Processing all four steps quickly is important, because more time remains for lead time, as well as time to adapt to the change measures. On the one hand, the need for change has to be identified fast. On the other hand, changes have to be executed rapidly to provide enough time to move to the new flexibility corridor. The sooner the need for change is identified and the correct change decision is applied, the softer and more controllable the change may be. Hence, fast and reliable decision making is an essential characteristic of holistic supply chain management concepts.

The following sub-chapters show how leveraging new technologies allows shortening the response time to a minimum while assuring valid planning decisions. Beneath the Real-World-Awareness technologies and the InfoBroker-concept presented in Chap. 2, further IT-based methods are proposed for fast decision support. The model based discrete event simulation [13] is the central tool in analyzing supply chains and evaluating possible change options.

3.1 Identification of Drivers of Change

According to WESTKÄMPER, drivers of change may be separated into internal and external classes [14]. Typical external drivers of change in complex distribution networks (e.g. an automotive supply network) are:

- Sudden changes in customer behavior
- Transportation delays
- New product of a competitor
- Legal changes
- Shortages of natural resources

Internal drivers of change include changes of the internal production capacity, changed strategic targets etc. It is common to all of them that the anticipated

Table 1 Scope and methods for getting information

Scope	Method	Information
Supply chain	InfoBroker	Material flow
Supply chain partners	InfoBroker + VDA-/EDIFACT-messages	Material flow + bilateral planning information
Planning domain	InfoBroker + VDA-/EDIFACT-messages + connection to operative systems	Material flow + internal planning information

demand profile *may* not fully match or even be out of frame of the previously planned capacity profile of the supply network. Note the formulation: even if the causes of change are transparent, their consequences are not clear.

In the "Identification" step, those drivers of change have to be identified and formalized. It should be distinguished between three scopes of information and appropriate methods how to get them. With the methods presented in Chap. 2, a constant and nearly real-time monitoring of drivers of change in the material flow within the whole supply chain is possible. Partners in the supply chain use sophisticated Advanced Planning systems that support standardized information exchange. Typically, in the automotive industry, VDA- or EDIFACT-messages are used. The transfer channels have to be monitored constantly by appropriate middleware. Finally, operative systems in the core planning domain [15] have to be checked permanently for new information. Table 1 shows which information is available through which method depending on the observed Supply Chain segment.

Often, all the aforementioned systems offer a notification concept. The system proposed in this paper offers, moreover, the possibility to register itself, as well as notifications upon the occurrence of relevant events. By leveraging these technologies, the time required for "Identification" is kept on the lowest possible level. Hence, the subsequent "Analysis" step may start without delay.

3.2 *Analysis of the Need for Change*

As previously mentioned, it has to be determined whether the need for change exists. Only if drivers of change shift the demand profile beyond the immanent flexibility-corridor [16] of the system, change options need to be identified and applied.

Up to now, the perception of drivers of change is distributed among several InfoBroker systems, in messages and databases of operative systems. If not already available in figures, all drivers of change have to be quantified and related to business objects as well as to a specific time span. For instance, signals indicating a change in the future market demand for a car manufacturer have to be translated into measures of volume and option quotas over time [17]. This kind of further formalization is an essential requirement in order to be able to implement at least partially an automated evaluation process.

All information that is relevant for planning has to be made available in an integrated information model. Being decomposed into a description of current and expected material flows, restrictions of planning domains, market behavior and product data; this model represents the complete planning-relevant knowledge about the supply chain. Thus, a fast and automated generation of this model paves the road to permanent planning [18].

The core analysis step itself is performed by the previously mentioned model-based simulation, which is a fast as well as holistic approach. Based on the integrated information model, simulation may deduce and analyze future system states. Appropriate reports or performance indicators may be extracted from simulation results, which are able to reveal bottlenecks, shortfalls or low utilization rates of resources.

In the final filter step, a human planner assesses these results and decides whether there is the need for change in the supply chain. Excessively small long-term misalignments may seldom cause any action, while short-term severe bottlenecks inevitably require changes to be made.

3.3 Planning with Virtual Experiment Fields

If a need for change has been identified, the set of possible decision alternatives shall be systematically identified and evaluated. A valid decision alternative is any combination of pre-defined change options.

A change option is a time-limited or unlimited and feasible measure to change planning-relevant parameters. Thus, the execution of change options may be planned in advance and their effect on the supply chain may be evaluated. Examples for change options are

- Raise of suppliers' capacity
- Influence on market demand by an advertising campaign
- Earlier arrival of material by using air cargo
- Shift of production to another factory (if the product has already been qualified for the production line according to ISO/TS 16949)

A set of change options to be applied on the system under investigation is the basis of an experiment. Several experiments developed through structured methods, each one forming the basis for an experiment, are called experiment plan.

Virtual Experiment Fields [19] form a model of relevant design aspects and aim at covering the entire scope of decisions. They provide a toolkit that supports the planner in making its decision on change options while achieving short lead times that meet the requirements of decision makers. The concept of Virtual Experiment Fields describes a structured approach for the design of experiment plans and for hedging the resulting scenarios. Methodological support is provided in particular by the model-based discrete-event simulation [20].

The planning phase comprises the four steps "define system boundaries", "define change options", "create experimental plan" and finally "evaluate experiment plan".

During the first step system boundaries have to be defined. They restrict the entities of the network that may be influenced in their behavior. Afterwards, the set of possible change options has to be restricted in a way that, on the one hand, only valid options remain and, on the other hand, enough room is left for experiments.

Based on the solution space of possible change options, an experimental plan has to be generated. By leveraging expert knowledge from a human planner or IT-based reasoning engines, reasonable combinations of change options among the set of possible change options have to be selected and configured. Each combination represents the description of an experiment and is stored in the experiment plan.

The procedure has similarities to iteration in evolutionary methods for optimization [21]. The main difference is the generation of new solution candidates. In classical heuristic approaches, new solution candidates are usually derived from existing solutions through generic operators. However, the procedure in the context of Virtual Experiment Fields focusses on leveraging expert knowledge. In an optimal implementation, expert knowledge is formalized as decision rules. They may as well be used in automated reasoning algorithms [22]. This technique may be applied to generate change scenarios fully automated without human interaction. If these rules or the reasoning algorithms do not exist, manual scenario generation is possible. For this purpose, appropriate user interfaces have to be provided.

After having defined all combinations of change options, simulation models for the evaluation have to be built up. This step comprises primarily the fusion of the base information with the configured change options. The result is a range of evaluable simulation models. Again, the model based discrete-event simulation provides the necessary tools to carry out the evaluation. The application of efficient, model-based methods, e.g. simulation and optimization, is crucial in order to guarantee fast processing [12].

3.4 Decision for the Best Set of Change Options

The evaluation of all scenarios is followed by the decision phase. Among all alternatives in the experiment plan, the scenario with the most promising evaluation results has to be identified for realization according to the preferences of the decision maker. These preferences may cover a wide field and are mostly multi-dimensional. The dimensions may be quantified with key performance indicators and clustered in technical, economic and ecological classes ([23, 24]). While the technological dimensions are based on thinking in quantities and times, the economic dimension asks for thinking in values. Ecological objectives gain more and more perception and are evaluated in terms of emissions [25].

If none of the experiments yields promising or satisfying results, a new, advanced Virtual Experiment Field with other system limitations and change options has to be defined and conducted through another evaluation process.

4 Application of the Concept

In this chapter, we will present how the concept of Virtual Experiment Fields has been implemented in a prototypical IT system. A vehicle manufacturer produces one of its models at two sites in Germany and Argentina. Suppliers of parts as well as complex assemblies are located in South America and Europe. Both sites receive material from suppliers on both continents. The transport is organized multimodal and dominated by scheduled vessels between Argentina and Germany. For faster delivery, air cargo is also possible to hedge against delays in transportation, and warehouses next to the production sites keep inventory for some days. This supply chain is depicted schematically in Fig. 3.

Due to the long transport lead times, dispatchers at the production and suppliers sites are facing complex decision problems. The amount of material requested from overseas has to be determined under uncertainty and depends on many factors. In order to be able to offer order flexibility to the customer, the production program is not fixed at the time of material disposition. Periodically, the sales department calls for additional orders to be placed in production. Due to the schedule, a small disturbance in one stage of the multimodal transport chain may severely affect all subsequent stages. Inventory costs shall be kept as low as possible, while ensuring the availability of material.

Drivers of change are primarily unexpected new orders, delays in the long and intercontinental supply chain and quality problems with suppliers. Among the possible change options are:

- Changing or amending orders
- Changing suppliers capacities
- Arranging extra transports

A Logistic Assistance System (LAS) was developed [26] to support all planning tasks under economic and ecological objectives. Generally, LAS provide support for recurring decision situations by giving transparency over the consequences of possible decision alternatives [27]. They can be seen as symbiotic compositions of decision-support systems and workflow management systems (cf. [28, 29]). LAS allow a tight cooperation between the cognitive and creative abilities of a human planner and the computing power of IT. Four functional modules constitute the LAS used in this case: an information management module with appropriate persisting and data access components, the simulation suite OTD-NET [13], a business logic engine, and a Web-based GUI module (see Fig. 4).

The basis for all actions supported by the system is the integrated information model. All methods introduced in Sect. 3.1 are extensively leveraged for data

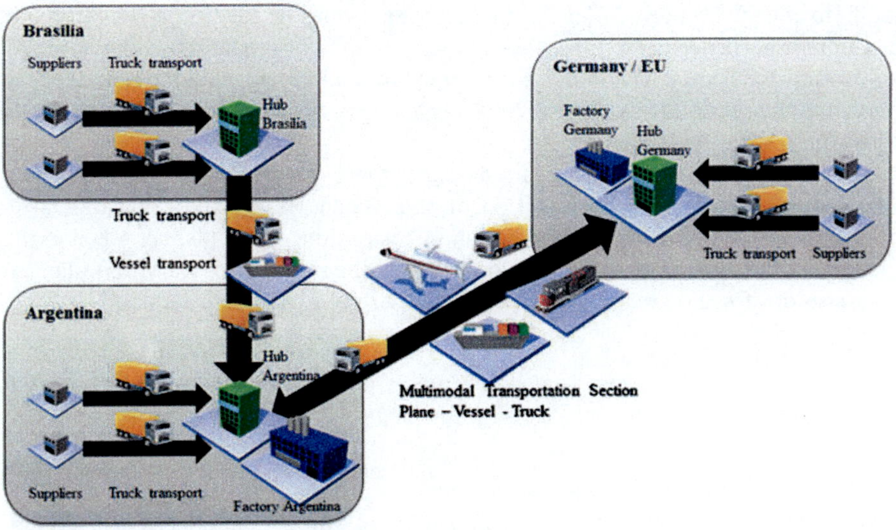

Fig. 3 Multimodal intercontinental supply chain

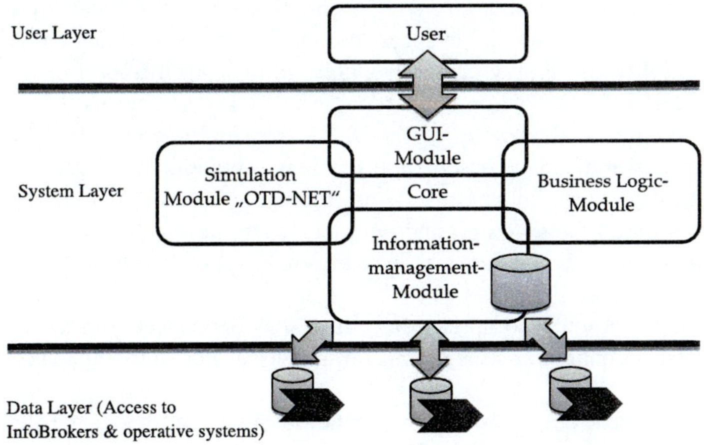

Fig. 4 Architecture of the logistic assistance system

collection. For instance, a connector to the web service of vesseltracker.com provides real time information about the position of vessels.

The concept of virtual experiment fields is an integral feature of the LAS. Drivers of change are constantly monitored and their quantitative effect may always update the information model. The simulation component OTD-NET provides an initial in-depth analysis of the unmodified scenario. In the case of any shortages, the system supports the structured generation of an experiment plan containing multiple change scenarios. For instance, business rules on lead time, severity and duration of bottlenecks facilitate the creation of different scenarios with additional air- or ship-transports. One of these rules may, in a simplified version, look like "IF (leadtime > 5w AND severity > 20 % AND duration > 1w) THEN [additional ship transport]".

An experimental plan is built and all experiments are evaluated by simulation. This method allows the consideration of ship schedules, capacity constraints of pre- and post-carriage and other dynamic restrictions. The planner decides to implement the experiment with the lowest overall costs. If necessary, the planner may also opt for the creation of an extended virtual experiment field.

5 Conclusion

High dynamics and complex decision situations impose new challenges in supply chain management. This paper presented "real time" approaches to support planning processes in such environments. Information about the current flow and the current position of goods on the supply chain can be provided through RFID, AIS and similar technologies on the hardware side, and Implementations of InfoBroker-systems on the software side. Information is—if allowed by the information provider—made available across company borders and may be used for an optimal control of the supply chain. The concept of Virtual Experiment leverages this information for an automated generation and evaluation of scenarios of decision alternatives. Reasoning methods, the domain knowledge of experienced human planners and model based time discrete simulation constitute the framework for this purpose. In contrast to traditional industrial practice, where the generation of partly validated production plans under new supply chain conditions takes up to 30 days, the method presented in this paper provides a fully validated result within minutes.

Future developments comprise the automated deduction of suitable change options based on characteristics of misalignments of demand, capacity and business targets.

The concept has been applied in the automotive industry and rolled out to control an intercontinental and multimodal supply chain.

References

1. Deiseroth, J., Weibels, D., Toth, M., Wagenitz, A.: Simulationsbasiertes Assistenzsystem für die Disposition von globalen Lieferketten. In: Rabe, M (ed.) Advances in Simulation for Production and Logistics Applications. 13. ASIM Fachtagung Simulation in Produktion und Logistik Berlin, pp. 41–50. IRB Verlag, Stuttgart (2008)
2. van Wyk, J., Baerwaldt, W.: External risks and the global supply chain in the chemicals industry. Supply Chain Forum: Int. J. **6**(1), 2–15 (2005)

3. McLaren, T., Head, M., Yuan, Y.: Supply chain collaboration alternatives: understanding the expected costs and benefits. Internet Res.: Electron. Networking Appl. Policy **12**(4), 348–364 (2002)
4. Dudek, G., Stadtler, H.: Negotiation-based collaborative planning between supply chains partners. Supply chain management and advanced planning. Eur. J. Oper. Res. **163**(3), 668–687 (2005). doi:10.1016/j.ejor.2004.01.014
5. Mentzner, J., Foggin, J., Golicic, S.: Collaboration—the enablers impediments and benefits. Supply chain management review **16**(September/October), 52–58 (2000)
6. Kuhn, A., Hellingrath, B., Hinrichs, J.: Logistische Assistenzsysteme—Entscheidungsunterstützung für eine robuste Logistik. In: ten Hompel, M (ed.) Software in der Logistik. Weltweit sichere Supply Chains—Anforferungen, Funktionalitäten und Anbieter in den Bereichen WMS, ERP, TMS und SCM. Huss, München (2008)
7. Lange, B.: Automobilindustrie wagt gemeinsame RFID-Wege. VDI-Nachrichten(13), http://www.autoran.de/fileadmin/autoran.de/data/Dokumente/Artikel_VDI_Nachricht_042011.pdf (2011)
8. Federal Ministry of Economics and Technology: RAN-based Automotive Network—Transparent and optimum management of processes in the motor-vehicle industry, http://www.autonomik.de/en/218.php
9. Hegmanns, T., Toth, M.: RFID-based Real Time Decision Support in Supply Chains. In: Massei, M., Frydman, C., McGinnis, M (eds.) The 10th international conference on modeling and applied simulation, Rom, 12–14 Sept (2011)
10. Eisenführ, F.: Rationales Entscheiden, 4th edn. Springer, Berlin (2003)
11. Kuhn, A., Klingebiel, K., Schmidt, A., Luft, N.: Modellgestütztes Planen und kollaboratives Experimentieren für robuste Distributionssysteme. In: Spath, D. (ed.) Wissensarbeit—zwischen strengen Prozessen und kreativem Spielraum, pp. 177–198. Gito, Berlin (2011)
12. Klingebiel, K., Winkler, M., Klaas, A., Laroque, C.: A Cross-Level Approach to Distribution Planning. In: Rykalova, Y. (ed.) Proceedings SpringSim. 2012 Spring Simulation Multiconference, 26–29 March (2012)
13. Wagenitz, A.: Modellierungsmethode zur Auftragsabwicklung in der Automobilindustrie. Dissertation, Technische Universität Dortmund. http://hdl.handle.net/2003/24408 (2007)
14. Westkämper, E.: Wandlungsfähige Produktionsunternehmen. Das Stuttgarter Unternehmensmodell. Springer, Berlin (2009)
15. Stadtler, H., Kilger, C (eds.): Supply chain management and advanced planning. Concepts, models, software, and case studies, 4th edn. Springer, Berlin (2008)
16. Hegmanns, T.: Dezentrales Planungs- und Prozesskonzept für ein kollaboratives Bedarfs- und Kapazitätsmanagement in Produktionsnetzwerken. Praxiswissen, Dortmund (2009)
17. Meyr, H.: Supply chain planning in the German automotive industry. OR Spectrum **26**(4), 447–470 (2004)
18. Westkämper, E.: Die Digital Fabrik. Kontinuierliche und partizipative Planung. In: Erfolg in Netzwerken, 1st edn., pp. 247–260. Springer, Berlin (2002)
19. Toth, M.: Real-Time Logistics for Adaptive Supply Networks. In: Katkalo, V.S., Ledyaev, A.P., Zierpka, D (eds.) Research and Education in Logistics and Supply Chain Management: the Current Situation and New Perspectives. Petersburger Dialogue, St. Petersburg, 14–16 Dec (2011)
20. Law, A.M., Kelton, W.D.: Simulation Modeling and Analysis, 3rd edn. McGraw-Hill, New York (2000)
21. Dréo, J., Pétrowski, A., Siarry, P., Taillard, E.: Metaheuristics for Hard Optimization. Methods and Case Studies. Springer, Berlin (2006)
22. Pearl, J.: Probablistic Reasoning in Intelligent Systems. Network of Plausible Inference. Morgan Kaufmann, San Francisco (1988)
23. Pfohl, H.-C.: Logistikmanagement. Konzeption und Funktionen, 2nd edn. Springer, Berlin (2004)
24. Keller, M.: Kennzahlenbasierte Wirtschaftlichkeitsbewertung der Integration von Unternehmen in Produktions- und Logistiknetzwerken. Praxiswissen, Dortmund (2010)

25. Gruden, D.: Umweltschutz in der Automobilindustrie. Motor, Kraftstoffe, Recycling. Vieweg + Teubner, Wiesbaden (2008)
26. Kuhn, A., Toth, M.: Assistenzsysteme für die effektive Planung logistischer Netzwerke. In: Scholz-Reiter, B. (ed.) Technologiegetriebene Veränderungen der Arbeitswelt, 1st edn, pp. 257–278. Gito, Berlin (2008)
27. Kuhn, A., Toth, M., Wagenitz, A.: Integrierte Versorgungsplanung im Rahmen der Digitalen Logistik. In: Schenk, M. (ed.) Digital Engineering—Herausforderung für die Arbeits- und Betriebsorganisation, pp. 175–194. Gito, Berlin (2009)
28. Turban, E., Aronson, J.E.: Decision support and expert systems. Management support systems, 5th edn., Upper Saddle River, Prentice-Hall International, NewJersy (1998)
29. Hollingsworth, D.: The Workflow Reference Model. Document Number TC00-1003. The Workflow Management Coalition, Winchester. http://www.wfmc.org/Download-document/TC00-1003-The-Workflow-Reference-Model.html (1995)

New Mechanisms in Decentralized Electricity Trading to Stabilize the Grid System: A Study with Human Subject Experiments and Multi-Agent Simulation

Sho Hosokawa and Nariaki Nishino

Abstract The Smart Grid concept has lately attracted attention because of the increase of decentralized electricity generators and the development of the information communication technology. In the smart-grid concept, mutual information exchange among suppliers and consumers can be achieved to balance and optimize the supply and demand of electricity, which is generally necessary for a grid system. Taking this background into consideration, the necessity for electricity trade by which small-scale consumers such as households buy and sell electricity is now advocated to realize further stability of the grid system. However, it is noteworthy that consumers are self-interested, which endangers the grid system stability. This study proposes new trading mechanisms applied in the electricity trade and evaluates them in terms of stability and social surplus in the market. We examine their validity using experiments with human subjects and multi-agent simulations.

Keywords Decentralized electricity trading · Trading mechanism · Multi-agent simulation · Human subject experiment

1 Introduction

In consideration of global warming and the steep increase of energy prices, various countries have been promoting the introduction of renewable energy generation

---

This contribution was previously published in Logistics Research (2012) pp. 123–131. DOI:10.1007/s12159-012-0092-y.

S. Hosokawa (✉) · N. Nishino
The University of Tokyo, 7-3-1 Hongo, Bunkyo-ku, Tokyo, 113-8658 Japan
e-mail: hosokawa@css.t.u-tokyo.ac.jp

K. Windt (ed.), *Robust Manufacturing Control*, Lecture Notes in Production Engineering, 341
DOI: 10.1007/978-3-642-30749-2_25, © Springer-Verlag Berlin Heidelberg 2013

modes such as photovoltaic (PV) power generation and wind power generation [1]. Because there is generally difficulty in storing the electricity generated using those means, it is necessary to balance demand and supply of electricity simultaneously to stabilize public electrical grid systems. Considering such characteristics of electricity, introduction of renewable energy generations into the current grid system makes it more difficult to stabilize the supply of electricity because some renewable energy generation modes involve output-power fluctuation. Using the smart-grid concept, mutual information exchange among medium-scale electricity suppliers, small decentralized suppliers, and consumers can be realized using information communications technology, which can balance and optimize the supply and demand related to electricity. Moreover, expansion of residential PV systems might enable electricity trade among even small-scale consumers such as households and might play a role in the further stabilization of the grids through household participation in the electricity trade market. Information communications technology helps consumers to give real-time information related to the balance of demand and supply, which is expected to achieve balanced trades by market principals and thereby increase of social surplus. In addition, a technology exists to enable electricity trade among small-scale consumers. Digital Grid, which was advocated by Abe [2], enables identification of who generates how much electricity by attaching information such as an address to units of generated electrical power.

To realize electricity trading, however, we must take several points into consideration. First, as described above, the amount of electricity supply and that of electricity demand must be balanced at any given time. Second, members of the grid are self-interested, meaning that their only purpose of trading electricity is to maximize their profit, which might endanger the grid system stability. Third, electricity is one of the most vital daily necessities which people almost always use incessantly. Therefore, trading mechanisms that compel consumers to follow a complicated process are not desirable.

Recently, the number of studies related to electricity trade is increasing as the introduction of renewable energy for electricity is promoted worldwide. For example, Rudkevich et al. [3] estimated electricity pricing under a market mechanism called Poolco, in which electric power companies bid to maximize their profit; Tanaka [4] simulated the Japanese wholesale electricity market as a transmission-constrained Cournot market. Above all, Vitelingum et al. [5] propose a mechanism based on a continuous double auction for electricity trade in which small-scale consumers can participate. However, few reports describe that kind of electricity trade. Moreover, even the mechanism proposed by Vitelingum et al. is not sufficiently simple for households when considering the features of electricity such as incessant daily use.

In this study, we define "Decentralized Electricity Trading" as "electricity trading in which small-scale consumers who possess their own generator participate not only as consumers but also as producers."

In these circumstances, we propose two new electricity trading mechanisms which entail simple procedures, and which can stabilize a system even when grid participants are self-interested. To evaluate the mechanisms, we conduct human

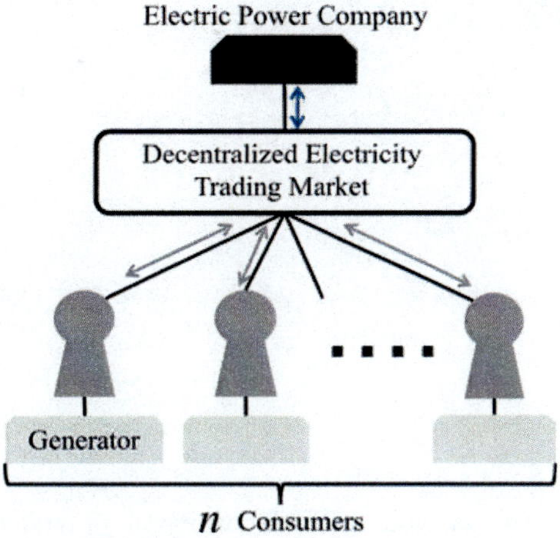

Fig. 1 Model of decentralized electricity trading

subject experiments and multi-agent simulations. A model of decision-making by human beings is constructed through subject experiments. We use it for multi-agent simulation as input data.

2 Modeling Decentralized Electricity Trading and Proposed Trading Mechanisms

2.1 Model of Decentralized Electricity Trading

We construct a model of decentralized electricity trading. As Fig. 1 shows, decentralized electricity trading consists of a market, one electric power company, and n consumers who have their own generator and who can generate electricity independently.

Decentralized Electricity Trading Market. As Fig. 1 shows, in decentralized electricity trading, the electric power company and all consumers trade the generated electricity in this market. This market has a trading mechanism that determines what kind of information consumers must send as an input to trade electricity and how much electricity is traded in the market. The market then determines how to distribute electricity to consumers and the electricity price based on its mechanism.

Electric Power Company. An electric power company exists in the model. Compared with consumers, this electricity trading company has much greater capacity to generate electricity. It takes responsibility for stabilizing the electrical grid system. This company executes actions of two types in decentralized electricity trading.

- In case too much amount of electricity is generated by consumers, the company purchases the excess electricity for constant price p^{min}.

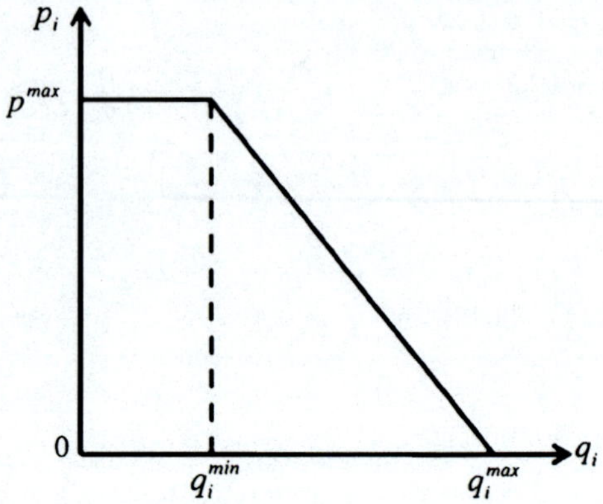

Fig. 2 Reservation price for electricity of consumer i

- In case too little electricity is generated by consumers, the company sells an amount to alleviate the shortage for constant price p^{max}.

Consumers in the market have no incentive to sell their electricity to other consumers for less than p^{min}, for which they can surely sell their electricity to the electric power company, and also have no incentive to buy other consumers' electricity for more than p^{max}. In this study, we assume $p^{min} = 0$ for simplicity. We designate p^{max} as the "electric power company's electricity sales price."

Consumers. Consumers send necessary information and electricity they generate to the market. Each of the consumers gains profits through the trade, which is determined by their own demand function for electricity and the amount of electricity they consume through the trade. The purpose of consumers is to maximize their profit.

Consumer's Reservation Price for Electricity. Each consumer has its own reservation price for electricity per unit. The reservation price for electricity of consumer i ($1 \leq i \leq n$) is determined by min-demand q_i^{min}, max-demand q_i^{max} and the reservation price for min-demand p_i^{max}. The min-demand means the minimum amount of electricity that consumer i consumes irrespective of electricity price. The max-demand means the maximum amount of electricity that consumer i can consume. We assume in this study that all consumers' reservation prices for the min-demand equal the electric power company's electricity sales price. We can write the reservation price as Eq. (1) in decentralized electricity trading. $p_i(q)$ represents the reservation price of consumer i for the amount of electricity q.

$$p_i(q) = \begin{cases} p^{max} & (0 \leq q < q_i^{min}) \\ -\frac{p_i^{max}}{q_i^{max} - q_i^{min}}(q - q_i^{max}) & (q_i^{min} \leq q \leq q_i^{max}) \end{cases} \quad (1)$$

This can be represented as Fig. 2.

Consumer's Demand Function. The demand is easily derived from Eq. (1). The potential maximum amount of electricity that the consumer wants to consume is determined according to Eq. (1) if a certain price is given. Figure 3 portrays consumer i's demand function $q_i(p)$ given price p.

Fig. 3 Demand function for electricity of consumer *i*

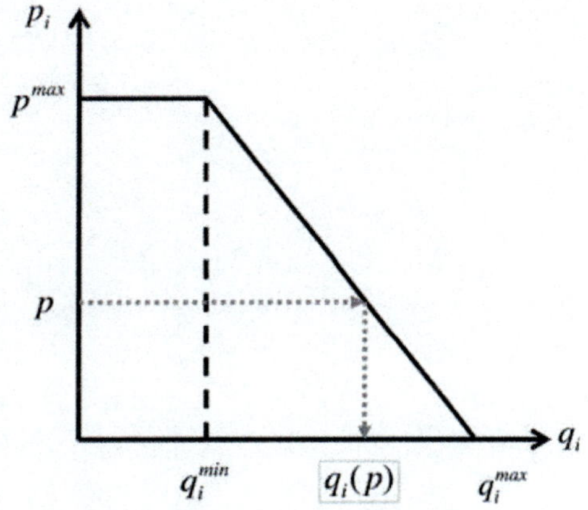

Consumer's Profit. Consumers gain profits through electricity trading. The consumer's profit is divisible into three elements as follows:

- Profit from consuming electricity
- Profit from selling electricity
- Payment for purchasing electricity

Consumer *i*'s total profit π_i can be calculated as Eq. (2) using these elements.

$$\pi_i = \int_0^{q_i^c} p_i(q_i)\, dq_i + \sum_j pq_{i,j}^s - \sum_i pq_{i,j}^c \tag{2}$$

The first term represents consumer *i*'s profit from consuming electricity. Here we assume that a consumer consumes all the electricity the consumer has purchased from the market and does not sell it to other consumers or store it. This profit is calculable with their demand function for electricity and the amount of electricity they consume q_i^c. This profit can be depicted as the colored area in Fig. 4.

The second term represents consumer *i*'s profit from selling the electricity the consumer generates, which is calculable with the electricity price in the market and the amount of electricity sold in the market. p represents the electricity market price, and $q_{i,j}^s$ represents the amount of electricity that consumer *i* sells to another consumer *j*.

The last term represents consumer *i*'s payment for purchasing electricity, which is calculable with the electricity price in the market, p, and the amount of electricity purchased by consumer *i* from consumer *j*, $q_{i,j}^c$.

Consumers' purposes for making their decisions in the decentralized electricity trading are to make this total profit as large as possible.

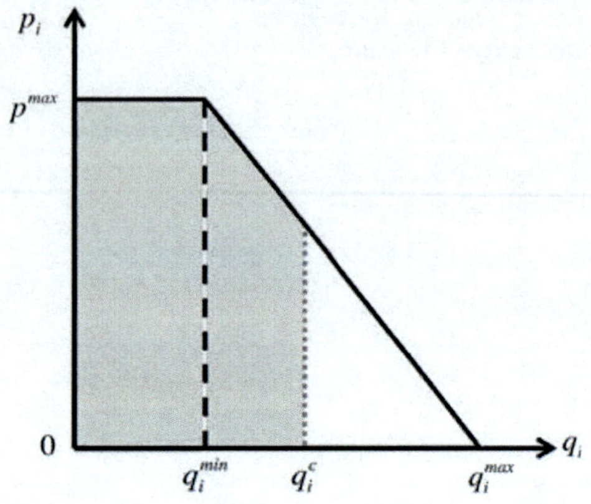

Fig. 4 Profit from consuming electricity of consumer *i*

2.2 Proposed Trading Mechanisms

We propose two new trading mechanisms applied to the decentralized electricity trading. These trading mechanisms are devised not only to make the social surplus larger, which is calculated as the sum of consumers' profit, but also to stabilize the grid system, which means that electricity trading under mechanisms can balance the demand and supply of electricity. The two mechanisms differ from each other in three points as shown below.

- Kind of input information
- Tradable amount of electricity
- Rule of determining the selling order

2.2.1 Mechanism 1: Aggregated Demand–Supply Mechanism

Input and the Amount of Electricity Sent to the Market. Each consumer must input an "Offer Price" in the market under Mechanism 1. The Offer Price is the price at which a consumer wants to sell the electricity the consumer generates. All consumers' electricity is traded in the market in this mechanism (Fig. 5).

Determining the Electricity Price. The electricity price is determined using an aggregate demand curve and aggregate supply curve. The aggregate demand curve is made from demand curves of all consumers, as shown in Fig. 6. The aggregate supply curve is made from the Offer Price and electricity sent from all consumers as shown in Fig. 7. Here, p_k^s and q_k^g respectively represent the Offer Price and generated electricity output of consumer *k*, who sends the *k*th cheapest Offer Price in the market.

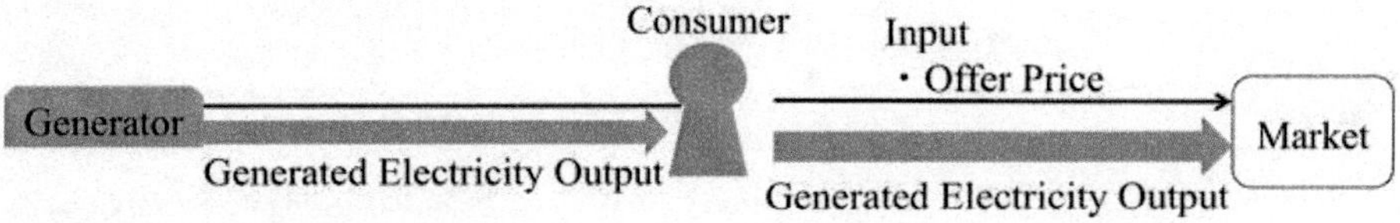

Fig. 5 Input and the amount of electricity sent to the market under mechanism 1

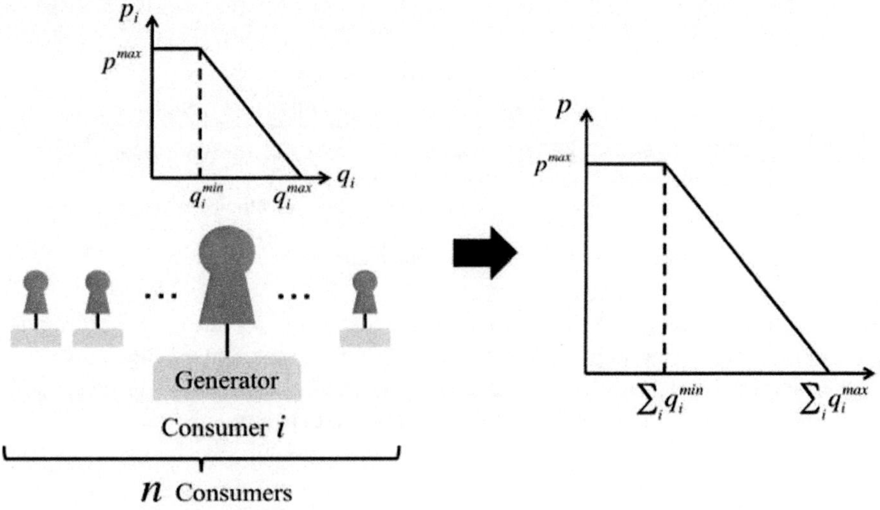

Fig. 6 How to make an aggregate demand curve under mechanism 1

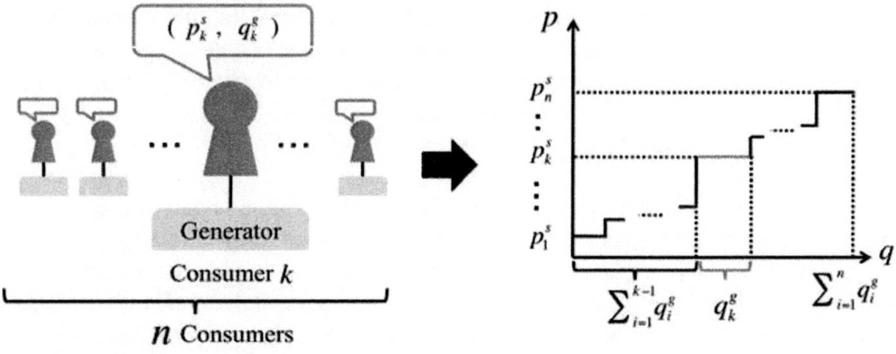

Fig. 7 How to make an aggregate supply curve under mechanism 1

The intersection point of the aggregate demand curve and the aggregate supply curve is determined as the electricity price in the market, which we call a "Trading Price." All the electricity is traded for that price in the market (Fig. 8).

Order of Selling Electricity. The order of selling electricity is determined by the consumers' Offer Price. The cheaper an Offer Price consumers input, the earlier

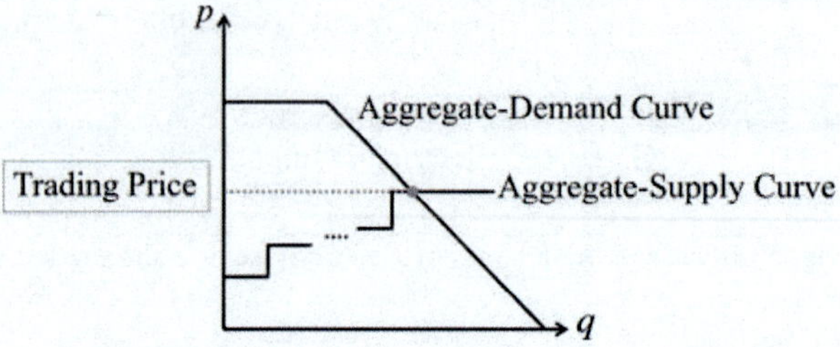

Fig. 8 Determining the trading price under mechanism 1

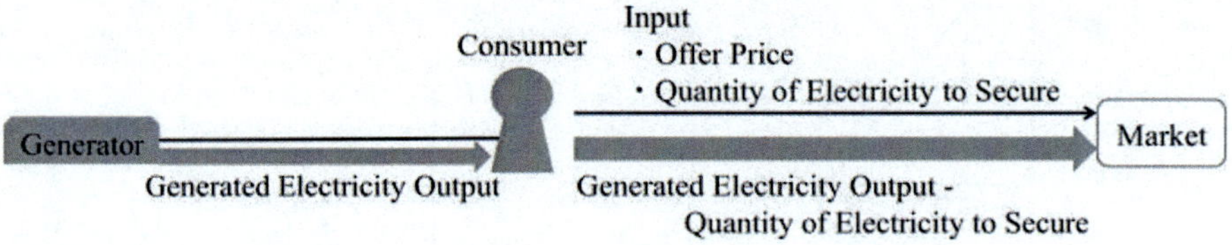

Fig. 9 Input and amount of electricity sent to the market under mechanism 2

they can sell their electricity in the market. For example, a consumer who inputs the cheapest Offer Price can sell electricity first, and a consumer who inputs the second cheapest Offer Price can sell electricity next. A cheaper Offer Price gives a lower probability of not selling all their generated electricity.

2.2.2 Mechanism 2: Residual Electricity Based Mechanism

Input and the Amount of Electricity Sent to the Market. Consumers must input an “Offer Price” and a “Quantity of Electricity to Secure” in the market under Mechanism 2. The Offer Price is the price at which a consumer wants to sell electricity that the consumer generates. The Quantity of Electricity to Secure is the quantity of electricity which a consumer wants to consume from the electricity that is generated. This quantity of electricity is not traded. The rest is traded in the market (Fig. 9).

How to Determine the Electricity Price. The electricity price is determined by the aggregate demand curve and aggregate supply curve from which secured amounts are removed. The aggregate demand curve is made from the demand curve. The Quantity of Electricity to Secure of all consumers is shown as Fig. 10. The aggregate supply curve is produced from the Offer Price, Quantity of Electricity to Secure, and electricity sent from all consumers as shown in Fig. 11.

The intersection point of the aggregate demand curve and the aggregate supply curve determines the Trading Price in the market. All electricity is traded with that price in the market (Fig. 12).

Order of Selling Electricity. The order of selling electricity is determined by the consumer’s Offer Price, as shown in Mechanism 1. The lower the consumers set their Offer Price, the earlier they can sell their electricity.

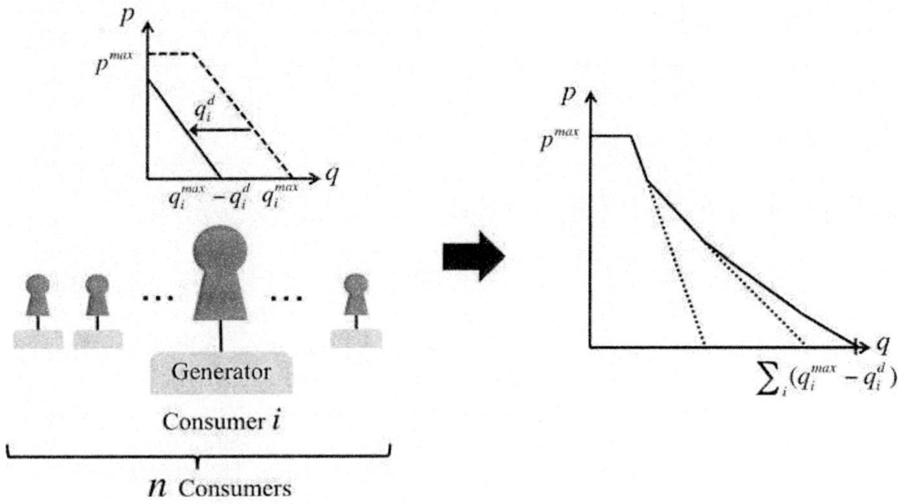

Fig. 10 How to make the aggregate demand curve under mechanism 2

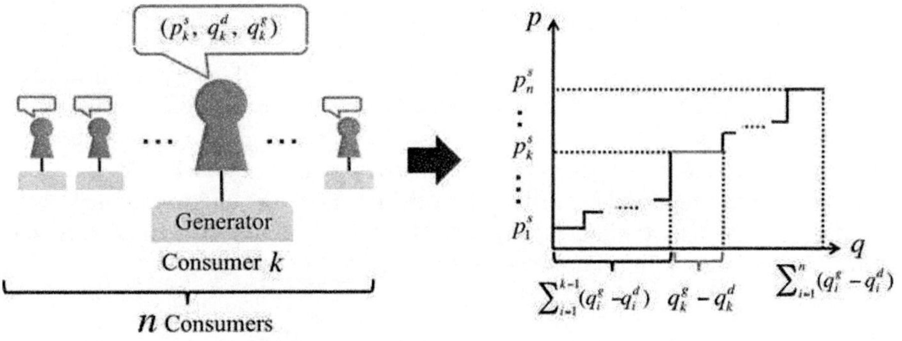

Fig. 11 How to make the aggregate supply curve under mechanism 2

Fig. 12 How to determine the trading price under mechanism 2

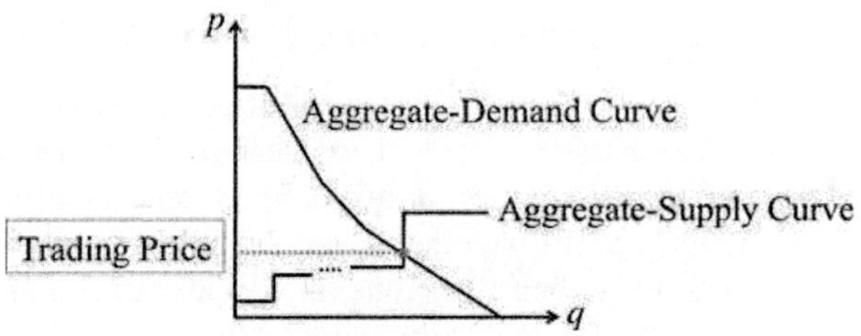

Table 1 Consumer parameters

	Type 1	Type 2	Type 3
Min-demand	10	20	30
Max-demand	100	100	100
Generated electricity output	120	80	60

3 Experiments with Human Subjects

We conducted experiments with human subjects to analyze how human beings make their decisions in the decentralized electricity trading under each of two mechanisms. Though perfect rationality is assumed in economic theory, human beings do not always make their decisions rationally, which should be considered when we evaluate how stable the mechanisms are. The experiments are based on the experimental economics methodology [6, 7]. Subjects were promised a monetary reward according to the payoff earned in experiments. The experiments were conducted with 54 subjects on December 3 and 14, 2011.

3.1 Experimental Settings

We fixed the number of consumers in the market as three. 27 subjects joined each day. Thereby, the subjects are divided into nine groups. Consumers of three types are assumed and are set to each subject respectively in a group. Each type has a different min-demand and generated electricity output, as Table 1 shows.

Subjects made their decisions based on their parameters and profits they were able to gain through trading electricity. They knew their own parameters and profits, but they could not know the others'.

3.2 Experimental Results

We were able to elicit models of decision-making by human subjects from the experiments. The model we elicited is the following.

- Subjects make their decisions based on decision change and profit change from the previous trade. The decision change shows whether consumers make values of input larger, smaller, or no change them; the profit change shows whether the consumer's profit becomes larger, smaller, or shows no change.
- Subjects make their decisions for the next trade based on a combination of their decision change and profit change from the previous trade. Whether they make the values of input larger, smaller, or do not change them for the next trade is determined stochastically according to a probability derived from the experiments.

Table 2 Consumer parameters

	A change in offer price from the previous trade	Next decision	Probability (%)
Became larger	Raised	Raise	17
		No change	33
		Lower	50
	Did not change	Raise	11
		No change	80
		Lower	9
	Lowered	Raise	25
		No change	54
		Lower	21
Unchanged	Raised	Raise	40
		No change	40
		Lower	20
	Did not change	Raise	5
		No change	72
		Lower	22
	Lowered	Raise	64
		No change	21
		Lower	14
Became smaller	Raised	Raise	8
		No change	8
		Lower	85
	Did not change	Raise	16
		No change	58
		Lower	27
	Lowed	Raise	75
		No change	15
		Lower	10

Tables 2 and 3 portray the probability elicited from the result of the experiments. These results are used in the next section as the decision-making model of agents.

4 Multi-Agent Simulation Considering Decision-Making by Humans

We conducted a multi-agent simulation in which agents' decisions were based on the model elicited in the experiments with human subjects in the former section. We evaluated the two mechanisms proposed in Sect. 2 in terms of stability of the grid system and social surplus.

Table 3 Model of decision-making by human beings under mechanism 2

A change in quantity of electricity to secure from the previous trade	Next decision on quality of electricity to secure	A change in offer price from the previous trade	Raised			Did not change			Lowered				
		Next decision on offer price	Raise (%)	No change (%)	Lower (%)	Raise (%)	No change (%)	Lower (%)	Raise (%)	No change (%)	Lower (%)		
Raised	Raise		0.0	0.0	14.3	0.0	23.8	0.0	10.5	5.3	0.0	Became larger	A change in profit from before last trade to last trade
	No change		0.0	14.3	57.1	4.8	47.6	0.0	10.5	36.8	10.5		
	Lower		0.0	14.3	0.0	0.0	14.3	9.5	15.8	10.5	0.0		
Did not change	Raise		0.0	0.0	0.0	0.0	4.9	2.4	4.2	0.0	4.2		
	No change		50.0	0.0	50.0	2.4	85.4	2.4	8.3	62.5	4.2		
	Lower		0.0	0.0	0.0	0.0	2.4	0.0	0.0	12.5	4.2		
Lowerd	Raise		0.0	0.0	33.3	0.0	20.0	0.0	6.7	6.7	6.7		
	No change		0.0	33.3	0.0	10.0	50.0	0.0	26.7	46.7	6.7		
	Lower		33.3	0.0	0.0	0.0	10.0	10.0	0.0	0.0	0.0		
Raised	Raise		11.1	11.1	11.1	0.0	33.3	0.0	0.0	0.0	0.0	Unchanged	
	No change		11.1	11.1	11.1	0.0	11.1	11.1	0.0	0.0	0.0		
	Lower		11.1	11.1	11.1	0.0	33.3	11.1	50.0	50.0	0.0		
Did not change	Raise		11.1	0.0	0.0	0.0	6.7	2.2	0.0	33.3	0.0		
	No change		11.1	44.4	33.3	7.8	72.2	3.3	33.3	16.7	0.0		
	Lower		0.0	0.0	0.0	0.0	7.8	0.0	16.7	0.0	0.0		
Lowerd	Raise		0.0	0.0	0.0	7.7	15.4	15.4	0.0	0.0.	0.0		
	No change		0.0	50.0	50.0	0.0	30.8	0.0	0.0	100.0	0.0		
	Lower		0.0	0.0	0.0	7.7	15.4	7.7	0.0	0.0	0.0		
Raised	Raise		8.3	8.3	0.0	8.3	16.7	0.0	0.0	0.0	50.0	Become smaller	
	No change		0.0	0.0	8.3	0.0	8.3	0.0	0.0	0.0	0.0		
	Lower		8.3	16.7	50.0	8.3	33.3	25.0	50.0	0.0	0.0		
Did not change	Raise		6.7	13.3	13.3	1.9	7.4	3.7	20.0	20.0	0.0		
	No change		0.0	13.3	40.0	3.7	68.5	9.3	20.0	20.0	0.0		
	Lower		0.0	0.0	13.3	0.0	5.6	0.0	0.0	20.0	0.0		
Lowerd	Raise		0.0	11.1	44.4	7.7	53.8	15.4	60.0	0.0	0.0		
	No change		0.0	0.0	33.3	0.0	23.1	0.0	0.0	0.0	20.0		
	Lower		0.0	11.1	0.0	0.0	0.0	0.0	0.0	0.0	20.0		

Table 4 Simulation results

	Mechanism 1	Mechanism 2
Variance of offer price	714.92	658.04
Variance of quantity of electricity to secure	–	280.53
Variance of trading price	30.76	19.88
Social surplus	604284.6	630106.3

4.1 Parameters

We set up the parameters used in the simulations as follows:

- The number of agents n is 100.
- Max-demand of each consumer is 100.
- Min-demand of each consumer is between 10 and 50 in intervals of 10.
- The generated electricity output of each consumer is between 10 and 200 at intervals of 10.
- The min-demands and electricity outputs are uniformly distributed, meaning that all consumers have a different set of min-demand and electricity output.

4.2 Simulation Results

Table 4 presents the simulation results. We use the variance of each consumer's decisions as an index of how stable the grid system under each of the mechanisms is. Variances are average values of 100 trials. Values of social surplus are the moving average values of the prior 200 steps in 100 trials. Each trial has 1000 steps.

As Table 4 shows, Mechanism 2 achieves smaller values of variance in both the Offer Price and Trading Price compared with Mechanism 1. The low variance indicates stability in the mechanism because it means consumers in the market do not change their values of input frequently. It can therefore be said that Mechanism 2 makes the electricity trading more stable than Mechanism 1 does. In addition, as Table 4 and Fig. 13 show, social surplus in the market under Mechanism 2 is larger. Moreover, it is apparent in Fig. 13 that the high social surplus is realized in early steps under Mechanism 2. We infer that consumers can obtain profits to some extent without fail because they are sure to consume some amount of electricity as the Quantity of Electricity to Secure under Mechanism 2. We conclude that Mechanism 2 is better than Mechanism 1 not only in terms of grid system stability but also in terms of social surplus.

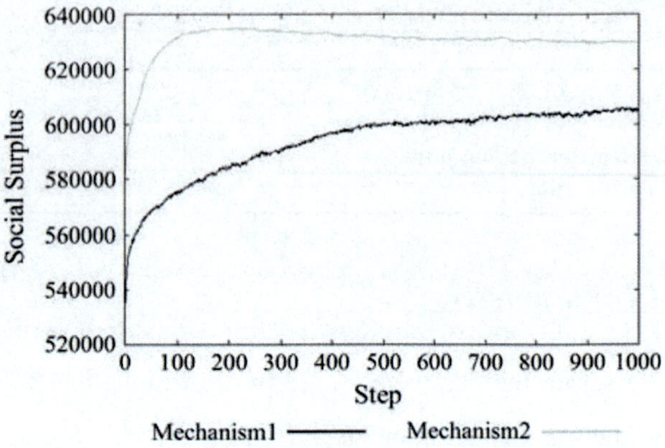

Fig. 13 Social Surplus transition of Mechanism 1 and mechanism 2

5 Conclusion

Electricity trade in which small-scale consumers such as households participate, which we call Decentralized Electricity Trading, is regarded as realizable in the near future. This paper proposes new trading mechanisms applied to decentralized electricity trading and is evaluated using an integrated approach with experiments using human subjects and multi-agent simulation. Results show that, when considering irrationality in decision-making by human beings, Mechanism 2, by which consumers secure the electricity they use beforehand and by which the rest is traded in the market, achieves a good result in terms of grid system stability and social surplus. Considering the fact that balancing demand and supply with robustness is desired to make large profit, which Makris et al. [8] mention in their research, we think the methods used in this research are also useful in manufacture.

References

1. Gross, R., Leach, M., Bauen, A.: Progress in renewable energy. Environ. Int. **1**(3), 105–122 (2003)
2. Abe, R., Taoka, H., McQuilkin, D.: Digital grid: communicative electrical grids of the future. IEEE Trans. Smart Grids **2**(2), 399–410 (2011)
3. Rudkevich, A., Duckworth, M., Rosen, R.: Modeling electricity pricing in a deregulated generation industry: the potential for oligopoly pricing in a Poolco. Energy J. **19**(3), 19–48 (1998)
4. Tanaka, M.: Oligopolistic competition in the Japanese Wholesale electricity market: a linear complementarity approach. In: RIETI Discussion Paper Series 07-E-023, pp. 1–32 (2007)
5. Vytelingum, P., Ramchurn, S.D., Voice, T.D., Rogers, A., Jennings, N.R.: Trading agents for the smart electricity grid. In: Proceedings of the Ninth International Conference on Autonomous Agents and Multiagent Systems, pp. 897–904 (2010)
6. Smith, V.L.: Experimental economics—Induced value theory. Am. Econ. Rev. **66**(2), 274–279 (1976)
7. Smith, V.L.: Microeconomic systems as an experimental science. Am. Econ. Rev. **72**(5), 923–955 (1982)
8. Makris, S., Zoupas, P., Chryssolouris, G.: Supply chain control logic for enabling adaptability under uncertainty. Int. J. Prod. Res. **49**(1), 121–137 (2011)

Decentralized Manufacturing Systems Review: Challenges and Outlook

Dimitris Mourtzis and Michalis Doukas

Abstract During the last three decades the economic landscape has abandoned its local characteristics and evolved into a global and highly competitive economy. The market demands towards high product variety, the low human labour costs in specific locations, the evolution of Information and Communication Technologies, and specific social and political forces are the principal reasons towards globalization. The main trend currently outlining the development of manufacturing paradigms is the ever-increasing tendency in the direction of decentralization of manufacturing functions towards decentralized entities. This has caused a fundamental reorganization process of the manufacturing organizations in order to cope with this trend. Several critical issues rise in the control and management of such organizations. These criticalities are further compounded by the need to achieve mass customization of industrial products, as this greatly complicates the manufacturing and supply activities. Moreover, the modalities for the configuration and implementation of each of the distributed manufacturing typologies are identified. The purpose of this paper is to specify the main trends, issues, and sensitive topics that characterize the behaviour and performance of these production systems. Based on this review, a discussion over existing production concepts is performed.

Keywords Decentralised manufacturing · Globalisation · Production concepts · Mass customisation

---

This contribution was previously published in Logistics Research (2012) pp. 113–121. DOI:10.1007/s12159-012-0085-x.

D. Mourtzis (✉) · M. Doukas
Department of Mechanical Engineering/Laboratory for Manufacturing Systems and Automation (LMS), University of Patras, Patras 26500, Greece
e-mail: mourtzis@lms.mech.upatras.gr

K. Windt (ed.), *Robust Manufacturing Control*, Lecture Notes in Production Engineering, DOI: 10.1007/978-3-642-30749-2_26, © Springer-Verlag Berlin Heidelberg 2013

1 Introduction: Globalization and Economical Facts

The economic landscape has drastically altered during the last three decades. The local economy has evolved into a global and highly competitive economy. Industries started to operate on a global basis, expanding the limits of their business. The export of finished goods to foreign markets has been the dominating theme in the international trade up to the 1990s, and gained even more attention the last decade. Moreover, location-specific factors such as low-cost labour and highly-skilled personnel in specific locations enabled the globalisation. Enterprises started to seek for fertile production environments into developing or developed countries [1]. A number of developments have fuelled the effectiveness of global production (Fig. 1). The advent of the Internet and the increasing computational power enabled globalization [2]. The widespread application of Information and Communications Technology (ICT) in the 1970s boosted the development of cooperative and collaborative structures [3, 4].

The transportation costs for intercontinental transports also keep dropping significantly. This allows manufacturers to distribute their products at dispersed production sites and markets in massive volumes. The amount of freight traffic kilometres presents a high annual growth rate and is envisioned to triple in the next 20 years [8]. The world Gross Domestic Product (GDP) grew at a Compound Annual Growth Rate (CAGR) of 5.1 % over the years 1990–2006. International trade has outmatched this trend with a CAGR of 7 %. The strong growth in trade volume further increased over the last 10 years. While for 1950–1992 the trade volume grew 1.5 times faster than the GDP, this ratio increased to 2.6 for the time period 1992–2008 [9]. The growth in trade volumes indicates that trade intensive production setups are of increasing importance for companies. In recent years, a greater share of companies source parts and components abroad, or re-import finished goods from their manufacturing plants in other countries. In such setups, production equipment and other capital goods are exported to the country of the manufacturing site. Thereby, the trade volume increases substantially compared to the trade paradigm dominant prior to the 1990s.

The manufacturing systems, in order to compensate with these rapid developments, are continuously evolving, leading to the future paradigm. Future manufacturing will be characterised by increased automation, high flexibility and modularity, focusing on seamless interoperability and environmental friendliness (Fig. 2).

2 Evolution of Manufacturing Paradigms

Manufacturing is the key driving force of the European economy. In 2010, 34 million people were employed in the EU-27 manufacturing sector, representing 15.9 % of the total employment. Indirectly (with related sectors and activities), manufacturing accounts for close to 50 % of the European economy [10, 11].

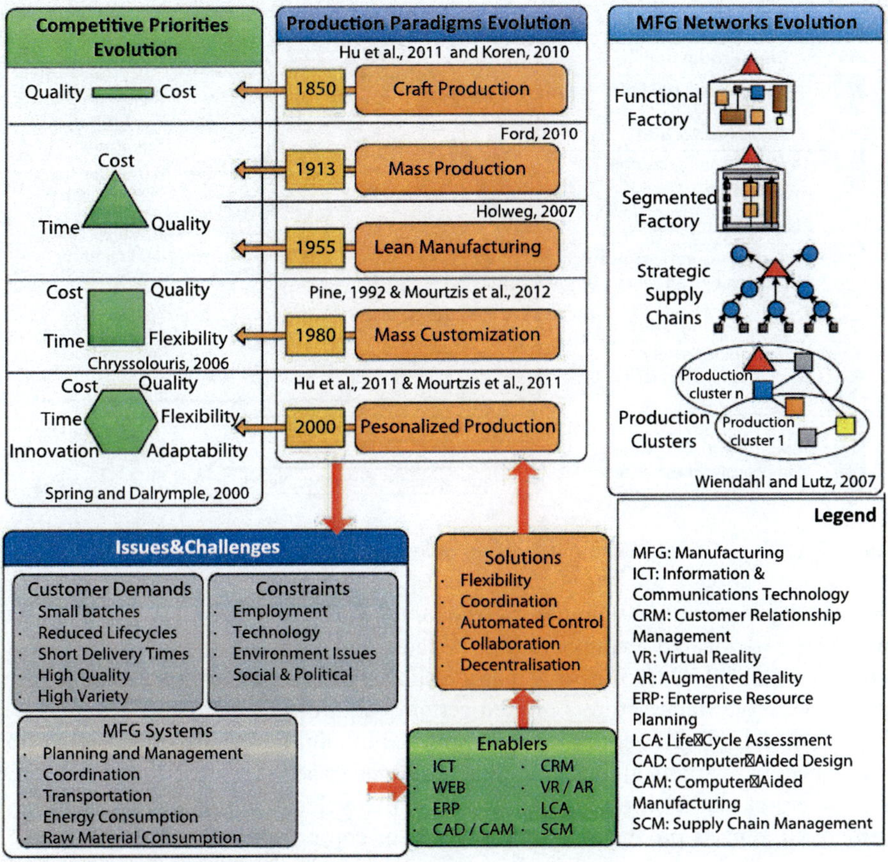

Fig. 1 Evolution of production paradigms, manufacturing networks and competitive priorities, issues generated and solutions [2, 5–7, 12–15, 20]

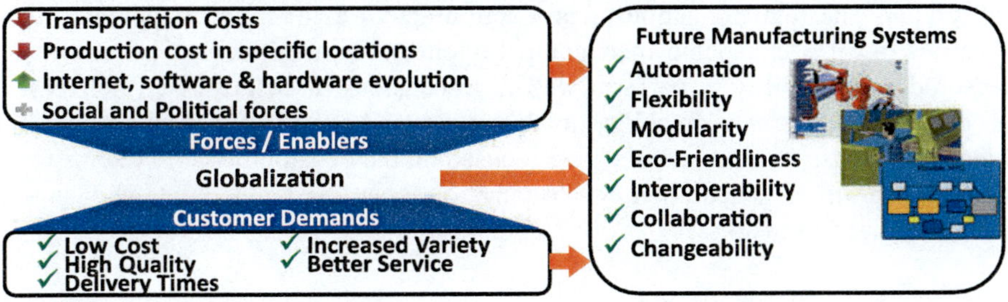

Fig. 2 Forces and enablers towards the manufacturing systems of the future

Since its birth two centuries ago, manufacturing has evolved through several paradigms, addressing the needs of market and society (Fig. 3). The first paradigm was "Craft Production" that focused on creating exactly the product that the customer requests [12, 13]. In the 1910s, "Mass Production" allowed low-cost

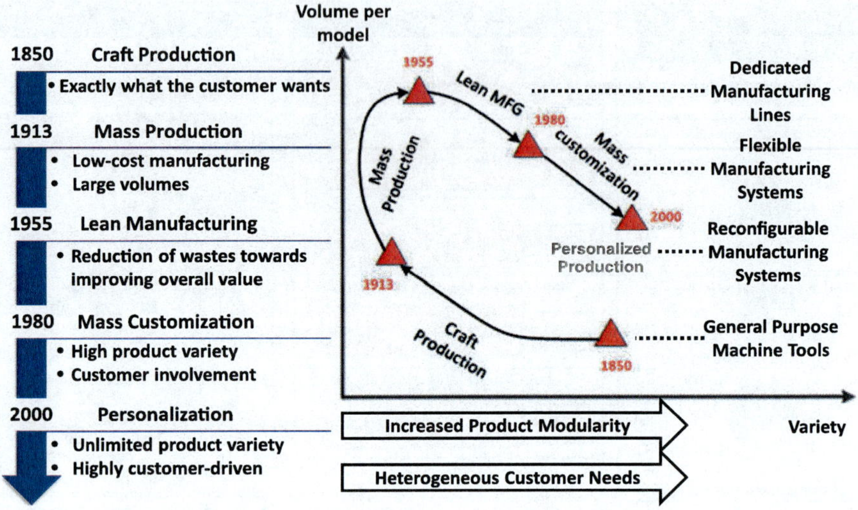

Fig. 3 Evolution of manufacturing paradigms and market needs (Adapted from [12, 13])

manufacturing of large volumes of products with limited variety, which was enabled by dedicated manufacturing systems [14]. In the late 1980s, "Mass Customization" (MC) [15] emerged as a response to consumer demands for higher product variety. Manufacturers offered certain variations of their standard product [12]. Nowadays, high product variety is offered by quite every industrial sector to heterogeneous markets around the globe, via web-based means [16].

In an era of market segmentation and short lifecycles, traditional manufacturing methods, like mass production, are incapable of coping with market demand, due to their rigidity and low responsiveness. They are being replaced by the MC paradigm. Decentralised manufacturing approaches replaced traditional centralized practices, showing their benefit in delivery times, transportation costs and agility [17]. The regionalization of production activities offers great potential to industries towards enhancing their competitiveness. Competitiveness is currently measured by the ability to perform well in dimensions of cost, quality, delivery, speed, innovation and adaptability to demand variations [18]. To achieve such objectives, industry and academia have focused on the development of systems for control, monitoring, scheduling, synchronization, coordination and data-exchange in decentralized networks [19].

3 Evolution of Cooperation Structures in Manufacturing

Cooperation amongst industries existed in the past, when companies operated in relatively stable market environment where reasonable forecasts were possible and adequately accurate. Inside that "deterministic" environment, optimisation was primarily focused on internal processes and manufacturing improvement [20]. However, the architectures of these information systems were fairly rigid. Production concepts that enable faster adaptation to changing market needs were developed over time. Cooperative structures with increased focus on flexibility started forming. Flexibility can be achieved internally through re-organization of structures and processes. To increase flexibility further, companies had to extend their sphere of influence to other companies, so that flexibility could be accomplished externally [21]. A differentiation between intra-firm and inter-firm production concepts with respect to the amount of flexibility becomes reasonable as presented in [22]. In addition, the choice of the building blocks of a cooperation network, the supply chain partners, is based on the analysis of their core competencies and their coherence with the network's needs and strategy. They are evaluated based on their uniqueness in the market, and their ability to provide a variety of products and services towards the satisfaction of the customer demands [20]. According to a study [23], company managers perceive "quality" as the most important attribute for a supplier. However, the same sample of managers assigned more weight to "cost" and "on-time delivery" attributes when actually choosing a supplier. Furthermore, due to increased legal and public environmental conscience, companies integrate environmental criteria into the supplier selection process. Humphreys et al. [24] proposed a framework of quantitative and qualitative environmental criteria that a company can consider during the supplier selection.

3.1 Supply Chain Management (SCM)

In the 1990s a fundamental transformation took place on the strategic level of the manufacturing domain [20]. Increased complexity led companies to the decision whether to produce or outsource, concentrating only on high added value procedures. Following that, companies started to outsource entire components and modules. The formation of strong bonds between the stakeholders took place, and the networks were linked by logistic companies. This resulted in the establishment of Supply Chains (SC). Inside a SC companies cooperate with suppliers over various tiers in order to improve business performance, by reducing the number of self-manufactured components and by substituting them by components from external partners [25]. Such SC networks are commonly led by one central organization, mostly the end-product manufacturer. The company's goal is to add value to its products as they pass through the SC [26], and integrate and coordinate the operational activities with decisions and actions of their external business partners [27]. Moreover, SCs allow the transportation of products to geographically dispersed markets in the correct quantities, with the correct specifications, at the correct time, and at a competitive cost [18]. Enterprise Resource Planning (ERP) systems were developed over time (Fig. 4) for management of internal and external resources, and for facilitating information flow.

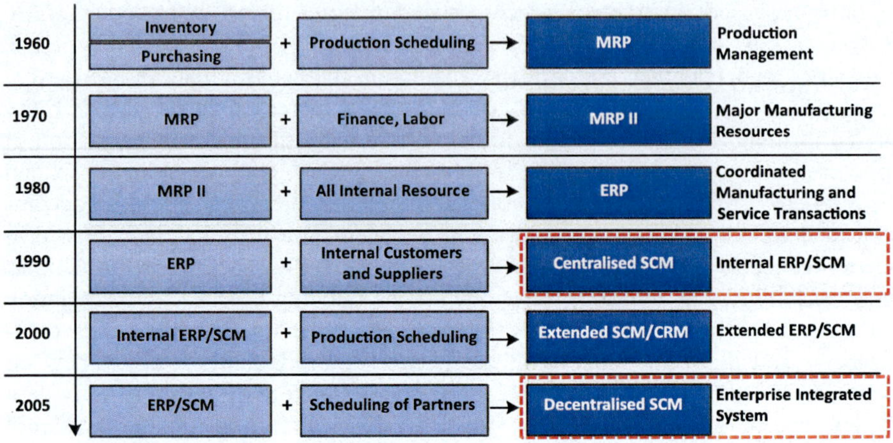

Fig. 4 Evolution of supply chain management and ERP (Adapted from [31])

3.1.1 Centralised Supply Chain

A centralized supply chain is regarded as one entity that aims at optimizing the system performance [26]. In the first half of the 20th century, large manufacturers began to study the emerging global supermarkets. They aimed at enhancing their production planning in terms of storing and self-stocking technics. Toyota in the framework of the Toyota Production System (TPS) developed the Kanban subsystem. Kanban aimed at controlling inventory levels, production and supply of components and in some cases even raw materials [28]. Many systems that follow the Kanban logic have been proposed, depicting the difficulties in applying the Kanban logic in real productive systems or on highly complex and decentralised systems [29]. Different production environments require different SC coordination mechanisms [30].

3.1.2 Decentralised Supply Chain

A decentralized supply chain differs from a centralized as the entities that comprise the network act independently in order to optimize their individual performance. Although firms throughout the globe realize that the collaboration with their supply chain partners can improve their profits, the decentralization of inventory and decision-making is often unrealistic [26, 30]. The need is presented to not only to coordinate the activities of the independent partners but also to align their objectives in order to achieve a common goal. For reducing purchase costs and attract a larger base of customers, retailers and OEMs are constantly seeking suppliers with lower prices. These suppliers however, may be located at long distances from the OEM sites and retailer distribution centres and stores [32]. A wider integration of the logistics into the supply chain is required [28], in order

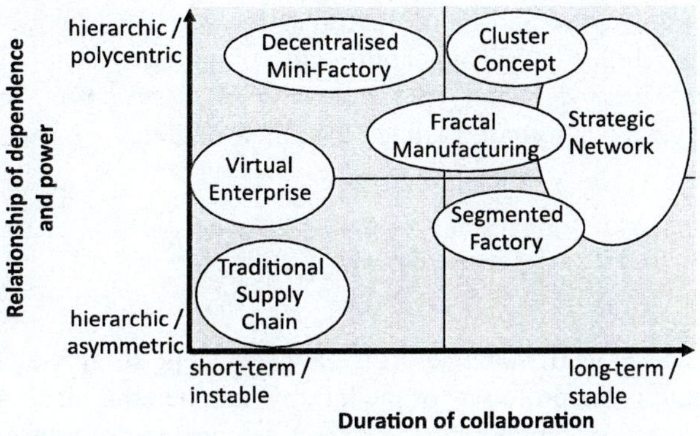

Fig. 5 Categorization of the different production and cooperation concepts (Adapted from [20])

to enhance the performance of the collaborative production network. Globalized transportations entail some risks. A significant proportion of the shipped products are susceptible to defects due to missing parts, misplaced products, or mistakes in orders and shipments [31]. Risk assessment and evaluation models have been proposed, such as chance constrained programming (CCP), data envelopment analysis (DEA), and multi-objective programming (MOP) [33].

4 Decentralized Manufacturing Concepts and Networks

Modularisation is a fundamental organizational principle for a successfully operating globalised and decentralised entity [34]. It involves the reforming of the organizational structure into small, manageable units (modules) on the basis of integrated, customer-oriented processes. These units have a decentralized decision-making authority and the responsibility for results. The organisational structures presented below are characterised by a high degree of modularity and non-hierarchic relationships (Fig. 5).

4.1 Segmented Manufacturing

The Segmented Factory (SF) is modularly organized in small, flexible, and decentralized structures that are self-responsible as well as market and human oriented. This organizational structure leads to a reduction of interfaces. Thus, coordination complexity and costs can be reduced, which is extremely important in decentralized organisations and for manufacturing modular products. Inside the company the segments pursue different competitive strategies. They may also act as customers and manufacturers towards the other segments, which results in very efficient final outcomes [35]. The distance between operational and strategic functions is small, so that information flow is frictionless. The modular architecture and the decentralisation of a SF provide the necessary structures for flexibility and changeability. In comparison to the Fractal Factory however, the Segmented Factory has relatively fixed structures because process stability and specialization are realized to a high degree.

4.2 Fractal Manufacturing

The Fractal Manufacturing (FM) concept comprises units, the so called fractals, and is the prototype of the internal and heterarchical organization [36]. A fractal is an autonomous unit, the objectives and performance of which can be described unambiguously. The FM is for many the European answer to lean production [37]. It has been practiced in many businesses and proven to be very successful [38]. The fractals are characterized by self-similarity, self-organization and self-optimization features. The constitution of fractals could be interpreted according to the systems theory, in a way that the interior relations are stronger than the exterior relations (flow of material, resources and information). In case of environment changes, the fractals adjust accordingly. They must fulfil the principle of vitality that is basically determined by their lifecycle: conception, realization, maturation, optimization, deterioration. Insufficient vitality results in stagnating or decreasing revenues and competitiveness. Therefore, a fundamental challenge is to constantly adjust to the exterior requirements [39]. The operative self-organization guarantees fast and flexible reactions and adjustments to changing customer orders. The tactical and strategic self-organization enables the fractals to adjust independently, and to cope with highly personalised orders.

The vitality and self-optimization characteristics significantly support the capability of changeability. Fractals can grow and shrink, so that the requirement of scalability and changeability is fulfilled. Additionally, they can separate, dissolute, and restructure, because they are build-up of smaller fractals that can be grouped differently, fulfilling thus the requirements of modularity. The fact that there are functional fractals supporting the others can be added as a further advantage, because in this way, every fractal can concentrate on its core competencies. Functional fractals focus on their supporting functions and producing fractals focus on producing. An additional strength is the strong communication and interaction network between the fractals. The weakness of the FM is the high coordinative complexity. There is no centralized strategic leadership, and consequently the fractals have to harmonize their objectives continuously. Bionic and Holonic are similar to FM concepts, but are differentiated by the biological or mathematical analogies that they draw characteristics from [40].

4.3 Decentralized Mini-factories

The concept of Decentralized Mini-factories (DMF) has specially been developed to support MC. It is supposed to bridge the gap between centralized production, decentralized distribution and customer contact. A DMF is a scalable, modular, and geographically distributed unit, located in proximity to the customer and connected to other DMFs. The DMF is able to perform distribution, maintenance and repair service as well as additional services [41]. A central supporting unit for all DMF assures the support with standard components, fundamental product developments and training for the employees. This central unit however, has no decision competencies. DMFs can be interpreted as a heterarchic inter-firm network. All the DMFs are independent even though they are all part of one company. The interaction and interdependency among the DMFs is extremely low compared to all the other networks.

An advantage that a DMF offers is the facilitation of acquisition of customer information. DMFs foster a better access to "sticky information", through direct interaction with the customer, during the product specification [42]. The proximity to the customer can tie the customer closer to the factory and make a repurchase more probable (economies of relationship) [43]. Economies of relationship describe the potential of cost reduction on basis of customer loyalty. Because of the small market it addresses, the complexity of a DMF is low and manageable. Another advantage lies in the low initial investment. A DMF can gradually adjust to the market requirements through scaling [44]. The internal organization however, is not determined at all. Therefore, the potential of changeability cannot be generally assessed and it highly depends highly on the internal organization of the DMFs. An application of the DMF concept can be found in South Africa at the Automotive Supplier Park [45].

4.4 The Strategic Network

Strategic Networks (SNs) are described by Jarillo as "long-term, purposeful arrangements among distinct but related for-profit organizations, that allows those firms to gain or sustain competitive advantage vis-à-vis over their competitors outside the network" [46]. The efficiency of a SN can be explained by the help of the transaction cost theory. The network is economically efficient if the costs of the extern partners plus transaction costs are lower than the costs of intern production. A necessary prerequisite for this collaboration is a high degree of trust. Such networks are especially profitable for young enterprises without many resources at their disposal. In sectors with high demand for changeability, flexibility and global competition, the SNs can strengthen competitiveness and help share the risk [47]. The "strength" of a SN is the focal leader; usually the end-product manufacturer. In case the leader or hub firm is missing, the network is called "regional network".

Typically, such networks are composed of small and medium enterprises that are often located close to each other.

The key factor of success for a SN is modularization; the modules and components have to be assembled finally to the end product. The end product manufacturer determines the optimal number of component suppliers, distributes the orders and coordinates the partners. These formal coordination mechanisms and the contractual ties limit the danger of opportunistic behaviour. The long-term cooperation of a SN provides a necessary stable production environment for supporting MC. MC aims for a mass market and longer lifecycles of a basic product design. In this production environment, the participating firms can develop their core competencies and specialize over time on the required market niches. Thus, a highly efficient in terms of scope economies network, develops. The danger of this mutual reliance is however, the strong dependencies of the firms. The firms may rely on the orders of the focal firm and do not interact directly. For this reason and due to the fact that the network is designed for long term cooperation, flexibility is decreased. MC however, requires a certain degree of flexibility, supporting the efficient standardization and stability.

4.5 The Virtual Enterprise

Virtual Enterprises (VEs) were developed in the 2000 s and presented a new approach to the sharing of tasks between the collaborators inside the supply chain. According to a broadly accepted definition [48], a VE is not a single corporation but a network of many corporations that are perceived by customers as one entity. VEs are set up in order to carry out a single project and after that the bonds between the partners are broken [49, 50]. The VE is extremely flexible and adjustable because its composition can be restructured very fast according to the requirements of a specific order. The problem for MC is however, that VEs are designed for small market niches. It is optimal to quickly exploit chances that occur for a short time. MC however targets mass markets and not niche markets. VEs are appropriate for fundamentally different orders. The broad range of firms with very different competencies is at disposal to realize any upcoming order, in order to serve personalisation. In order to achieve this, collaborating enterprises try to utilise the capacities and competencies of their partners, as there is no constant end-product manufacturer. This is important for the development of a long term learning relationship and economies of relationship. Therefore, it is possible to conclude life-cycle contracts for certain products. However, this does not equal the potential for establishing stable and standardized processes in a long-term cooperation for several products and product generations. The short-term collaboration and the location-independent cooperation provide high incentives for opportunistic behaviour. At the same time, the need for trust is very high, because there is no central coordination which is a problematic contradiction. A main characteristic of VEs is the mutual use of inter-organisational information systems. Typical examples of VE application are low-tech products with very short life-cycles like textile and fashion retailing industries [51] and the construction sector [52].

4.6 The Cluster Concept

The Cluster Concept (CC) is basically an extension of the VEs. Inside a cluster we find a heterarchical network of OEMs, end users, suppliers and information, machinery, resources and materials that are needed for the operation of such a network. The difference between the CC and other production networks is that the different stakeholders may use the same infrastructure, share identical customers and/or skills bases. Moreover, clusters can include research institutes and the government. Similarly to regional networks, there are regional agglomerations of companies, mostly specialized on one business sector. They provide advantages for both the region and the participating network. Inside the cluster, different (regional networks) can be involved, i.e. the cluster itself is composed of smaller clusters. Typically such a cluster includes large parts of the value chain and is vertically integrated. Clusters often can be found within automobile manufacturing. A prominent example is the motorsports cluster around Oxford in south England, with approximately 200 highly specialized small and medium companies [53]. Other applications of the CC can be found in Canadian maritime industry [54] and Scottish electronics sector [20, 55].

5 Discussion and Conclusions

The review of the existing production concepts is based on a set of Key Performance Indicators (KPIs). The selection of the KPIs is defined on a strategic level. The conclusions support the selection of the most suitable production network structure for the realisation of decentralised production that serves a MC model.

Chryssolouris [2] states that “in general, there are four classes of manufacturing attributes to be considered when making manufacturing decisions: cost, time, quality and flexibility”. However, it is not possible to simultaneously optimize all of them because they partially contradict each other. It is rather important to find the optimal trade-off between all of them. For MC, the attributes of time and cost are emphasized. Quality became more important for the German and Japanese manufacturing. Moreover, “flexibility will become a major competitive weapon for the manufacturing industry” [2]. Decentralisation of production offers many advantages towards supporting today’s turbulent MC and personalisation environment. MC is nevertheless a promising strategy providing many opportunities and chances. However, at the same time its realization is highly demanding. Mass production is so successful, because it can significantly reduce complexity. Complexity is extremely high in MC due to the many variations that disturb the

Table 1 Applicability of concepts regarding the KPIs, for decentralised mass customisation

KPI	Strong applicability	Medium applicability	Weak applicability
Complexity & modularization [2, 34]	Segmented factory, strategic network	Fractal factory, virtual corporation	Mini-factories
Interaction [56]	Mini-factories, fractal factory, strategic network	Virtual corporation, segmented factory	
Economies of scale, scope & integration [57]	Strategic network, segmented factory	Mini-factories, virtual corporation, fractal factory	
Changeability [19]	Mini-factories, virtual corporation, fractal factory	Strategic network, segmented factory	

smooth function of the manufacturing systems. Therefore, it is a crucial requirement to master variety and to reduce complexity in order to lower costs and increase flexibility. A key element in complexity reduction is modularization and decentralization of decision making. For the coordination of these decentralized units it is important to build upon a system of intensive interaction [56]. Communication and the exchange of information and knowledge are highly valuated in decentralised organisations. In this context, customer proximity is crucial for all customized work because it starts with the customer (customised order) and ends with the customer (delivery and after-sales services). For MC it is an essential KPI not to only make use of Economies of Scale, but also to exploit the potentials of Economies of Scope and Integration in order to improve the performance in the dimension of costs [57]. Finally, there is the requirement that changeability; flexibility and the responsiveness, have to be maintained [19]. Apart from that, time-to-market (quick responsiveness) and customization are relevant for customer friendly customization. These concepts support the attributes of flexibility and time. In Table 1, the level of applicability of the examined production concepts for the defined KPIs is summarized.

The strengths-weaknesses analysis reveals that the examined concepts are generally applicable for the decentralised production of MC products from a strategic point of view. This is attributed to the fact that they were developed against the background of new challenges in manufacturing similar to the challenges of MC. All of the concepts exhibit a relatively high degree of decentralisation. Another conclusion that can be drawn is the fact that a certain degree of hierarchy is also beneficial for MC. The least hierarchic and most flexible network, the VE also shows the most disadvantages of all concepts. Similarly, the weaknesses of the Fractal Factory can be weighted stronger than the weaknesses of the Segmented Factory. The reason is that MC does not need to be extremely flexible, because the operative flexibility and individualization is confined to a limited space of specification [58]. Furthermore, hierarchic structures decrease flexibility but help to increase the degree of standardization of processes and interfaces.

This is very important to guarantee for Economies of Scale so that production costs can be lowered. The long-term development of a learning relationship between the end-product manufacturer and the consumers, contributes further to the importance of a strategic leadership.

The currently imposed environmental regulations consist of further constraints towards the realisation of decentralised manufacturing. The carbon emissions have to be kept under control. Moreover, the digitalisation of manufacturing is an enabler for the transition from labour-intensive setups towards knowledge-based and automated manufacturing structures. The traditional structure of industrial practice is based on capital and labour; it is evident that the future needs call for structures based on knowledge and capital [59].

Acknowledgments This work has been supported by the EC funded project "A Web-based Collaboration System for Mass Customization – e-CUSTOM" (260067).

References

1. Abele, E., Elzenheimer, J., Liebeck, T., Meyer, T.: Reconfigurable Manufacturing Systems and Formable Factories—Globalization and Decentralization of Manufacturing, 1st edn, pp. 4–5. Springer, NY (2006)
2. Chryssolouris, G.: Manufacturing systems: Theory and Practice, 2nd edn. Springer Verlag, New York (2005)
3. Lazonick, W.: Business Organisation and Competitive Advantage: Capitalist Transformations in the Twentieth Century, Technological Enterprise in A Historical Perspective, pp. 119–163. Oxford University Press, Oxford (1991)
4. Grazia, D., Santangelo, G.D.: The impact of the information and communications technology revolution on the internationalisation of corporate technology. Int. Bus. Rev. **10**(6), 701–726 (2001)
5. Mourtzis, D., Doukas, M., Michalos, G., Psarommatis, F.: A web-based platform for distributed mass product customization: conceptual design. DET 2011. 7th International Conference on Digital Enterprise Technology, pp. 604–613. Athens, Greece, (2011)
6. Holweg, M.: The genealogy of lean production. J. Oper. Manage. **25**(2), 420–437 (2007)
7. Spring, M., Dalrymple, J.: Product customisation and manufacturing strategy. Int. J. Oper. Prod. Manage. **20**(4), 441–467 (2000)
8. Airbus Global Market Forecast 2011–2030: URL: http://www.airbus.com/company/market/forecast/cargo-aircraft-market-forecast/, (2011)
9. Mongobay, using EIA data: URL http://rainforests.mongabay.com/09-carbon_emissions.htm, (2009)
10. Eurostat High-tech statistics: URL http://epp.eurostat.ec.europa.eu/statistics_explained/index.php/High-tech_statistics
11. Manufuture-EU: URL: www.manufuture.org, (2011)
12. Hu, S.J., Ko, J., Weyand, L., El Maraghy, H.A., Kien, T.K., Koren, Y., Bley, H., Chryssolouris, G., Nasr, N., Shpitalni, M.: Assembly system design and operations for product variety. CIRP Ann. Manuf. Technol. **60**(2), 715–733 (2011)
13. Koren, Y.: The global manufacturing evolution, 1st edn. Wiley, NY (2010)
14. Ford, H.: My Life and Work—An Autobiography of Henry Ford. Greenbook Publications, LLC., Libiya (2010)

15. Pine, B.J.: Mass Customization: The New Frontier in Business Competition. Harvard Business School Press, Boston (1992)
16. Thirumalai, S., Sinha, K.: Customization of the online purchase process in electronic retailing and customer satisfaction: an online field study. J. Oper. Manage. **29**(5), 477–487 (2011)
17. Leitao, P.: Agent-based distributed manufacturing control: a state-of-the-art survey. Eng. Appl. Artif. Intell. **22**(7), 979–991 (2009)
18. Mourtzis, D., Papakostas, N., Makris, S., Xantakis, V., Chryssolouris, G.: Supply chain modelling and control for producing highly customized products. CIRP Ann. Manuf. Technol. **57**(1), 451–454 (2008)
19. Wiendahl, H.P., El Maraghy, H.A., Nyhuis, P., Zäh, M.F., Wiendahl, H.H., Duffie, N., Brieke, M.: Changeable manufacturing—classification, design and operation. CIRP Ann. Manuf. Technol. **56**(2), 783–809 (2007)
20. Wiendahl, H.P., Lutz, S.: Production in networks. CIRP Ann. Manuf. Technol. **51**(2), 573–586 (2002)
21. Takeishi, A.: Bridging inter and intra-firm boundaries: management of supplier involvement in automobile product development. Strateg. Manag. J. **22**(5), 403–433 (2001)
22. Hilpert, U.: Regional Innovation and Decentralization: High-Tech Industry and Government Policy. Routledge, London (1991)
23. Verma, R., Pullman, M.E.: An analysis of the supplier selection process. Omega **26**(6), 739–750 (1998)
24. Humphreys, P.K., Wong, Y.K., Chan, F.T.S.: Integrating environmental criteria into the supplier selection process. J. Mater. Process. Technol. **138**(1–3), 349–356 (2003)
25. Shapiro, J.F.: Modelling the supply chain, 2nd edn. Duxbury Press, Pacific Grove (2001)
26. Frayret, J.M., D'Amours, S., Montreuil, B., Cloutier, L.: A network approach to operate agile manufacturing systems. Int. J. Prod. Econ. **74**(1–3), 239–259 (2001)
27. Chase, R.B.: Production and Operations Management: Manufacturing and Services, 8th edn. McGraw-Hill, San Francisco (1998)
28. Jayaram, J., Das, A., Nicolae, M.: Looking beyond the obvious: unravelling the Toyota production system. Int. J. Prod. Econ. **128**(1), 280–291 (2010)
29. Junior, M.L., Filho, M.G.: Variations of the Kanban system: literature review and classification. Int. J. Prod. Econ. **125**(1), 13–21 (2010)
30. Mourtzis, D.: Internet based collaboration in the manufacturing supply chain. CIRP J. Manuf. Sci. Technol. **4**(3), 296–304 (2011)
31. Jacobs, F.R., Weston, F.C.: Enterprise resource planning (ERP)—a brief history. J. Oper. Manage. **25**(2), 357–363 (2007)
32. Kim, H., Lu, J.C., Kvam, P.H., Tsao, Y.C.: Ordering quantity decisions considering uncertainty in supply-chain logistics operations. Int. J. Prod. Econ. **134**(1), 16–27 (2011)
33. Wu, D., Olson, D.: Supply chain risk, simulation and vendor selection. Int. J. Prod. Econ. **114**(2), 646–655 (2008)
34. Yoo, J., Kumara, S.R.T.: Implications of k-best modular product design solutions to global manufacturing. CIRP Ann. Manuf. Technol. **59**(1), 481–484 (2010)
35. Wiendahl, H.P., Scholtissek, P.: Management and control of complexity in manufacturing. CIRP Ann. Manuf.Technol. **43**(2), 533–540 (1994)
36. Koestler, A.: The Ghost in The Machine, 1st edn. Macmillan, USA (1967)
37. Sihn, W.: Fractal businesses in an e-business world. In: 8th International Conference on Concurrent Enterprising. Rome, Italy, (2002)
38. Warnecke, H.J.: The fractal company: a revolution in corporate culture. Springer Verlag, New York (1993)
39. Warnecke, H.J., Huser, M.: Lean production. Int. J. Prod. Econ. **41**(1–3), 37–43 (1995)
40. Tharumarajah, A.: Comparison of the bionic, fractal and holonic manufacturing system concepts. Int. J. Comput. Integr. Manuf. **9**(3), 217–222 (1996)
41. Reichwald, R., Stotko, C., Piller, F.: Distributed Mini-Factory Networks as a Form of Real-Time Enterprise: Concept, Flexibility Potential and Case Studies. Springer, New York (2005)

42. Hippel, R.: Sticky information and the locus of problem solving: Implications for innovation. Manage. Sci. **40**(4), 429–439 (1994)
43. Zaeh, M.F., Wagner, W.: Factory planning modules for knowledge sharing among different locations. IFIP Int. Fed. Inf. Process. **160**, 239–251 (2005)
44. Reichwald, R., Piller, F., Jaeger, S., Zanner, S.: The customer centric enterprise: advances in mass customization and personalization, 1st edn, pp. 51–69. Springer, Berlin (2003)
45. Automotive Supplier Park: URL http://www.supplierpark.co.za/AboutLocation.aspx, (2007)
46. Jarillo, J.C.: On strategic networks. Strateg. Manag. J. **9**(1), 31–41 (1998)
47. Blecker, T., Neumann, R.: Knowledge Management and Virtual Organizations, 1st edn. Idea Group Publishing, Hershey (1999)
48. Davidow, W.H., Malone, M.S.: The virtual corporation: Structuring and revitalizing the corporation for the 21st century, 1st edn. Harper Collins Publishers, New York (1992)
49. Martinez, M., Fouletier, P., Park, K., Favrel, J.: Virtual enterprise organisation, evolution and control. Int. J. Prod. Econ. **74**(1–3), 225–238 (2001)
50. Browne, J., Zhang, J.: Extended and virtual enterprise—similarities and differences. Int. J. Agile Manage. Syst. **1**(1), 30–36 (1999)
51. Wang, W.Y.C., Chan, H.K.: Virtual organization for supply chain integration: two cases in the textile and fashion retailing industry. Int. J. Prod. Econ. **127**(2), 333–342 (2010)
52. Rezgui, Y.: Role-based service-oriented implementation of a virtual enterprise: a case study in the construction sector. Comput. Ind. **58**(1), 74–86 (2007)
53. Saward, J.: New owners in Formula 1, URL: http://joesaward.wordpress.com/2011/10/03/new-owners-in-formula-1/, (2011)
54. Doloreux, D., Shearmur, R.: Maritime clusters in diverse regional contexts: The case of Canada. Mar. Policy **33**(3), 520–527 (2009)
55. Cieminski, G., Carrie, A.: Production planning and control and exchange of information in the Scottish electronics industry. In: Proceedings of IFIP WG 5.7 Working Conference, pp. 113–122. Tromo. Norway, (2000)
56. Wildemann, L.: Alliances and networks: The next generation. Int. J. Technol. Manage. **15**(1–2), 96–108 (1998)
57. Piller, F.T., Möslein, K.M.: From economies of scale towards economies of customer interaction: Value creation in mass customization based electronic commerce. In: 15th Bled Electronic Commerce Conference, pp. 214–228. TU München, (2002)
58. Fogliatto, F.S., da Silveira, G.J.C., Borenstein, D.: The mass customization decade: an updated review of the literature. Int. J. Prod. Econ. **138**(1), 14–25 (2012)
59. MANUFUTURE, A Vision for 2020, URL: http://www.fp7.org.tr/tubitak_content_files/270/ETP/ManuFuture/ManufutureVision2020.pdf (2011)

Environmental Impact of Centralised and Decentralised Production Networks in the Era of Personalisation

Dimitris Mourtzis, Michalis Doukas and Foivos Psarommatis

Abstract The current trend of globalisation and decentralisation of the production activities has created a series of environment related issues. The increase of transportation distances, the escalated consumption of natural resources, and toxic emissions are among the generated challenges. Additionally, the manufacturing complexity, due to high product variety leads to increased energy consumption. Nevertheless, natural resources are limited and emission levels must be kept under the limits. This paper presents a methodology, implemented through a software tool, for the investigation of the environmental impact caused by centralised and decentralised manufacturing networks, under heavy product customisation. Simulation models of automotive manufacturing networks were developed, utilising real life industrial data, for the investigation of the impact of the production networks under highly diversified product demand, on environmental aspects. Multiple user-defined criteria have been used for the evaluation of the environmental footprint, including CO_2 emissions and energy requirements in terms of depletion of natural resources. This paper aims at identifying optimal configurations of centralised and decentralised production networks, characterised by reduced energy requirements, low consumption of natural resources and reduced toxic emissions.

Keywords Decentralised manufacturing · Environmental impact · CO_2

D. Mourtzis (✉) · M. Doukas · F. Psarommatis
Department of Mechanical Engineering/Laboratory for Manufacturing Systems and Automation (LMS), University of Patras, 265 00, Patras, Greece
e-mail: mourtzis@lms.mech.upatras.gr

K. Windt (ed.), *Robust Manufacturing Control*, Lecture Notes in Production Engineering, 371
DOI: 10.1007/978-3-642-30749-2_27,

1 Introduction

The continuously increasing customer demands towards higher product variety and uniqueness has multiplied the complexity of designing and planning optimal production networks to serve the market's needs [1]. Towards trying to compete, industries throughout the world have increased their outsourcing activities, aiming at reduced costs and times [2]. Adding to that, the globalisation of manufacturing activities entails increased transportation distances, very short delivery times and high quality standards, factors that create further disturbances in the operation of manufacturing systems [3, 4]. Moreover, strict environmental regulations comprise an additional constraint to the manufacturers, such as trying to comply with the directives that emerge from the Kyoto Protocol [5]. The decentralisation of production comes with an increase of the required processes, both direct and indirect. Results revealed that the location of suppliers is a significant factor that can alter the raw material's embodied energy as different locations use electricity generated from various combinations of energy sources. The careful selection of suppliers combined with the reduction of road transportation for supplying the high quantities of raw materials over significantly long distances can reduce the environmental impact [6]. The forecasts for future energy consumption indicate a deceleration of 1.6 % at the growth of primary energy consumption up to 2030, compared to 2.0 % p.a. the previous 20 years, and a growth of 0.7 % p.a. energy consumption per capita, which is approximately the same rate as it has been since 1970. Moreover, as seen in the charts (Fig. 1), the transportation and manufacturing sectors lead the growth of final energy consumption, especially in rapidly developing economies. The 2030 projection indicates that the industrial sector accounts for 60 % of the projected growth of final energy demand [7].

The demand for goods has increased and so has the demand for natural resources and energy. However, since resources and energy are finite, new ways of producing more with less ought to be found [8]. Several life cycle studies conducted in previous research work have a tendency to assess the environmental impact of products during their use phase, focusing less on the manufacturing stage and transportation of both final product as well as components [9]. The decisions taken during the manufacturing phase can indirectly impact the use phase of the product. By increasing the precision during the manufacturing of products, can lead to longer life and greater efficiency [10]. The environmental effect of transportation activities in a make-to-order environment is of high importance. Postponement strategies have been proposed for environmentally efficient transportation [11]. A decision framework was proposed and tested in a pilot case in 2010 for analysing the impacts of transportation policies on social system, environmental issues, and energy [12]. In addition, studies have focused on reducing the managing discrete facilities in greener ways [13]. Weinert et al. in 2011 presented a methodology based on the representation of production operations as segments of specific energy consumption for each operating state of the production equipment [14]. The method integrated energy-efficiency criteria with evaluation and decision processes during

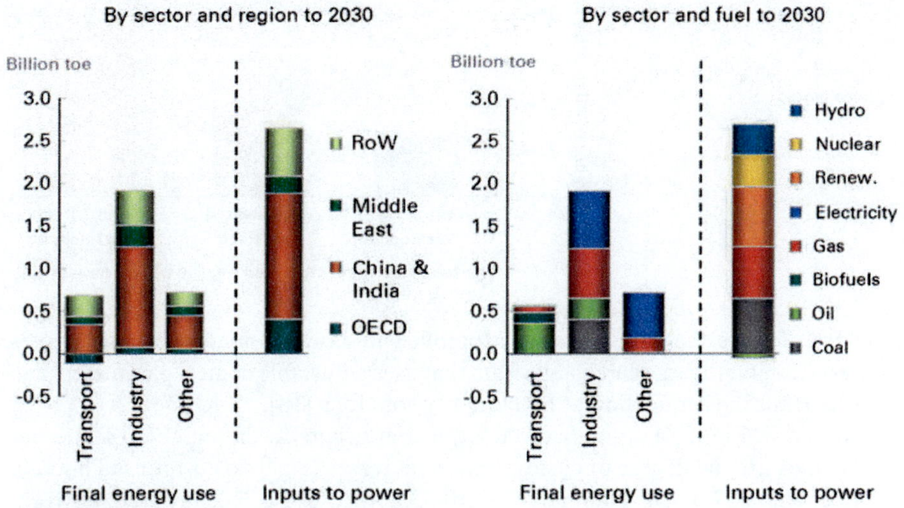

Abbreviations:
OECD: Organisation for Economic Co-operation and Development
RoW: Rest of World
TOE: Tones of Oil Equivalent

Fig. 1 The growth of energy consumption by sector [7]

production system planning and scheduling, and its application in a case study emphasised the need for system-wide approaches that are based on accounting of energy consumption on machine level. Towards the protection of natural resources and the preservation of the ecological environment initiatives for energy-saving, resource-efficiency and environmental sustainability, must be promoted, as well as the development of the 'circular' economy through recycling [15]. The Gross Energy Requirement (GER) and Global Warming Potential (GWP) are key environmental indicators [16] that European companies only recently started to calculate. The CO_2 nevertheless, is the main contributing factor of the greenhouse effect and reducing its emissions is a critical goal.

2 Environmental Impact Assessment

The presented research work consists of a methodology for the efficient identification of environmentally friendly manufacturing and transportation schemes, aiming at supporting the planning and configuration of supply chains, in a mass customisation environment [17, 18]. The generation of alternative manufacturing and transportation schemes is performed using an exhaustive search method and the evaluation of the schemes is carried out against multiple user-defined criteria, with adjustable weight factors. The four steps of the decision making process can be summed up to the following: (a) Create Alternatives, (b) Select Criteria, (c) Evaluate Alternatives

Fig. 2 Number of alternatives vs. network configuration and degree of customisation

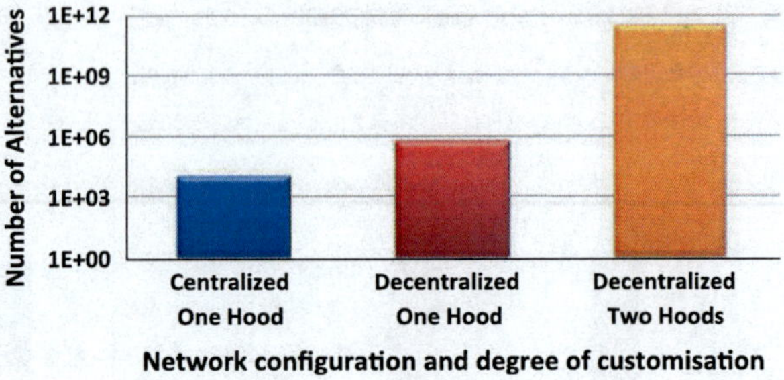

and (d) Calculate the Utility Value. Moreover, in case the number of alternatives increases, an intelligent search algorithm that uses adjustable control parameters can be utilised for the acquisition of high quality solutions [18].

As seen in Fig. 2, a combinatorial explosion occurs in the number of alternatives in the case the degree of customisation increases (e.g. two customised hoods). They become 287×10^9 alternatives from 535×10^3 for a decentralised network, thus obstructing the use of an exhaustive search, due to the required computation time [17].

2.1 Centralised and Decentralised Network Configuration

The customised product under investigation is a car hood that is produced in four variants. The supply chain for the production and assembly of the hood components comprises OEM plants, Suppliers and Dealers all cooperating in order to fulfil the customer orders. The centralised supply chain is modelled in such a way that assembly tasks can be performed only by the OEM at one of the owned plants. After the final hood assembly is performed, the OEM delivers the product to the sales representatives (dealers) so that it can be sold to a customer. In a decentralised scenario however, the assembly of the hood and the customisation activities can be carried out by suppliers or even dealers. The decentralised network modelling, allows final assembly operations or special works (e.g. application of the warp cast carbon) to be performed at a dealer or a supplier site, close to the customer [17]. Different actors are capable of different operations and charge different prices for them.

2.2 Criteria

The criteria used for the quality measurement of the alternative manufacturing and transportation schemes are the Energy Consumption and CO_2 Emissions. The values of these two metrics are calculated as the sum of the values required for the

Table 1 Tasks, supply chain partners, resources and processing time for the customised car hood case study

Task	Supply chain partner	Work centre	Recourse	Processing time (min)	Description
T_{HF}	P_1	WCP1_1	RP1_1_1	0.7	Hood frame (HF) production
	S_6	WCS6_1	RS6_1_1	1	
T_{HS}	S_3	WCS3_1	RS3_1_1	0.5	Hinge support (HS) production
	S_4	WCS4_1	RS4_1_1	1	
	S_7	WCS7_1	RS7_1_1	1.5	
T_{LS}	S_3	WCS3_1	RS3_1_2	0.5	Lock support (LS) production
	S_4	WCS4_1	RS4_1_2	0.8	
	S_5	WCS5_1	RS5_1_1	1	
	S_6	WCS6_2	RS6_2_1	0.7	
T_{EXC}	P_1	WCP1_1	RP1_1_2	1	External cover (EXC) production
	P_2	WCP2_1	RP2_1_1	0.6	
T_{WCC}	S_1	WCS1_1	RS1_1_1	2	Wrap cast carbon (WCC) production
	S_2	WCS2_1	RS2_1_1	1.6	
	S_5	WCS5_2	RS5_2_1	2	
	S_7	WCS7_2	RS7_2_1	1	
T_{BAS}	P_1	WCP1_1	RP1_1_3	15	Assembly of the basic hood (BAS)
	P_2	WCP2_1	RP2_1_2	20	
T_{AWC}	S_1	WCS1_1	RS1_1_2	20	Application of the wrap cast carbon to the hood (AWCC)
	S_5	WCS5_2	RS5_2_2	30	
	D_1	WCD1_1	RD1_1_1	480	
	D_2	WCD2_1	RD2_1_1	320	
	D_3	WCD3_1	RD3_1_1	250	
T_{ORN}	S_2	WCS2_2	RS2_2_1	30	Ornament (ORN) production
	S_3	WCS3_2	RS3_2_1	32	
	S_5	WCS5_3	RS5_3_1	40	
	S_7	WCS7_3	RS7_3_1	20	
	S_4	WCS4_2	RS4_2_1	25	
T_{AOR}	S_1	WCS1_2	RS1_2_1	18	Assembly of the ornament to the hood (AOR)
	S_6	WCS6_3	RS6_3_1	20	
	D_1	WCD1_2	RD1_2_1	30	
	D_2	WCD2_2	RD2_2_1	40	
	D_3	WCD3_2	RD3_2_1	35	

T: Task, P: Plant, S: Supplier, D: Dealer, WCP: Plant work centre, RP: Plant resource

production and assembly for each of the hood components and their transportation within the supply chain. Upon entering the resources into the software tool, the Watt specifications and the processing time of each resource are stored in the database and are used for the calculations (Table 1). For each process, the Energy Consumption is measured in Joules, and the values are summed together with the transportation energy requirements for each alternative scheme. The energy requirements for transportation processes take into consideration the covered distance and the fuel consumption of each truck. The formula for the calculation of the Energy Consumption (1) is:

$$EC = EC_T + EC_P = \sum_{r=1}^{R}(D_r * TC) + \sum_{k=1}^{K}(Pt_k * RW_k) \quad (1)$$

where:

EC the sum of Energy Consumption for the manufacturing scheme (Joule),
EC_T the sum of the Energy Consumption due to transportation activities (Joule),
EC_P the sum of the Energy Consumption for all the processes (J),
D transportation distance covered (Km),
r the number transportation roots ($r \in \mathbb{N}/r = 1, 2, \ldots R$),
TC the Energy Consumption per kilometre (J/km) [19],
Pt production time (sec),
k the task of a job ($k \in \mathbb{N}/k = 1, 2, \ldots K$),
RW the Watts of the resource responsible for task k (J/s).

The CO_2 Emissions (2) value is calculated by the distance travelled and the emission of CO_2 per kilometre (km):

$$\mathrm{CE} = \sum_{r=1}^{R}\frac{G * D_r}{N} \quad (2)$$

where:

CE Carbon dioxide (CO_2) Emissions (gr of CO_2),
D transportation distance (Km),
G CO_2 Emissions (gr/Km) [19],
N number of products that the truck is carrying ($N \in \mathbb{N}$).

Moreover, the developed software tool provides the ability to evaluate the alternative schemes against classic criteria, such as Cost, Lead time, Resource Availability, Reutilisation and stochastic indicators, namely Annual Production Rate and Flexibility.

2.2.1 Aggregation of Criteria Values

The decision between the alternative manufacturing schemes requires a normalisation of the values of each criterion as described in [20]. Afterwards, a decision matrix is used for the selection among the alternative schemes. The rows of the matrix represent the possible alternatives and the columns the evaluation criteria (Fig. 4). The matrix contents are the values of the criteria of each alternative. The cardinal preference (utility value) is calculated using a sum of weighted criteria normalised to the sum of one. The alternatives with the highest utility value are the most preferable.

3 Software Tool Implementation

In order to test the functionality and performance of the methodology, a prototype software tool has been designed and implemented using Unified Modelling Language Diagrams (UML) in an object-oriented programming language, using the .NET Framework^{TM} (Fig. 3). To ensure fast data retrieval and respect data integrity constraints, a Relational Database Management System (RDBMS) has been implemented using the Oracle 9i Database. The workstation used for performing the experiments was equipped by an Intel^{TM} i7 3.4 GHz processor, with 8 GB of RAM. The tool interface comprises user-friendly Graphical User Interfaces (GUIs) for performing the required data entry, for configuring the control parameters of the intelligent search algorithm, and for visualizing the results. The user of the software tool is provided with the ability to select between the exhaustive search and intelligent search algorithm functionalities, and define the search parameters. If the intelligent algorithm is selected for the evaluation, the tool can generate any number of alternatives upon request and present their performance in the form of bar charts. The resource assignments and operations designated for each product component and subassembly are stored in database tables. Moreover, the tool has the capability of automatic generation of Discrete Event Simulation (DES) models utilizing an integrated commercial simulation software suite. The user can designate the demand profile for the examined product, and then the manufacturing and supply alternatives are evaluated against this demand profile. The architecture of the developed software is presented in Fig. 4.

4 Industrial Case Study

The products, resources and dataset (eco-profile for materials, cycle times etc.) used in the presented case study for the environmental impact calculations are coming from a European automotive manufacturer. The car hood comprises six components, two of which are optional customisation additions. The customisable components are offered by the OEM to the customers. The basic components of the hood are the external hood cover, the hood frame, the hinge support and the lock support, whereas the customisation options include an ornament and a wrap cast carbon. These components are either manufactured internally at the OEM or they are outsourced to Suppliers. The assembly processes can be performed at the OEM plants or in some cases e.g. application of the wrap cast carbon, at the facilities of the suppliers or even the dealers at different cost, time and environmental impact. The processes required for the production and assembly are depicted in Fig. 5. The transportation activities are also modelled as processes, taking into account the distance between the supply chain actors, the average speed of the truck, the fuel consumption per litre of fuel, and the CO2 Emissions per litre

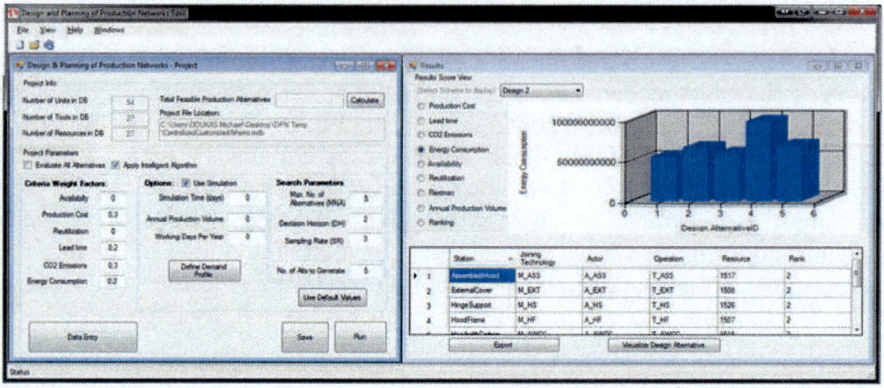

Fig. 3 Software tool interfaces and visualisation of results in chart form

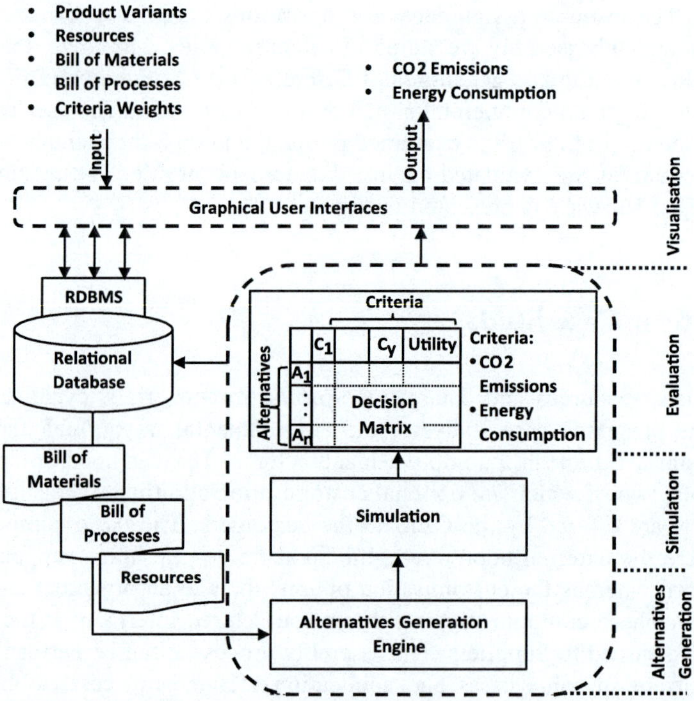

Fig. 4 Architecture of the software tool

of fuel for the calculation of the CO2 Emissions. The resource characteristics include the processing time, setup time, Energy Consumption, Mean Time Between Failure (MTBF) and Mean Time To Repair (MTTR).

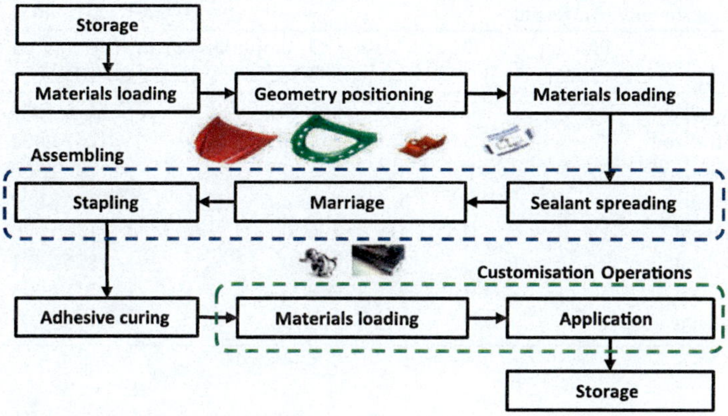

Fig. 5 Bill of processes (*BoP*) for the fully customised hood

5 Results and Discussion

A series of computer simulation experiments have been carried out for the evaluation of the environmental impact of the degree of customisation and for the alternative network configurations. The dataset used for the experiments is presented inside Table 1. The first column contains the task required for the manufacturing and assembly of the hood components, the second column contains the supply chain partner that can perform the task, the third column represents the work-centres that comprise a set of similar resources or resources that are utilised for performing a job, the fourth column includes the resources that can perform the tasks and, finally, the fifth column includes the processing time of the task.

The parameters of the conducted experiments are included inside Table 2. The first column is the number of the experiment. The second column describes the network configuration, namely "Centralised" and "Decentralised" network. The third column "Product Variant" determines the degree of customisation. The four variants used in the experiments were:

1. **Product variant 1 (L1)—Basic Hood:** This variant is a non-customised product, comprising four components, namely the external cover, the hood frame, the hinge support and the lock support.
2. **Product variant 2 (L2a)—Ornament:** This variant is customised by the customers and comprises the basic hood assembly and the ornament option.
3. **Product variant 3 (L2b)—Wrap Cast Carbon:** This variant is customised by the customers and comprises the basic hood assembly and the wrap cast carbon option.
4. **Product variant 4 (L3)—Ornament + Wrap Cast Carbon:** This variant is fully customised comprising the basic hood assembly and the wrap cast carbon and ornament options.

Table 2 Experiments and results

No.	Production network	Product variant	W_{CO2}	W_{EC}	CO_2 emissions (gr of CO_2)	Energy consumption (MJ)
1	Decentralised	1	0.5	0.5	490000	11760,000
2	Decentralised	2	0.5	0.5	515600	12374,402
3	Decentralised	3	0.5	0.5	722800	17347,206
4	Decentralised	4	0.5	0.5	748400	17961,609
5	Centralised	1	0.5	0.5	490000	11760,000
6	Centralised	2	0.5	0.5	761200	18268,801
7	Centralised	3	0.5	0.5	738000	17712,001
8	Centralised	4	0.5	0.5	904400	21705,602

The fourth and fifth columns contain the weight factors (W_{CO2} and W_{EC}) of the used criteria that are used to express the objectives of the evaluation. The weight factors get values between 0 and 1, according to the relevant importance of the criterion. Finally, the last two columns contain the calculated values for the Energy Consumption and the CO_2 Emissions.

The first observation regards the impact of the degree of customisation on the CO_2 Emissions and on Energy Consumption. As seen in the diagrams in Figs. 6a and b and 7 a and b, the CO_2 Emissions and the Energy Consumption increase as the required product components, and consequently the product complexity, increase. For the decentralised case (Fig. 6), the production of the fully customised hood (basic hood + ornament + wrap cast carbon) yielded 34.52 % more CO_2 emissions, and required 34.53 % more energy. For the centralised network, the production of the fully customised hood yielded 45.82 % more CO_2 emissions, and required 45.8 % more energy consumption. In addition, for the production of the non-customised hood in a decentralised network, the CO_2 Emissions and the Energy Consumption are 32.2 and 32 % respectively lower compared to the production of the hood with the wrap cast carbon customisation.

Additionally, by combining the diagrams above (Figs. 6 and 7) the comparison between the centralised and decentralised production network configurations can be visualised more clearly in Fig. 8a and b. The centralised scenario values are depicted by the blue exponential trend line and the decentralised by the red. Both lines display a positive correlation; the environmental impact increases as the degree of customisation increases. Moreover, a deviation is observed between the trend lines as the degree of customisation increases. The decentralised production network displays significantly less average CO_2 Emissions and Energy Consumption, compared to the centralised network for the production of the same hood variant.

Finally, by comparing the total environmental impact (aggregation between the CO_2 Emissions and Energy Consumption values) for the decentralised scenario we can see a direct relation between the quality (utility value) of the manufacturing and transportation schemes versus the degree of customisation (Fig. 9). The values that appear in the diagram are normalised and the higher the utility value, the less

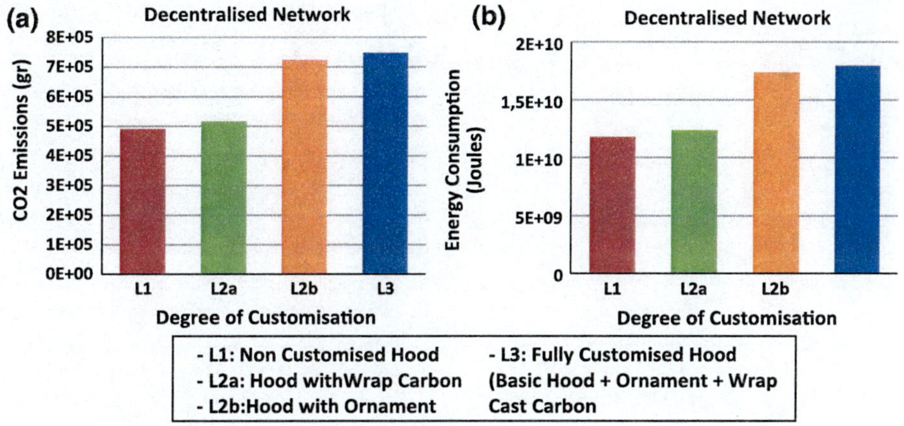

Fig. 6 **a** CO_2 emissions vs. degree of customisation for the decentralised network configuration, for the best alternatives. **b** Energy consumption vs. degree of customisation for the decentralised network configuration, for the best alternatives

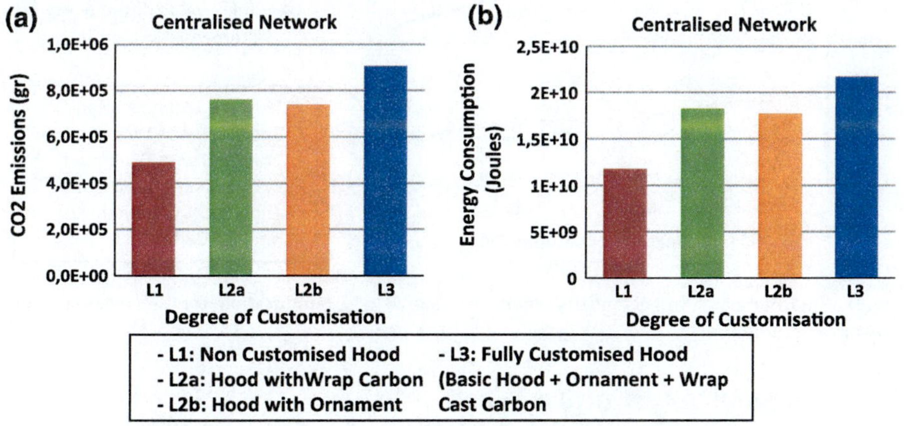

Fig. 7 **a** CO_2 emissions vs. degree of customisation for the centralised network configuration, for the best alternatives. **b** Energy consumption vs. degree of customisation for a centralised network configuration for the best alternative

the environmental impact. The blue line represents the utility value of the best solution (less environmental impact) and the red line represents the utility value of the worst alternative (high environmental impact). The utility value of the best alternative compared to the utility value of the worst has an average difference of 29.75 %.

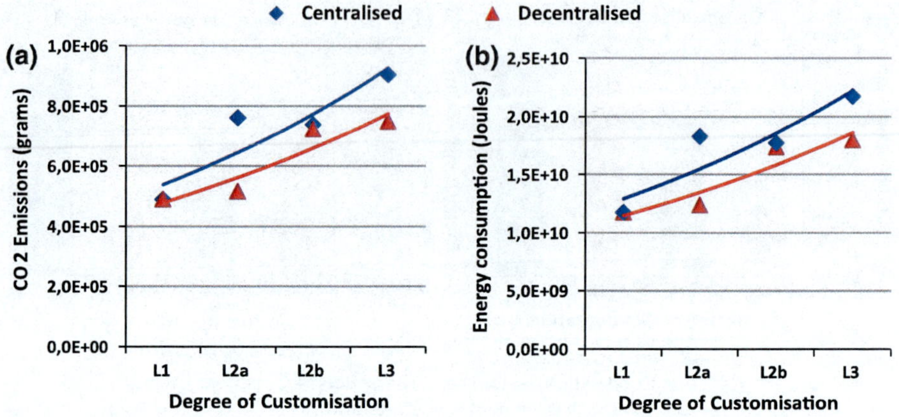

Fig. 8 **a** Comparison of CO_2 emissions of centralised vs. decentralised networks. **b** Comparison of energy consumption of centralised vs. decentralised networks

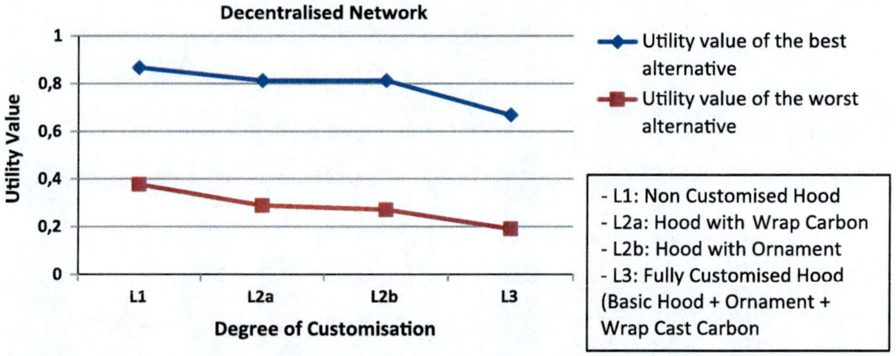

Fig. 9 Environmental impact utility value vs. degree of customisation for the decentralised network configuration

6 Conclusions and Future Work

The presented methodology can support the decision makers during the configuration of the supply network in an eco-friendly way, taking into consideration reduced environmental impact in terms of Energy Consumption and CO_2 Emissions. The significant variation in the quality of the alternatives for each product variant, indicates the necessity of using the evaluation mechanism during the decision making process. The results revealed a direct relation between the degree of customisation and the environmental impact. The results depicted that the higher the customisation degree of a product, the higher the CO_2 emissions and the energy requirements for its production and transportation. Especially in the case of the fully customised hood, the impact was almost double than in the case of the basic, non-customised product for the centralised and the decentralised network configurations. The addition of the customisation options (ornament and wrap cast carbon) imposes a number of additional processes and transportation routes that increase the environmental impact of the final product. Additionally, the decentralised network configuration displayed reduced CO_2 Emissions and Energy Consumption values compared to the centralised, for the manufacturing and transportation of the same product variant. The constraints in a centralised network allow assembly operations to be performed only at an OEM plant, thus leading to a limited number of alternative manufacturing and supply schemes. However, some of the excluded schemes are of high quality with respect to environmental indicators, which are reduced due to the decreased transportation distance [17].

Future work will focus on extending the capabilities of the methodology, incorporating the selection of different types of materials and transportation means. Moreover, additional environmental indicators will be included for the evaluation of the alternative schemes, such as environmental impact of diverse raw materials, toxic emissions, and eutrophication indicators. A commercial Life Cycle Assessment (LCA) software suite will be integrated to the developed tool for the detailed simulation of the environmental impact of the processes, materials and transportation. Finally, apart from the environmental indicators, the criteria of Cost, Lead time, Quality and Flexibility will be taken into account for the evaluation of the alternative manufacturing and transportation schemes.

Acknowledgments This work has been supported by the EC funded project "A web-based collaboration system for mass customisation—e-CUSTOM".

References

1. Paralikas, J., Fysikopoulos, A., Pandremenos, J., Chryssolouris, G.: Product modularity and assembly systems: An automotive case study. CIRP Ann. Manuf. Technol. **60**, 165–168 (2011)
2. Tolio, T., Urgo, M.: A rolling horizon approach to plan outsourcing in manufacturing-to-order environments affected by uncertainty. CIRP Ann. Manuf. Technol. **56**, 487–490 (2007)
3. Chryssolouris, G.: Manufacturing Systems—Theory and Practice, 2nd edn. Springer, New York (2005)
4. Gunasekaran, A., Ngai, E.W.T.: The future of operations management: An outlook and analysis. Int. J. Prod. Econ. **135**, 687–701 (2012)
5. Kyoto Protocol. http://unfccc.int/kyoto_protocol/items/2830.php(2005)
6. Kara, S., Ibbotson, S.: Embodied energy of manufacturing supply chains. CIRP J. Manuf. Sci. Technol. **4**, 317–323 (2011)
7. BP. Energy Outlook 2030. http://www.bp.com/liveassets/bp_internet/globalbp/STAGING/global_assets/downloads/O/2012_2030_energy_outlook_booklet.pdf (2010)
8. Chryssolouris, G., Papakostas, N., Mavrikios, D.: A perspective on manufacturing strategy: Produce more with less. CIRP J. Manuf. Sci. Technol. **1**, 45–52 (2008)
9. Helu, M., Vijayaraghavan, A., Dornfeld, D.: Evaluating the relationship between use phase environmental impacts and manufacturing process precision. CIRP Ann. Manuf. Technol. **60**, 49–52 (2011)

10. Diaz, N., Helu, M., Jayanathan, S., Chen, Y., Horvath, A., Dornfeld, D.: Environmental analysis of milling machine tool use in various manufacturing environments. IEEE International Symposium on Sustainable Systems and Technology. Arlington (2010)
11. Yang, B., Yang, Y., Wijngaard, J.: Impact of postponement on transport: an environmental perspective. Int. J. Logistics Manage. **16**, 192–204 (2005)
12. Ülengin, F., Kabak, O., Önsel, S., Ülengin, B., Aktaş, E.: A problem-structuring model for analysing transportation–environment relationships. Eur. J. Oper. Res. **200**, 844–859 (2010)
13. Eagan, P.D., Joeres, E.: The utility of environmental impact information: a manufacturing case study. J. Cleaner Prod. **10**, 75–83 (2002)
14. Weinert, N., Chiotellis, S., Seliger, G.: Methodology for planning and operating energy-efficient production systems. CIRP Ann. Manuf. Technol. **60**, 41–44 (2011)
15. Jovane, F., Yoshikawa, H., Alting, L., Boër, C.R., Westkamper, E., Williams, D., Tseng, M., Seliger, G., Paci, A.M.: The incoming global technological and industrial revolution towards competitive sustainable manufacturing. CIRP Ann. Manuf. Technol. **57**, 641–659 (2008)
16. OECD Key Environmental Indicators. http://www.oecd.org/dataoecd/20/40/37551205.pdf(2008)
17. Mourtzis, D., Doukas, M., Psarommatis, F.: A multi-criteria evaluation of centralised and decentralised production networks in a highly customer-driven environment. CIRP Ann. Manuf. Technol. Vol 61. (2012) (To be published)
18. Michalos, G., Makris, S., Mourtzis, D.: An intelligent search algorithm-based method to derive assembly line design alternatives. Int. J. Comput. Integr. Manuf. **25**, 211–229 (2012)
19. EPA. www.epa.gov(2010)
20. Chryssolouris, G., Dicke, K., Lee, M.: On the resources allocation problem. Int. J. Prod. Res. **30**, 2773–2795 (1992)

Innovative Approaches for Global Production Networks

Global Footprint Design: An Evolutionary Approach

Günther Schuh, Till Potente, Daniel Kupke and Rawina Varandani

Abstract The continuous growth of production networks has led to a threefold complexity issue for multinational companies. Firstly, it lies in the tremendous number of design options taking into regard all product groups and their production processes which need to be allocated to the amount of existing or new production sites. The second complexity issue is characterized by a short amount of time available in companies for highly important decisions. The third complexity problem lies in the complexity evaluation within the production network taking into account product, manufacturing and organisational structures. These three challenges are addressed within the scope of an approach which avails itself of a digital tool using interactive computing methods. While the complexity of the solution space is handled through a mathematical optimization, visual components help to understand and analyze the given solution through interactive computing. The identification of complexity drivers within production networks constitutes the final challenge.

Keywords Production networks · Complexity · Global production

G. Schuh (✉) · T. Potente · D. Kupke · R. Varandani
Laboratory for Machine Tools and Production Engineering (WZL), RWTH Aachen University, Steinbachstr. 19, 52074, Aachen, Germany
e-mail: G.Schuh@wzl.rwth-aachen.de

T. Potente
e-mail: T.Potente@wzl.rwth-aachen.de

D. Kupke
e-mail: D.Kupke@wzl.rwth-aachen.de

R. Varandani
e-mail: R.Varandani@wzl.rwth-aachen.de

K. Windt (ed.), *Robust Manufacturing Control*, Lecture Notes in Production Engineering,
DOI: 10.1007/978-3-642-30749-2_28,

1 Introduction

Multinational corporations have created heterogeneous manufacturing structures by jumping after market opportunities, opening new production sites in low-cost countries to harness low personnel cost levels, following their customer or OEM and aiming for quick-wins through acquisitions. The continuous growth of production networks has led to an increasing number of planning objects which need to be integrated into the planning scope [1]. While there may be a debate about the actual increase in labor costs in China and uncertainty about the development in currency exchange rates, material prices or tax regulations—one thing manufacturers seem to agree on is that complexity within global production networks will become an even greater issue over the next decade than it is today [2]. This paper addresses the complexity issues which need to be assessed by multination corporations in the Chap. 2. It then demonstrates how existing methods can be linked with innovative approaches to meet current challenges in "Global Footprint Design"—commonly known as global production network design. For this purpose existing methods in the field of Operations Research are presented and complemented by the latest research results in complexity management in manufacturing in Chap. 3. Chapter 4 presents the result of the research: a complexity-oriented approach aiming at solving the conflict of the three complexity issues addressed earlier. A tool using a genetic algorithm serves to handle the complexity of the solution space, while interactive computing methods are used to facilitate the decision making process for CEOs. A model presenting the key complexity drivers serves to address the third complexity issue. While the validation for the first two aspects of the complexity-oriented approach is presented in Chap. 4 as well, the work on the structural complexity in managing production networks is yet to be validated. The work is concluded in Chap. 5.

2 Challenges

The following chapter serves to describe the initial situation for global footprint design. It also addresses three key issues for which this paper attempts to find a profound solution.

2.1 Initial Situation

Today's global manufacturing landscape is characterized by the dilemma between two central dichotomies: the need for diversification in order to penetrate new markets and serve global customer needs as well as the realization of economies of scale for leveraging cost potentials. To ensure a company's competitiveness within this area of conflict the need for a fundamental optimization of the global production network becomes apparent [3]. Companies have different approaches to the optimization process. Many do not use a green field approach or real mathematical optimization techniques but rather conduct multiple projects through which they implement incremental changes. The fundamental structure of a production network is thus not questioned [4]. Furthermore, the decision-making process is given too minor a role. Companies concentrate on trusting their intuition rather than employing techniques in order to prepare and provide knowledge to the decision makers in an integral way [5]. This paper aims at providing a solution to the following three key issues met when designing or redesigning global production networks.

2.2 Key Issues in Global Production Network Design

Given the described initial situation, the first challenge lies in the tremendous number of design options within a production network taking into regard all product groups and their production processes which need to be allocated to the existing or new production sites [6]. Most companies use a manual planning approach, i.e. by intelligence and experience of the planner. Various forms of allocation of production resources are developed and evaluated with spreadsheet programs. Usually, some scenarios are developed for the production network, and then evaluated for cost. If necessary, an iterative process takes place in order to find the best solution. By nature, in this way only a very small number of possible scenarios can be considered and evaluated. Furthermore, additional considerations, such as the consideration of alternative value chains or equipment for a product, are out of scope because the spreadsheet method cannot master this level of complexity. The consequence is that the probability of finding the best possible solution is not very high. With a limited number of products, short supply chains and a one-to-one correspondence between process and machine the manual planning method may deliver sufficient results. However, once new as well as alternative production processes for a product group are regarded and the value chain may be distributed over different production sites in order to achieve the best possible utilization of the global production system, the size of the solution space increases exponentially. In this case, a manual planning is no longer appropriate [3]. The described problem may be summarized as the first key issue: complexity of the solution space.

Many of the various production sites in global production networks have been set up or acquired empowered by individual decisions—managerial decisions. Hence, the second complexity issue is characterized by a short amount of time available in companies for highly important decisions such as the closing or opening of a site and the demand for a focused summary of all the important information required to take these decisions, as shown in Fig. 1. A study shows that while a normal person makes about 70 choices per day, CEOs make many more. On top of daily life decisions CEOs make about 250 decisions more per week. However, given the short amount of time available, 50 % of a CEO's

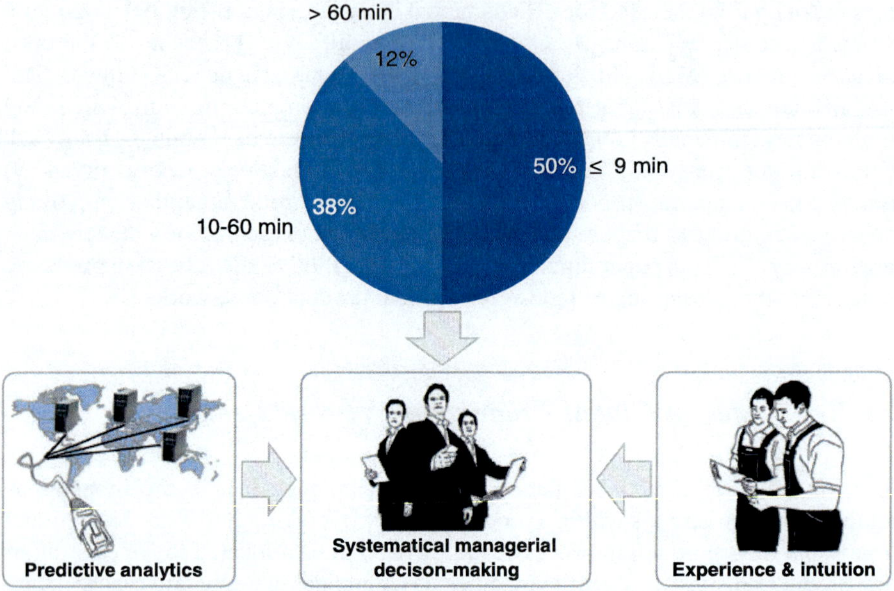

Fig. 1 Managerial decision-making process [7]

decisions are made in nine minutes or less. Only about 12 % of the decisions require an hour or more time [7]. This makes it apparent that facts and data, on which sound decisions should be based, need to be processed through predictive analytics. Information must be analyzed and prepared in a way that decision makers can process it directly visually [8]. Experience and intuition from experts and CEOs themselves complete the required set of information. It is apparent that the reduction of complexity in the decision-making process becomes a more important factor as companies grow.

The third complexity problem, which multinational companies today see as a real future challenge, lies in the complexity evaluation within the production network taking into account product, manufacturing and organizational structures. When companies are asked about the main complexity drivers within their production network, most of them name the product variety first [9]. Complexity issues are mostly linked to the topic of product complexity, while the complexity of production structures, especially in a global network, has not yet received much attention by researchers.

3 Approaches for Global Production Network Design

This chapter provides an overview of the development of technical solutions to solve problems with large solution spaces. The different methods belonging to the field of Operations Research are presented. Furthermore, approaches considering factors of complexity are presented.

3.1 Operations Research in Global Production Networks

The design of global production networks deals with a great number of variables (number of plants, facilities and resources, possible locations etc.) and possible states of those variables. There have been some developments to solve the problem of designing global production networks from an Operations Research perspective, especially in the field of supply chain management. A short overview of methods and approaches developed in the field of Operations Research are presented below.

Vidal and Goetschalckx [10] present a critical review of strategic production–distribution models with emphasis on global supply chain models. Geoffrion and Graves [11] develop an approach to optimize multi-commodity single-period production–distribution systems by a modified Benders Decomposition. In 1978, Geoffrian et al. [12] published a status updated with some evolved ideas about the use of decomposition techniques. The so-called PILOT model is another approach brought up by Cohen and Lee. It is designed to be a deterministic, periodic, mathematical program to minimize costs that extends the model of Geoffrion and Graves. The extensions include for example opening and closing of plants [13].

A model for locating international plants is presented by Hodder and Dincer. It includes many factors and the model results into a large-scale Mixed-integer programming (MIP) model that is difficult or impossible to solve, even by using an approximation procedure. The model covers many financial, but few production related factors [14].

Brown and Olson present a mathematical framework including three algorithms for row factorization: one for generalized upper bound rows, one for pure network rows and one for generalized network rows [15]. The authors apply their algorithms to two kinds of production–distribution systems, the ODS (similar to the systems used by Geoffrion and Graves [11]) and the DEC, which was developed by Arntzen et al.. The approach of Arntzen et al. [16] is based on a multi-period, multi-commodity model to optimize a global supply chain. The model focusses rather on designing and locating manufacturing plants than on characteristics of a supply chain.

A generic model for the strategic design of production–distribution systems is presented by Goetschalckx et al. In addition to their model, they develop algorithmic components that significantly reduce the calculation times compared to commercial solvers based on standard MIP solutions [17, 18].

The more variables one takes into account, the more complex the problem becomes. At some point, it is nearly impossible to find the optimal solution using linear, integer or non-linear programming (Hillier and Lieberman [19]). Nevertheless, in many cases it is necessary to find a best-fitting solution and therefore approaches that include Lagrangian relaxation, linear programming based heuristics and met heuristics are utilized (Melo et al. [20]).

Pirkul and Jayaraman present a heuristic procedure called *Planwar* for logistic purposes that is based is based on the results of a Lagrangian relaxation and finds a feasible solution for a given distribution problem in the SCM context.

Metaheuristics include genetic algorithms that lead to best-fitting solutions relatively fast and with little effort [21]. Canel et al. [22] developed an algorithm that is divided into three phases and aims to finds a best-fitting solution for capacitated, multi-commodity, multi-period and multi-stage facility location problems. Keskin and Üster [23] describe genetic algorithms used in the design process of global production networks: the scatter search and the tabu search algorithms, first introduced by Glover [24]. Both approaches convince due to short run times on large-scale problems.

The overview shows that there has been a lot of research in developing optimization methods for global network design. Some of the approaches focus rather on the design of logistics than production networks. However, the basic approach is the same for both. Most approaches, however, are not very user-friendly due to the complexity of the optimization.

3.2 Complexity Analysis Within Global Production Networks

Complexity management is a field which is commonly associated with the reduction of product complexity. While there are many approaches in this research field, there has been a progress on evaluating complexity in manufacturing systems in recent years.

ElMaraghy et al. use a hybrid mathematical model consisting of three components in order to measure complexity for single products. It combines the absolute quantity of information, the diversity of information and effort to achieve the required result. In other words, it takes product features and specifications into account while the process complexity analysis focuses on the tools, equipment and operations used to manufacture it. The operational complexity analysis considers the cognitive and physical effort associated with the tasks related to a product/process combination. By the use of a code-system the complexity for manufacturing systems can be classified [25].

Papakostas et al. introduce a simulation-based approach for measuring the complexity of manufacturing systems. Their work aims at uncertainty in production, statistical analysis and simulation-based solution methods. A manufacturing execution complexity index is introduced. This number links the intrinsic structure of the production system and the uncertainty related to the operations of the system. The figure heavily depends on the workload profile of workers. Therefore, a set of manufacturing models, characterized by different production configurations and part routings, is simulated and evaluated through a series of experiments, employing diverse workload patterns. The results are used for determining the sensitivity of a manufacturing system to workload changes, measuring the complexity of a manufacturing system and discussing measures to control the complexity of a manufacturing system based on the proposed simulation-based approach [26].

Vrabic et al. offer an approach using computational mechanics in order to quantify complexity in manufacturing. The computational mechanics approach towards complexity in manufacturing systems is based on the hypothesis that the more difficult it is to predict a process, the more complex it is [27, 28].

Zhang proposes a function for cellular manufacturing systems. In this study the relationship between complexity and utility in a manufacturing system are researched. Based on the analysis of characteristics of cellular manufacturing systems, using both, the static entropy model and the operational entropy model, complexity in manufacturing systems is investigated [29].

The overview shows that there has been quite a progress in the research on complexity in manufacturing; however, an approach which clearly focusses on the complexity, moreover the management complexity of a global manufacturing network is yet to be described.

4 Global Footprint Design: A Complexity-Oriented Approach

This chapter describes the technical solution designed to optimize global production networks enhanced by visual components for usability and comprised information handling. Furthermore, it provides an overview of complexity drivers in production networks laying the ground for a complexity-oriented Global Footprint Design.

4.1 Technical Solution

In order to cope with the challenges identified in Sect. 2.2, the Laboratory for Machine Tools and Production Engineering (WZL) has developed a software tool that applies a method based on Operations Research, a genetic algorithm belonging to the group of evolutionary algorithms, to optimize production networks. This software tool, named "OptiWo", consists of two key components: the "optimizer" on the one hand, which avails itself of a genetic algorithm to calculate a cost-optimal Global Footprint within a defined solution space and a "data viewer" on the other hand. The optimizer uses a genetic algorithm which can be parameterized depending on the size of the solution space in order to facilitate the search of an optimal solution. The key parameters are population size, defining the number of individuals per generation, the mutation rate, describing the frequency of transformation of individuals and finally, the number of generations. The operation of the genetic algorithm can be described as evolutionary. With each new generation the genetic algorithm calculates the fitness value of a configuration of the production network based on a fitness function. Following the motto "survival of the fittest" only the configuration with the best fitness value survives, until a stopping criterion is reached.

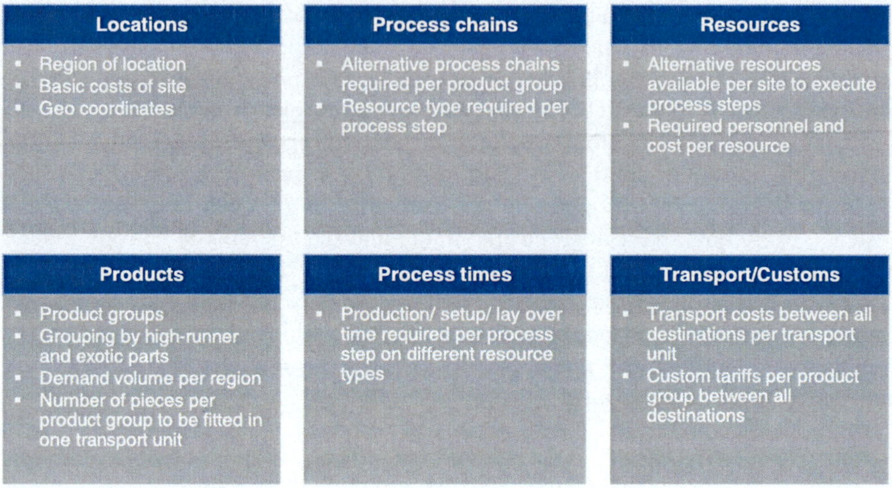

Fig. 2 Spredsheet based input data

The optimizer uses a common data interface by means of six.csv (comma separated value) spreadsheets. The spreadsheets need to be filled by the user and hold all information necessary to model a global production network, as shown in Fig. 2.

Generally, the user is not limited regarding the data complexity—he can model both existing and future production networks, with alterations regarding product groups, process chains and production locations. The user can model any amount of product groups with their specified demand volume per region. For each product group at least one process chain needs to be defined, elucidating the technologies required to produce a certain product group. If desired, alternative process chains can be defined for the same product group.

This option also allows including new technologies which do not exist yet but are most likely to be used in the near future. The software will then also consider the split of the production of a product on different process chains. Transport steps can be foreseen at any point within a process chain; however, they need to be specified before running OptiWo. The software will then calculate the total landed costs for different configurations of production networks in order to find the cost-optimal solution. The cost calculation includes all cost types shown in Fig. 3. This cost model was developed with experts from the manufacturing industry in order to ensure its viability in the industry. Variable costs are considered via machine hour rates and multiplied by the amount of hours used on a machine. Fixed costs are calculated per site separately. Following cost types are included in the calculation of the overall landed costs per year: Direct and indirect labor, depreciation of machines, building, intra-logistics, energy, maintenance, consumables, allocations, custom duties, transportation and raw material. A productivity factor, that can be defined not only for each site but also for each resource separately, is used to increase the time required in certain countries to execute a particular task.

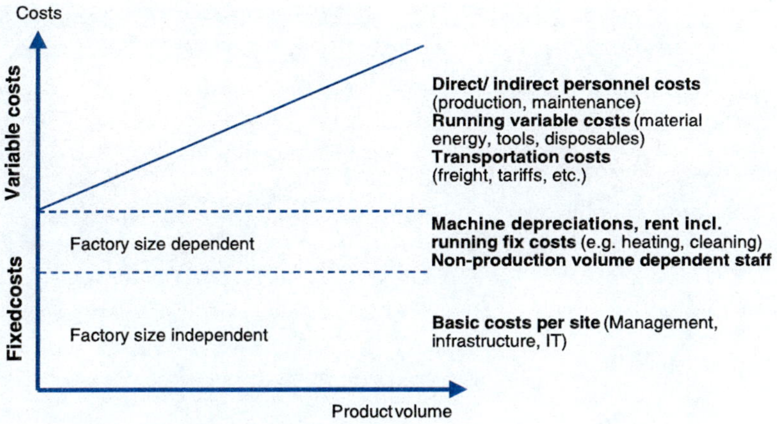

Fig. 3 The cost model builds the basis for the calculation of total landed costs

4.2 *Visualization as a Key to Understanding*

While the genetic algorithm used for the calculation of the total landed costs and their minimization plays a major role when preparing the decision information, the visualization of the solution appears just as important in order for a CEO to understand the key findings. Therefore, the software solution OptiWo developed at the Laboratory for Machine Tools and Production Engineering in Aachen uses innovative visualization elements. Visualization enables a high information density and thus enables CEOs to quickly understand complex data and relationships.

A so-called data viewer which is web-based provides different visualization elements to display the results. Two key elements are the World Map and the Resource Map which provide central information for management in highly compressed form. The World Map displays the distribution of production capacities (in production hours) is on a world map to give an overview of a configuration calculated through OptiWo, as shown in Fig. 4.

The Resource Map is based on the concept of a tree map which is commonly used in the financial sector. It visualizes the size of all production sites within the global production network, the amount of machines per resource type for each site and finally their utilization rate through a color code, as shown in Fig. 5. In this figure, the size of the rectangles represents the number of machines of a certain resource type, whereas the color of the rectangles represents the average utilization of those machines, ranging from red (low utilization rate) to green (high utilization rate). Additionally, the rectangles representing the machines are aggregated into a larger rectangle whose size represents the size of the production site. A user interface input mask allows an easy-to-use and quick highlighting of different production sites and resource types—the rectangles that match the search criteria are then highlighted.

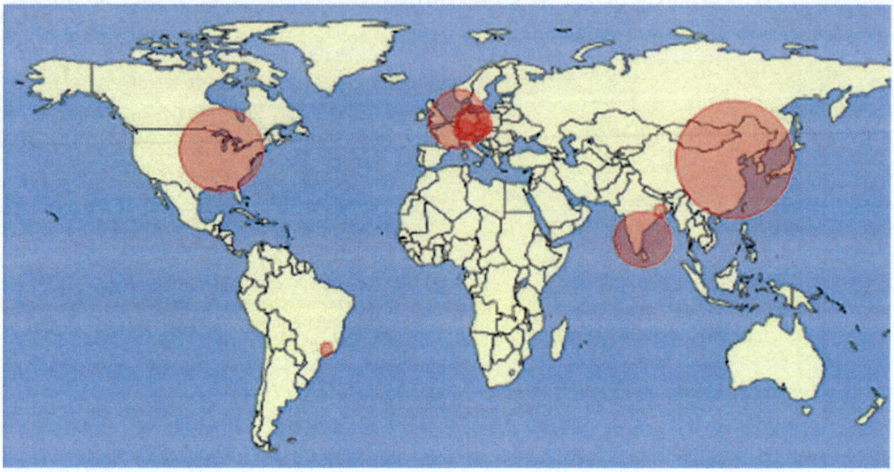

Fig. 4 World map

Finally, a Manual Modeller and a Sensitivity Check are integrated as visual components. The Manual Modeller assists the user in changing the distribution of products onto different production sites. The impact on costs and delivery times of each change in the configuration can directly be monitored. The Sensitivity Check supports the user in testing the robustness of a particular solution. A number of parameters such as productivity factors, exchange rates, wages and material costs, may be varied to view the financial implications of a change in the parameters.

4.3 Complexity Drivers Within Global Production Networks

In order to address the third issue presented in the Sect. 2.2 an overview of complexity drivers identified in the management of global production networks is given in Fig. 6. While there has been a significant research in the field of complexity management in manufacturing in general, which was presented in Sect. 3.2, the research on key complexity issues in the context of global production is not very broad. It rather focusses on logistical aspects, e.g. uncertainty in wage, currency exchange rates and oil price development. However, the main focus of this paper is to provide a first insight on complexity drivers in the management of global production networks, see Fig. 6.

The next steps in the research at the Laboratory for Machine Tools in Aachen aim at specifying the relationships between these drivers and quantifying them in order to measure complexity in global production networks. The mathematical equation can then be formed into a fitness function that can be implemented within the software solution OptiWo described in Sect. 4.1. This enables an optimization not only towards the total landed costs but with respect to the increasing complexity faced by global manufacturers.

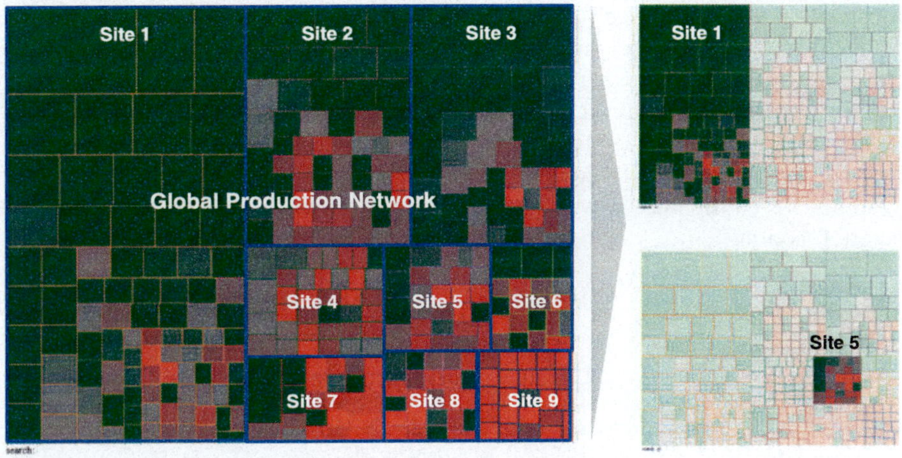

Fig. 5 Resource map

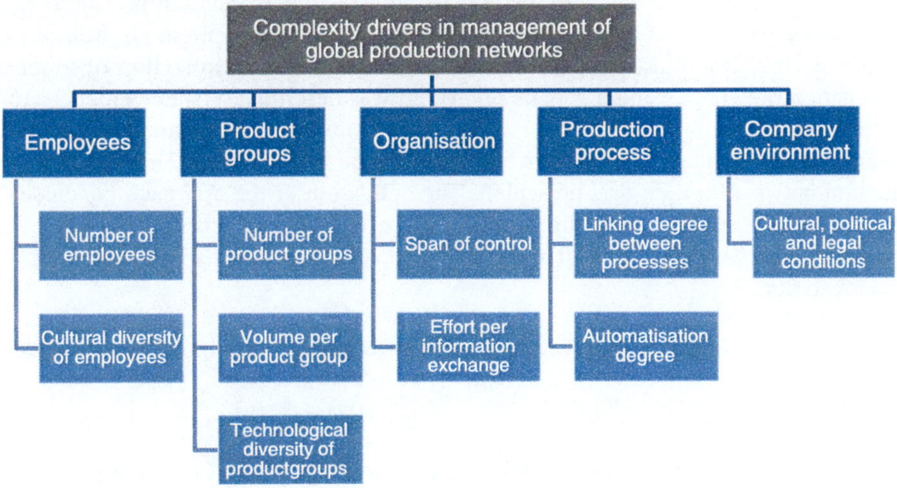

Fig. 6 Complexity drivers in management of global production networks

4.4 Validation

The presented approach was applied for the optimization and design of a global production network in the manufacturing industry. A project was conducted with a business unit of a world leader in electronic and energy technology. 240 product groups each requiring about 3–8 technological process steps to be performed on 2–3 different production resources at one of ten production sites were in the project scope. While the first task was to define an ideal production network without any restrictions, trying to identify the main direction for future efforts, a future network considering restrictions such as existing resources at a particular site was finally defined. On the way there, discussion rounds with CEO and CFO of the business unit were required to derive useful scenarios and finally develop a scenario which can be implemented within the next five years. Through this approach the use of the tool OptiWo was embedded into a complete project approach. While the main focus in the project was to derive a cost-optimal solution, limiting delivery time factors were considered as well and defined as a boundary condition within the fitness function.

5 Conclusion

The presented work addresses three types of complexity issues in global production networks. While an overview of existing methods were given for the field of operations research and complexity analysis in this context. Different methods used to optimize production networks, mostly on cost-basis, and quantify complexity, however as discovered mainly in site-specific manufacturing systems, were discussed. An innovative approach, using visualization methods from different sectors, was presented: a technical solution for the optimization of global production networks using a genetic algorithm was described, visual elements and their benefit in facilitating managerial decisions were introduced and finally a first overview of complexity drivers, which need to be evaluated within global production network design, was presented. The validation of the approach concludes the research. Further investigations in the field of complexity assessment in global production network design need to be made and concluded through a mathematical manifestation.

References

1. Schuh, G., Gottschalk, S., Wesch, C.: Potentials in computer-aided factory design—negotiation-based top-down-bottom-up approach. In: De Silva, A.K.M., Harrisson, D.K. (eds.) Proceedings of the 20th International Conference on Computer-Aided Production Engineering; Cape, Glasgow, 6–8 June 2007
2. Capgemini: Manufacturing in 2020—Envisioning a Future Characterised by Increased Internationalisation, Collaboration and Complexity, joint study conducted by Capgemini and IDG Global Solutions (2008)
3. Abele, E., et al.: Global Production—A Handbook for Strategy and Implementation. Springer, Heidelberg (2008)
4. Schuh, G., Nöcker, J., Varandani, R., Schwartze, J., Schilling, R.: Dealing with the need for flexibility and economies of scope in global production network design. In: 1st WGP Congress, Berlin (2011)
5. Ariely, D.: Predictably Irrational. HarperCollins, New York (2009)
6. Hübner, R.: Strategic Supply Chain Management in Process Industries: An Application to Speciality Chemicals Production Network Design. Springer, Heidelberg (2007)

7. Iyengar, S.: How to make choosing easier. http://www.ted.com/talks/sheena_iyengar_choosing_what_to_choose.html. Accessed 13 Mar 2012
8. Ariely, D.: Dan Ariely asks, are we in control of our own decisions, EG Conference, Monterey, Dec 2008, http://www.ted.com/talks/lang/en/dan_ariely_asks_are_we_in_control_of_our_own_decisions.html, (2009)
9. WZL. Study on management of value-added chains, (2011)
10. Vidal, C.J., Goetschalckx, M.: Strategic production-distribution models: A critical review with emphasis on global supply chain models. Eur. J. Oper. Res. **98**(1), 1–18 (1997)
11. Geoffrion, A.M., Graves, G.W.: Multi-commodity distribution system design by Benders decomposition. Manage. Sci. **20**(5), 822–844 (1974)
12. Geoffrion, A.M., Graves, G.W., Lee, S.J.: Strategic Distribution System Planning: A Status Report, Studies in Operations Management, pp. 179–204. Elsevier, Amsterdam (1978)
13. Cohen, M.A., Lee, H.L.: Strategic analysis of integrated production-distribution systems: Models and methods. Oper. Res. **36**(2), 216–228 (1988)
14. Hodder, J.E., Dincer, M.C.: A multifactor model for international plant location and financing under uncertainty. Comput. Oper. Res. **13**(5), 601–609 (1986)
15. Brown, G.G., Olson, M.P.: Dynamic factorization in large-scale optimization. Math. Program. **64**, 17–51 (1994)
16. Arntzen, B.C., Brown, G.G., Harrison, T.P., Trafton, L.L.: Global supply chain management at digital equipment corporation. Interfaces **25**(1), 69–93 (1995)
17. Goetschalckx, M., Nemhanser, G., Cole, M.H., Wei, R., Dogan, K., Zang, X.: Computer aided design of industrial logistic systems. In: Proceedings of the Third Triennial Symposium on Transportation Analysis (TRISTAN III), Capri, Italy, pp. 151–178 (1994)
18. Goetschalckx, M., Cole, M.H., Dogan, K., Wei, R.: A generic model for the strategic design of production-distribution systems, version 1.0. Georgia Institute of Technology; submitted for publication, (1995)
19. Hillier, F., Lieberman, G.: Introduction to Operations Research. McGraw Hill, New York (2005)
20. Melo, M., Nickel, S., Saldanha-da-Gama, F.: Facility location and supply chain management—a review. Eur. J. Oper. Res. **196**(2), 401–412 (2009)
21. Pirkul, H., Jayaraman, A.: Multi-commodity, multi-plant, capacitated facility location problem: Formulation and efficient heuristic solution. Comput. Oper. Res. **25**, 869–878 (1998)
22. Canel, C., Khumawala, B.M., Law, J., Loh, A.: An algorithm for the capacitated, multi-commodity multi-period facility location problem. Comput. Oper. Res. **28**, 411–427 (2001)
23. Keskin, B.B., Üster, H.: Meta-heuristic approaches with memory and evolution for a multi-product production/distribution system design problem. Eur. J. Oper. Res. **182**(2), 663–682 (2007)
24. Glover, F.: Heuristics for integer programming using surrogate constraints. Decis. Sci. **8**, 156–166 (1977)
25. ElMaraghy, H.A., Kuzgunkaya, O., Urbanic, R.J.: Manufacturing systems configuration complexity, intelligent manufacturing systems (IMS) centre. CIRP Ann. Manuf. Technol. **54**(1), 445–450 (2005)
26. Papakostas, N., Efthymiou, K., Mourtzis, D., Chryssolouris, G.: Modelling the complexity of manufacturing systems using nonlinear dynamics approaches. CIRP Ann. Manuf. Technol. **58**, 437–440 (2009)
27. Vrabic, R., Butala, P.: Assessing operational complexity of manufacturing systems based on statistical complexity. Int. J. Prod. Res. **50**(14), 1–13 (2011)
28. Vrabic, R., Butala, P.: Computational mechanics approach to managing complexity in manufacturing systems. CIRP Ann. Manuf. Technol. **60**, 503–506 (2011)
29. Zhang, Z.: Modeling complexity of cellular manufacturing systems. Appl. Math. Model. **35**(9), 4189–4195 (2011)

Part IV
Process Optimization and Strategic Approaches Towards Robustness

Evaluation of Production Processes Using Hybrid Simulation

Norbert Gronau, Hanna Theuer and Sander Lass

Abstract Changing market conditions, variable customer demands and growing customer requirements are some reasons for producing companies to create flexible and adaptable processes and to fulfil the customer demands in a high quality. For this reason it may be beneficial to change the production system from a centralized towards a decentralized production management approach. It is of high importance to figure out the best mix of centralized and decentralized production control for every company separately, while at the same time ensuring that the process continues running. Comprehensive analyses often turn out to be time-consuming and expensive. Especially small and medium sized enterprises have to avoid these side-effects. This article presents a method for the fast and well-founded evaluation of the best mix of decentralized and centralized production control by using autonomous technologies.

Keywords Decentralized production control · Hybrid simulation · Autonomous technologies · Robust production control

N. Gronau · H. Theuer (✉) · S. Lass
Chair of Business Information Systems and Electronic Government, University of Potsdam, August-Bebel-Str, 14482 Potsdam, Germany
e-mail: hanna.theuer@wi.uni-potsdam.de

N. Gronau
e-mail: norbert.gronau@wi.uni-potsdam.de

S. Lass
e-mail: sander.lass@wi.uni-potsdam.de

K. Windt (ed.), *Robust Manufacturing Control*, Lecture Notes in Production Engineering, 401
DOI: 10.1007/978-3-642-30749-2_29, © Springer-Verlag Berlin Heidelberg 2013

1 Introduction

Due to various changes of production conditions during the last years—e.g. variable customer demands, growing customer requirements as well as a shift from a seller to a buyer market—global production networks are faced with several exigencies. It is important that companies cogitate about new methods and evaluate those toward the achievements of objectives in consideration of their processes.

Decentralized production control by the use of autonomous technologies seems to be an adequate method to deal with the current requirements on production processes. It would be pointless to make generally applicable statements towards the best degree of decentralized and centralized production control. It is of high importance to figure out the best mix for every company separately, while at the same time ensuring that the manufacturing process continues. Often comprehensive analyses turn out to be time-consuming and costly. Especially small and medium sized enterprises try many times to avoid these efforts. The distribution of such concepts or methods in practice is low. For reinforcing of competitiveness it is important to use modern autonomous technologies. Therefore it is necessary to create a method with low effort to ensure the consideration of all relevant parts of a production system.

The authors perform a research project ("Leistungsfähigkeitsbeurteilung unabhängiger Produktionsprozesse"—LUPO) [1] with the aim of creating a possibility for the fast and well-founded evaluation of the best mix of decentralized and centralized production control by using autonomous technologies [2]. Therefore they develop a hybrid simulation environment that combines the advantages of computer simulation and physical model factory. Production processes are mapped, recreated, simulated and analyzed towards their suitability of decentralized structures. Due to a high flexibility regarding the construction of the simulation environment and the possible integration of miscellaneous autonomous technologies, it is possible to analyze a high variety of scenarios. The integration of a Manufacturing Execution Systems (MES) ensures the consideration of the state-of-the-art information technology.

This article presents firstly, the concept of the simulation environment and secondly the integration of information technology. The third part introduces a key figure for the evaluation of decentral and autonomous technologies. This enables an analysis of benefit and applicability for concrete production processes. The article ends with an outlook for further research activities.

2 Hybrid Simulation Environment

For the evaluation of autonomous production processes it is of high importance to use a suitable method for modeling, simulation and analysis. The most important requirements for the simulation of autonomous production processes regarding the objectives mentioned above are:

- High flexibility of processes
- Quick set up
- Possibility to reproduce physical issues, e.g. antenna orientation

Common methods that are used in coherence with production processes are the Digital Factory and the Model Factory. The analysis of both show that they have their strengths and weaknesses. A combination of the two approaches would fit the requirements of the simulation the best way. The following sections shortly describe both simulation methods.

2.1 Digital Factory

Digital Factory is a planning approach, which is used in product development for modeling suitable versions for future objects in order to visualize and perform analyses. The objective is the optimization of product relevant structures, processes and resources [3, 4]. Real time monitoring and planning support are connected to one system [5] and establish a shared database for all product relevant software systems. Modeling a Digital Factory is often expensive. Despite decreased costs of information technology systems small and medium sized enterprises forego the use of this kind of planning. In the automotive industry this planning approach is used successfully [6]. A digital mockup of a product is used to analyze ergonomic issues and functionality, possible construction methods or mere visualization [7]. Its strength lies within shortening of duration for product development and design at unchanged costs [8]. An optimization of production processes primarily occurs during product development and focuses on the product itself.

But in the early phases of designing manufacturing processes there are no integrated factory and logistic planning methods that allow a structured comparison of different alternatives based on a set of various criteria [6]. The introduction and use of the different technologies in production process are marginally considered. With regard to the stated aims above, physical systems appear only as data provider. In summary, it allows simulation without the use of hardware components through digital mockup, if there is an adequate implementation of a physical model within the software. Otherwise, it has to be implemented. This possibly causes high effort compared with a hardware variant and results in software tools with high complexity. The strength of Digital Factories is the product development. With regard to the given objectives Digital Factories cannot be applied without extending their concepts.

2.2 Model Factory

A model is a simplified representation of a planned or existent system build up to reduce complexity [7]. A Model Factory represents concrete production processes in a simplified way under lab conditions. There is no standard or universally agreed definition for the term Model Factory which is mainly used for educational and teaching purposes, for example at the RWTH Aachen [9] or the Technical University Darmstadt [10]. The components in a Model Factory are a physical implementation of their real counterpart and works the same way e.g. machines or production islands. In most cases the specialized model elements have a small field of operations. Their application is limited to simple products and scenarios. Due to the inflexibility of model factories, the analysis of new ideas and concepts are restricted to cases with similar usage. The limitation to a concrete production process impede the use for different production situations and prevents the implementation of the model into various processes. The evaluation of alternative scenarios is difficult. The obtained results subsequently have to be transferred from this special application to other applications. To sum up, a Model Factory allows a quick realization of physical problems with low effort, but it has a low degree of flexibility concerning different production scenarios. Thus, it is not an adequate solution for the objectives stated in the introduction.

2.3 Hybrid Simulation

In the Hybrid Simulation environment there are physical models for the relevant production objects. They are used for the representation of the existing system by deploying a combination of software and hardware components. For every single part of simulation, the most appropriate way of simulation can be identified. The implementation of original equipment in the digital model, opens the possibility of testing physical effects e.g. detection rate, field intensity or antenna pointing of AutoID-elements with minimal effort [11]. Neither the purely physical nor the purely digital simulation can offer this advantage. The hybrid approach is suitable for the realization of the started objectives.

The LUPO simulation environment consists of the work piece and machine tool demonstrators as well as transport lines that connect the various machine tool demonstrators. The demonstrators with their ability to communicate in different ways and the flexible transport system do provide an effortless integration of hardware components into the overall system. The software is designed for a quick integration of sensors and other devices using standard communication protocols. The hardware section provides the interfaces for an easy connection. The system supports the integration of hardware-components by design. This is an important advantage compared with sole software models which are supplemented by some hardware parts. For an investigation of receiving characteristics in a RFID scenario, for example, it is not sufficient to connect merely a reader device, but in addition it is necessary to realize moved work pieces with a kind of conveyer. A cost intensive construction of further hardware parts is imperative for good results. Thus, the presented approach avoids these efforts.

A demonstrator consists of a box which is configured with the parameters of a certain production object. The interaction of demonstrators allows the setup and

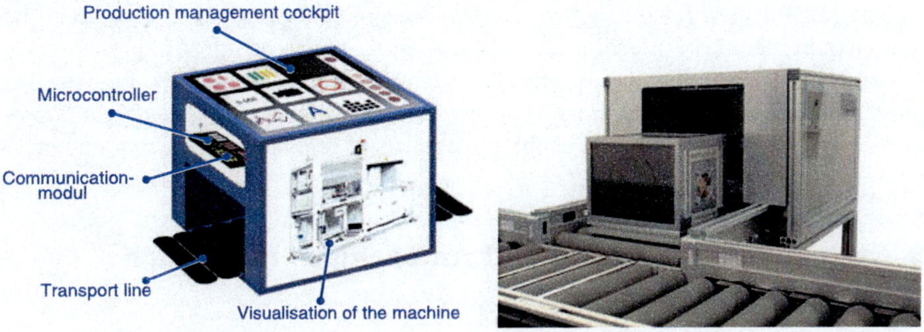

Fig. 1 Machine tool demonstrator—schema and original

simulation of a whole production process. Relevant environmental information is delivered for input by various sensors. Some interface and communication modules allow the connection of different types of sensors and enable the interaction with other components. Thus, it is possible to configure demonstrators and complete them with further pieces of hardware. Figure 1 illustrates the setup as a matter of principle as well as a picture of the existing demonstrator. The illustrations of the parts to be worked on are displayed as a 2D or 3D model on both sides of the demonstrator. The monitoring display is on the top side reporting the relevant product, process and job information. All of this information is up to date at every time during the simulation. The diverse machine center demonstrators are aligned by transport lines. To ensure a high level of flexibility and adaptability to the simulation environment transport, several line elements like switch plates, circular shelves as well as entry points and gates are used. Various factory layouts with sequences, parallelism or repetition can be represented. For process control of the simulated production processes, it is necessary to use corresponding software tools (described in the following section).

2.4 Integration of Information Technology

The LUPO simulation environment distinguish between two types of software: the LUPO operation system (LOS) and the MES. The tasks of the LOS are the control of the different types of demonstrators and the transportation lines as well as the internal communication of demonstrators. Additionally it provides the user interface that enables the configuration of demonstrators and the input of relevant process parameters.

It distinguishes between the simulation operation level and simulation application level. All technical adjustments for the configuration of demonstrators are made in the simulation operation level. In the simulation application level the concrete process is displayed. All settings that are necessary for the configuration of the real process are made.

The MES is used for the control and analyses of processes. An exchange of the default parameters of machine center and work piece demonstrator has to be ensured. This includes the production plan, relevant-related key figures like process time, set up time or transport times as well as data on possible breakdowns of the simulated process. This enables a control of relevant process parameters.

3 Evaluation of Decentral and Autonomous Production Processes

Autonomous technologies are one possibility for the realization of a decontrol controlled production. Therefor In the following the focus is on this kind of production control. It is necessary to have a suitable method for the documentation and analyses of simulation runs of autonomous processes at the LUPO hybrid simulator. There should be an opportunity for creating an easy overview, whether a process acts autonomously or not. A comparison of different processes towards their degree of autonomy is possible. A modeling method had to be selected for the documentation of different scenarios being simulated in the hybrid simulation environment. In accordance with criteria different methods, e.g. Value Stream Design (VSD) and Event driven Process Chain, were compared.

The most important criteria for the selection are:

- Suitability for production processes
- Good opportunity to evaluate and compare various production processes towards relevant objectives
- Basis for discussion
- Consideration of special requirement of autonomous technologies.

The comparison revealed that Value Stream Design is the most suitable method for the aims of the LUPO simulation environment. This method fulfills the first four of the requirements mentioned above.

3.1 Value Stream Design

Originally designed by Rother and Shook for mass production in automotive industry, the VSD method took on significance even for small batch production in recent years as for many companies Lean Production came into the focus of attention [12]. By distinction of value-adding (non waste) and non-value-adding (waste) processes it is a simple way to analyze the current situation of a production towards lean aspects. Sources of waste can be discovered—the basis for improvements is given. Based on the findings, different production scenarios for improvement can be compared and analyzed. Therefore VSD and Lean Production are a good combination for long lasting improvements [12]. For modeling, the method offers a clearly defined set of symbols that considers different properties of a supply chain such as production processes, inventory, customer, supplier and material flow. Furthermore relevant key data (e. g. lead times, waiting times, set-up times, number of persons at one process, stock) are mapped. Information flow is also of interest but the focus of this method is on material flow [12, 13].

The relation of value-adding and total process time is called Lean Index. It is expressed in x:y (x to y with x = value-adding times, y = total process time). The more similar both numbers are, the less (time) waste can be found in the production. A Lean Index of 1:1 presents a perfect piece flow with no waiting times for the products, and complete adjusted cycle times. Nowadays the Lean Index in many companies is bigger than $1{:}y > 100$ [12, 13].

3.2 Extended VSD

The original Value Steam Design method does not consider the special requirements for the analyses of processes using autonomous technologies. As this is of high importance for the analyses of production processes in the LUPO simulation environment it is necessary to extend the method by details of the information flow. By an additional documentation of data which is relevant for autonomous control, traceability increases. Furthermore a reproducibility of the process is given. For the process evaluation an index that measures the degree of autonomy is introduced. Since Value Stream Design is easy to understand and to practice directly at the workflow without great effort, the extension should be likely.

Set of Symbols In order to create an easy overview as to whether a process acts autonomously or not the symbolism of processes has to be extended. An autonomous process is marked with a black triangle on the right corner. During Value Stream Mapping the author of the diagram can decide whether a process is autonomous or not and mark the process, if necessary.

Figure 2 shows an example of a Value Stream Map of a production site with three processes. While Process A and B and non-autonomous controlled, Process C its autonomous controlled. The supplier delivers the goods to a production supermarket. From there Process A takes them in. After finishing the procedures at Process A the goods are given into a FIFO line to Process B. Another FIFO line connects Process B and Process C. After Process C is completed all goods are put into stock where they have to wait for a certain time. From there the products are delivered to the customer. All customer and supplier processes are like a black box. They exist but no further information is known. External processes are similar. The only figure that is known is the total process time for the external process [13].

Data Dictionary For the reproducibility of the autonomous process it is necessary to document all data that is relevant for process execution. This includes all data that is exchanged between production objects which are involved in the process as well as data that the process needs to decide how to act. The relevant

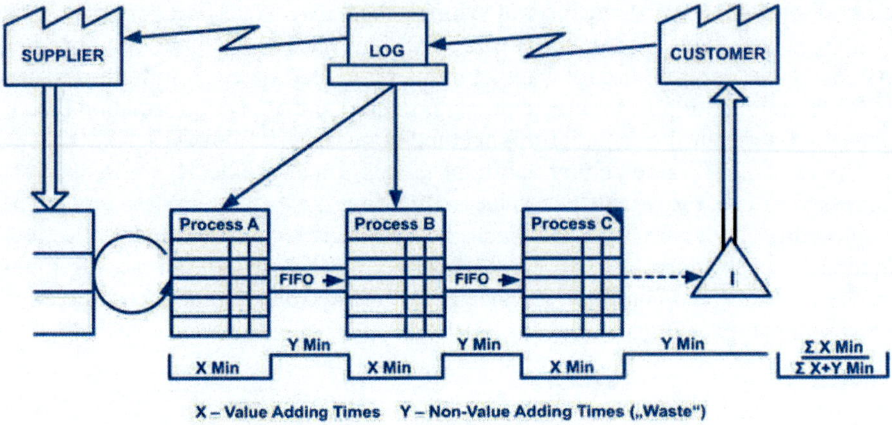

Fig. 2 Example of a value stream map with three processes connected by FIFO lines

Process Data	Information Flow Data	Product Data
Predefined rules stored	Data-On-Tag or Data-On-Network	
Set-up time matrix	Used technology of data exchange	Type of product
Amount of different products beeing worked on the specific process	Frequency of data exchange	Relevance (express or not)
	Amount of data per exchange	Planned completion data
Process times for different products beeing worked on the process	Mission critcal index - what happens in case where the data needed is not availbale	Additional information

Fig. 3 Relevant data for data dictionary in extended value stream design

data can be divided into three super classes: process data, information flow data and product data.

Process data is specific for the process. It includes all information that is necessary to enable the process to make decisions on its own. Information flow data specifies the data exchange of the production objects at the process (e.g. process and product). This data is necessary to rebuild the technological settings of the process. Product data specify the product that is worked on in the process. Process data and information flow data require particular values. As there may be various products in one specific process product data are of Boolean type. It is necessary to know what product data is exchanged without a concrete definition. Relevant data may be (but is not limited) as shown in Fig. 3:

Autonomy Index For the evaluation of value streams with autonomous technologies the introduction of a key performance figure is necessary. To underline the interest it is named Autonomy Index. It specifies the degree of autonomy used at the value stream and thereby give the possibly to compare different production systems. In coherence with the Lean Index the Autonomy Index should clarify the amount of autonomy in comparison to the whole value stream. When defining the index the basis for the comparison has to be specified. There are a number of possibilities:

- number of autonomous processes: number of all processes
- autonomous controlled process time: total cycle time
- quantity of autonomous data: total quantity of data

Due to high importance of data exchange in autonomous production control the decision was made in favor of the third possibility. The Autonomy Index AI is calculated as shown in Eq. 1. With an orientation to the Lean Index AI is noted as *DEaut: DEall.* The range is between 1:X (X means number higher 1) and 1:1. The low AI the higher the degree of autonomy at the value stream. For documentation purpose AI is written down on every Value Stream Map.

$$AI = \frac{\sum_{i=1}^{n} F_i \cdot A_i}{\sum_{j=1}^{m} F_j \cdot A_j} = \frac{DE_{aut}}{DE_{all}} \qquad (1)$$

with:

AI: Autonomy Index
DE_{aut}: total amount of autonomous data exchange
DE_{all}: total amount of data exchange
F: frequency of data exchange
A: average amount of data volume per exchange
$i \epsilon I$ with I: amount of autonomous data exchanges
$j \epsilon J$ with J: amount of all data exchanges
$I \subset J$
n: number of autonomous data exchanges
m: number of all data exchanges

It has to be considered that the amount of data exchange does not include all exchanged data but only the relevant. For example not the file size of a picture is relevant (as it may vary strongly depending on the kind of file, e.g. jpg, tiff or bmp), as well as the resolution or quality. Instead the relevant data, e.g. color or size of the product, has to be measured.

3.3 Evaluation

The calculation of AI enables an evaluation of the correlation of the Autonomous Control and another key figure, e.g. total process cost or Lean Index, in a specific value stream. Based on this information it is possible to generate statements about the concrete benefit of decentral production control. It is possible to compare different mixture of decentral and central production control configurations and thereby gain knowledge about the applicability of decentral production control by autonomous production processes for this specific process.

The graphical representation results in a scatter plot since there are different ways of achieving the same value of DE_{aut}. Additionally same values of DE_{aut} can result in different grades of the regarded key figure. The plot may indicate the best degree of Autonomous Control for the considered value stream.

It is possible to analyze which processes have major or minor impact on the decision towards an autonomous control. Based on this, a cost-benefit analysis can be indicated. The reciprocals of both data are put into a scatter plot. An example is depicted in the following chapter.

4 Case Study

This section provides an example for the usage of extended Value Stream Mapping in connection with simulation of production processes in the LUPO laboratory. The analyzed production consists of five processes (process A to E). The production sequence is predefined and identical for all products produced. The process can be classified into two sections that are linked by an interim storage. This storage is the point of product individualization also called order decoupling point. The first section consists of process A and B. Both produce non-individual intermediate products. Products handled at process B are put in an interim storage. There are two variants produced in this section. Processes C to E produce customer individual products. Intermediate products are taken from the storage and then handled in process C. There are three variation possibilities in process C and D, two in process E. All of those possibilities are combinable, so that there are $2 \times 3 \times 3 \times 2 = 36$ variants of the end product.

To change from one to another variant process setups are necessary. As set up times vary from initial state to target state, there are setup matrixes for all five processes. Process times differentiate from process to process as well as from variant to variant. For satisfaction of the customer requirements it is of major importance that the right product is manufactured at the right time. In the case analyzed the delivery performance deteriorated and stock rose. It is for this reason that all processes should be reconsidered. Additionally to the mentioned problems it should be analyzed how to deal with express orders, that ensure a highly shortened delivery time to customers. At the current state all five processes are central controlled. While there is a push control installed at the first section there is a pull control at the second section. The produced amount of both basic variants in the first section is planned due to a sales forecast. The production program for the next week of the second section is planned due to concrete customer orders. Difficulties arose due to missing intermediate products. All five processes with their relevant characteristics have been recreated and simulated at the LUPO laboratory.

Table 1 Characteristics of processes

Process	Average amount of data volume per exchange	Frequency of data exchange
A	1	3
B	2	3
C	3	3
D	3	3
E	2	3

After the validation of simulation results with the existing processes variations of the type of control are made. In addition to a completely centralized or decentralized control mixed control concepts are analyzed. Therefore the first section is set decentralized controlled while the second is central controlled and in a second trial the other way around. Substancial differences are perceived at lead times and stock building. For each set-up a value stream map was created and a value stream analysis performed. Lean and Autonomy Index were calculated. As second quantity the Lean Index is used exemplarily.

Assuming that as well the average amount of data volume per exchange as the frequency of data exchange are identical in all scenarios the relevant data for the scatter plot can be determined with Table 1 and 2:

The reciprocals of both data are put into a scatter plot. The evaluation of the correlation between Autonomy and Lean Index is shown in the scatter plot in Fig. 4. The numbers refers the belonging scenario.The scatter plot indicates that there is no correlation between Autonomy and Lean Index in the specific production analyzed.

It is obvious that the lowest Lean Index is realized with AI = 1:1,5. This situation occurs if the first section is controlled central and the section decentralized controlled. The worst Lean Index is achieved by a decentral controlled first section and a central controlled second section (AI = 1:3,67). The change of a complete central production (AI = 1:∞) to a complete decentralized production (AI = 1:1) only causes marginal differences in consideration of the Lean Index. For all decentral controlled processes a supplementary analysis is realized. Its results are recorded at the data dictionary. This enables a later reproduction of the process and therefore the usage of identified advantages in the real production. Additionally the extended value stream maps enable a well-founded discussion with company internal and external persons. The best mix of central and decentral controlled production has been determined. The problems of the original process mentioned above were reduced to a minimum. Due to a well arranged and completely documentation regarding lean production aspects, a successful implementation of simulation results in real production processes is provided.

Table 2 Characteristics of scenarios

Scenario	AI	LI
1 (completely decentral controlled)	1:1	1:50
2 (completely central controlled)	1:∞	1:45
3 (first section decentral, second section central controlled)	1:3,67	1:70
4 (first section central, section decentral controlled)	1:1,5	1:10

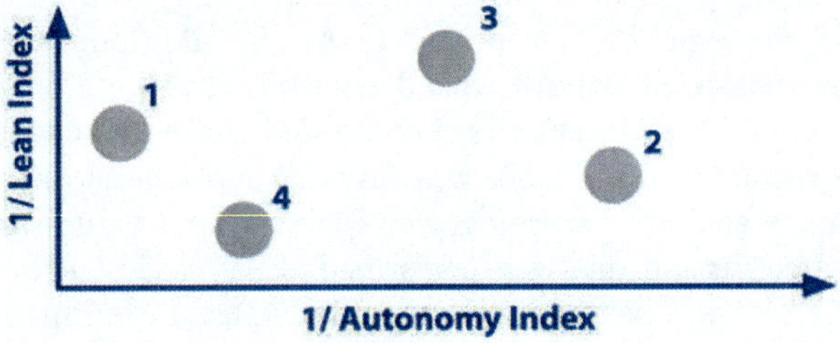

Fig. 4 Scatter plot for the analysis of the correlation of autonomy and lean index

5 Outlook

After the installation of the MES as well as the finalization of the construction of the hybrid simulation environment, it has to be analyzed how to record data relevant for the analysis presented from the MES. It would be possible to compare the results of several processes simulated. A template for the data dictionary is necessary. On this account a number of different autonomous processes have to be analyzed firstly. All relevant information will be extracted and documented separately. A comparison allows the recognition and filtering of recurring data. With the creation and comparison of numerous scatter plots regularities are worked out. It is examined whether it is possible to define rules regarding specific industries or manufacturing techniques.

Acknowledgments The project upon which this publication is based is funded by the German Federal Ministry of Economics and Technology (BMWi) under the project number: 01MA09018A. This publication reflects the views of only the authors.

References

1. Lupo-Projekt. http://www.lupo-projekt.de
2. Lass, S., Theuer, H., Gronau, N.: A new approach for simulation and modeling of autonomous production processes. In: Proceedings of the 45th Hawaii International Conference on System Sciences (HICSS 2012), Maui, Hawaii, pp. 1247–1256. Jan 2012

3. Association of German Engineers (VDI) : VDI-Guideline 4499—Digital Factory—Basics, (Verein Deutscher Ingenieure). Beuth, Berlin (2008)
4. Muhs, D., et al.: Roloff/Matek—Machine Elements, Wiesbaden, Germany, p. 9 (2007) (in German)
5. Westkämper, E.: Management of production in a turbulent environment. In: PPS Management 12, 2. Berlin, p. 70 (2007) (in German)
6. Scholz-Reiter, B., Lütjen, M.: Digital factory—basic approaches of an integrated design of products and processes, In: Industrie Management 25, 1. Berlin, p. 19 (2009) (in German)
7. Freund, G.: Development of a Methodical Procedure for Implementation of Digital-Mock-Up- Technologies in Automotive Product Construction, Freiberg (2004) (in German)
8. Syska, A.: Production Managment: Important Methods and Concepts for Today's Production, Gabler Wiesbaden, (2006) (in German)
9. IRT Homepage: IRT Modellfabrik. http://www.irt.rwth-aachen.de/institut/ausstattung/irt-modellfabrik/?type=99
10. Institut für Produktionsmanagement, Technologie und Werkzeugmaschinen|TU Darmstadt. http://www.prozesslernfabrik.de/
11. Gronau, N., Theuer, H., Lass, S., Nguyen, V.: Productivity evaluation of autonomous production objects. In: Proceedings of the 8th IEEE International Conference on Industrial Informatics. Osaka, Japan, p. 751 July 2010
12. Rother, M., Shook, J.: Learning to See: Value Stream Mapping to Add Value and Eliminate MUDA. Lean Enterprise Institute, Cambridge (1999)
13. Erlach, K.: Value Stream Design: The way to a Lean Enterprise. Springer, Berlin (2007) (in German)

Robust Manufacturing Through Integrated Industrial Services: The Delivery Management

Horst Meier and Thomas Dorka

Abstract Industrial Product-Service Systems (IPS^2) can be used to provide robust manufacturing. To support the providers of IPS^2 in the management of a provider network and in planning of resources for the IPS^2 delivery, the cloud-based IPS^2-Execution System (IPS^2-ES) is proposed in this paper. By analyzing the responsibilities the IPS^2 provider has to accept in his role, the functionalities the system has to provide can be derived. The requirements therefore include possibilities to manage a network of suppliers, the flexible handling of fluctuations in the network, potentials to adapt to different suppliers, customers or partners automatically and offer strategic and operational real-time planning of resources (e.g. by a genetic algorithm and using cloud resources). Not least, the integration of a system which is able to monitor and control the IPS^2 delivery on customer premises is necessary to provide highest production control flexibility for the IPS^2 customer.

Keywords Industrial Product-Service System (IPS^2) · IPS^2-Execution System · Cloud Computing

1 Introduction

Robustness has become highly relevant in the manufacturing industry because production downtimes have a big effect on the manufacturers' business. To ensure continuous production as needed, the production processes of the manufacturers have to be robust as well as the manufacturing systems that are used. Thus, the

H. Meier · T. Dorka (✉)
Ruhr-Universität Bochum, Bochum, Germany
e-mail: dorka@lps.ruhr-uni-bochum.de

H. Meier
e-mail: meier@lps.ruhr-uni-bochum.de

K. Windt (ed.), *Robust Manufacturing Control*, Lecture Notes in Production Engineering, 415
DOI: 10.1007/978-3-642-30749-2_30, © Springer-Verlag Berlin Heidelberg 2013

manufacturing systems are one of the keys to success. To be able to leverage the machine as required, the manufacturer has to take care of several factors that are valuable for him, e.g. the functionality of the system, the availability of the system or the number of produced goods in a given timeframe. Industrial Product-Service Systems (IPS^2) represent a new way to provide manufacturing systems that offer the required customer value. Instead of buying a manufacturing system and ordering services, the manufacturer now buys a system with integrated industrial services. Thus, the customer purchases the required value (availability, functionality or result) from an IPS^2 provider instead of taking the responsibility for himself. Hence, the IPS^2 provider is responsible for the manufacturing system. In turn, this gives the manufacturer the opportunity to shift his focus to optimization of his production processes. Due to the transfer of responsibility for the manufacturing system to the IPS^2 provider, he has to ensure the delivery of the customer value. For this, service delivery in networks and detailed planning for IPS^2 are required. After giving an overview of the delivery of IPS^2, this paper describes which responsibilities the IPS^2 provider has to take and how he can be supported in their management. Based on this, available software systems for supporting the provider in his duties are reviewed and a concept for a new software system is proposed.

2 State of the Art

Industrial Product-Service Systems (IPS^2) are a way of providing customer value in an industrial context in contrast to selling products and offering services separately [1], [2]. Beyond that, an IPS^2 is not only the sum of product and service shares, but rather an inseparable integration of these shares in planning, development and provision. A definition of IPS^2 is as follows [3]:

> An Industrial Product-Service System is characterized by the integrated and mutually determined planning, development, provision and use of product and service shares including its immanent software components in Business-to-Business applications and represents a knowledge-intensive socio-technical system.

Earlier fundamental research has shown that providers of IPS^2 can differentiate from competitors in the market by delivering extended customer value [4]. Challenges during the delivery are already addressed by research on concepts like Provider Networks, IPS^2 Networks and IPS^2 Delivery Networks [5] or on the complex adaptive IPS^2 resource planning [6]. However, the support tools and systems to provide IPS^2 are not sufficient yet [7].

2.1 IPS^2 Service Delivery in Networks

During the operation phase of an IPS^2, the provider has to ensure the customer value agreed upon in the contracted business model. Therefore, several delivery processes have to take place. These delivery processes can be maintenance

Fig. 1 Dynamics of the networks for IPS^2

procedures, repair tasks and even the training of workers or technology upgrades. All processes have several requirements that have to be met to execute them. Among them can be specially skilled workers, dedicated tools or third-party spare parts.

An IPS^2 provider cannot necessarily fulfill all these requirements. Hence, he has to work in a network with other companies to carry out the delivery processes [8]. Each company in the network provides the contracted resources, e.g. service technicians, spare parts, tools or Industrial Product-Service Modules (IPSM), which the provider can then utilize. Among these companies, there is also the customer of the IPS^2, who can especially provide local resources, such as compressed air or electricity. The whole network is defined as the Provider Network (PN) and includes the provider himself, the customer and third-party companies [4].

Since every IPS^2 offered by a provider can be different, the associated processes and therefore the required resources can be different. Thus, every IPS^2 might need a slightly different subset of the PN. This IPS^2 specific network is called IPS^2 network [4]. For each of the processes for an IPS^2, only resources of some of the companies in the IPS^2 network are involved. The network of companies involved in the delivery of a process is called the IPS^2 delivery network (IPS^2 DN, [4]).

The aforementioned networks are dynamic (see Fig. 1), as they change over time. Partners can be added to or removed from the PN. Furthermore, if the contracted business model of an IPS^2 is changed, the required processes and therefore the required resources change. This in turn has an effect on the IPS^2 network.

2.2 Adaptive IPS^2 Planning

To determine the required capacities of resources in the PN and to schedule all processes that have to be executed for the different IPS^2 of a provider, the strategic capacity planning and the operational resource planning methods are used. Strategic capacity planning is the key to create an initial delivery plan and ensure sufficient capacities throughout the life cycle of the IPS^2. Whenever an unplanned event (such as a broken machine part) requires a quick response, operational resource planning needs to take place to include a new process (e.g. a repair or a part replacement) in the delivery plan [6].

Strategic capacity planning ensures that the resources needed to execute the IPS^2 processes in the future can be provided on time and simultaneously protects the provider from overstocking. Peaks in the use of different resources are covered by partners in the PN and allow the provider to cope with these special demands while keeping reasonable amounts of manpower and stocks over time.

Based on the strategic guidelines, the initial delivery plan is created by the scheduling of delivery processes and the corresponding assignment of resources. To execute a delivery process for an IPS^2, several resources are needed. Therefore, each delivery process has a list of requirements that need to be fulfilled to carry it out, e.g. special skills, tool types or spare part types. To fulfill these requirements, different human resources (e.g. service technicians), tools or spare parts are available in the PN and can be assigned to the different delivery processes [4].

Operational resource planning on the other hand is the change of a delivery plan based on requirements that arise during the operation of an IPS^2. A new delivery process has to be included into the delivery plan and hence, the assignment of resources and the schedule of the processes have to be recalculated [4].

To provide sufficient degrees of freedom for both strategic capacity and operational resource planning for IPS^2, resources have to be available from one or more partners in the IPS^2 network. Several factors have an influence on which resource should be used in a delivery process. These factors include limited availability of workers or IPS^2, logistics and travel from one delivery location and different skills of workers. If multiple partners are in the PN and multiple IPS^2 have to be supported in possibly several countries, a highly complex planning problem arises, which has to be solved. The principle of simultaneous creation and consumption of services (simultaneity, [9]) requires, that each process has to be delivered in time.

One solution to solve this problem using a heuristic algorithm is suggested by [4]. Due to the complexity of the heuristic algorithm, high computing power is needed to help support all of the above factors. Cloud computing can help provide the required computing power [10].

2.3 Cloud Computing

Since Amazon brought the first elastic cloud computing service online [11], the use of cloud computing has spread to several companies. It is also described as “the” IT trend topic in [12]. The following definition of cloud computing can be found [13]:

> Cloud computing is a model for enabling ubiquitous, convenient, on-demand network access to a shared pool of configurable computing resources (e.g., networks, servers, storage, applications, and services) that can be rapidly provisioned and released with minimal management effort or service provider interaction. This cloud model is composed of five essential characteristics, three service models, and four deployment models.

The definition of the five essential characteristics and the three service models are further described in [13]. The deployment models are (1) private cloud, i.e. the cloud system is exclusively used by one company, (2) community cloud, in which several organizations can form a community and share a cloud system, (3) public cloud, where generally the open public can use the cloud, and (4) hybrid cloud, which is a composition of two or more of the above with data interfaces between the cloud systems [13].

Due to the possibilities cloud computing provides and the flexibility it offers, many software systems are already available in the cloud. Among these are office products [14] or ERP systems [15].

3 Responsibilities of IPS^2 Providers During the Operation Phase

After an IPS^2 has been planned, developed and implemented by a provider, he is still responsible for the IPS^2 during the operation phase. In this phase his main duty is to ensure the operation of the provided IPS^2 as agreed upon in the contracted business model. He is in charge of representing the IPS^2 as a whole to the customer in the manner of "one face to the customer".

The required delivery processes for a proper operation are already developed and are described in the IPS^2 product model. However, the allocation of adequate resources is still needed for the execution of the processes of an IPS^2 at a designated time and location and is part of the responsibilities of the provider. To avoid resource shortages, the provider has to plan the different resource capacities accordingly, e.g. by stocking resources or including additional partners in his PN. The qualitative and quantitative decisions as to which and how many resources are stocked, as well as the question of whether resources shall be provisioned or produced by the IPS^2 provider or another party, etc. have therefore to be made and answered strategically. This also ensures the flexibility needed for resource planning.

The provider also has to take care of network fluctuations, i.e. a partner leaving or a new partner joining a PN and the connected changes to strategic planning. The network itself can be distributed across several different countries. This allows for supporting IPS^2 globally by using network partners in the vicinity of the location of each IPS^2, e.g. on customer premises.

When partner resources are involved in the execution of a process, the provider has to ensure that these partners have access to associated process descriptions and details. Additionally, some parties of an IPS^2 DN involved in a process delivery might have to exchange information. The provider has to make sure that communication in the IPS^2 DN is supported in a structured and extensible way and all required data can be exchanged.

Whenever an unexpected event occurs at an IPS^2, the provider has to take care of this incident. Based on the contracted response times or availabilities, new processes need to be scheduled in real-time to deliver the promised customer value. Hence, a system that detects incidents and is able to monitor and control the IPS^2 and the execution of delivery processes has to be integrated. This system can then offer valuable information about each IPS^2 and its delivery processes to the IPS^2 provider.

Throughout the operation phase, the provider has to arrange the collection of data for knowledge extraction. This extraction can be either fully automated or supported by an adequate system. For example, accumulated occurrence of incidents and related process scheduling together with the original operation schedule can give valuable information for the redesign of processes and therefore need to be accessible.

Across all tasks and interaction, the provider has to ensure that all data and information exchanged under his administration is handled securely and vconfidentially. Especially information about human resources and details about IPS^2 customers and partners should be handled with best care and attention.

Not least, the provider has to supervise the operation of all IPS^2 offered and keep track of the information of processes. This includes the coordination of delivery processes and the network management for potentially hundreds of IPS^2, so that all customer needs can be met.

To summarize, the provider has to accept the following responsibilities:

- Represent the IPS^2 to the customer
- Assign resources to processes
- Ensure sufficient resource capacities and plan resources strategically
- Manage fluctuations in the global provider network
- Make process descriptions and details available to partners involved in a process
- Support communication in IPS^2 DN and the exchange of required information
- Allow for short response times to incidents
- Provide information about incidents and the operation schedule
- Manage data securely and confidentially
- Supervise the operation of all IPS^2

Fulfilling all the aforementioned demands without proper tool support is not possible. Especially when global partner networks are involved and multiple IPS^2 have to be maintained, the complexity of the requirements become even more visible. As a consequence, a tool or a set of tools offering support for the duties of the provider has to be selected or developed. Thus, available software systems have to be analyzed with regard to their applicability.

4 Available Software Systems for the IPS^2 Provider

There are several software systems available on the market with focus on the industrial production and service sector. Among these are planning, process data acquisition and machine monitoring systems. Examples are Enterprise Resource Planning systems (ERP), Production Planning and Scheduling systems (PPS, also

Production Planning and Control), Advanced Planning Systems (APS), Customer Relationship Management systems (CRM), Service Management Systems (SMS) and Manufacturing Execution Systems (MES) as well as systems for personnel time tracking, maintenance planning and Quality Management (QM) [16] and [17] give a good overview over products currently available on the market and their features.

The systems have their special field of application. For example, SMS are tools to schedule and plan service assignments, whereas MES are used for detailed manufacturing planning and organization. The former are often included in or connected to a CRM systems, the latter are typically connected to an ERP system, which represents the master system for data. The high interconnection between the available solutions is obvious.

Among the aforementioned software solutions, ERP systems seem to provide the highest support for an IPS^2 provider. However, they do so only within limits. Especially detailed planning is often excluded from ERP systems to avoid negative effects on the system performance [18]. Hence, the functionality of an ERP system alone is not sufficient, because planning is one of the key features needed for the provisioning of IPS^2.

The other existing applications are either focused mainly on products (e.g. MES and PPS focus on production systems), or centered on services (like SMS), do not include the management of a network of partners integrating the customer or are only built to be used by one of the involved parties. Even if they support resource planning, the strong interaction with partners in dynamic networks is not taken into account. In contrast to the conventional approaches, product and service shares as well as planning and network management for IPS^2 are tightly coupled and interdependent in an IPS^2. Thus, they need to be managed simultaneously as a whole, involving all participants in the PN, including the customer.

Hence, the responsibilities that the IPS^2 provider has to take can not be supported by a single system of the ones mentioned above [10], [4]. Even a landscape of those systems would at least need some new features to be implemented. Therefore, the specific requirements of IPS^2 need to be covered by a new system that incorporates special functionality. Because of the complex structures, a landscape of the existing systems has to be enriched by a new tool to fill the gap for an effective support of the IPS^2 provider. This also allows for using cloud computing without changing existing software solutions.

5 Requirements and Concept for an IPS^2-Execution System

As explained in the chapter above, already established software systems are not capable of supporting the special needs of IPS^2 providers during the operation phase of IPS^2. Thus, a new system needs to be developed that can be integrated with the other system through loose coupling. The responsibilities of the provider as well as the consideration of the integrated service and product shares are the special requirements of IPS^2 that have to be covered. Hence, the IPS^2-Execution

System (IPS^2-ES) is proposed. Due to the special requirements of IPS^2 the following demands arise for this IPS^2-ES.

The responsibilities of the provider can be divided into three categories: First, there are duties that are not connected to a specific IPS^2 but are general in nature: assigning resources to processes, ensuring sufficient resource capacities and planning resources strategically, managing fluctuations in the global network, and providing information about incidents and the operation schedule. All these are mainly connected to the PN or to the development of IPS^2 managed by the provider.

Furthermore, there are obligations connected to a specific IPS^2 and the corresponding customer: representing the IPS^2 to the customer, allowing for short response times to incidents, and supervising the operation of all IPS^2. These are influenced by the business model and the connected contract for the IPS^2 [19].

On a more detailed level, the responsibilities are connected to delivery processes: making process descriptions and details available to partners involved in a process, and supporting communication and exchange of required information in IPS^2 DN. These duties are of special interest for the partners. In all three categories, the last liability of the provider has to be supported: managing data securely and confidentially.

5.1 Levels of the Execution System

Information management for the IPS^2 networks is complex and needs enabling technology, e.g. access control [20]. Hence, to represent the aforementioned categories and to have a basis for access control, three interdependent levels in the IPS^2-ES are characterized, on which it has to support the IPS^2 provider. These levels are the provider level (PL), the IPS^2 level (IL) and the delivery process level (DPL). The PL supports functionality needed for all IPS^2 offered by the provider, the IL supports each individual IPS^2 and the DPL supports the execution of different delivery processes. This partition is also reflected by the availability of data. While an IPS^2 customer can only see the data connected with his specific IPS^2 on the IL, the provider needs to have access to a broader set of data. Therefore, the PL provides an overview over all IPS^2 offered. On the DPL, a third party supplier that takes part in a delivery process for a customer can get information about that specific process and the involved parties. Hence, the different levels present different views for the IPS^2-ES users and reflect their unique requirements to the system. They also represent the responsibilities of the provider and are related to the different networks for IPS^2: provider network, IPS^2 network and IPS^2 delivery network (see Fig. 2).

On the PL, sufficient resource capacities to deliver all contracted IPS^2 customer value have to be ensured. Therefore, a network of partners needs to be established and maintained. The mechanisms introduced in [5] have to be supported by the software.

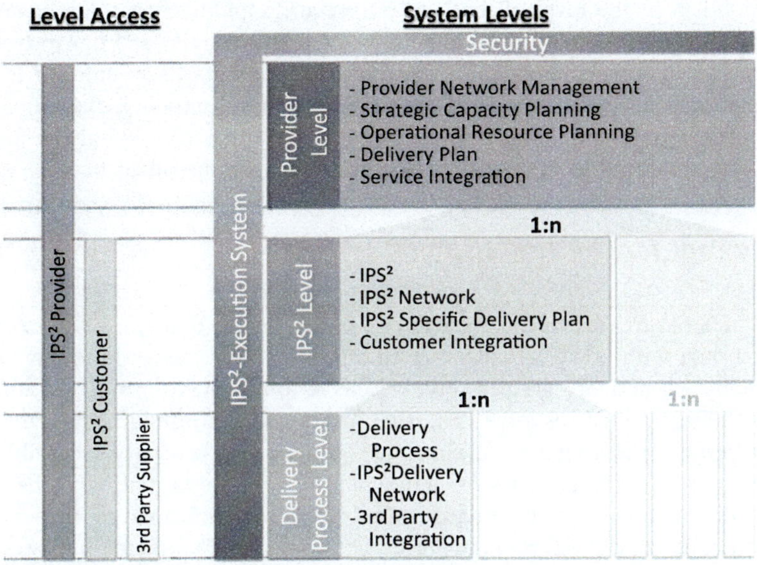

Fig. 2 System levels of the IPS^2 Execution System and role access to these levels

Strategic capacity planning as explained in [21] presents a compass that provides decision support that can be directly used to initiate required preproduction processes and storage. Based on that, resource planning can be executed as presented in [4]. The generated information, e.g. the deliver plan and the occurrence of incidents, has to be made available for storage and future analysis and has to be sent out to network partners to promote transparency of upcoming tasks.

In contrast to the PL, which allows management of the PN and supports planning methods, the IL is mainly an information system. Still, it has to trigger the PL whenever response to an incident reported by one of the IPS^2 is required, so that resource planning can be executed again. This is done to fulfill the shortly arising needs of one IPS^2 while providing standard support for the others. Apart from that, information about the IPS^2, its operation and its projected delivery plan has to be presented to the customer and to the provider.

Lastly, the DPL is the stage at which details of a special delivery process are exchanged and clarified. Process descriptions and details have to be made available to the involved parties, i.e. the provider, the customer and third-party suppliers, as long as possible prior to their execution. On this platform, automated communication as well as required manual data exchange and personal contact have to be provided. For example, the ordering of the production of required spare parts can be automatically triggered. During and after the process execution, the data needs to be accessible for a specific time to include more sophisticated features like customer feedback on process and personnel quality.

Independent of the level of the IPS^2-ES, security and access rights have to be taken into account. All data have to be handled confidentially and has to be made sure that the customer of one IPS^2 can only see the data he is entitled to. Thi-s excludes especially data about IPS^2 of other customers and the included partners as well as data about the PN as a whole. Also, third parties should only be allowed to see data connected to the delivery processes they are involved in.

5.2 Concept for the IPS^2-ES

The IPS^2-ES has to present the exclusive software system that is used to exchange operation data with the IPS^2 customer to allow for "one face to the customer". Additionally, all network partners involved have to be able to provide availability data for their provided resources and they need to be informed about when and where to provide these resources. Thus, the system has to be easily accessible by a user interface, provide services for automated data requests and have the possibility to actively connect to external systems for pushing information. Based on the strategic parameters, the IPS^2 network can be initiated. Contracted partners, who are capable of providing resources for the offered IPS^2, need to be included in the network as well as the IPS^2 customers and the IPS^2 provider himself. Data for personal contact as well as automated data exchange need to be collected and maintained. An automatic coupling of the IPS^2-ES with the software systems used by the network partners (e.g. ERP, MES, SMS, etc.) has to be targeted. This might also mean the integration of existing service interfaces that are used to automatically initiate processes like logistics. Additionally, a connection to an agent system that monitors and controls the IPS^2 and its delivery processes has to be established [22]. For that, an open and dynamic architecture needs to be implemented that reconfigures and optimizes itself whenever needed. This also enables the system to handle fluctuations of the network easily.

The management of the network has to focus on the following tasks: handling the integration and retirement of network partners, the partner communication (automatically as well as personally) and the building of IPS^2 networks and IPS^2 DN. The building of networks is closely connected to the resource planning and is therefore covered by the algorithm used there. To support the integration of new partners, standardized service registries can be leveraged. All configuration data needed to connect to the systems of the partners has to be made available so that the IPS^2-ES can use required partner interfaces. This implies that the adapters to connect to those interfaces have to be available. Again, the openness of the system, e.g. by integrating a plug-in system, can allow for easy extension of the IPS^2-ES by adding new adapters.

IPS^2 can be provided to customers all over the world. Hence, the need to provide resources for required processes in different locations arises. The communication between the partners therefore needs to cross borders and has to be available over a global network like the internet. To avoid security concerns when using the internet as a communication medium, secure connections, e.g. using virtual private networks, can be used without limiting the given potentials.

The most computing resource intensive part of the IPS^2-ES is the resource planning system. While acceptable delivery plans need to be created or changed in short time, the algorithm to handle this is using complex evolutionary algorithms from metaheuristics. Especially the evaluation of possible solutions is very computing-intensive [23]. However, while no planning takes places, no computing resources are needed. This fits ideally to the flexibility of cloud computing. Resources can be allocated when needed and freed when unused. However, especially when using a public cloud system, several security aspects have to be considered. For example, depending on the location of the cloud system, different laws are applicable. Laws like the United States Patriot Act [24] grant authorities the access to all data stored on servers located in the U.S, which conflicts with the european law for protection of personal data [25]. Using private cloud systems or using data anonymization can help overcome these issues, but still other security aspects have to be taken into account.

6 Conclusion and Outlook

To support the provision of robust manufacturing systems, the key elements on the organizational level of IPS^2 have been introduced. Then, the responsibilities that have to be accepted by an IPS^2 provider have been described. Based on these duties, available software systems have been analyzed to find software tools to support the provider. Since none of the systems provides the required functionalities, the requirements for developing an IPS^2-ES have been outlined. Therefore, the different levels of the IPS^2-ES have been introduced as provider level, IPS^2 level and delivery process level. These levels also show the accessibility of data for the different roles.

Then, a concept for the IPS^2-ES has been given. The functionalities the system has to offer are: an easily accessible user interface, services for automated requests, information push capability, automatic coupling with other software systems via adapters, an open and dynamic architecture, self-optimization and self-configuration, the usage of standardized service registries, availability over the internet, secure data management and using cloud computing to provide effective real time planning. Through the support of these tasks, IPS^2 can be provided effectively and robust manufacturing is possible.

Still, the architecture of the software IPS^2-ES has to be further developed and services for use in the system as well as interfaces have to be designed. The implementation of a planning algorithm capable of using cloud computing is needed to provide essential functionality. Especially a focus on integration with existing software has to be set. An IPS^2-ES prototype for demonstrating the value of the system by offering the required support has to be realized. If these developments are given, the proposed concepts can be used to ultimately provide robust manufacturing systems.

Acknowledgments The authors would like to thank the German Research Foundation (Deutsche Forschungsgemeinschaft, DFG, www.dfg.de) for funding their research within the project Transregio 29 "Engineering of Industrial Product-Service Systems" (www.tr29.de).

References

1. Aurich, J.C., Clement, M.H. (eds.): Produkt-Service Systeme: Gestaltung und Realisierung. Springer, Berlin (2010)
2. Durugbo, C., Bankole, O.O., Erkoyuncu, J.A., Tiwari, A., Alcock, J.R., Roy, R., Shehab, E.: Product-service systems across industry sectors: future research needs and challenges. In: Sakao, T. (ed.), Industrial product-service systems (IPS^2): Proceedings of the 2nd CIRP IPS^2 Conference [2010, Linköping, 14–15 April] 535–542. Linköping University, Linköping (2010)
3. Meier, H., Roy, R., Seliger, G.: Industrial product-service systems—IPS2. CIRP Ann. Manuf. Technol. **59**(2), 607–627 (2010)
4. Meier, H., Funke, B.: Resource planning of industrial product-service systems (ips^2) by a heuristic resource planning approach, In: Sakao, T. (ed.) Industrial product-service systems (IPS^2): Proceedings of the 2nd CIRP IPS^2 Conference [2010, Linköping, 14–15 April]. Linköping University, Linköping (2010)
5. Meier, H., Völker, O.: Aufbau- und ablauforganisation zur erbringung hybrider leistungsbündel. In: Meier, H., Uhlmann, E. (eds.) Integrierte Industrielle Sach- und Dienstleistungen: Vermarktung, Entwicklung und Erbringung hybrider Leistungsbündel, pp. 137–161. Springer, Berlin (2012)
6. Meier, H., Funke, B.: Planungsmethoden für den betrieb hybrider leistungsbündel. In: Meier, H., Uhlmann, E. (eds.) Integrierte Industrielle Sach- und Dienstleistungen: Vermarktung, Entwicklung und Erbringung hybrider Leistungsbündel, pp. 163–190. Springer, Berlin (2012)
7. Sakao, T., Berggren, C., Björkman, M., Kowalkowski, C., Lindahl, M., Olhager, J., Sandin, J., Sundin, E., Tang, O., Thollander, P., Witell, L.: Research on services in the manufacturing industry based on a holistic viewpoint and interdisciplinary approach. In: Hesselbach, J., Herrmann, C. (eds.), Functional Thinking for Value Creation: Proceedings of the 3rd CIRP International Conference on Industrial Product Service Systems, 27–32. Springer, Berlin (2011)
8. Becker, J., Knackstedt, R., Pfeiffer, D. (eds.): Wertschöpfungsnetzwerke: Konzepte für das Netzwerkmanagement und Potenziale aktueller Informationstechnologien, 1st edn. Physica-Verlag, Heidelberg (2008)
9. Fitzsimmons, M.J., Fitzsimmons, J.A.: Service management: Operations, strategy, and information technology, 5th edn. McGraw-Hill/Irwin, Boston (2006)
10. Meier, H., Funke, B., Dorka, T.: Cloud computing für eine integrierte leistungssteuerung. Ind. Manage. **2012**(1), 49–52 (2012)
11. Hof, R.D.: Jeff Bezos' Risky Bet, Business Week, 13 Nov. (2006). http://www.businessweek.com/stories/2006-11-12/jeff-bezos-risky-bet
12. Spath, D., Weiner, N., Renner, T., Weisbecker, A. (eds.): Neue Geschäftsmodelle für die Cloud entwickeln: Methoden, Modelle und Erfahrungen für "Software-as-a-Service" im Unternehmen. Fraunhofer Verlag, Stuttgart (2012)
13. Mell, P., Grance, T.: The NIST definition of cloud computing recommendations of the national institute of standards and technology. NIST Spec. Publ. **145**(6), 1–7 (2011)
14. Microsoft Corporation: Office365. http://www.office365.com Accessed 19 Mar 2012
15. Eggert, S.: ERP in der cloud: wie können sich cloud-anwender rechtlich absichern? ERP Manage. **7**, 34–36 (2011)
16. Theuer, H., Leo, A.K.: IT in der produktion. Productivity Manage. **16**(4), 37–45 (2011)
17. Al-Scheikly, B.: Marktübersichten. http://www.it-production.com/markt Accessed 29 Feb 2012

18. Roth, F.: Die grenzen klassischer ERP-systeme. In: Jacob, O. (ed.) Xpert.press, ERP Value, pp. 61–74. Springer, Berlin (2008)
19. Rese, M., Meier, H., Gesing, J., Boßlau, M.: HLB-Geschäftsmodelle: Partialmodellierung zur Systematisierung von Geschäftsmodellen "Hybrider Leistungsbündel" (HLB) (Deutsch), wt Werkstattstechnik online. **101**(7/8), 498–504 (2011)
20. Periorellis, P., Parastatidis, S.: Task-based access control for virtual organizations, In: Guelfi, N., Reggio, G., Romanovsky, A. (eds.) Lecture Notes in Computer Science, vol. 3409, Scientific Engineering of Distributed Java Applications: 4th International Workshop, FIDJI 2004, Luxembourg-Kirchberg, Luxembourg, Nov 24–25, 2004, Revised Selected Papers, pp. 38–47. Springer, Berlin (2005)
21. Krug, C.M.: Framework zur strategischen Kapazitätsplanung hybrider Leistungsbündel. Aachen: Shaker (2009). ISBN:978-3-8322-9351-2
22. Uhlmann, E., Geisert, C., Raue, N., Stelzer, C.M.: Automatisierungstechnik für erbringungsprozesse hybrider leistungsbündel. In: Meier, H., Uhlmann, E. (eds.) Integrierte Industrielle Sach- und Dienstleistungen: Vermarktung, Entwicklung und Erbringung hybrider Leistungsbündel, pp. 245–263. Springer, Berlin (2012)
23. Luke, S.: Essentials of Metaheuristics: A Set of Undergraduate Lecture Notes. Lulu, Raleigh (2009). http://www.cs.gmu.edu/~sean/book/metaheuristics/ . ISBN:978-0-557-14859-2
24. United States Government: Public Law 107—56—Uniting and Strengthening America by Providing Appropriate Tools Required to Intercept and Obstruct Terrorism: USA PATRIOT ACT (2001)
25. Europäisches Parlament: Richtlinie 95/46/EG des europäischen Parlaments und des Rates vom 24. October 1995 zum Schutz natürlicher Personen bei der Verarbeitung personenbezogener Daten und zum Datenverkehr (1995)

Enhancements of a Logistic Model to Improve the Time Synchronicity of Convergent Supply Processes

Sebastian Beck, Friedrich Gehler and Peter Nyhuis

Abstract The assembly plays a key role in ensuring the competitiveness of industrial enterprises. From a logistics perspective, one of the central challenges with assembly processes is simultaneously and punctually supplying the necessary parts at the date required. In doing so, the goal is to satisfy the customers' desire for short throughput times and high schedule reliability, thus keeping inventory related costs as low as possible. The content of this article is the transfer of the idea of the supply diagram to the assembly process. With the help of the assembly throughput diagram and SCHMIDT's method using finite process elements to create the supply diagram it is possible to build two new diagrams integrating the point of assembly start and the assembly end into consideration. These diagrams are similar to the supply diagram but focus the successive process and provide complete different opportunities of interpretation. Besides the visualisation of order sequence interchanges or capacity flexibility it is possible to evaluate the quantity of disrupted WIP and regular WIP of an assembly process. The use of the two models supports analysing the logistic performance like the schedule reliability and reveals shortcomings in the concerned period.

Keywords Schedule reliability · Model · Assembly

S. Beck (✉) · P. Nyhuis
Institut für Fabrikanlagen und Logistik,
Leibniz Universität Hannover, Hannover, Germany
e-mail: beck@ifa.uni-hannover.de

P. Nyhuis
e-mail: nyhuis@ifa.uni-hannover.de

F. Gehler
Gesellschaft für Technologie Transfer mbh,
Hannover, Germany
e-mail: friedrich.gehler@gtt-online.de

K. Windt (ed.), *Robust Manufacturing Control*, Lecture Notes in Production Engineering, DOI: 10.1007/978-3-642-30749-2_31,

1 Introduction

Currently, producing industrial companies face a host of tasks and changing framework conditions. Both the growing international competition and the increasing prices for raw materials caused by resources becoming more scarce, as well as the differentiation of the customer requirements, pose major challenges to companies. The change from the supply-driven market towards a demand-driven market in recent years has caused far-reaching changes in many industrial companies [1]. Thus, companies find it difficult to distinguish themselves from the competition in terms of product prices and product quality and thereby achieve competitive advantages.

Many companies attempt to stand out from the competition by their logistical characteristics and move into the centre of the customers' attention by short delivery times and high delivery reliability. Since today's customer behaviour is characterised by individual requests for multiple product variants, companies are expected to be able to react flexibly to changing needs of customers. Companies can and must react to that by advancing the customer order decoupling point before assembly, in order to be able to flexibly create the marketable customized product with the assembly process [2].

Here the confluence of material and information flows in assembly entails a high degree of complexity and high control requirements; thus high logistic performance of the production management is needed to supply assembly on schedule and within budget. The situation is aggravated by the fact that in assembly, as the final stage of the value chain, all delays from previous process stages accumulate and thus directly affect the schedule situation for assembly [3]. At the same time, the system of objectives for logistics must take into account the actions required in the area of conflict between high performance and low costs in logistics [4]. This explains why over the last few years assembly has gained in importance relative to production.

In order to meet the diverse requirements referred to, production planning must find ways and means to collect the large quantity of data converging in assembly, to filter them quickly and expediently and to make them accessible to the viewer. To this end, models are suitable that generate meaningful performance figures and so allow for an overview of the progress of the production processes that generates transparency, thereby providing the basis for adapted planning processes and situation-optimized decisions [5].

Here the use of models can be effective only if they identify trends and deviations on the basis of planned versus actual status comparison and thus allow to determine whether intervention is necessary or not.

2 Existing Models for the Assembly Process

2.1 The Assembly Throughput Element

For the collection and presentation of the complex logistical situation in assembly, the descriptive model for assembly throughput elements by Schmidt is available [6].

Figure 1 shows the basic idea of the assembly throughput element (ATE). Here, three supply paths converge into assembly. In the ATE, the time shares of the individual sub-processes of a working process are plotted over a time axis. This is measured universally in time units, since the choice of the time dimension (shop calendar days, hours etc.) depends on the accuracy of the feedback from the assembly considered. The time shares are: Queuing after calling or after processing, respectively; transport; queuing before assembly; set-up; and assembly.

The key data of the ATE are the individual dates which correspond to the feedback from production data acquisition PDA. These are essentially the completion dates of the supply paths and the start and end dates of the actual assembly work. Apart from a few master data, such as the value of components or the working time of the assembly processes, they form the basis for the calculation of the key figures derivable from the ATE and presented in Fig. 1.

2.2 The Assembly Throughput Diagram

Based on the ATE, Schmidt created the assembly throughput diagram. It plots several assembly orders to an assembly system, weighted by value. Here the three aforementioned supply processes contribute to each assembly process with their individual sizes and values. The assembly orders are shown in Fig. 2, above, chronologically arranged in a row. From this diagram for the analysis of logistic relationships four curves can be derived, presented in Fig. 2.

The input curve represents the supply processes of all assembly orders under consideration, cumulated over time. Here, the currently most recently received supply order forms the completing order and thus a section of the completing curve amounting to the value of the assembly order or of the sum of the corresponding supply orders, respectively. The starting points of the assembly operations (set-up and assembly) over time form the assembly start curve. The course makes it clear that only one assembly system is considered, as the starting point of one assembly process corresponds to the endpoint of the previous assembly process. The fourth curve, the output curve, is formed by the assembly completion time points and the values of the assembly processes.

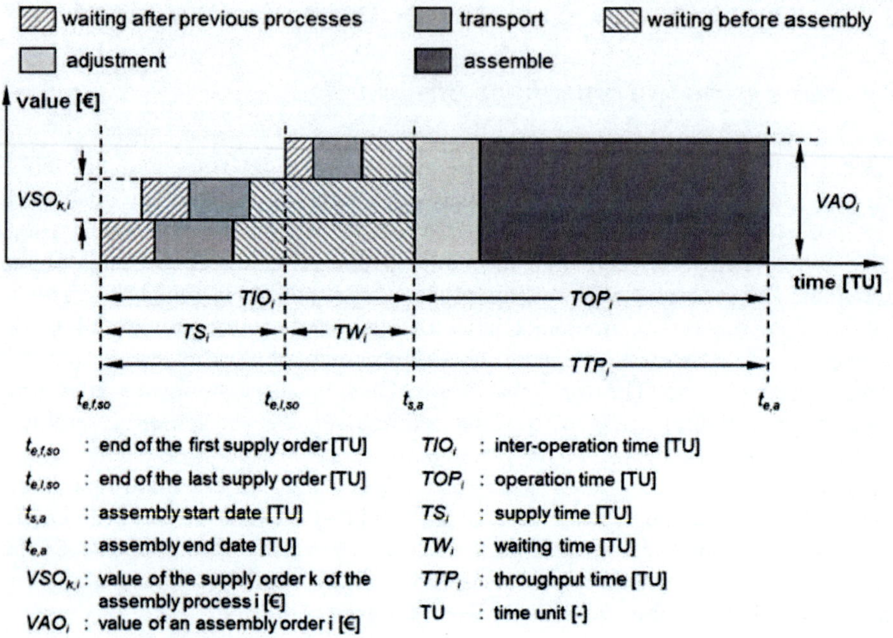

Fig. 1 The assembly throughput element [6]

2.3 The Supply Diagram

Building on the development of the supply diagram by Nyhuis, Nickel and Busse [7], Schmidt opts for deduction of the supply diagram using the assembly throughput diagram.

Figure 3 depicts the procedure resulting from this model concept for constructing the value weighted Supply Diagram. The starting point is the 'target-actual comparison' via the value related assembly throughput element oriented Assembly Throughput Diagram, presented in Part A. Here, six generic assembly throughput elements can be seen along with the dates they are required on. In Part B, the assembly throughput elements are converted into a standardized depiction with regards to the date required. In Part C, the supply orders are first sorted according to their input lateness and weighted with their values before being cumulatively plotted in a curve (standardized input curve). For the chronological first three in-going supply orders, this is made visible with the dashed line. In contrast to conventional models of Supply Diagrams, the last supply orders provided for an assembly operation are also monetarily considered.

Following that, the completed assembly operations along with their values are sorted according to their standardized completion dates—that is the lateness of the last supply orders—and cumulatively plotted in a graph. The result is the Supply Diagram depicted in Part C with the standardized input curve and standardized completion curve.

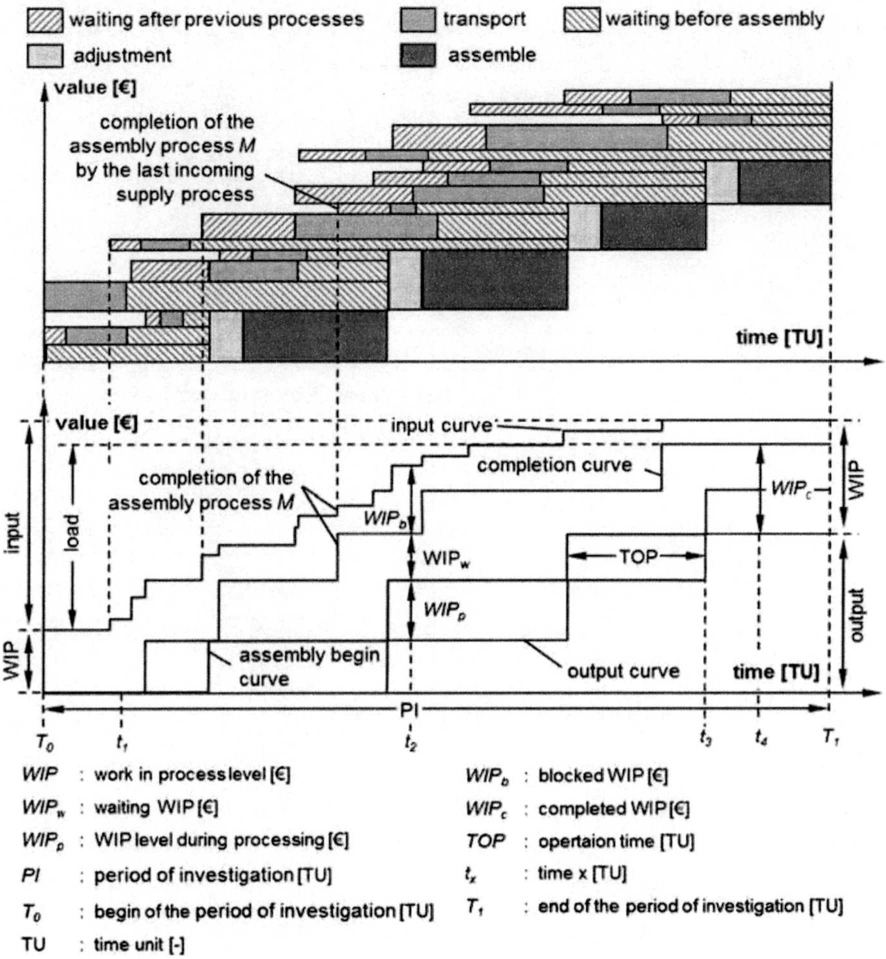

Fig. 2 The assembly throughput diagram [6]

Part D depicts the idealized representation of the Supply Diagram and the typical s-shaped progression of the supply curve. The area enclosed by the two curves represents the quantity of disrupted WIP waiting before the assembly system. This inventory corresponds to the quantity of already supplied components which could not yet be processed, since not all of the materials required for the assembly job have been completely supplied.

This Supply Diagram provides a very clear description of the schedule situation of an assembly's supply. It describes the extent to which the supply processes have processed the needed components on schedule and provided them to the subsequent assembly for further processing. From the perspective of logistics, provision is the better, the closer both curves run to each other, and the steeper and closer to the demand date they are. The area enclosed by the two typically

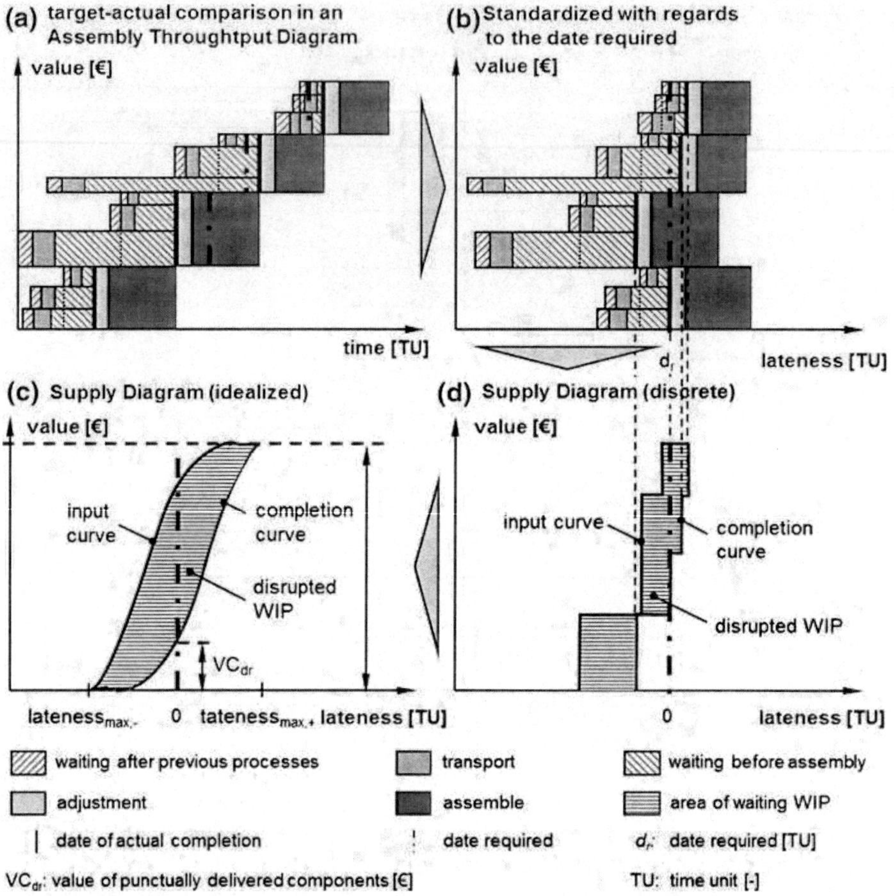

Fig. 3 Deriving the supply diagram from the assembly throughput diagram

S-shaped curves represents the quantity of disrupted WIP waiting before the assembly system. This inventory corresponds to the quantity of already supplied components which could not yet be processed, since not all of the materials required for the assembly job have been completely supplied. This area is also a good measure of the current logistical quality of provision and should be kept as small as possible.

2.4 Possibilities of Further Development

So far, however, it is not evident yet to which extent the individual components were delivered in time to the place of assembly, and to what extent the final assembly process could be completed according to schedule. Therefore, so far the supply diagram allows to visualize the schedule reliability of a part of the assembly process or of the provision of the components only. Therefore, the following two sections transpose the modelling system of the supply diagram to the assembly process, thereby expanding it. This promises contemplation of inter-process cause-effect relationships between provision and assembly and analysis of the logistics performance of this process sequence [8]. For the extension of the supply diagram, the system of generation of the supply diagram is used with the help of the assembly throughput diagram.

3 The Inter-Operation Time Diagram

3.1 Creation of the Diagram

First, the possibility will be described of including the start of assembly, as the process step directly following provision, into the contemplation. In relation to the assessment of the punctuality of the start of assembly, the scheduled (required) start of assembly represents the reference date for each assembly order. In the first step, the required and actual dates of each order are identified and highlighted in the assembly throughput diagram (cf. section A of Fig. 4).

The actual dates are represented by a solid line, the required dates by a dashed line. A Comparison of these two lines reveals that they do not match for every order and indicates a deviation of the actual from the required dates. In order to plot the lateness of all actual assembly start dates considered during the investigation period and present them clearly for comparison, the required assembly start date for each order will be defined as the reference value and set to a deviation value of zero on the X axis. This is in contrast to the modelling of the supply diagram, where with the demand date of individual components another reference date is chosen. In the second step, the assembly throughput elements of all orders of one period under consideration are standardised for the demand date (cf. section B). Graphically, this corresponds to a horizontal shift of an assembly throughput element by the amount of the deviation of the actual from the required start date on the X axis. Now the time axis no longer represents a continuous time line, but a temporal deviation from the set deadline. The standardization performed reveals at a glance whether there has been any lateness relative to the demand date.

In the next step, the assembly orders are sorted by the respective lateness of assembly begin. The individual orders remain weighted by their value in monetary units (e.g. €). Both the completion and the assembly start curve are created by weighting all orders and plotting them cumulated according to their temporal deviation over the X axis. For the temporally first three incoming supply processes, the transition from Step 2 to Step 3 is indicated by the dashed lines (sector C). In the interest of clarity, from this time on presentation of the transportation and queuing times will be omitted.

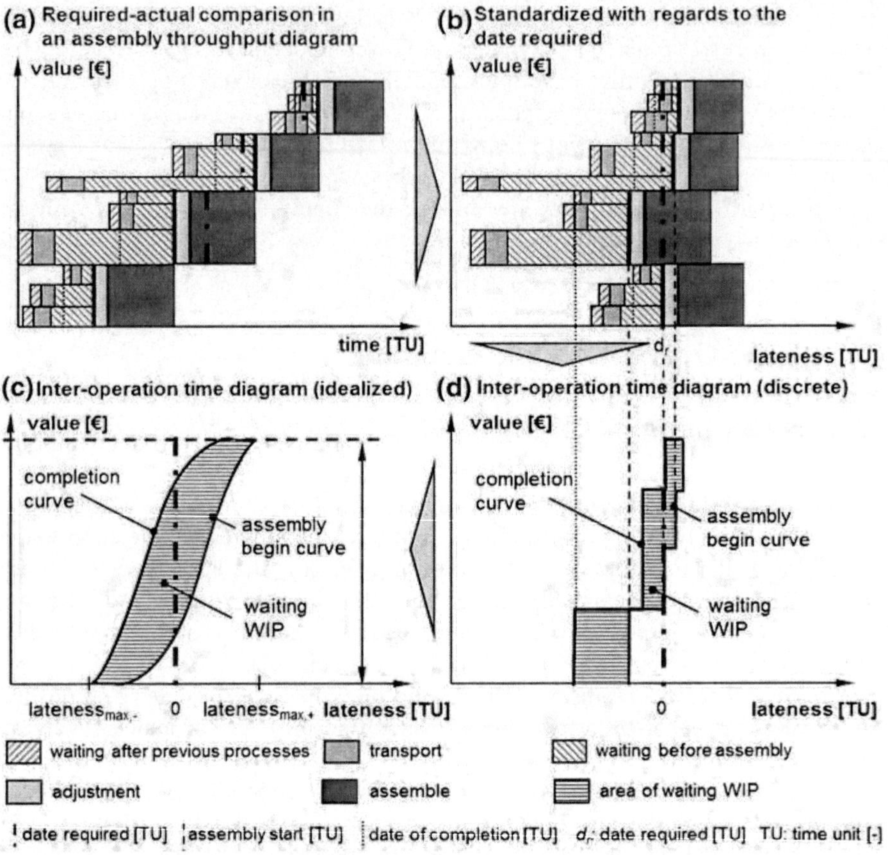

Fig. 4 Deriving the inter-operation time diagram from the assembly throughput diagram

The presentation of the diagram based on temporal elements allows interpretation of the area between the completion and assembly start curves. This area is known as waiting work in progress (WIP). In the present study it comprises all orders whose supply processes have been completed and are ready for assembly, but whose assembly has not begun yet.

The completion curve in the inter-operation time diagram is generated from the differences from the required start of assembly, unlike the completion curve of the supply diagram, which is produced from the difference of the date of the component provided last to the demand date. The completion curve of the supply diagram and the completion curve of the inter-operation time diagram are therefore not identical. Furthermore, the two curves of the inter-operation time diagram do not begin and end, in contrast to the supply diagram, in the same point. Relative to the assembly start curve, the completion curve is shifted leftwards approximately by the value of the transitional period. However, the horizontal distance cannot be interpreted exactly as inter-operation time. Due to the ascending sorting of the completion and assembly start dates by temporal deviations, the dates lose their relationship to each other.

If the definitions of the input or completion time points according to SCHMIDT are taken into account, these denote the ends of their respective predecessor processes [6]. The assembly start date denotes the beginning of the set-up, which together with the time for assembling forms the assembly process (cf. Fig. 1). Between the completion and assembly start date, there is thus the inter-operation time with the time segments 'Queuing after predecessor process', 'Transportation' and 'Queuing before assembly'. The presentation of the completion curve in combination with the newly developed assembly start curve is therefore called, in analogy to the supply diagram, an inter-operation time diagram.

In individual cases, there may be inevitably high transportation time shares, as well as process-inherent, organizational or technical queuing times. In these cases, at the beginning of the inter-operation time the assembly order cannot be called completed yet. The area of the waiting WIP will then consist not exclusively of assembly orders which are already available for further processing. Therefore, in such cases it is expedient to supplement the completion with an assembly capacity curve which is shifted, relative to the completion curve, by inevitable transport and queuing times. If there are no inevitable transport and queuing times, the area between the completion and input curves already represents the WIP level of assembly orders waiting for processing. Such waiting WIP causes capital and complexity costs. Small WIP buffers are nevertheless necessary in an assembly system to prevent undesired interruptions of material flow. Excessive WIP levels, however, must be avoided.

3.2 Application of the Inter-Operation Time Diagram

In addition to the examination of the WIP waiting before an assembly station, the inter-operation time diagram opens some further possibilities for analysis. Comparison of the courses of the assembly start and completion curves allows conclusions regarding any sequence permutations. A steeper course of the assembly start curve compared to the completion curve indicates delays of prematurely completed orders as well as accelerations of belated orders. In addition, from the inter-operation time diagram a number of key figures for the analysis of the reliability of the assembly start can be derived (cf. Fig. 5). For any schedule reliability value, a statement on the share of already initiated assembly orders (or still queued orders) in the full order volume can be made. In Fig. 5, this is exemplified for the demand date.

In analogy, for any schedule reliability value the portion of completed assembly orders on the overall input can be determined. This allows deducing the extent to which a delayed start of assembly is justified by provision not being complete yet.

Juxtaposition of an inter-operation time diagram with a supply diagram of the same study period provides further insights. A comparison of the WIP areas of the

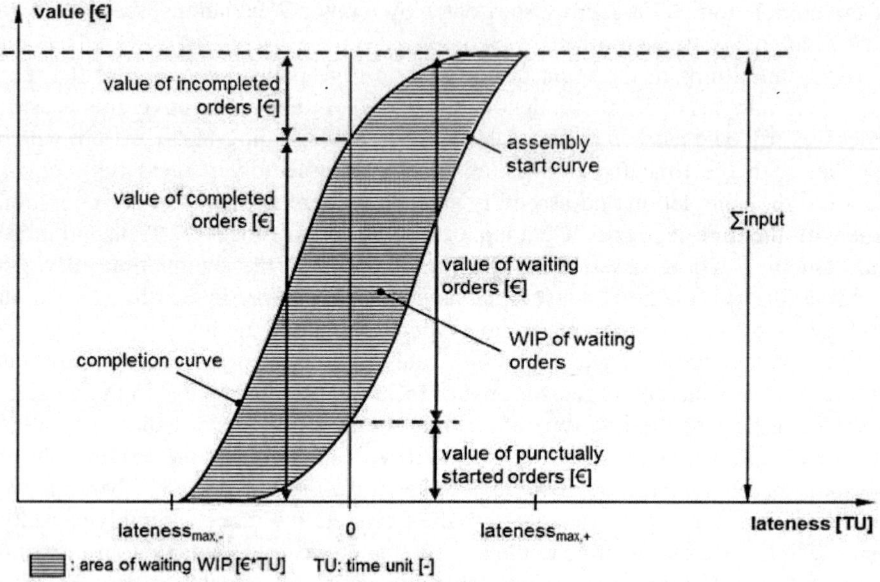

Fig. 5 Key figures of the inter-operation time diagram

supply diagram and the assembly start diagram provides qualitative statements about the ratio of the areas of both WIP types. Thus, conclusions regarding the WIP (blocked WIP and waiting WIP) can be drawn, and simultaneously causes for longer throughput times identified. If the blocked WIP area in a supply diagram exceeds the area of the waiting WIP in a inter-operation time diagram, this indicates insufficient logistical coordination in provision. A significantly higher waiting WIP than blocked WIP level clearly indicates an excessive WIP buffer of orders waiting before the assembly station under consideration. Comparison of a supply diagram with a inter-operation time diagram requires identity of the investigation periods and of the assembly stations analysed.

4 The Operation Time Diagram

4.1 Creation of the Diagram

The development of the inter-operation time diagrams revealed how the assembly start curve in combination with the completion or assembly capacity curve opens additional sources of information for schedule reliability and WIP analysis. Inclusion of the concluding time point of assembly, the end of assembly, allows to subsequently focusing on the entire process chain, thereby further extending the information base.

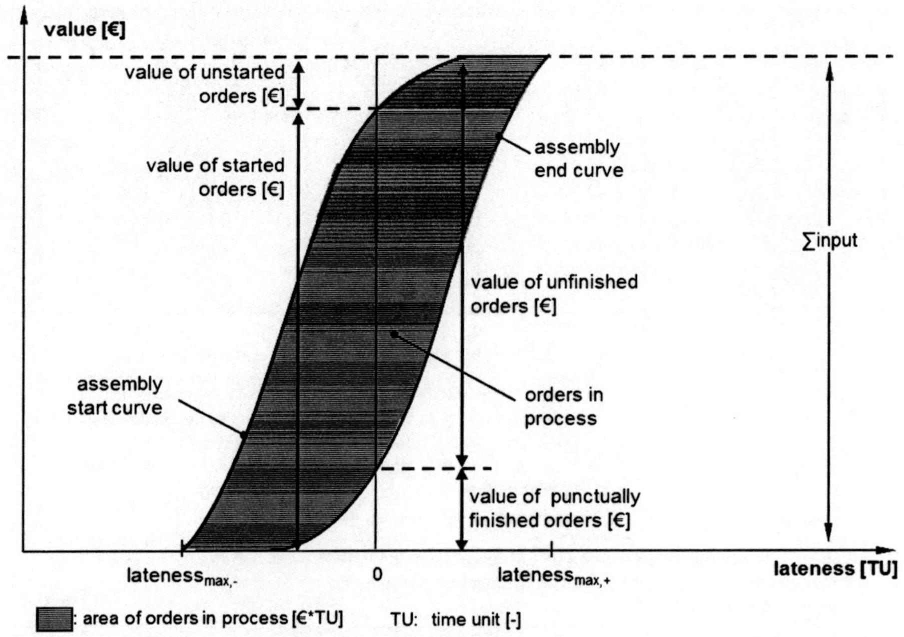

Fig. 6 Key figures of the operation time diagram

The operation time diagram is produced in analogy to the procedure for the inter-operation time diagram. For the operation time diagram, the reference point used for standardisation of the orders is provided by the scheduled end date of the assembly processes under consideration.

The input events are to be assessed, as in the preparation of the supply and assembly start diagrams, with the amount resulting from summation of the values of the individual components. This neglects the actual increase in the value of an assembly order and its associated product through the use and expenditure of material, resources and time. All the curves of the provision, inter-operation period and operation time diagrams will therefore have the same height, supporting comparability of the diagrams. In order to be able to clearly present the schedule reliability of assembly and thus a prominent point in the value chain, the assembly completion diagram is limited to the presentation of two curves: in this case the assembly start and completion curves.

4.2 Application of the Operation Time Diagram

The operation time diagram likewise offers further possibilities for the interpretation of the assembly process and its logistical efficiency. In analogy to the handling and inter-operation time diagrams, at any schedule compliance value the processing

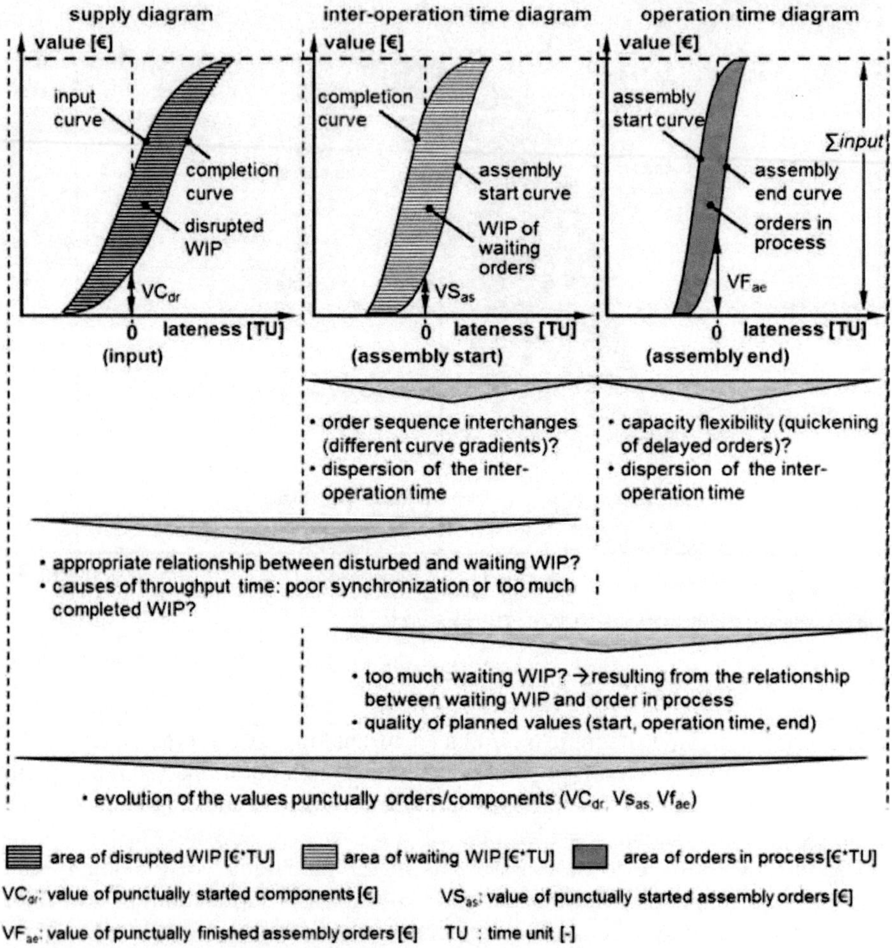

Fig. 7 Interpretations of the assembly sub-process diagrams

state (not yet completed, already completed) of the present order volume can be assessed (cf. Fig. 6).

The area bounded by the assembly start and completion curves represents the total amount of WIP in processing accrued during the investigation period. In addition, there are some possibilities for interpretation from the comparison of the geometry of the two curves. If the course of the assembly start curve is significantly steeper than that of the assembly completion curve, this indicates existing capacity flexibility. Belated orders must be accelerated. A prerequisite for this statement is, however, that the order values have not fluctuated overly during the study period. Highly fluctuating work contents may also lead to different curve slopes. In such cases no clear interpretation is possible.

A juxtaposition of the operation time diagram with the inter-operation time diagram immediately upstream in the process chain can also provide additional insights. However, only a comparison of the area of the waiting WIP with that of the orders in process allow a statement regarding its absolute level.

5 Summary

Usually an assembly process represents the final stage of a process chain. Since the schedule reliability of the last process affects the punctuality to the customer in a great measure, the controlling of scheduled assembly dates has main importance. The diagrams presented provide an instrument which allows further insight into the causes of schedule deviations at assembly processes. The following Fig. 7 summarizes the potentials.

With the help of the inter-operation time diagram as well as the operation time diagram it is possible to visualize order sequence interchanges and significant key figures. Especially the analyse of the development of the three models in the process sequence (supply, inter-operation and operation) opens up new chances to find potential to improve the logistic performance. Besides the evaluation of the present level of WIP for each process the comparison of the diagram shows the development of the lateness of the orders. The consideration of the left curve of each diagram provides an indication in which extent a more punctual further processing would have been possible.

References

1. Lotter, B.: Einführung. In: Lotter, B., Wiendahl, H.-P. (eds.) Montage in der industriellen Produktion, pp. 1–9. Springer, Heidelberg (2006)
2. Nof, S.Y., Wilhelm, W.E., Warnecke, H.J.: Industrial Assembly. Chapmann & Hall, London (1997)
3. Reinhart, G., Cuiper, R., Loferer, M.: Die Bedeutung der Montage als letztes Glied in der Auftragsabwicklung. In: Reinhart, G.(eds.) Montage-Management—Lösungen zum Montieren am Standort Deutschland. München. pp. 7–12,TCW Transfer-Centrum, München. (1998)
4. Nyhuis, P., Wiendahl, H.-P.: Fundamentals of Production Logistics—Theory, Tools and Applications. Springer, Berlin (2009)
5. Hopp, W. J.: Spearmann, M. L.: Factory physics. Irwin, Chicago (2008)
6. Schmidt, M.: Modellierung logistischer Prozesse der Montage. Berichte aus dem IFA, Garbsen (2011)
7. Nyhuis, P., Nickel, R., Busse, T.: Controlling der Materialverfügbarkeit mit Bereitstellungs-diagrammen. ZWF—Zeitschrift für wirtschaftlichen Fabrikbetrieb, vol. 5, pp. 265-268. Carl Hanser Verlag, München (2006)
8. Beck, S., Schmidt, M., Nyhuis, P.: Modeling converging Material Flows in the Supply Chain. Proceedings of the 21st International Conference on Production Research (ICPR), Stuttgart (2011)

Self-Optimizing Decision-Making in Production Control

Günther Schuh, Till Potente, Sascha Fuchs, Christina Thomas, Stephan Schmitz, Carlo Hausberg, Annika Hauptvogel and Felix Brambring

Abstract This paper deals with the concept for self-optimizing decision-making in production planning and control. The concept is based on a value stream that provides real-time production data. This data enables a qualified decision regarding production planning and control. Practice has shown that production systems with a high production process complexity—such as job shop production with low volume production—are difficult to control automatically. Therefore, employees have an important role to play but need to be supported regarding their decision-making. The goal is to highlight relevant decisions and put them into the correct context. An unconventional and interactive illustration that abandons

G. Schuh · T. Potente · S. Fuchs · C. Thomas · S. Schmitz · C. Hausberg (✉) ·
A. Hauptvogel · F. Brambring
Laboratory for Machine Tools and Production Engineering (WZL),
RWTH Aachen University, Steinbachstr. 19 52074 Aachen Germany
e-mail: c.hausberg@wzl.rwth-aachen.de

G. Schuh
e-mail: g.schuh@wzl.rwth-aachen.de

T. Potente
e-mail: t.potente@wzl.rwth-aachen.de

S. Fuchs
e-mail: s.fuchs@wzl.rwth-aachen.de

C. Thomas
e-mail: c.thomas@wzl.rwth-aachen.de

S. Schmitz
e-mail: st.schmitz@wzl.rwth-aachen.de

A. Hauptvogel
e-mail: a.hauptvogel@wzl.rwth-aachen.de

F. Brambring
e-mail: f.brambring@wzl.rwth-aachen.de

K. Windt (ed.), *Robust Manufacturing Control*, Lecture Notes in Production Engineering,
DOI: 10.1007/978-3-642-30749-2_32,

classic numerical key performance indicators helps to derive the correct decisions. Varying levels of detail regarding the depicted data allow the user to "zoom" in or out of the state of his production system. By support of simulation and visualization tools, the aim of this paper is to present a concept for self-optimizing decision-making in production control in order to help user making the right decision.

Keywords Self-optimizing · Decision-making support · Production planning and control · Tool for visualization

1 Introduction

The dilemma of production planning and control is to achieve high process efficiency, low throughput times and good planning confidence in spite of a turbulent environment which can be caused—amongst others—by an increasing product variety and a growing individualization of demands [1]. Today's offered solutions in production planning and control are numerous. For most problems, large numbers of tools and methods following different philosophies are available in the industrial environment [2]. Supply Chain Management (SCM) [3], Enterprise Resource Planning (ERP) and Manufacturing Execution Systems (MES) [4] are examples of the various means to keep the growing complexity under control. In particular, the growing popularity of Advanced Planning and Scheduling Systems (APS) shows that many companies see a necessity to improve their capability of mastering high dynamics in processes and demand. The basic idea of these systems is to use real-time feedback from production systems to continuously adapt the production schedule to any kind of change, induced by turbulence [5].

These solutions are often promised to be holistic approaches, but only work stable under certain conditions which are not transparent to the user. Complex interdependencies have the effect of coupling processes although they have statistical variations, so that decision makers are left with an insolvable amount of choices [6]. This lack of decision-making support is the main reason for the low acceptance of these systems within shop floor environments. The system's algorithms are very complex and results can change daily, so that the workers are confronted with non-reproducible situations. On the other hand, also simple decision heuristics as proclaimed in many lean approaches are not capable of handling the complexity found within production control sufficiently. The main challenge in production control on the one side is to fulfill the classic logistic targets defined by Wiendahl [7] such as low stocks, short throughput times and a high adherence to delivery dates, and on the other side—following Goldratt's argumentation—[8] to maximize the throughput at a minimum of operating costs, despite any turbulence.

The environment that this paper focuses on is a classic job shop manufacturing. Due to a strong customer orientation resulting in lots of product variants, the customer decoupling point is positioned at the beginning of the value chain. This results in multi-stage manufacturing processes where several machines take part in the value adding process, which leads to long delivery times. As the shop floor consists of hundred or more machines with some of them being redundant, orders can take numerous routes through the production.

The aim of this paper is to enhance transparency of production control for user in order to support him within decision-making process. All necessary data from production should be prepared in an intelligent way so that the user understands important information at a glance in order to make the right decisions.

2 Challenges

Especially in machine and equipment manufacturing and small and medium sized batch production are confronted with the problem to handle orders with different lot sizes and a variety of different scales. Highly varying product design induced by customer demand volatility produces high variance and dynamics in the production processes [9]. Furthermore, work stations and resources are allocated by their technological working process, which is why such a production is also called function-oriented [10]. Within this environment, the complexity of material flows is characterized by the connectivity (diversity of dependency) and variety (diversity of elements) [11] of objects within the production system. Production structures with more than 70 capacities show very complex material flows [12]. For example, the analysis of a company from the machinery and equipment industry showed that for example 8,000 manufacturing orders were produced in 4,000 different routes. This led to strongly intersecting material flows. As a result the future capacity impact of newly released orders was not predictable. Any form of planning and controlling became unstable [13]. A large amount of production orders due to crossing material flows increases the effort necessary for coordination and transport within the system. It also increases work in process (WIP) because of long exposure times from unfinished goods. Hence, working capital is comparatively high.

Furthermore, changes in order sequences cause a shifting work load regarding up and downstream stages of production [14]. Unfortunately, the optimization of order sequences is challenged by the law of combinatorics: in case of five work processes at five different resources there are already $2{,}49 \times 10^{49}$ alternatives of possible order sequences [15].

The complexity of material flows also results in confusion over the location of bottlenecks. It has been recognized that bottlenecks occur to be highly dynamic as they can shift between different resources. This phenomenon, often described as shifting bottlenecks, is well known and often published, so that a considerable amount of scientific approaches deal with that topic [16, 17]. Shifting bottlenecks emerge because of high product variance combined with alternating demands and general changes of the order mix. They significantly influence the overall logistical performance of a production system [18]. Goldratt recommends the identification of existing bottlenecks that determine the throughput of the production system. They demand an increased attention and fair control effort in order to avoid losses of throughput. According to Lödding, a guideline to design an effective production control is needed to consider the capacity utilization of bottleneck resources [19]. Schuh agrees that sophisticated tools of capacity planning and control are necessary, such as time capacity adjustment for several resources [9].

2.1 Approaches in Production Control

Existing approaches in production control can be divided into two philosophies: deterministic and rule-based approaches. This classification can be characterized by their level of detail. Deterministic approaches deliver a meticulous production plan with exact starting times for each manufacturing order. On the other hand rule-based approaches fulfill the production control tasks based on local and decentralized state variables [20]. This difference can also be shown within the need for IT support. Where deterministic approaches need high processing power to meet the requirements of the process complexity, rule-based approaches normally do not require such support and can be executed via the shop floor and planning staff.

The growing individualization and stronger stochastic effects in job shop manufacturing have a negative effect on the quality of deterministic results. Especially the high deviation of customer demand leads to continuously changing guidelines in production control, e.g. regarding the sequence of processing customer orders. In addition, disturbances cannot be considered within a proper time. Rule-based approaches are much more robust against these dynamic influences. They can be exchanged quickly regarding the actual preferential targets and support the worker's knowledge and process understanding as they leave some open space for individual decisions which might be necessary due to disturbances or other unforeseeable situations. The main reason for criticism is their lack regarding the consideration of delivery dates. They prioritize orders rather against each other than in connection to the target delivery date. Especially in complex environments occurring in job shop manufacturing the prediction of due-dates is very important though.

The approach developed in this paper will combine the advantages of both approaches, delivering processing support for handling the large variety of data produced by the huge amount of orders as well as enabling an integration of simple rules which support the decision-making process of the staff involved. The key feature of this approach is on the one hand to develop a modular production control in order to adapt production to dynamic market demand. On the other hand, there is a need for a new way of visualization of data and the generation of correct information in order to support operator within is decision-making process while configuring production control. This will allow the workers to capture the current situation on which they can rely when making decisions.

2.2 *Approaches for Visualization Concepts*

The term visualization refers to the graphical representation of data sets and information [21]. Often, it also simultaneously denotes the process of producing representations of any given data by encoding its inherent information in charts and graphs [22]. In contrast to series and tables of data, the graphical representation of data ideally allows a quick overview as well as broader analyses [23]. Therefore, visualization is a tool to analyze data by facilitating the discovery of hidden statistical connections within datasets and a way to effectively communicate the discovered results to others [24].

For some decades now, the rising availability of computational power has made it possible to accomplish these tasks by generating interactive computer graphics which can literally be updated with newly gathered data in real-time [21, 25]. Technically, the fulfillment of Shneiderman's Visual Information Seeking Mantra to give an "Overview first, zoom and filter, then details on demand" poses no difficulties for today's IT infrastructure [26]. However, the translation of this claim into an adequate visualization tool is a very problem-specific task because the quality of any data visualization depends heavily on the purpose it is designed for [24]. This is true for any visualization, whether the primary goal may be providing a rough overview, allowing an in-depth analysis or presenting results [24].

The possibilities for collecting data in production control are virtually endless. Due-dates, lead times, adherence to schedule, demand and supply of resource capacities and WIP are just a few examples of possibly interesting data and computable Key Performance Indicators (KPI). Numerical representation of this data still has its value in undergoing in-depth analyses, but visualizing the data enormously facilitates performance benchmarking, controlling production and highlighting critical developments before they impact manufacturing processes [23, 27].

Nevertheless, the analysis of feedback data from production takes still high effort. First, employees do not know which data to look at. Second, visualization of data is often static and brings no new findings. It is still difficult for employees to get all dependencies of production control tasks. There is still a need for an innovative visualization, which clusters all important information sufficiently. The user should be able to grasp all needed information at one glance.

3 Concept of Self-Optimizing in Production Control

The challenge for the economic planning in manufacturing companies is to dynamically find the right balance between a detailed planning and the direct design of the value stream [13]. This conflict can be solved by an adaptive system that adjusts its objectives depending on the situation [28]. Today's static planning and control processes in organizational and technological solutions must therefore be replaced by self-optimizing, decentralized control loops [29]. However, the current developments in automation technology show that this is not yet possible even with comparatively simple facts. Hence in many cases the human being still has an important role. The aim is to create a self-optimizing system that serves the people as a decision support through visualization. For this purpose, the term "self-optimization" is defined in the following, before the concept of modular software architecture for a detailed scheduling system and the intelligent visualization is discussed.

3.1 Definition of Self-Optimizing in Production Control

According to the definition of the SFB 614 (Collaborative Research Centre of German Research Foundation—DFG) self-optimizing systems have the ability to react independently and flexibly to changing environmental conditions, interventions of the user or actions of the system [30]. Self-optimizing systems optimize their behavior through independent learning. In a closed system, self-optimization describes the recurring execution of the following three actions:

- A continuous analysis of the current situation, where the regarded situation includes the current state of the system and all performed observations of the environment. The most important part of this first step is the analysis of the fulfillment of the objectives.
- The next step is the determination of the system's objectives. These objectives can be selected out of programmed objectives or generated by the system not regarding the implemented possibilities.
- The last step's goal is to adapt the system's behavior. Changed objectives require an adaptation of the system's behavior. This can be achieved by adapting the parameters and by adapting the structure of the system [31].

The self-optimizing process leads, according to changing influences, to a new state.

Today the flexibility of automated systems is limited. These systems require explicit specifications that define clear instructions for each possibly occurring situation. This leads to high costs in the implementation of automated systems. However, today's production systems demand flexible processes and procedures and are characterized by a high degree of complexity [32]. If the number of possible situations, which are regarded and implemented in the system, is n, today's automation systems would not be able to react to the situation $n + 1$. However, an adaptive self-optimizing system has to be able to react properly in the situation $n + 1$. Therefore, self-optimizing systems have to be equally intelligent as human beings [33]. It follows that human beings must be seen as a central component of modern production systems. The concept, which also regards the human being, is presented below.

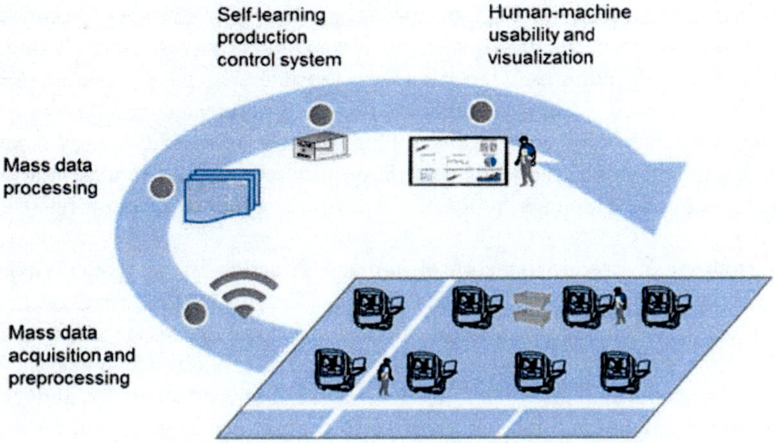

Fig. 1 Concept of self-optimizing decision-making in production control

3.2 Concept

The aim of this concept is to provide a high resolution, adaptive production control based on cybernetic support systems and intelligent sensor technology. In order to increase the efficiency of production, production control systems should support operators, who have to make decisions regarding production control, by using high resolution data in an ideal way (cf. Fig. 1).

The generation of high resolution data is done by using simple and inexpensive sensors. To ensure a quick and easy interchangeability of different sensor actuator elements, an identification and implementation of modular functional blocks is provided. These function modules are communicating and acting autonomously in smart objects and subsystems. It should be recognized, that data granularity can be adapted to the particular level of observation. The interoperable operational capability of components and sub systems is enabled by provision of standardized interfaces to the higher-level IT systems and the design of the modular function blocks. After generating the data, IT architecture can be developed in order to support the cybernetic-physical production systems. Therefore, the measured data are first stored centrally. Then, data can be further processed by selective access provided to (human) operators.

The processed data is transferred via a dedicated interface to a modular structured control system. The developed software system for production control is offering a new opportunity to manage high resolution production data while meeting individual requirements efficiently. The architecture of this control system is based on modular control elements in order the enhance usability and transparency of production control. Each control method represents a control module within the IT system. For instance, different sequencing rules can be mapped by different modules which can be chosen by operator. The definition and selection of control principles can be done by the plug and play-principle.

The feedback data like throughput, throughput times, WIP, operation times, setup times or resource utilization gathered from production can be evaluated with regard to occurring patterns. Depending on the detected patterns, control alternatives are proposed which are validated with a connected simulation model. Then, the user has the ability to adapt production control by interchangeable control elements using the plug-and-play principle. In particular, the interchangeable control elements represent control modules for order release and sequencing.

The analysis of production control and the selection of control modules are realized by an innovative way of visualization. The aim of this visualization is to highlight relevant issues for operator's decision-making processes and to provide necessary information in a context so that decision makers can make the right decision. An important aspect by the interaction between operator and control system is the degree of automation. In order to support operators to understand and apply the control system, an ergonomically designed software system is required. The key question is which information is needed for operators about previous (semi-) autonomous decisions of the control system in order to react in critical situations quickly and with as few errors as possible. During increasing operation time of the control system, the system is able to provide better and better control alternatives while learning from previous decisions about production control. Therefore, this modular production control system can be defined as a self-optimizing system.

The explained concept for high resolution production control offers the following advantages:

- A direct connection between the real production and software by realization of a cyber-physical production system
- High resolution data generation using simple, low-cost sensors
- High transparency of production and of the used production control principle
- Integration of participating software systems including embedded systems
- Adaptive use of high resolution data
- Support of understanding and application for operators by intelligent visualization software and ergonomic software design.

4 Detailing of Monitoring Concept

The elements of the proposed monitoring system for production control and their functional relationships are depicted in Fig. 2.

Its heart and main functionality is the visualization, where two possible views on the production process can be distinguished: the resource view and the job view. The differentiation between these two views allows a more thorough analysis, especially in complex production situations [20]. Possible visualizations for

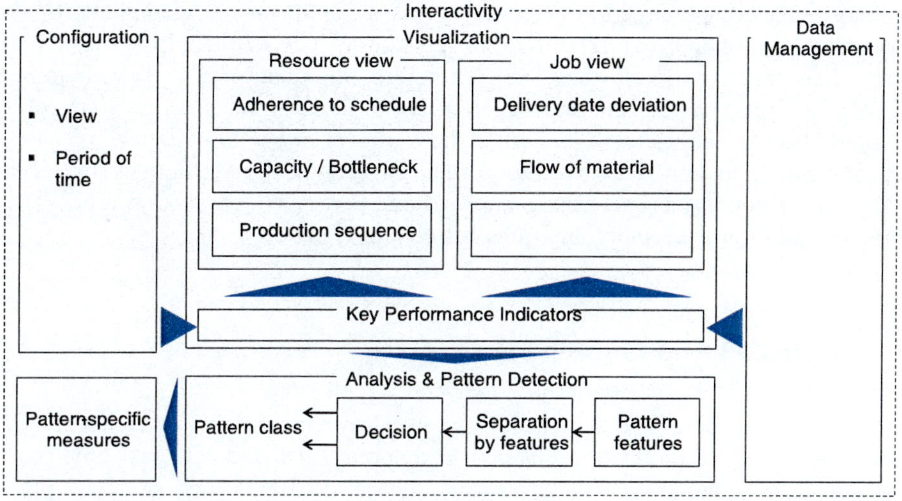

Fig. 2 Elements of the monitoring system

the resource view are the adherence to schedule of the individual production systems, the utilization of the capacities of these production systems and identification of the system which is representing the bottleneck as well as the adherence to the projected production sequence. Within the job view, possible visualizations include the job's delivery date deviations and the flow of material through the production area.

With these visualizations, it should not only be possible to control current and past developments, but also to make predictions about possibly uprising problems in the near future. Therefore, besides the desired view of the visualization, the monitoring system should also allow the user to configure the examined period of time. Depending on the user's choice, the necessary data is pulled by the Data Management module which is the groundwork for calculating the Key Performance Indicators (KPIs). The KPIs themselves are the basis for the visualization. Therefore, the choice of the visualization is obviously closely connected to the gathered and available production data.

The ultimate goal of the whole monitoring process is to improve current production processes by identifying suitable measures to remedy problems which are currently occurring or might do so in the near future. These measures have to be specifically aimed at improving production goals, e.g. to increase the adherence to schedule for the production jobs. For this purpose the root causes of the goal deviations need to be identified. A first step for doing so is to analyze the visualizations and trying to detect patterns within the data. Pattern recognition denotes the allocation of patterns to predefined pattern classes and can principally be done visually or mathematically/statistically. The main problem of this classification process is the definition of specific pattern features which distinguish one class from others [34].

Due to the complexity of production processes and the many different questions that may arise while investigating goal deviations, it is impossible to specify a procedure which fits in all cases [35]. Therefore, the monitoring tool needs to be highly interactive and allow the user to navigate different levels of data consolidation. Usually, the user would start at the top level to get an overview over the general situation and then start to further investigate by drilling down [36]. This can be done by concentrating on a specific part of the production process, a product group or even individual jobs for certain clients.

5 Conclusion and Outlook

This paper has described a concept for a self-optimizing tool that assists in decision-making in production planning and control. It is fed with real-time production data, either from simple sensors (such as RFID-sensors) in a real production environment or from a simulation model.

The tool is mainly designed for a job shop manufacturing environment with a high level of production complexity. Its goal is to highlight relevant decisions to the user (e.g. production planners) and to put the available decision options into the correct context, so that the user can identify the best production control configuration regarding the existing conditions (product mix, capacity load, resources etc.).

Whereas existing approaches of computer-based tools for the support of production planning and control can be divided into two philosophies—deterministic and rule-based, where deterministic approaches deliver a meticulous production plan and need high IT-processing power to meet the requirements of the process complexity and where rule-based approaches fulfill production control tasks based on local and decentralized state variables and therefore can be easily executed via the shop floor and planning staff—the tool described in this paper combines the advantages of both approaches, delivering processing support for handling the large variety of data produced by the huge amount of orders as well as enabling an integration of simple rules which support the decision-making process of the staff involved.

Its IT architecture is based on a control circuit that not only monitors the target state of the target parameter, but at the same time adapts controller parameters to the observed changes. This embodies adaptive behavior and therefore allows for self-optimizing actions. Amongst others, the tool contains a module for pattern detection that allows to systematically analyze the user's choices and to match them with the underlying decision task so that for future issues the user can chose from a set of preselected options.

Regarding the visualization design of the tool, several approved design guidelines have been incorporated, such as the precept of effectiveness which stands for enabling the viewer to intuitively understand the meaning or message of the portrayed data, or the precept of expressiveness which stands for presenting the data as unaltered as possible. The outcome is an unconventional and interactive visual illustration that abandons classic numerical key performance indicators. Varying levels of detail regarding the depicted data allow the user to "zoom" in or out of the state of his production system.

The tool will be implemented at the Campus Demo-Factory at RWTH Aachen University as well as at the consortium partners of the respective research project.

Acknowledgments The new concept of self-optimizing in production control as well as the described implementation is being investigated by the Laboratory for Machine Tools and Production Engineering (WZL) within the publicly funded excellence initiative "Integrative production technology for high wage countries" at RWTH Aachen University in cooperation with the DFG (Deutsche Forschungsgemeinschaft, German Research Foundation).

References

1. Jones, D.: Creating Lean Solutions 2, pp. 17–28. Lean Management Summit, Aachen (2005)
2. Schuh, G., Lenders, M., Nussbaum, C., Kupke, D.: Design for changeability. In: ElMaraghy, H. (ed.) Changeable and Reconfigurable Manufacturing Systems. Springer, London (2008)
3. Milberg, J., Neise, P., Organizational Design of Supply Chains, WGP, Production Engineering XIII/2:181–186 (2006)
4. Valckenaers, P., Van Brussels, H.: Holonic manufacturing execution systems. Annals of the CIRP **54**(1), 427–432 (2005)
5. Zijm, W.H.M.: Towards intelligent manufacturing planning and control systems. J. Spectrum Springer **22**(3), 313–345 (2000)
6. Schwartz B., The paradox of choice: why more is less (2004)
7. Wiendahl, H.-P.; Load-oriented Manufacturing Control; Springer (1995)
8. Goldratt, E., Cox, J.: The Goal: Excellence in Manufacturing. North River Press, Croton-on-Hudson (1984)
9. Schuh, G., Potente T., Fuchs S.: Shifting Bottlenecks in Production Control., In: Hoda, A., ElMaraghy (Hrsg.): Enabling Manufacturing Competitiveness and Economic Sustainibility. Proceedings of the 4th International Conference on Changeable, Agile, Reconfigurable and Virtual Production, (CARV 2011), Montreal, pp. 505–511 (2012)
10. Dipl.-Wi.-Ing. Matthias Bornhäuser.: Reifegradbasierte Werkstattsteuerung Heimsheim: Jost-Jetter-Vlg; 1. Auflage. Zugl. Stuttgart, Univ., Diss. (2009)
11. Lücke, O.: Methodische Nutzung der betrieblichen Lernfähigkeit. Essen: Vulkan-Verlag (Schriftenreihe des IWF) Zugl. Braunschweig, Techn. Univ., Diss. (1999)
12. Hirschberg, A. G.: Verbindung der Produkt- und Funktionsorientierung in der Fertigung. München: iwb Forschungsberichte; 137. Zugl. München, Techn. Univ., Diss. (2000)
13. Brecher, C.: Integrative Production Technology for High-Wage Countries. Springer, Berlin (2011)
14. Adam, D.: Produktions-Management 9, überarb. Aufl., Nachdr., Wiesbaden, Gabler (2001)
15. Hahn, D., Laßmann, G.: Produktionswirtschaft: Controlling industrieller Produktion, 3. Aufl., Heidelberg, Physica-Verl. (1999)
16. Nakata, T., Matsui, K., Miyake, Y., Nishioka, K.: Dynamic bottleneck control in wide variety production factory. IEEE Trans. Semicond. Manuf. **12**(3), 273–280 (1999)
17. Roser, C., Nakano, M., Tanaka, M., Shifting Bottleneck Detection; Proceedings of the Winter Simulation Conference, p.1079–1086 (2002)
18. Goldratt, E.: The unbalanced plant. In: 24th APICS annual international conference, Boston, Mass., p. 195–199. Falls Church, USA (1981)

19. Lödding, H.: Dezentrale Bestandsorientierte Fertigungsregelung. Düsseldorf: VDI-Verlag, 2001. (Fortschritt-Berichte VDI: Reihe 2; 587) Zugl. Hannover, Univ., Diss. (2001)
20. Wiendahl, H.-P.: Betriebsorganisation für Ingenieure. 7., aktualisierte Aufl. München: Hanser (2010)
21. Chen, C.: Information visualization. Wiley Interdisciplinary Rev. Comput. Stat. **2**(4), 387–403 (2010)
22. Cleveland, W.S., McGill, R.: Graphical perception and graphical methods for analyzing scientific data. Science **229**, 828–833 (1985)
23. Wiendahl, H.-P.: Modelle und Systeme des Produktionscontrollings. In: Eversheim, W., Schuh, G. (eds.) Betriebshütte—Produktion und Management, Teil 2, 7, völlig neu bearb. Auflage. Springer, Berlin (1996)
24. Schumann, H., Müller, W.: Visualisierung—Grundlagen und allgemeine Methoden. Springer, Berlin Heidelberg (2000)
25. Sackett, P.J., Al-Gaylani, M.F., Tiwari, A., Williams, D.: A review of data visualization: opportunities in manufacturing sequence management. Int. J. Comput. Integr. Manuf. **19**(7), 689–704 (2006)
26. Shneiderman, B.: The eyes have it—a task by data type taxonomy for information visualizations. Proceedings of the IEEE Symposium on Visual Languages, pp. 336–343 (1996)
27. Sackett, P.J., Williams, D.K.: Data visualization in manufacturing decision-making. J. Adv. Manufact. Syst. **2**(2), 163–185 (2003)
28. Zaeh, M., Reinhart, G., Ostgathe, M., Geiger, F., Lau, C.: A holistic approach for the cognitive control of production systems. Adv. Eng. Inform. **24**(3), 300–307 (2010)
29. Scholz-Reiter, B., Höhns, H.: Selbststeuerung logistischer Prozesse mit Agentensystemen. In: Schuh, G. (Hrsg.): Produktionsplanung und -steuerung. Grundlagen, Gestaltung und Konzepte. 3. Auflage, Springer, Berlin (2006)
30. Frank, U. et al.: Definition der Selbstoptimierung. Arbeitspapier des Sonderforschungsbereichs 614: Selbstoptimierende Systeme des Maschinenbaus. Revision 1.105. Paderborn (2004)
31. Gausemeier, J., Frank, U., Donoth, J., Kahl, S.: Specification technique for the description of self-optimizing mechatronic systems. Research in Engineering Design, vol. 20(4), pp. 201–223. Springer, London (2009)
32. Schmitt, R., Beaujean, P.: Selbstoptimierende Produktionssysteme. ZWF 102, 9, Carl Hanser, München, pp. 520–524 (2007)
33. Possel-Dölken, F., Herfs, W., Kempf, T., Brecher, C.: Kognitive Automatisierung—Szenarien und Ansätze für die selbstoptimierende Steuerung von Abläufen. atp—Automatisierungstechnische Praxis 49, 11, p. 50–60 (2007)
34. Niemann, H.: Methoden der Mustererkennung. Akad. Verl.-Ges, Frankfurt (1974)
35. Ullmann, W.: Controlling logistischer Produktionsabläufe am Beispiel des Fertigungsbereichs. Düsseldorf: VDI-Verlag (Fortschritt-Berichte VDI, 2, 311). Zugl.: Hannover, Univ., Diss. (1994)
36. Reichmann, T.: Controlling mit Kennzahlen: Die systemgestützte Controlling-Konzeption mit Analyse- und Reportinginstrumenten. 8.,überarb. underw. Aufl. München: Vahlen (2011)

Robust Solution Approach to CLSP Problem with an Uncertain Demand

Wilhelm Dangelmaier and Ekaterina Kaganova

Abstract In this paper, we consider a production planning problem where demand for production is not exactly known, and only its lower and upper bounds are provided. Section 1 comprises an introduction to the production planning field, whereas the detailed problem statement is given in Sect. 2. Modeling aspects are contained within Sect. 3. The mathematical model of capacitated lot-sizing problem (CLSP) was considered as a basis. A demand uncertainty is included into the mathematical model by means of an uncertainty set D, which consists of corresponding demand uncertainty intervals in each planning period. We describe the solution approach in Sect. 4. The robust optimization techniques were chosen for solving, which allowed to construct the solution immunized against uncertainty. A computational example, comparison with the solution from stochastic optimization approach and analysis of obtained results are encompassed in Sect. 5. Conclusions and directions of future research are presented in Sect. 6.

Keywords Production planning · Demand uncertainty · Robust optimization

1 Introduction

In times of growing industries and rapid development of production technologies, production planning becomes increasingly important. It has a crucial influence on a company's competitiveness, since a reduction of costs by several percentage points

W. Dangelmaier (✉) · E. Kaganova
University of Paderborn, Fuerstenallee 11, Paderborn 33102, Germany
e-mail: wilhelm.dangelmaier@uni-paderborn.de

E. Kaganova
e-mail: kaganova@hni.uni-paderborn.de

K. Windt (ed.), *Robust Manufacturing Control*, Lecture Notes in Production Engineering,
DOI: 10.1007/978-3-642-30749-2_33, © Springer-Verlag Berlin Heidelberg 2013

provides a distinct competitive advantage in the market. However, the optimization of production processes in manufacturing faces great practical difficulties.

First of all, we must bear in mind that many real-life production systems are of high complexity. It means that all existing connections between system components cannot be considered and included into the corresponding mathematical optimization model. Nevertheless, with some kind of simplification, a relative majority of production systems can be described with optimization problems, well-known in the literature: ELSP, CLSP, DLSP, GLSP, PLSP, MLLP, etc. [1].

Secondly, production planning systems are affected by uncertainty. In reality, the considered data are usually uncertain. This means that information is not exactly known at the time the problem is being solved. Uncertainty can be classified into two main groups [2]:

- Environmental uncertainty;
- System uncertainty.

The first group includes, for example, demand or supply uncertainty, whereas quality uncertainty, production lead time or failure of production system uncertainty belongs to the second group.

Why is it so important to treat uncertainty in the problem data and how can we do this? Due to the fact that even a very small uncertainty in the data can make a constructed production plan infeasible or extremely expensive, uncertainty has to be definitely taken into account while planning a production. Otherwise, manufacturing companies risk going bankrupt or losing important customers when neglecting estimation or implementation errors. In the literature three main approaches dealing with uncertainty can be distinguished:

- Sensitivity Analysis [3];
- Stochastic Optimization [4];
- Robust Optimization [5, 6].

All three approaches can be applied to uncertain production planning problems, but they need a different initial knowledge about uncertain parameters and also treat uncertainty in different ways. Sensitivity analysis actually ignores data uncertainty and only provides information to which extent a proposed solution is still optimal. Stochastic optimization approaches require a probabilistic knowledge about the data and give the best solution according to the most expected data realization. Robust optimization approaches are based on the worst-case analysis and therefore do not require any probabilistic data.

We would like to concentrate on the robust modeling and optimization of production system, which is affected by specified demand uncertainty. In particular, the demand uncertainty is given by upper and lower bounds and no other information about demand is provided. This situation is often obtained in practice, when a company can only assume the value of customers demand (e.g. in a bakery or ski rental shop) or when changing demand is allowed on a fixed percentage by the contract. The robust optimization approach allows to guarantee the maximal value of total costs for the manufacturer and does not need any probabilistic data.

The presented research focuses on several challenges. The first goal is to model a production system, concurrently considering specified demand uncertainty. A constructed mathematical model should reflect all real-life system restrictions and has to be robust against uncertainty. Secondly, we have to solve a created model with an appropriate computational time and to guarantee an upper bound of total costs. Another goal is to improve the production plan, considering information from previous periods, such as product amounts which have already been produced and sold. This information allows to identify the current system statement and to correct the production plan for future periods.

All the goals mentioned above are achieved and a computational example is additionally presented. The comparison of the obtained solution with the solution provided by the stochastic optimization approach as well as with the best-possible solution is done. In spite of existing conservativeness, our research results show that the proposed solution approach can be applied in practice, concurrently providing significant benefits. In relation to the best-possible solution, the financial losses are also negligible.

Section 2 presents the problem statement and the main requirements on a model. Section 3 "Modeling" describes the constructed CLSP model and the developed, modified version of the model—the robust production planning model. In Sect. 4, we describe a solution approach for the proposed model. Section 5 "Computational Example" encompasses the data and results of a computational experiment, as well as a comparison of our model with the stochastic optimization approach and an analysis of the results. We conclude by providing directions for future research in Sect. 6.

2 Problem Statement

We consider the production planning problem which is affected by demand uncertainty. Specifically, the demand uncertainty is described in a following way: for each planning period only upper and lower bounds of possible demand are provided. No other information about the demand such as the expected value or probability distribution over this interval is available. The actual demand realization for the current period is available at the end of the previous period. For example, if one planning period is a day of the week, then the exact information about the customer demand for Monday is available only on Sunday evening, whereas the information about the demands for Tuesday, Wednesday, Thursday and Friday is only given by upper and lower bounds (see Fig. 1). In summary, information about demands is updated at the end of each planning period, and this process is repeated in a serial fashion manner until the end of the total planning horizon.

The main requirement for the possible solution of the problem is the robustness against uncertainty. The proposed production plan should be feasible for each realization of demand sequence. In addition, the upper border for the total costs should be provided. Furthermore, not only robustness of the solution should be achieved, but also all other main actual requirements for the production system should be fulfilled:

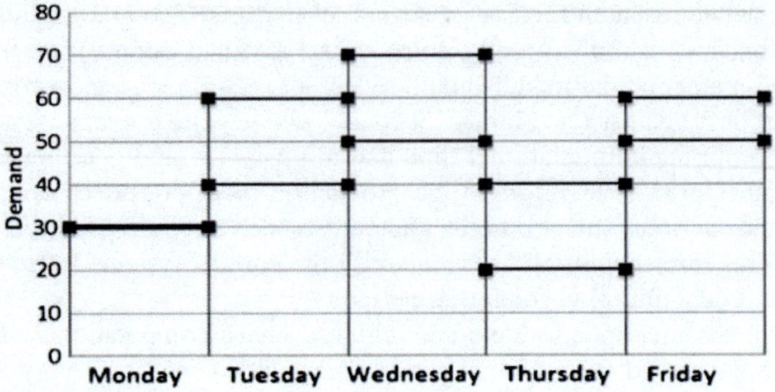

Fig. 1 Demand information

- Different production costs (normal and overtime working shifts) should be considered;
- Fixed setup costs in each of normal and overtime working shifts should be considered;
- Backlogging (production for current demand in later time slots) is allowed with some punishment;
- All demands should be satisfied;
- Production capacity limits in normal and overtime shifts should not be exceeded;
- Total setup, production and holding costs should be minimized.

3 Modeling

3.1 CLSP Model

Capacitated lot-sizing problem (CLSP) is one of the important problems in the production planning, which has been studied extensively in the literature and whose model is implemented in industries [1, 7].

In order to describe our production system, we will use the CLSP mathematical optimization model as a basis with an extension to allowed backlogging.

Indices:

$j = 1 \ldots M$—products,

$t = 1 \ldots N$—periods.

Data:

d_{jt} demand of product j in period t,

k_{jt} production capacity available for product j at normal working time slot of period t,

w_{jt} production capacity available for product j at overtime slot of period t,
c_{jt} production cost for product j at normal working time slot (per unit) of period t,
ov_{jt} production cost for product j at overtime slot (per unit) of period t,
h_{jt} holding cost for product j (per unit and per period) in period t,
p_{jt} penalty per unit of product j for the backlogged demand in period t,
s_t setup costs at normal working time slot of period t,
sov_t setup costs at overtime slot of period t,
I_{jo} initial stock of product j

Decision variables:
x_{jt} quantity of product j to be produced at normal working time slot in period t,
y_{jt} quantity of product j to be produced at overtime slot in period t,
z_{jt} backlogged demand of product j to be produced at overtime slot in period t,
I_{jt} positive inventory of product j at the end of period t,
Q_{jt} amount of product j at the end of period t (could be negative meaning backlogged demand and positive meaning the positive inventory),
b_t binary variable, which is equal to 1 when $x_{jt} \geq 0$ in period t and 0 otherwise,
bov_t binary variable, which is equal to 1 when $y_{jt} \geq 0$ in period t and 0 otherwise

Objective function:

Objective function is the function of costs, which should be minimized. It includes production costs at normal and overtime slots as well as holding, backlogging and setup costs summarized over products and periods.

$$min \sum_{t=1}^{N} \sum_{j=1}^{M} \left(c_{jt}x_{jt} + ov_{jt}y_{jt} + h_{jt}I_{jt} + p_{jt}z_{jt} + s_t b_t + sov_t bov_t\right) \quad (1)$$

Restrictions:

The restrictions presented below reflect all main requirements of a production system.

$$Q_{j1} = I_{j0} + x_{j1} + y_{j1} - d_{j1}, \forall j \in \{1...M\} \quad (2)$$

$$Q_{jt} = Q_{j,t-1} + x_{jt} + y_{jt} - d_{jt}, \forall t \in \{2...N\}, \forall j \in \{1...M\} \quad (3)$$

$$I_{jt} \geq Q_{jt}, I_{jt} \geq 0, \forall t \in \{1...N\}, \forall j \in \{1...M\} \quad (4)$$

$$z_{jt} \geq -Q_{jt}, z_{jt} \geq 0, \forall t \in \{1...N\}, \forall j \in \{1...M\} \quad (5)$$

$$x_{jt} \leq k_{jt}b_t, \forall t \in \{1...N\}, \forall j \in \{1...M\} \quad (6)$$

$$y_{jt} \leq w_{jt}bov_t, \forall t \in \{1...N\}, \forall j \in \{1...M\} \quad (7)$$

$$b_t \in \{0,1\}, bov_t \in \{0,1\}, \forall t \in \{1...N\} \tag{8}$$

$$x_{jt} \geq 0, y_{jt} \geq 0, \forall t \in \{1...N\}, \forall j \in \{1...M\} \tag{9}$$

Restrictions (2) and (3) are the so-called balance restrictions: the product amount, which is presented in the inventory at the end of period *t*, is equal to the product amount in the inventory at the preceding period plus the production volume at the normal and overtime slots minus the demand at the existing period.

Restrictions (4) and (5) show the connection between the positive inventory at the end of the planning period, the value of backlogged demand and the value of the generalized variable, which shows the positive or negative stock at the end of one period.

Restrictions (6) and (7) express the upper bounds on the possible production amount in each planning period and also include a binary setup variable, which allows taking corresponding setup costs into account.

Finally, restrictions (8) and (9) show the non-negative or binary nature of the appropriate variables.

However, the classical CLSP deterministic mathematical model often does not allow describing the production system and environment parameters precisely. In particular, in situations where the production system is affected by some kind of data uncertainty, the classical deterministic CLSP model cannot be used. Therefore, we should reformulate the constructed CLSP mathematical model in order to include uncertain demands and to make the model immunized against uncertainty.

3.2 Robust Production Planning Model

One of the goals of the presented research is to extend the classical CLSP mathematical model in order to treat the demand uncertainty.

First of all, we define an uncertainty set *D*. Since for each product *j* in the time period *t* the upper bound d_{jt}^{max} and lower bound d_{jt}^{min} of an uncertain demand are given, we can state that the demand d_{jt} belongs to the uncertainty interval $[d_{jt}^{min}, d_{jt}^{max}]$. In a next step, we consider a set of corresponding uncertainty intervals for each *j* and *t* which defines the uncertainty set *D*. This kind of uncertainty is often called "box uncertainty" and could be imagined in a space as a polyhedron [8].

In our case, the resulting uncertain linear optimization problem is comprised of instances of the CLSP mathematical model (1)–(9). It is parameterized by the uncertain demand running through the uncertain set *D*.

Since we want to get a robust solution immunized against uncertainty, we optimize the worst possible case. It means we try to find such a demand trajectory, belonging to the uncertainty set *D* that gives the maximal possible total costs; subsequently, we optimize the production plan according to this demand trajectory.

Therefore, we need to change the initial goal function in an appropriate way.

Objective function:

The objective function is the function, which minimizes the maximal possible costs over an uncertainty set *D:*

$$\min \max_{d_{jt} \in D} \sum_{t=1}^{N} \sum_{j=1}^{M} (c_{jt}x_{jt} + ov_{jt}y_{jt} + h_{jt}I_{jt} + p_{jt}z_{jt} + s_t b_t + sov_t bov_t)$$

Restrictions:

The restrictions presented in the previous subsection should not be changed; they rather have to be satisfied to each demand realization from the uncertainty set *D*. It means that for each d_{jt} from *D* the proposed model should be feasible.

$$\forall d_{jt} \in D :$$

$$Q_{j1} = I_{j0} + x_{j1} + y_{j1} - d_{j1}, \forall j \in \{1 \ldots M\} \quad (10)$$

$$Q_{jt} = Q_{j,t-1} + x_{jt} + y_{jt} - d_{jt}, \forall t \in \{2 \ldots N\}, \forall j \in \{1 \ldots M\} \quad (11)$$

$$I_{jt} \geq Q_{jt}, I_{jt} \geq 0, \forall t \in \{1 \ldots N\}, \forall j \in \{1 \ldots M\} \quad (12)$$

$$z_{jt} \geq -Q_{jt}, z_{jt} \geq 0, \forall t \in \{1 \ldots N\}, \forall j \in \{1 \ldots M\} \quad (13)$$

$$x_{jt} \leq k_{jt}b_t, \forall t \in \{1 \ldots N\}, \forall j \in \{1 \ldots M\} \quad (14)$$

$$y_{jt} \leq w_{jt}bov_t, \forall t \in \{1 \ldots N\}, \forall j \in \{1 \ldots M\} \quad (15)$$

$$b_t \in \{0, 1\},\ bov_t \in \{0, 1\},\ \forall t \in \{1 \ldots N\} \quad (16)$$

$$\mathrm{x}_{jt} \geq 0, \mathrm{y}_{jt} \geq 0, \forall t \in \{1 \ldots N\}, \forall j \in \{1 \ldots M\} \quad (17)$$

The proposed model reflects all the real-life requirements of a production planning system and is immunized against demand uncertainty, since we optimize the worst possible case.

4 Solution

In this section, we investigate how the proposed model can be solved. Due to the fact that our uncertainty set *D* is computationally tractable, our model is computationally tractable as well. However, we cannot just solve the model for each possible demand trajectory realization, because the number of instances we have to consider grows exponentially with the growth of model parameters and the length of uncertainty intervals.

Therefore, we will exploit a well-known maximum principle from the convex analysis: *"Let f be a convex function, and let C be a nonempty polyhedral convex set contained in dom f. Suppose that C contains no lines, and that f is bounded above on C. Then the supremum off relative to C is attained at one of the (finitely many) extreme points of C."* [9]. In our model, an objective function is a linear function and, therefore, it is convex. The uncertainty set D is also a convex polyhedral. All additional conditions take place as well, thus, the maximum principle is applicable in our case. By this reason, the inner maximum of the objective function is achieved on some combination of extreme demand values. Making intelligent transformations of our mathematical model we can analyze only extreme demand values instead of complete uncertainty set D and end up with an extended linear model.

In order to write our model as a standard linear optimization problem, we consider an additional variable R and represent our objective function in the following way:

$$\min R \tag{18}$$

$$\sum_{t=1}^{N}\sum_{j=1}^{M}\left(c_{jt}x_{jt} + ov_{jt}y_{jt} + h_{jt}I_{jt} + p_{jt}z_{jt} + s_t b_t + sov_t bov_t\right) \leq R \tag{19}$$

Adding the modified goal function and the presented restrictions to the (10)–(17) and considering this model for all possible combination of extreme values $d_{jt,}^{min}\ d_{jt}^{max}$, we will get the final mathematical optimization model. The appropriate parameterization of restrictions was done by adding additional variables and using the following equivalence:

$$a_0 + \sum_{j=1}^{M}\sum_{t=1}^{N}\left(a_{jt}d_{jt}\right) \leq 0, d_{jt} \in \left[d_{jt}^{min}, d_{jt}^{max}\right] \Leftrightarrow \begin{cases} a_{jt}d_{jt}^{min} \leq b_{jt} \\ a_{jt}d_{jt}^{max} \leq b_{jt} \\ a_0 + \sum_{j=1}^{M}\sum_{t=1}^{N} b_{jt} \leq 0 \end{cases} \tag{20}$$

The constraint of a general form in (20) is represented by a system of linear inequalities in variables a_0, a_{jt} and additional variables b_{jt}. In general, a_0 and a_{jt} could be not just the single variables, but rather the linear functions of such problem variables that are certain. This procedure is quite general and is also used in synthesis of linear controllers theory [5]. Making the transformations and putting all constructed systems of inequalities together, we obtain the resulting system of linear constraints.

The obtained model relates to the class of linear optimization problems and an appropriate solution could be calculated with the help of any available linear optimization software. The created production plan is feasible for each possible demand trajectory belonging to the uncertainty set D and, therefore, is immunized against uncertainty. Variable R provides the value of the objective function in the worst-case. Thus, we can guarantee the upper bound of total costs for a manufacturer.

5 Computational Example

In this section, we consider a computational example in order to analyze the results of the proposed method. To evaluate the effectiveness of the proposed model and solution, undoubtedly, a set of real-life examples with an appropriate degree of differentiation is needed. We use the proposed example here only for the educational purposes.

5.1 Data

We consider the production of two different products ($M = 2$) and the total planning horizon is one week ($N = 7$). The positive stock at the end of the last planning period ideally should be equal to zero. Thus, the corresponding holding costs in this period are relatively high. This value could be considered as the punishment or utilization costs of the remaining items. All model parameters are presented in Table 1. The numeric values of the demand given by the upper and lower bounds are given in Table 2.

For the first product, the upper and lower bounds of demand are defined by 15 % difference on a given value of demand (taking into account rounding to an integer. Due to the fact that the demand can be changed by 10 % from the initial value, borders for the second product are constructed. In Fig. 2, the borders for the demand are shown graphically. The growth of the demand for the first product is obtained up to the fifth planning period, but then customer's demand starts to decrease. Vice versa, demand for the second product decreases at the first periods and increases at the end of the planning horizon.

We will consider four possible scenarios for the actual demand realization (Table 3):

- Scenario 1: demand always takes the lowest possible value;
- Scenario 2: demand always takes the highest possible value;
- Scenario 3: demand alternates between maximal and minimal values;
- Scenario 4: demand takes minimal values at the first three periods, an average value at the fourth period and maximal value at the last three periods.

5.2 Solving the Model

Information about the demands is updated at the end of each planning period, and we want to update our model in order to include this information. Thereby, at the beginning we solve[1] the problem with seven planning periods, where demand is exactly known only for the first period and only upper and lower bounds are given for other periods.

---

[1] All models were solved with the help of IBM ILOG CPLEX Optimizer software.

Table 1 Example data

Variables		Periods 1	2	3	4	5	6	7
k_{jt} (\$)	j = 1	70	70	70	70	70	70	70
	j = 2	25	25	25	25	25	25	25
w_{jt} (\$)	j = 1	20	20	20	20	20	20	20
	j = 2	12	12	12	12	12	12	12
c_{jt} (\$)	j = 1	100	120	100	120	100	120	100
	j = 2	70	50	70	50	70	50	70
ov_{jt} (\$)	j = 1	150	180	150	180	150	180	150
	j = 2	100	70	100	70	100	70	100
h_{jt} (\$)	j = 1	2	2	2	2	2	2	300
	j = 2	3	3	3	3	3	3	100
p_{jt} (\$)	j = 1	200	200	200	200	200	200	800
	j = 2	130	130	130	130	130	130	500
s_t (\$)	j = 1	3	3	3	3	3	3	3
	j = 2	3	3	3	3	3	3	3
sov_t (\$)	j = 1	3	3	3	3	3	3	3
	j = 2	3	3	3	3	3	3	3
I_{jo} (*items*)	j = 1	0						
	j = 2	0						

Table 2 Values of upper and lower bounds for demand (items)

		Periods 1	2	3	4	5	6	7
j = 1	Lower bound	47	48	53	64	68	57	49
	Upper bound	63	66	71	86	92	77	67
j = 2	Lower bound	32	31	23	15	20	27	32
	Upper bound	40	37	29	19	24	33	40

However, we implement this production plan only for the first period. Then we solve the new model with only six planning periods and updated demand information, where the initial data is formed according to the implementation results of the previous production plan. This process repeats in a serial fashion manner up to the end of the planning horizon.

All in all, we consider seven models with the planning horizon equal from seven to one period correspondingly, solve these models with extreme values of demand from the uncertainty set D and each time realize the production plan only for the first period. Summarizing the costs for each of implemented periods, we will get the total value of costs.

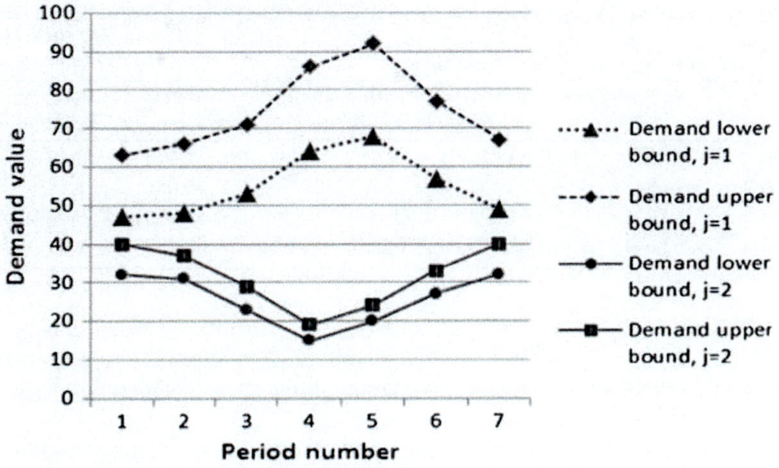

Fig. 2 Values of *upper* and *lower* bounds for demand

Table 3 Considered scenarios of demand realization (items)

		Periods						
		1	2	3	4	5	6	7
Scenario 1	j = 1	47	48	53	64	68	57	49
	j = 2	32	31	23	15	20	27	32
Scenario 2	j = 1	63	66	71	86	92	77	67
	j = 2	40	37	29	19	24	33	40
Scenario 3	j = 1	47	66	53	86	68	77	49
	j = 2	32	37	23	19	20	33	32
Scenario 4	j = 1	47	48	53	75	92	77	67
	j = 2	32	31	23	17	24	33	40

5.3 Analysis of Results

In order to evaluate the obtained results, we have considered the so-called "offline solution" and the solution given by the stochastic approach. Under the offline solution we understand the best-possible solution for the problem, which could be constructed when the exact demand realization is known at the beginning of the planning process and does not change [10, 11].

Talking about the stochastic approach, we should use probabilistic information about uncertain model parameters [4]. Since the demand distribution over the uncertainty interval is not given, it seems quite natural to propose that all demand values have the same probability. Therefore, we propose that the demand has a uniform discrete distribution. Under this assumption, we can easily calculate the demand expectation in each period and use it as model parameter. Considering stochastic programming, we also solve the set of production planning problems and update each one with the exact demand data.

Table 4 Obtained results: value of the objective function provided by different solution approaches and percentage difference between obtained and optimal solutions (RO—Robust Optimization, SO—Stochastic Optimization)

	Offline ($)	RO ($)	Difference RO and offline (%)	SO ($)	Difference SO and offline (%)
Scenario 1	52,636	53,016	0,72	52,737	0,19
Scenario 2	73,841	73,841	0	75,006	1,58
Scenario 3	60,995	61,248	0,41	61,025	0,05
Scenario 4	62,644	62,768	0,20	63,497	1,36

The total values of costs and percentage differences between the obtained solutions are given in Table 4.

For the presented four scenarios of demand realization, the solution provided by the stochastic approach was better than the solution provided by the robust approach in two cases. The maximal difference between the solutions provided by these two approaches is 0.53 %. However, in two scenarios where the quality of robust optimization solution was better, we have the maximal difference of 1.58 % that is three times higher.

Another important difference between the robust and stochastic approaches is the fact that the robust optimization approach can guarantee the upper bound for the objective function. In other words, we can guarantee that the total costs will not exceed the value of 73,841$ for any possible demand realization. In opposite, the expected value of the total costs is 62,243$ for the stochastic optimization. This value is not guaranteed, and in the worst case the total costs are 75,006$, which is 20.5 % higher.

Moreover, if we consider the production planning problem where backlogging is not allowed, the solution provided by the stochastic optimization approach may become infeasible.

6 Conclusions and Future Work

The proposed solution approach can be applied in practical production planning issues. In spite of existing conservativeness (a worst-case solution), it has significant benefits. The comparison with the stochastic optimization approach shows that the solution from the robust optimization is better in certain cases. The financial losses are also negligible in relation to the best-possible solution.

It should be noted that the computational example from the previous section cannot be used for the detailed analysis of the proposed approach. Therefore, implementation of the proposed approach on a set of real-life examples with different datasets and a variety of demand realization scenarios should be considered as one of possible directions for the future research.

It would be also interesting to investigate the applicability of the proposed approach for different kinds of production planning problems, for instance with and without backlogging, with and without setup costs, with different production and setup costs structure, etc. Problems with small and big bucket could also be considered.

Another direction for the future research is the application of the proposed approach for a rolling horizon and the determination of the "price of robustness" in comparison to the offline solution in this case.

References

1. Karimi, B., Fatemi Ghomi, S.M.T., Wilson, J.M.: The capacitated lot sizing problem: a review of models and algorithms. OMEGA, Int. J. Manag. Sci. **31**, 365–378 (2003)
2. Ho, C.: Evaluating the impact of operating environments on MRP system nervousness. Int. J. Prod. Res. **27**, 1115–1135 (1989)
3. Saltelli, A., Ratto, M., Andres, T., Campolongo, F., Cariboni, J., Gatelli, D., Saisana, M., Tarantola, S.:Global sensitivity analysis. The primer (2008)
4. Shapiro, A.: Stochastic programming approach to optimization under uncertainty. Math. Program. Ser., B **112**, 183–220 (2008)
5. Ben-Tal, A., Nemirovski, A.: Robust convex optimization. Math. Oper. Res. **23**(4), 769–805 (1998)
6. Beyera, H.G., Sendhoff, B.: Robust optimization—a comprehensive survey. Comput. Methods Appl. Mech. Eng. **196**, 3190–3218 (2007)
7. Quadt, D., Kuhn, H.: Capacitated lot-sizing with extensions: a review. 4OR, **6**(1), 61–83 (2008)
8. Ben-Tal, A., El Ghaoui, L., Nemirovski, A.: Robust Optimization. Princeton University Press, Princeton, New Jersey (2009)
9. Rockafellar, R.T.: Convex Analysis, pp. 342–346. Princeton University Press, Princeton, New Jersey (1972)
10. Borodin, A., El-Yaniv, R.: Online Computation and Competitive Analysis. Cambridge University Press, Cambridge, UK (2005)
11. Krumke, S.: Online Optimization, Competitive Analysis and Beyond. Technical University, Berlin (2001)

Evaluating Lead Time Standard Deviation with Regard to the Lead Time Syndrome

Mathias Knollmann and Katja Windt

Abstract Extending lead time standard deviation is a key figure that directly influences the due date reliability of a production process. Extending or reducing the planned lead time when trying to improve the due date reliability, does not only change the mean lead time, but also strongly affects the value of the lead time standard deviation. This connection is also associated with the Lead Time Syndrome of production control, which serves as a discussion framework. The aim of this paper is to investigate the lead time standard deviation influencing variables. As a result, various triggers of standard deviation will be discussed.

Keywords Lead time syndrome · Lead time standard deviation · Disturbances · Due date reliability · Production planning and control

1 Introduction

Fluctuations in production processes and inappropriate high work in process levels are only two reasons for a low proportion of orders that are produced on time and thus resulting in a low due date reliability. In order to finish all orders before their due date, a frequent response is to add a 'safety lead-time' to absorb uncertainty in time [1, 2]. With fixed due dates, this reaction implies that orders are released

M. Knollmann (✉) · K. Windt
Jacobs University Bremen, Campus Ring 1,
Bremen 28759, Germany
e-mail: M.Knollmann@jacobs-university.de

K. Windt
e-mail: K.Windt@jacobs-university.de

K. Windt (ed.), *Robust Manufacturing Control*, Lecture Notes in Production Engineering,
DOI: 10.1007/978-3-642-30749-2_34, © Springer-Verlag Berlin Heidelberg 2013

earlier, which increases the workload in the processes [3]. Therefore, the work in process level rises and lead times[1] get longer and more erratic. Finally, this circle of mistakes leads to a lower due date reliability than previous and demands for further measures. This chain reaction was first discovered by Mather and Plossl [4] in 1977 and is referred to as the 'Lead Time Syndrome' (LTS) of production control [5].

Even today's manufacturing facilities deal with interactions and effects of the LTS without knowing the syndrome itself [6]. Previous research on the LTS [6], using the logistic operating curve theory by Nyhuis and Wiendahl, has shown the great impact of the lead time standard deviation on both the effects and the occurrence of the LTS. While high lead time standard deviation represents a consequence, it also represents a trigger of the LTS. The aim of this paper is to question the reasons of lead time standard deviation in the context of the LTS.

To ensure this goal, in a first step the following chapter provides a more detailed description of the LTS and previous research, including an example of the LTS in its steady state. In Chap. 3, the main impacts and reasons for lead time standard deviation in the context of the LTS are outlined. Thereby, the approach of this paper provides a different perspective in order to reveal new due date reliability potentials in production processes, proceeding from the line of argumentation of the LTS.

2 The Lead Time Syndrome of Production Control

Due date reliability is a key performance indicator that directly affects customers. Therefore, direct and indirect measures are taken in production processes to improve it. Moreover, due date reliability is one of the four targets of production control [3], which are short lead times, low work in process (WIP), high capacity utilization and high due date reliability. It is calculated as follows [7, 8]:

$$DDR = \frac{number\ of\ orders\ with L_l \leq L \leq L_u}{total\ number\ of\ orders} \times 100 \tag{1}$$

DDR due date reliability [%]
L_l lower limit of due date tolerance period
L lateness [Shop Calendar Days]
L_u upper limit of due date tolerance period

As these targets are conflicting (e.g., to ensure appropriate capacity utilization a higher WIP is necessary, which induces higher lead times) one has to define a primary goal (e.g. appropriate lead times) to position its processes by means of the

[1] The terms 'throughput time' and 'lead time' are often used synonymously. According to the investigated 'Lead Time Syndrome' only the term lead time is used for simplification.

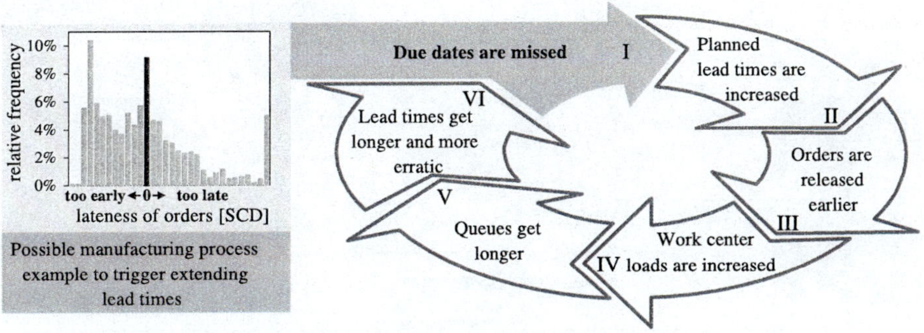

Fig. 1 Lead time syndrome of production control (adapted from [4, 5])

so-called 'logistical positioning' [7, 9]. Unfortunately, the powerful application of logistical operating curves [3] is still not well known in industry. Thus some 80 % of all production planners use safety lead-times as one measure to improve their due date reliability [1]. With longer order lead times disturbances get more likely, which can be defined as an event, "which affects a planned resource movement in such a way that a deviation from plan occurs" [1]. Therefore, the lead time standard deviation increases and due date reliability decreases. This example characterizes only one possible way to trigger the LTS, thus showing its topicality. The phenomenon was firstly described by Mather and Plossl [4] and can be summarized as follows (see Fig. 1) [5]: If the due date reliability is low (step I), it seems sensible to increase planned lead times (II), because it seems as if prior planned lead times were set too short to produce on time. Due to the applied backward scheduling, the orders are released earlier (III) and the workloads at the work stations increase (IV). Therefore, the WIP increases (V) and both the mean values and the standard deviations of lead times increase (VI).

The performed measure to improve the due date reliability leads to deterioration, which reinforces the problem in the end. Ultimately, an increased number of rush orders is necessary to deliver urgent (high prioritized) orders in time, which results in a low schedule reliability. In theory, this leads to a vicious circle, which continues until the lead times reach a very high level [3, 5].

The described 'high level' represents the worst case steady state of the LTS. It can be observed in real production processes when orders are released directly after their acceptance in order to meet due dates and to guarantee a high capacity utilization. The resulting condition (see left part Fig. 1 [6]) is accompanied with a high mean and standard deviation of lead times, high work in process levels and a very low due date reliability.

There is a large volume of published studies using the LTS line of argumentation to introduce, e.g., different production planning and scheduling techniques or to investigate logistical coherences [3, 5, 7, 10–12]. Selçuk et al. [13] initially demonstrated that the update frequency of planned lead times triggers uncontrolled production system states with a high mean and standard deviation of lead times.

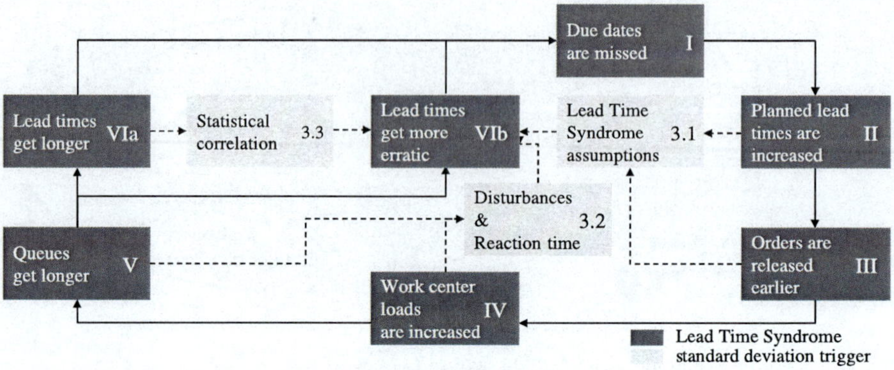

Fig. 2 Triggers of lead time standard deviation in the Lead Time Syndrome

In a more formal study, they [14] investigated subsequently the effects of frequently planned lead time updates under use of a two-dimensional Markov process. In the context of studying the impact of human and organizational factors on planning stability, Moscoso et al. [15] introduced the term 'planning bullwhip' that subsumes planning instabilities such as the LTS. The formal derivation and evaluation of the LTS line of argumentation by Knollmann and Windt [6] revealed on the one hand that fundamental assumptions of the LTS are still rarely investigated while, on the other hand it pointed out the lead time standard deviation as an essential element and trigger of the LTS.

Taking up this direct link between increasing lead times and its standard deviation, a further investigation is necessary. Moreover, knowing that the LTS can be avoided or mitigated most reliably with low and stable lead times [3], the high degree of cross-linking of the variables has to be shown in a first step, thus leading into a renewed understanding of the LTS.

3 Impacts and Reasons for Standard Deviation in the Lead Time Syndrome

Knollmann and Windt [6] investigated the Lead Time Syndrome through mathematical equations acc. to the logistics operating curve theory by Nyhuis and Wiendahl and derived that the occurrence of lead time standard deviation is both a consequence and a trigger of the LTS, thus a key figure which directly influences the due date reliability of a production process. Not only the LTS line of argumentation, but rather other triggers such as statistical correlations, disturbances and reaction times strongly influence the resulting lead time standard deviation in the last step of the LTS (VIb). These triggers are shown in Fig. 2 and will be evaluated in the following Sects. 3.1–3.3.

3.1 Raising the Lead Time Standard Deviation with New Planned Lead Times

The ninth basic law of production logistics [3] points out that higher lead time standard deviations inevitably leads to uncertainties in order scheduling. In order to mitigate the standard deviation, a safety lead time is often added in practice, which leads into the LTS cycle. Knollmann and Windt [6] derived the following equation and showed that in dependency of the adaption of the planned lead times (safety factor Δ)—in the second step of the LTS (see Fig. 2)—the resulting lead time standard deviation can be calculated as follows:

$$TL_{s,new} = \sqrt{\frac{1}{n-1}\sum\nolimits_{i=1}^{n}\left[(TL_i + TL^{\Delta}_{pl,i}) - (TL_m + TL^{\Delta}_{pl,m})\right]^2} \tag{2}$$

TL_s lead time standard deviation
n number of orders i
m mean value
pl planned value

This equation includes the line of argumentation of the LTS, thus including an order specific planned lead time adjustment $TL^{\Delta}_{pl,i}$. It reveals that the more the order specific $TL^{\Delta}_{pl,i}$ differ from one another, the higher the resulting lead time standard deviation will become:

$$TL_{s,new} = \sqrt{\frac{1}{n-1}\left[\sum\nolimits_{i=1}^{n}(TL_i - TL_m)^2 + \sum\nolimits_{i=1}^{n}\left(TL^{\Delta}_{pl,i} - TL^{\Delta}_{pl,m}\right)^2 + 2\sum\nolimits_{i=1}^{n}(TL_i - TL_m)(TL^{\Delta}_{pl,i} - TL^{\Delta}_{pl,m})\right]} \tag{3}$$

$$= \sqrt{TL^2_{s,old} + TL^{\Delta 2}_{s} + 2COV(TL_i, TL^{\Delta}_{pl,i})} \tag{4}$$

Moreover, Eq. (4) shows that if TL^{Δ}_{pl} is constant for all orders, the lead time standard deviation remains unchanged (with $TL^{\Delta}_{s} = 0$ & $(TL^{\Delta}_{pl,i} - TL^{\Delta}_{pl,m}) = 0$ *for* $i = 1..n$). Thus, in dependency of the way of adaption (e.g., random values or a multiplication factor) $TL_{s,new}$ increases or decreases. As $TL^{\Delta}_{s} \geq 0$ Eq. (4) gives the impression that the minimum new lead time standard deviation is the old lead time standard deviation (with $TL_{s,new} = TL_{s,old} + 0 + 0$). But the idea of order specific lead time adaptions in Eq. (2) offers the possibility to standardize the lead times, thus decreasing the new lead time standard deviation. This effect can be seen in the term of the Covariance, which can also be negative ($COV\left(TL_i, TL^{\Delta}_{pl,i}\right) \in R$). Therefore, the consequence of a planned lead time adaption with safety factors on

Fig. 3 Effect of earlier order releases under use of backward scheduling

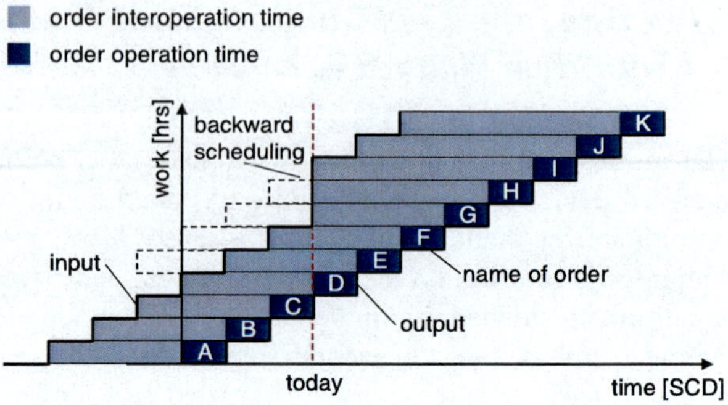

the value of the lead time standard deviation depends strongly on the adaption technique and can have either positive or negative effects on its final value.

As shown above, the impact of the lead time standard deviation directly results from the planned lead time adaption in the second step of the LTS. In addition, the applied backward scheduling in the LTS line of argumentation also influences the resulting lead time standard deviation.

Figure 3 shows the throughput diagram of a simple production process (constant work content; no disturbances) with the orders *A*–*K*. Increasing the lead times of the future orders *E*–*K* leads—under use of backwards scheduling—to an earlier order release, as due dates are fixed. Since the orders *E*–*H* had to be released in the past, their earliest possible order release is today. Therefore, the resulting order lead times of *G*, *H* and *I* differ from each other. Thus, the event of earlier order releases leads to a short term increase of the lead time standard deviation even in this simple example. In the dynamic environment of production processes, this short term increase might lead to further fluctuations within the process chain, which could lead to increasing lead time standard deviation in the long term.

3.2 Disturbances and Causes of Uncertainty in Production

During production, disturbances such as operation disruptions and unstable demand patterns may result in deviations from the original production schedule. According to Wiendahl [5], it is the responsibility of production control to mitigate the causes of uncertainty in production and to react to disturbances. Thus, disturbances in production processes inevitably lead to uncertainties in planning processes and thereby to an increasing discrepancy between planned and actual data. Disturbances are by themselves deviations of planned to actual data on the one hand and incidents that induce these deviations on the other hand [16]. Some of these disturbances are listed in Fig. 4, which affect the performance of production processes in a quantitative, qualitative, and a temporally way [17]. The first two elements result for example through rework in a temporally influence, thus all disturbances contribute to the measured due date reliability.

Production facilities	Human resources	Material	Information processing	Order processing
▪ Machine capacity shortages ▪ Processing errors (scrap/rework) ▪ Machine breakdown ▪ Unexpected maintenance	▪ Absenteeism ▪ Qualification ▪ Processing errors (scrap/rework) ▪ Flexibility (see envelope curves)	▪ Material shortages ▪ Poor supplier delivery performance ▪ Material defects ▪ high work in process level	▪ Inaccurate process feedback ▪ Wrong planning parameters ▪ Data errors ▪ Forecasting problems	▪ Sequence deviation through unexpected rush orders ▪ Fixed batch sizes ▪ Priority rules ▪ Order cancellations

Fig. 4 Abstract of disturbances and causes of uncertainty in production (based on [1, 16, 17, 22, 24])

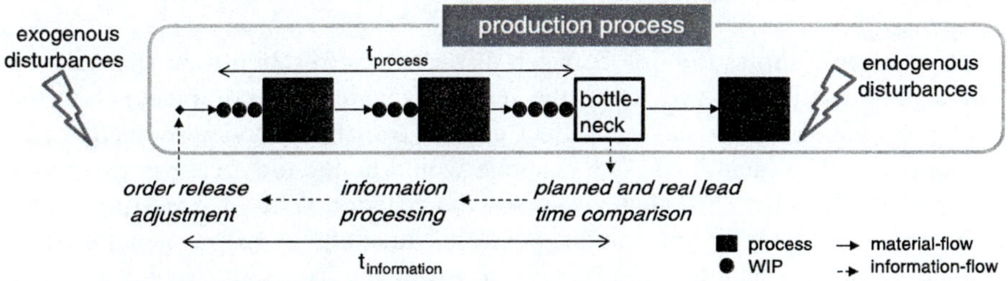

Fig. 5 Dead time of adjustments in production processes

Production planners in manufacturing processes execute safety actions that serve the purpose to avoid delays and prevent the propagation of these disturbances. For example LINDAU [1] observed that safety stocks (71.4 %), safety capacities (92.9 %), safety lead times (78.6 %), and overplanning (64.3 %) are most commonly used in the planning systems. Moreover, orders are, e.g., accelerated (85.7 %) or subcontracted (92.9 %) manually.

Thereby, introducing a safety lead time already raises the possibility of disturbances, thus leading to renewed measures. In addition, safety lead times are often added in order to minimize forecasting problems, as "forecasting accepts the stochastic variability and the inevitable problems it will cause in production" [18]. The example of safety lead times shows how easily the measures that are taken to weaken disturbances and to improve due date reliability lead to triggers of the LTS. Especially in the dynamic environment of production logistics, disturbances have a strong influence if they occur after a process monitoring, before improvement measures take effect. This system's reaction time problem introduces another key variable, the time t, of the lead time standard deviation in the context of the LTS. Figure 5 takes the disturbance topic one step further, bringing together disturbances, the reaction time, and the LTS in one example of the adjustment of a process in order to relieve a bottleneck system.

Before a process can be relieved properly, the current situation of the process chain has to be monitored in a first step at time t_x. Therefore, recent data has to be collected and compared with previously planned data. To derive an adjustment strategy, the data has to be processed and interpreted, which altogether takes the time $t_{information}$. After releasing new measures (e.g., new order release strategy), it takes the time $t_{process}$ until they reach the bottleneck system. This time occurs, as only new incoming orders can be released in a different way (see also Fig. 3), which will not directly affect the bottleneck system (depending on the WIP level). In total, the reaction time, until a suitable measure reaches the bottleneck, is the sum of $t_{information}$ and $t_{process}$. The longer this reaction time gets, the more likely endogenous disturbances (arise in the process itself, such as machine breakdowns) or exogenous disturbances (arise outside the process, such as material shortages) will be. These disturbances affect the running processes and possibly lead to a renewed adjustment of the process at t_{x+1} before the previous adjustment has reached the bottleneck system and thus taking effect at the bottleneck ($t_x + t_{reaction} > t_{x+1}$).

This example shows the link between disturbances, reaction time and the lead time standard deviation and raises the question, which other influences—in the context of disturbances—have an effect on the lead time standard deviation and how they could be quantified. One example would be the tool of envelope curves of capacity flexibility [19] that visualizes the reaction time of operating times. Therefore, the limit of maximum possible fluctuations of in general slow responding process chains could be derived under the use of this tool.

3.3 Statistical Correlation Between Mean and Standard Deviation

The value of the lead time standard deviation is on one hand "determined by the applied sequencing rule, the work in process level and the distribution of the work content" [3], while on the other hand the statistical correlation between mean and standard deviation demands for a further examination of the link between theory (statistics) and reality (production logistics).

In statistics, the mean $\overline{X}$ and standard deviation S of a random sample $x_1, \ldots, x_n$ is defined as $\overline{X} = \frac{1}{n}\sum_{i=1}^{n} X_i$ and $S = \sqrt{\frac{1}{n-1}\sum_{i=1}^{n} (X_i - \overline{X})^2}$ [20]. From the LTS point of view, the question is whether or not a higher mean value affects the value of the standard deviation. Thus, multiplying each X_i with a factor $B \in R$, the following new mean $\overline{X}_B$ and standard deviation S_B result:

$$\overline{X}_B = \frac{1}{n}\sum_{i=1}^{n} (X_i \cdot B) = B \cdot \overline{X} \tag{5}$$

$$S_B = \sqrt{\frac{1}{n-1}\sum_{i=1}^{n}([X_i \cdot B] - [\overline{X} \cdot B])^2} = \sqrt{B^2 \frac{1}{n-1}\sum_{i=1}^{n} (X_i - \overline{X})^2} = S \cdot B \tag{6}$$

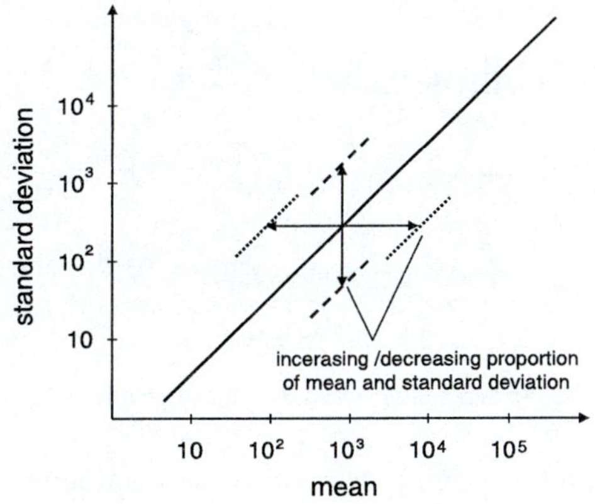

Fig. 6 Statistical correlation between mean and standard deviation (example values)

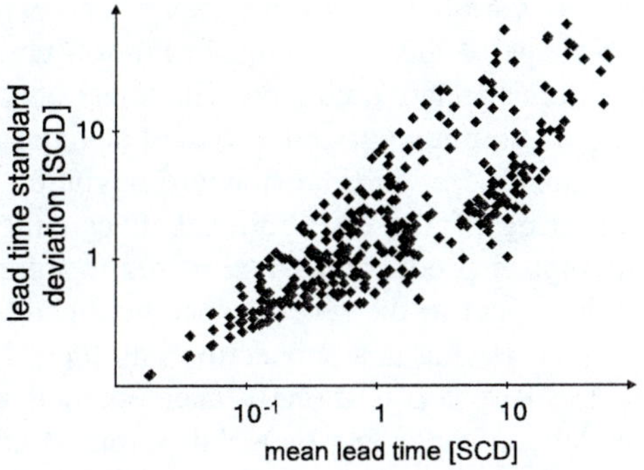

Fig. 7 Correlation of the mean and standard deviation of lead times in real production systems

cThese equations demonstrate that increasing the mean with a factor of B leads to an increase of the standard deviation with the same factor. This connection is known in statistics under the term ‘important rules for variance’ of random variables [21]. The effect is exemplary shown in Fig. 6. Based on a random proportion of mean to standard deviation, the slope of the straight line characterizes the mathematical connection between both variables. In the given logarithmic scale, the slope of the straight line does not change with increasing/decreasing proportions or example values of the variables.

This logical statistical connection can be compared with the experience of production planners that an increasing lead time leads to a higher lead time standard deviation. Therefore, we investigated production system feedback data from five different manufacturing companies, which are located in different logistical fields (i.e., process industry, tool and mounting, and sampling). These random companies offer the possibility to produce a representative distribution, which is shown in Fig. 7. It shows the average production system lead times compared with their individual lead time standard deviation[2] (logarithmic scale). The interpretation of this distribution firstly reveals an obvious correlation between

[2] The investigation periods of the feedback data differ from 3 month to 1 year.

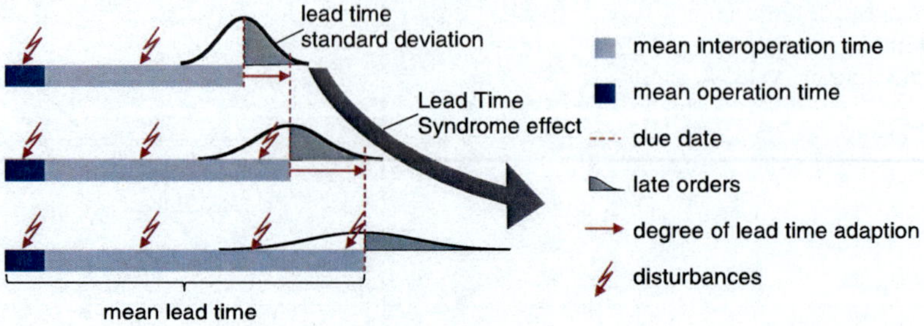

Fig. 8 Schematic representation of standard deviation caused by lead time adaptions

both variables. Moreover, the distribution lies in a corridor that can be defined through a lower/upper border. What counts for this paper is that the statistical link of both variables exists in practice. The upper bound describes a straight line from the shape of the correlation in Fig. 6, which can be interpreted as the maximum possible standard deviation. The lower bound of the distribution can be interpreted as the minimum possible standard deviation, which can be explained, e.g., with an existing operation time standard deviation [3, 22]. One approach to interpret the scattering of the distribution between the upper and lower border can be the attempt of production coordinators to minimize the lead time standard deviation with respect to the process specific mean lead time level.

Transferring this connection into the LTS steps, the increasing mean lead time in the last step is directly accompanied with an increasing lead time standard deviation. Depending on the slew rate of the mean value, a predictive value of the standard deviation can be derived. Figure 8 depicts the described correlation in the context of the LTS. Thereby, trying to produce late orders in time, lead times are increased to a certain degree. Thus, for example disturbances get more likely, which leads to an increased lead time standard deviation (described in Sect. 3.2) and demands for longer lead times once again. To absorb the increasing standard deviation, the degree of adaption rises with every LTS-loop.

The statistical correlation of both variables and moreover the correlation in real production systems raises the question, whether the schematic representation of the LTS in Fig. 8 can be transferred into a formula, calculating an estimated value of the lead time standard deviation for a striven mean lead time. Potentially, the logistic operating curve theory [5, 23] offers an intermediate step between Figs. 6 and 7, as it deals with, e.g., differing lead time values.

3.4 Aggregation of the Different Triggers

The effects that are shown in Fig. 2 and have been described in the Sects. 3.1–3.3 are not independent from another. For example the order specific lead time adjustment $\Delta TL_{pl,i}$ in Sect. 3.1 reflects partly the described effects of higher mean lead times in Sect. 3.3, which can be observed in practice. Equation (7) aggregates schematically all described effects. Therefore, the resulting new lead time standard deviation consists of the old lead time standard deviation and the magnitudes of influence. Moreover, these variables are overlapping and influencing each other such as the disturbance 'sequence deviation', which also depends on the applied scheduling technique, work center loads, etc. Thus, further research is necessary to define exact values and to derive a more detailed formula.

$$TL_{s,new} = TL_{s,old} \pm TL^{\Delta}_{s,scheduling} \pm TL^{t}_{s,disturbances} \pm TL^{B}_{s,statistics} \qquad (7)$$

$\mathrm{TL}^{\Delta}_{\mathrm{s,scheduling}}$ lead time standard deviation that depends on the underlying scheduling technique and the safety factor Δ

$\mathrm{TL}^{\mathrm{t}}_{\mathrm{s,disturbances}}$ lead time standard deviation that results from disturbances and the system reaction time *t* on adaptions

$\mathrm{TL}^{\mathrm{B}}_{\mathrm{s,statistics}}$ lead time standard deviation that reflects the statistical correlation of mean and standard deviation; multiplication factor *B*

4 Conclusion and Outlook

The aim of this paper was to point out different factors that influence the lead time standard deviation. It has been shown that several triggers affect the value of the lead time standard deviation, thus demonstrating the topicality of the LTS coherences. The lead time standard deviation is the result of different logistical effects (e.g., backward scheduling) on the one hand and on the other hand adversely affects production processes (e.g., disturbances).

The current research was not specifically designed to investigate each factor in a concluding analysis, but rather to raise further research questions in this topic. Therefore, a further investigation of the statistical correlation in Sect. 3.3 has to be processed (e.g., the allocation of the different industry segments in the distribution, a mathematical description of the upper/lower boundary, the derivation of underlying reasons such as dispatching rules and statistical methods to describe shape and correlation). Further research might also explore Eq. (7) of Sect. 3.4. The exact values of the variables in this equation—in the context of the LTS—are rising questions of importance, especially as each of the described coherences has a strong influence on the resulting lead time standard deviation in production processes.

The results of this study indicate that the introduction of valid logistical methods should be aspired in order to reduce the lead time mean and standard deviation. Thus, this research will serve as a base for future studies with the aim to define a composition of countermeasures that will work against the causes and effects of the LTS elements, to exploit new due date reliability potentials.

Acknowledgments The research of Katja Windt is supported by the Alfried Krupp Prize for Young University Teachers of the Alfried Krupp von Bohlen und Halbach-Foundation.

References

1. Lindau, R.A., Lumsden, K.R.: Actions taken to prevent the propagation of disturbances in manufacturing systems. Int. J. Prod. Econ. **41**, 241–248 (1995)
2. Ould-Louly, M.-A., Dolgui, A.: The MPS parameterization under lead time uncertainty. Int. J. Prod. Econ. **90**, 369–376 (2004)
3. Nyhuis, P., Wiendahl, H.-P.: Fundamentals of Production Logistics: Theory Tools and Applications. Springer, Berlin (2009)
4. Mather, H., Plossl, G.: Priority fixation versus throughput planning. Production and inventory management: J. Am. Prod. Inventory Control Soc., APICS. 3rd Q., pp. 27–51 (1977)
5. Wiendahl, H.-P.: Fertigungsregelung: Logistische Beherrschung von Fertigungsabläufen auf Basis des Trichtermodells. Hanser Verlag (1997)
6. Knollmann, M., Windt, K.: Exploitation of Due Date Reliability Potentials—Mathematical Investigation of the Lead Time Syndrome, 3rd International Conference on Dynamics in Logistics (LDIC). In print, p.12 (2012)
7. Lödding, H.: Verfahren der Fertigungssteuerung. Springer, Berlin (2008)
8. Windt, K., Nyhuis, P., Kolev, I., Gebhardt, N., Eilmann, J., Fronia, P., Bertsch, S.: Improving Due Date Reliability of Steel Mills: Identification of Punctuality Potentials, Innovation in Product and Production, 21st International Conference on Production Research (2011)
9. Nyhuis, P., Wiendahl, H.-P.: Logistic Production operating curves—basic model of the theory of logistic operating curves. CIRP Ann., Manufact. Technol. **55**, 441–444 (2006)
10. Wiendahl, H.P.: Betriebsorganisation für Ingenieure. Hanser Verlag (2008)
11. Schuh, G.: Produktionsplanung und -steuerung: Grundlagen, Gestaltung und Konzepte. Springer (2006)
12. Breithaupt, J.-W., Land, M., Nyhuis, P.: The workload control concept: theory and practical extensions of load oriented order release. Prod. Plann. Control **13**, 625–638 (2002)
13. Selçuk, B., Fransoo, J.C., De Kok, A.G., DeKok, A.: The effect of updating lead times on the performance of hierarchical planning systems. Int. J. Prod. Econ. **104**, 427–440 (2006)
14. Selçuk, B., Adan, I.J., de Kok, T.G., Fransoo, J.C.: An explicit analysis of the lead time syndrome: stability condition and performance evaluation. Int. J. Prod. Res. **47**, 2507–2529 (2009)
15. Moscoso, P., Fransoo, J., Fischer, D., Wäfler, T.: The planning bullwhip: a complex dynamic phenomenon in hierarchical systems. In: Fransoo, J.C., Wäfler, T., Wilson, J. (eds.) Behavioral Operations in Planning and Scheduling, pp. 159–186. Springer, Berlin (2011)
16. Patig, S.: Ansatzpunkte und Rechnerunterstützung des produktionsorientierten Störungsmanagements. Fakultät für Informatik, Otto-von-Guericke-Universität Magdeburg (1999)
17. Schwarz, F.: Störungsmanagement in Produktionssystemen. Shaker (2004)
18. Kingsman, B.G.: Input/output backlog control and dynamic capacity planning in versatile manufacturing companies. In: Christer, A.H., Osaki, S., Thomas, L.C. (eds.) Stochastic Modelling in Innovative Manufacturing, pp. 97–122. Springer, Berlin (1997)
19. Breithaupt, J.-W.: Controlling production dynamics—managing uncertainties with automatic production control. Int. J. Prod. Res. **38**, 4235–4246 (2000)
20. Hoang Pham: Basic Statistical Concepts. In: Pham, H. (ed.) Springer Handbook Of Engineering Statistics, pp. 3–48. Springer, Berlin (2006)
21. Kozak, A., Staudhammer, C., Watts, S.: Introductory probability and statistics: applications for forestry and natural sciences. CABI Publishing (2008)
22. Ludwig, E.: Modellgestützte Diagnose logistischer Produktionsabläufe. VDI Fortschritt-Berichte (2) 362 (1994)
23. Yu, K.-W.: Terminkennlinie: Eine Beschreibungsmethodik für die Terminabweichung im Produktionsbereich. Fortschritt-Berichte VDI (2001)
24. Koh, S.C.L., Saad, S.M.: Managing uncertainty in ERP-controlled manufacturing environments in SMEs. Int. J. Prod. Econ. **101**, 109–127 (2006)

An Integrated Approach: Combining Process Management, Organizational Structure and Company Layout

Günther Schuh, Till Potente, Fabian Bachmann and Thomas Froitzheim

Abstract When production engineers consider productivity and efficiency, much attention has been paid to optimizing production processes. Numerous approaches and software tools have been developed in order to enhance the optimization of production processes like Proplan, ARIS-Toolset or aixperanto. The corresponding intention of these approaches is to present, analyze and optimize business processes with a great focus on the involvement of the people imbedded in the processes. Hardly any of these approaches integrate the organizational structure of a company like different departments or diverging hierarchical levels. Neither of these approaches include the company layout for example the local office layout. The innovative goal of this paper is to develop a new interdisciplinary model calculating the productivity of a process by using the two factors identification and communication which are highly influenced by the organizational structure and the company layout. The idea of the model is to address the weak points on the interfaces in a process by considering the influence of the mentioned factors and to show the possible increase of productivity. The model will be validated by a case study conducted by the WZL.

Keywords Process management · Organizational structure · Company layout

G. Schuh · T. Potente · F. Bachmann (✉) · T. Froitzheim
Laboratory for Machine Tools and Production Engineering (WZL) of RWTH Aachen University, Aachen, Germany
e-mail: F.Bachmann@wzl.rwth-aachen.de

G. Schuh
e-mail: G.Schuh@wzl.rwth-aachen.de

T. Potente
e-mail: T.Potente@wzl.rwth-aachen.de

T. Froitzheim
e-mail: T.Froitzheim@wzl.rwth-aachen.de

K. Windt (ed.), *Robust Manufacturing Control*, Lecture Notes in Production Engineering,
DOI: 10.1007/978-3-642-30749-2_35,

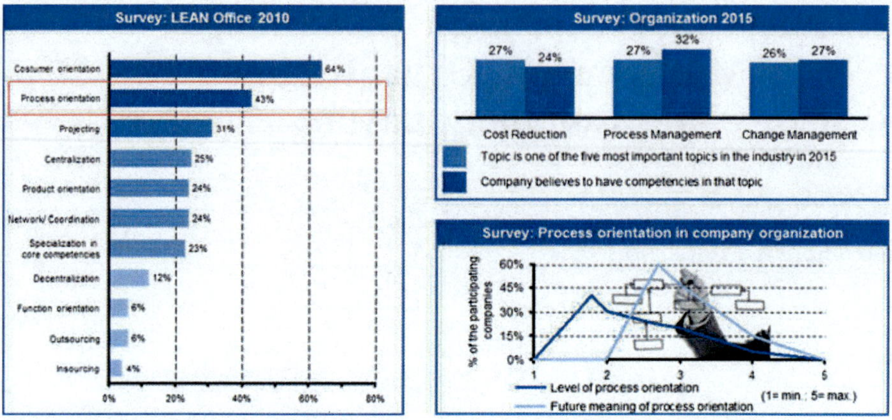

Fig. 1 Process management—Key factor for success in the industry [1–3]

1 Introduction

Process management is a key theme in the industry. The current interest in process management is rated as very high by more than 40 % of the participants taking part in a survey by Fraunhofer IPA [1]. Moreover process management is ranked among the five most important strategic themes for companies [2]. Despite this high interest in process management and its strategic priority among industrial firms, the implementation and understanding of process orientation is not yet fully established [3] (Fig. 1).

The general task of process management is the process-oriented design of business processes as well as their optimization concerning quality, time, costs and customer satisfaction in order to enhance the companies' competitiveness [4]. A business process consists of a sequence of activities generating benefit for internal or external customers. It does not end at the borders of organizational units but crosses different departments within the organizational structure [5]. Traditionally companies are organized by functional units in which consistent activities are combined to create a high degree of specialization and clear areas of responsibility in order to guarantee efficient activities within these units [6]. Disadvantages of functional organizational structures are their lacking focus on customer needs, inefficient and nontransparent cross unit operations and redundancies [7, 8].

In contrast to that, process-oriented organizations adjust their organizational structure along their process chains. These process chains in turn are designed according to the needs of internal and external customers. Thus from a process- and customer-oriented perspective companies' specific processes can be seen as the conceptual framework which should determine the flow of material and information. In process-oriented organizations functional units are widely replaced by processes. Furthermore, process owners or process managers, who are responsible for the process quality, are defined. As a result hierarchies and interfaces are reduced, whereby

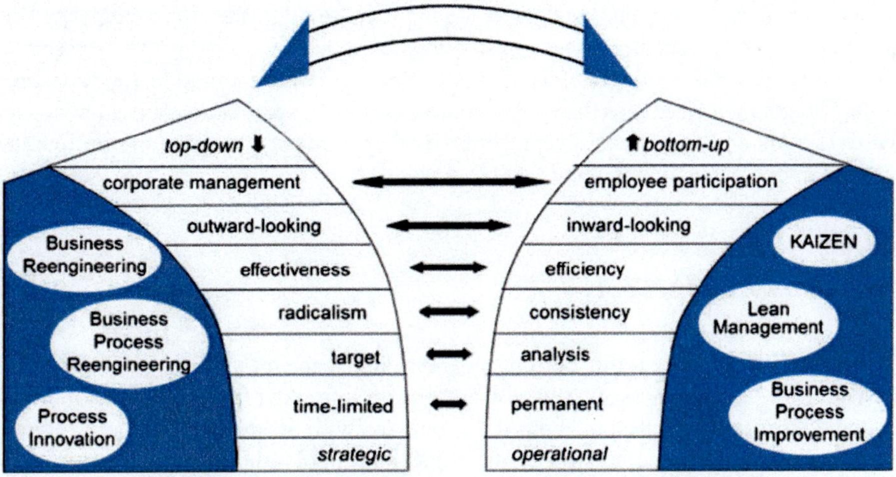

Fig. 2 Top-down and bottom-up approaches to implement process-orientation [14]

employees gain a better understanding of the process chain, coordination effort decreases, the company can react more flexible to changes in customer needs and thus the productivity increases too [8–10]. However, the realization of process orientation requires a high amount of staff and financial effort. For a sustainable turnaround from a functional to a process-oriented company, it is crucial to reach all employees and make them familiar with a process-oriented way of thinking and acting [7, 11, 12].

2 Implementation of Process Orientation: Overview of Approaches and Tools

2.1 The Top-Down and Bottom-Up Approach

In order to realize the potential efficiency benefits and improve customer satisfaction through process-orientation two main approaches can be distinguished: firstly, a bottom-up oriented approach which aims at constant and incremental improvements and secondly, the top-down approach with the main idea of a radical process redesign [13, 14] (Fig. 2).

The term Business Process Reengineering was established by Johansson in the early 1990s. Later, Hammer and Champy defined it as a fundamental rethinking and a radical redesign of business processes in order to achieve dramatic improvements in key performance indicators such as costs, quality, services and process time. Business processes with a relatively high contribution to the added value are generally redesigned by the top-down principles, by gradually detailing the superior strategic goals. In contrast, the bottom-up approach starts with an analysis of the current process characteristics and the identification of major inefficiencies at the operating level. The design of ideal processes and the implementation of mechanisms for constant process improvements are also part of this approach. One of the most famous synonyms for this kind of process improvement is the Japanese term Kaizen, which stands for "turn for the better". Lean Management with its various tools and frameworks represents another concept for different continuous improvement mechanisms [10].

2.2 Visualization Tools

To get an overview of the current process landscape of a firm or one single division, the top-down as well as the bottom-up approach uses different kinds of visualization tools. With their help the processes with all their interdependencies, information, material paths, processing times and involved parties are mapped, so that project teams gain a first transparent overview and deeper understanding of the actual happenings between order inflow and delivery.

Up to now several business process modeling tools have been developed. By paying attention to the resource, functional and organization view, it is possible to model the business processes and to provide transparency. Three business process modeling tools, which are frequently used in practice and during projects conducted by the Laboratory of Machine Tools and Production (WZL) of RWTH Aachen University, are presented below:

- Aris-Toolset
- Proplan
- Aixperanto

The **ARIS-toolset** is an analysis and modeling software for the realistic depiction, analysis and optimization of a company's organizational and operational structure. It focuses especially on the application of integrated information systems as a support of business processes. The analysis is generally structured by the five perspectives: organization, functions, performance, data and control mechanisms. In order to guarantee a consistent and complete depiction of the relevant process steps and characteristics, the ARIS architecture contains a special process modeling language called event driven process chain. It characterizes processes mainly by their corresponding functional units and interdependencies concerning time and content. A further aspect of the process analysis with ARIS is the identification and classification of process inefficiencies [13, 15, 16].

The **Proplan** visualization tool-set uses 14 standardized process elements to depict the process landscape of a firm. These process elements are divided into 8 direct and 6 indirect elements. The former are used to characterize process steps which directly add value to the finished product, like a manufacturing process. Whereas the latter are used to depict process steps with less contribution to the value creation, like transportation processes. Furthermore, intuitive understandable standard symbols are used to facilitate and accelerate the depiction process.

Besides the depiction of the process, Proplan supports the evaluation of the process' weaknesses [14].

Aixperanto gives the possibility to illustrate and analyze business processes as a whole. The main aspect of this visualization tool is the integration of the affected employees from the beginning. Thus, it is possible to create an awareness of the problems in business processes. As a result, the employees are capable to find weak points and suggest possible improvement measures. Therefore, easy symbols and understandable colors are used to help the employees recording their business process. In addition both qualitative and quantitative parameters are analyzed to influence the employees understanding of the business processes. These parameters are described very simple and distinct, e.g. by using the green–yellow–red logic taken from the traffic light. Thus, well and bad performing business processes are immediately distinguished from each other. Kaizen flashes mark the identified weak points and the different business units are placed in different swim lanes in order to see interfaces between the units at first sight [14, 17].

The comparison of the three different visualization tools show, that ARIS is especially suitable for process documentation due to the five different layers. Proplan is more suited to analyze business processes, whereas aixperanto serves as a supportive tool for business process optimization.

3 Problem Specification

Besides the goal of visualizing the current process characteristics all three tools support the user in the identification and analysis of current weaknesses and the definition of specific activities which are needed to reach a target process. In order to get a vision of such an optimal target process the identification of current inefficiencies is crucial. Process weaknesses, as e.g. dissipation of resources or bottlenecks can be found generally within a specific area of responsibility or at the interfaces between different departments.

For an improvement of process quality within a closed area of responsibility current literature offers varied standardized measures like illustrated in Fig. 3 [14]. Typical improvement measures are for instance the elimination or standardization of specific process sections. 5S-activities like a disciplined adherence to cleanness at a working station contribute can also be assigned to this group of improvement measures.

The weak points at the interfaces between different departments and hierarchy levels are not well described in literature. It is simply advised to reduce these interfaces or, if possible, to standardize them by implementing specialized IT-systems [18].

A set of optimization projects carried out by WZL of RWTH Aachen University show that local improvements in single areas of responsibility are much easier to realize than across the linked units. Since a local optimization of process characteristics does not guarantee an improvement of the whole process chain of a

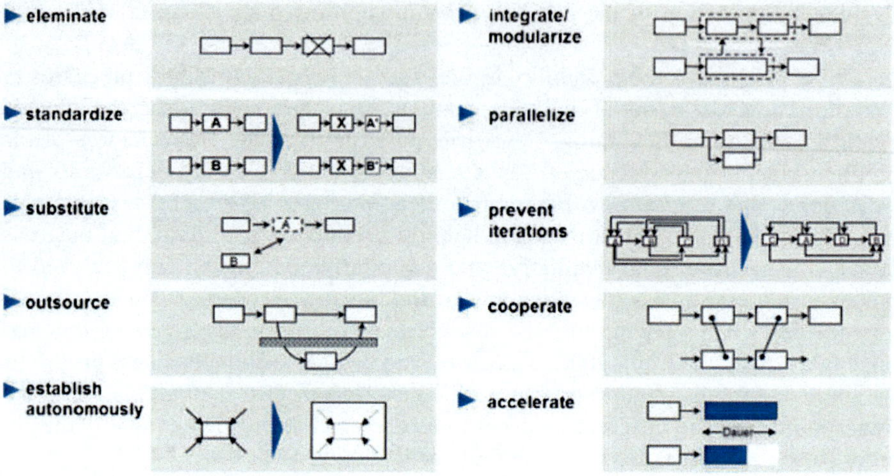

Fig. 3 Standard measures to improve business processes [14]

firm or in some cases even provoke new weaknesses in following process steps, a lot of improvement projects do not reach the aspired target process conditions. The missing sustainability of process reengineering projects is also documented by several specific surveys [19, 20]. According to them about 60 % of realized optimization projects do not achieve their target objectives. Remaining interfaces between departments can be identified as the crucial challenge for the productivity of the processes. Very often problems at the interfaces of departments can be traced back to insufficient communication between the involved employees or to interpersonal problems [21]. The reason for most communication difficulties along the process chain lays in the functional organization structure of a firm and the resulting strict division of labor among the departments. The flow of communication is split due to hierarchies in a vertical direction and due to department interfaces in a horizontal direction [22].

Putting communication processes in the focus of business process reengineering as a new approach of reaching sustainable improvements it becomes necessary to modify the original definition of business processes as "a collection of activities that take one or more kinds of input and create an output that is of value to the customer". Therefore, business processes should be seen from the perspective of the communication theory [23]. By this approach business processes are abstracted to a set of linked communicational operations. A communicational operation in turn consists of three elements: information (difference, which changes the condition of a system), message (intended action of the sender) and understanding (meaning of information and message are determined by the addressee) [24].

By this approach the productivity of business processes is mainly determined by the firms' communication patterns. As a result the set of the above mentioned classical process optimization measures should be expanded by two dimensions which are highly relevant for the communication patterns which develop within a company. This is firstly the configuration of the organizational structure, which addresses the previously mentioned vertical dimension of communication due to hierarchy. The second dimension, addressing especially the horizontal flow of communication can be seen in the technical distance of employees respectively the company layout.

4 Optimization Approach

In order to increase the process productivity the WZL developed a new approach with special considerations of communication patterns. In addition to the standard measures, two new main components have to be considered: the company layout and the organizational structure.

Thereby, the company layout means the spatial environment of a company like the relationships between rooms, spaces and other physical features as well as the spatial arrangement of the companies' stuff. Up to now the company layout is important during the company planning process in order to achieve for example good material flows or the best utilization of space [25].

The organizational structure is a framework within an organization arranges its lines of authority, the communication and the task allocation. It can be structured in different ways like a functional, process or matrix organization depending on the company objectives. It also determines which individuals get to participate in which decision-making processes [26]. An explanatory model was created in order to quantify the influences.

4.1 Explanatory Model

The key idea of the explanatory model is to focus on the increase of productivity over time. The model does not claim to be a holistic approach and to cover all eventualities. It rather tries to use as less parameters and variables as possible in order to guarantee a simple usage and feasibility in practice. The model is based on the interaction between an actor (employee) and a resource or a process. An algorithm is applied to calculate how likely it is that ratio potentials will be used. However, to calculate the potential change in productivity it is necessary to repeat the calculation for the entire organization structure involved in the tasks and activities concerning the resource or process. From a recent benchmarking study conducted by the WZL in 2011 about production systems, it can be derived that an increase of process productivity is based on the experience of the employees and the two main factors [27]:

- Identification and
- communication.

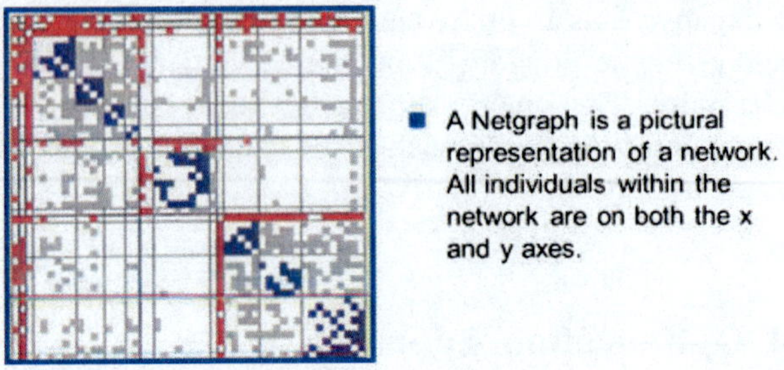

Fig. 4 Communication along the organization hierarchy [29]

The explanatory model showed below based on the findings of ALLEN and HENN and on the results of the benchmarking study allows quantifying the explained effects. The correction factors used in the model are based on experiences from the benchmarking project and serve the balance of the three different terms.

$$\frac{dP}{dt} = \underbrace{[(OH + FH)]}_{\text{Identification}} + \underbrace{[(OF + KF)]}_{\text{Communication}} *B \tag{1}$$

$\frac{\mathrm{dP}}{\mathrm{dt}}$: increase of productivity over time; B: experience; OH: organizational effect; FH: span of control; OF: technical exchange probability; KF: effect of distance

4.1.1 Identification

Identification, which means the creation of a 'us-feeling' between employees and the company, is based on two effects: the organizational effect and the span of control.

The organizational effect represents the synchronism of the objectives to be achieved e.g. different departments can be accounted for disjunctive goals. The basic idea is that the longer the distance between two persons involved in the process the more the objectives and targets differ. The influence of the organizational structure postulated GREENBAUM already in 1983: "the change in structure would significantly increase the effectiveness or communication processes" [28]. An empirical analysis of this postulate has been accomplished by ALLEN and HENN [29]. (Fig. 4)

The Netgraph shows a company division with three departments. All individuals of the division are on both the x and y axes. The purple, red and blue filled-in squares indicate intradepartmental communication between two people, the grey ones represent interdepartmental communication. The department placed in the center is also located in a separate building. Thus the Netgraph considers the organizational hierarchy within this company division. The evaluation shows that the two divisional leaders communicate mostly with the department leaders but also with different people throughout the three departments. However, a decreasing level of communication is seen with the department located in a separate building. The Netgraph also shows that the three department leaders mostly communicate with people within their own departments. Thus, it becomes apparent that organizational structures affect communication between people in an organization [29]. Therefore the organizational distance is measured in integer variables counting the number of hierarchy changes needed to get for example from the actor to the resource. The decisive variables and constants that need to be addressed are listed and named below.

$$\text{Organizational effect}: OH = k_1 \frac{1}{x^2} \tag{2}$$

Correction factor $k_1 = 0{,}25$; distance of hierarchy X = var.

The span of control represents how good objectives and principles are transferred from management or foreman to the employees. In this context, the more employees in the organizational structure the more difficult it becomes to transfer goals and principles. This has an impact on the identification and the effectiveness of the employees. Therefore the functional relationship is showed in the following.

$$\text{Span of control}: FH = k_2 \frac{1}{m} \tag{3}$$

Correction factor $k_2 = 0{,}125$; employee m = var.

4.1.2 Communication

The communication term in the overall equation is based on two effects: technical exchange probability between colleagues and the effect of spatial distance.

One of the main strengths of functional units is the exploitation of economies of scale within a department. This effect can be basically traced back to the technical exchange probability between colleagues. Consequently, the increasing number of employees with the same area of activities lead to a higher performance of the individual employee. As a result of a growing communication complexity the mentioned effect has a certain saturation effect. Once a certain department size is exceeded, the performances of the employees' don't rise further (Fig. 5 right side). Also, account must be taken to the fact that an improvement of efficiency due to technical exchange is only possible if a spatial distance between them is granted [29]. This effect is about as strong as the identification and is the reason why a functional organizational structure in many cases leads to better results than a process orientation. The important variables and the functional relationship are described below.

$$\text{Technical exchange probability}: \; OF = \left(k_3 * EK^{\frac{\ln(c)}{\ln(2)}}\right) * \left(k_4 * i^{(r)^N}\right) \tag{4}$$

correction factor $k_3 = 0{,}2$; correction factor $k_4 = 0{,}8$; communication rate c = 0,65; middle distance to employee EK = var.; initial exchange i = 0,2; exchange decrease r = 0,6; employee N = var.

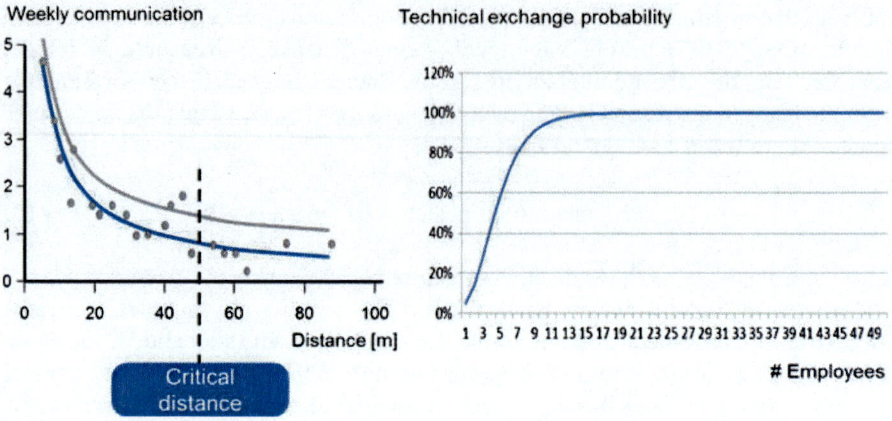

Fig. 5 Critical distance and technical exchange probability [29]

The effect of spatial distance describes the influences of the distance between two persons involved in the process. Therefor the spatial distance of the employee to the place of action is highly crucial. The calculation of the effect of distance is based on the empirical findings of ALLEN and HENN who showed that the communication frequency among employees in an organization highly correlates with spatial distance between them [29]. Left side of Fig. 5 indicates the asymptotic relationship between weekly communication frequency and spatial distance between the interacting employees. The curve is characterized by a steep negative slope up to about 20 m of distance and a fast approximation to an asymptotic level between zero and one communication events per week starting at about 50 m of distance. Besides the absolute distance between employees, the probability of communication also depends on special restrictions of the building. Thus it is of importance for communication activities, whether colleagues are arranged at a common floor or wing of a building or in different buildings. [29] (Fig. 5).

The relationship between physical location of employees and their communication among each other has been investigated by BOUTELLIER. He says: "office arrangement does influence communication and therefore can be used as a tool to reach higher productivity" [30]. STRYKER comes to a similar result and identifies a correlation between office layout and "team face 2 face interaction" [31]. The functional relationship is described below.

$$\textit{Effect of distance} : KF = k_5 * EA^{\frac{\ln(c)}{\ln(2)}} \tag{5}$$

Correction factor $k_5 = 0{,}2$; physical Distance EA = var.; communication rate $c = 0{,}65$

4.1.3 Experience

Finally the experiences gained by the employees make a contribution to improve the productivity over time. In this context, economies of scale have a high impact too, as a high wealth of experiences enhances productivity. This approach considers the positive effect that a function-oriented structure has on specialization in certain processes or resources. The equation to calculate the economics of scale is described below.

$$economics\ of\ scale: B = 1 - \left(k_6 * \left(\frac{f}{ma}\right)^{\frac{\ln(L)}{\ln(2)}}\right) \tag{6}$$

correction factor $k_6 = 1$; operations per year f = var.; employee in charge of plant ma = var.; communication rate L = 0,8

5 Validation: Case Study Within the Process Industry

5.1 Goal and Project Approach

The WZL has carried out a project with a client within the process industry at one German site, where the described explanatory model has been used. The initial situation in the plant showed that the department of industrial engineering was placed too far away from the manufacturing machines. Therefore it was analyzed which extent would be an improvement to integrate the department of industrial engineering into the single product lines and to put them under the control of the person in charge for the machine management, which means implementing an process-oriented structure.

The single decisive variables were taken up by the WZL and the involved employees together during several workshops for every machine and organizational changes. Thus, the equation for calculating the change in productivity could be used. Table 1 shows the values recorded for a single plant as an example.

5.2 Results

The generated results of the project show the increase of productivity for the current process or a function-oriented structure by using the overall equation based on identification and communication as well as experiences on the one hand. On the other it shows the increase of productivity for the new structure (Table 2).

Based on the 'increase of productivity' it is possible to calculate costs and possible revenues. Thus the comparison of the achieved results can help the

Table 1 Decisive variables

	x	m	N	EK	EA	f	ma	S
New structure	2	100	2	250	50	200	1	200
Old structure	4	6	6	20	150	200	4	50

Table 2 Results

	Old structure	New structure
Identification	3.65	6.38
Organizational effect	1.56	6.25
Span of control	2.08	0.13
Communication	3.77	2.12
Technical exchange	2.88	0.36
Effect of distance	0.89	1.79
Experiences	71.26	81.84
Increase of productivity	5.31	6.95

management to decide whether the company structure should be adjusted by implementing process orientation or to keep the function-oriented structure.

6 Conclusion and Critical Reflection

Process management is one of the key themes in the industry. The general task is the process-oriented design of business processes as well as their optimization in order to enhance the companies' competitiveness. Unfortunately a lot of improvement projects do not reach the aspired target process conditions. The remaining interfaces between departments can be identified as the crucial problem. Moreover, these problems can be traced back to insufficient communication between the involved employees due to hierarchies in a vertical direction and due to department interfaces in a horizontal direction.

A new model was presented focusing on communication patterns and linking the process productivity to the organizational structure and the company layout. The model gives the possibility to calculate the process productivity in different organizational and layout situations. Therefore the two main factors identification and communication have been identified. Identification can be operationalized on the one hand by using an organizational effect influenced by the number of hierarchy changes and on the other hand by considering the span of control in the organizational structure. Communication as the second factor is based on the technical exchange probability between colleagues as well as the spatial distance between two persons involved in the process. In addition to the described factors the experiences gained by the employees make a contribution to improve the productivity. On this basis a decision to select the most efficient system can be made.

The described optimization approach was so far verified by only one case focusing on the production at which positive results could be achieved. Hence it should be applied in further projects within production processes in order to approve the mathematical manifestation for calculating the productivity over time. Furthermore it would be interesting to see how the developed formula works on administrative processes or on projects. Thus a universally valid formula could be created.

References

1. Westkämper, E. et al.: Lean Office 2010: Erfolgsfaktoren der Lean Implementierung in indirekten Unternehmensbereichen. Fraunhofer Verlag, Stuttgart (2011)
2. Roghé, F. et al.: Organisation 2015—Wie werden die organisatorischen Herausforderungen während der derzeitigen Krise und in der Zukunft bewältigt. In: Future of Organization, vol. 78, No. 5, pp. 201–208. Schäffer Poeschl Verlag, Stuttgart (2006)
3. Projekt Prozessorientierung in der Unternehmensorganisation, LMU (2009-2012)
4. Gaitanides, M., et al.: Prozeßmanagement: Konzepte. Umsetzungen und Erfahrungen des Reengineering. Hanser, München (1994)
5. Koch, S.: Einführung in das Management von Geschäftsprozessen. Springer, Berlin (2011)
6. Much, D.: (Auftragsabwicklung und Produktionsplanung und -steuerung) Harmonisierung von technischer Auftragsabwicklung und Produktionsplanung und -steuerung bei Unternehmenszusammenschlüssen. Dissertation, p. 15. RWTH Aachen (1997)
7. Eversheim, W.: Prozessorientierte Unternehmensorganisation. Springer, Berlin (1996)
8. Gaitanides, M.: Prozessorganisation. Entwicklung, Ansätze und Programme des Managements von Geschäftsprozessen. Vahlen, München (2007)
9. Draft, R.L.; Murphy, J.; Willmott, H.: Organization theory and design. Cengage Learning EMEA (2010)
10. Schmelzer, H.J. Sesselmann, W.: Geschäftsprozessmanagement in der Praxis: Kunden zufrieden stellen, Produktivität erhöhen, Wert steigern, vol. 6, p. 75. Carl Hanser, München (2008)
11. Binner, H. F.: Prozessorientierte TQM-Umsetzung. 2. Aufl., Carl Hanser Verlag, München (2002)
12. Knuppertz, T. et al.: Umfrage Status Quo Prozessmanagement 2010/2011. (2011)
13. Schuh, G. et al.: Strategien und Management produzierender Unternehmen, pp. 371–372. Springer, Berlin (2011)
14. Schuh, G.: Change Management—Prozess strategiekonform gestalten. Springer, Berlin (2006)
15. Hippner, H., et al.: Grundlagen des CRM—Strategie, Geschäftsprozesse und IT-Unterstützung. Gabler Verlag, Wiesbaden (2011)
16. Seidlmeier, H.: Process Modeling with ARIS: A practical Introduction. Vieweg, Wiesbaden (2004)
17. Boos, W. et: Effizienz in Geschäftsprozessen durch partizipative Gestaltung steigern. In: ZWF, vol. 12, pp. 888–892. Carl Hanser Verlag, München (2008)
18. Greiling, M., Dudek, M.: Schnittstellenmanagement in der integrierten Versorgung. Kohlhammer GmbH, Stuttgart (2009)
19. Bartscher: Veränderungen erfolgreich managen: Ein Handbuch für interne Prozessbearbeiter. Haufe-Lexware, Freiburg (2011)
20. Freitag, M., et al.: Projektkommunikation: Strategien für temporäre soziale Systeme. Springer, Wiesbaden (2011)

21. Pfeifer, T., Schmitt, R.: Handbuch Qualitätsmanagement. Carl Hanser, München (2007)
22. Ahlers, G.M.: Organisation der Integrierten Kommunikation: Entwicklung eines prozessorientierten Organisationsansatzes. Gabler, Wiesbaden (2006)
23. Luhmann, N.: Einführung in die Systemtheorie. Carl-Auer, Heidelberg (2004)
24. Quandt, T., Scheufele, B.: Ebenen der Kommunikation. VS, Wiesbaden (2011)
25. Erickson, T.: From interface to interplace: The spatial environment as a medium for interaction. In: Spatial Information Theory. Springer, Berlin (1993)
26. Baligh, H.H.: Organization structures: Theory and design, analysis and prescription. Springer, Berlin (2006)
27. Schuh, G.; Potente, T.; Nöcker, J.; Bachmann, F.; Völker, M.: Konsortial-Benchmarking "Production Systems". WZL der RWTH Aachen (2011)
28. Greenbaum, H.H. et al.: Organizational structure and communication processes: A study of change. In: Group and Organization Management. SAGE, New York (1983)
29. Allen, J., Henn, G.W.: The Organization and Architecture of Innovation: Managing the Flow of Technology. Elsevier Inc, Burlington (2007)
30. Boutellier, R. et al.: Impact of office layout on communication in science-driven business. In: R&D Management, vol. 38, Blackwell Publishing Ltd, Oxford (2008)
31. Stryker, J.B. et al.: Facilitating face-to-face communication in high-tech teams. In: Research-Technology Management, vol. Jan–Feb, pp. 51–56 (2012)

Design and Quality Control of Products Robust to Model Uncertainty and Disturbances

Beata Mrugalska

Abstract As increasingly today more and more complex products are designed, manufactured or maintained. Later changes to design solutions are costly or impossible to come up with the suggestions for consideration and application by designers, engineers, manufacturers. Thus, nowadays mathematical models are widely used in the process of product design and control. This paper provides a comprehensive overview of a new concept of modelling in these both stages. In particular, a new method of product design robust to disturbances existing in a technological process is presented. The application of mathematical modelling and methods of parameter estimation enabled formulating an approach in which parameters and acceptable tolerance of manufacturing product parameters are calculated. Moreover, the proposed approach can be easily extended and applied to the product quality control.

Keywords Quality control · Parameter estimation · Product design · Robust modelling

1 Introduction

Over the past two decades the rapidity of change in industrial practices has been remarkable and astonishing. The development of technologies has facilitated continuous development of products. Nowadays the products are more and more modern, functional and reliable in exploitation process. The cycle of their study and development is much shortened than it used to be. In the consequence, the

B. Mrugalska (✉)
Poznań University of Technology, Faculty of Engineering Management,
ul. Strzelecka 11, 60-965 Poznań, Poland
e-mail: beata.mrugalska@put.poznan.pl

K. Windt (ed.), *Robust Manufacturing Control*, Lecture Notes in Production Engineering, 495
DOI: 10.1007/978-3-642-30749-2_36, © Springer-Verlag Berlin Heidelberg 2013

products appear systematically in shorter periods of time on the market [1]. Moreover, the process of design of products is aimed at decrease of economical loss caused by insufficient quality of manufactured products. In most cases the effort of constructors is focused on such a design of parameters in manufacturing processes that the product, which characterizes the best quality, is elaborated [2]. This goal is usually achieved by the identification of the most significant process parameters and choice of their optimal values which allow to maximize quality of process and product, respectively. In order to achieve it, the development of new methods of product design, which are based on mathematical model, has been noticed [3]. These approaches are based on the choice of parametric structure of the model, which reflects expected traits or behaviour of designed product, and parameter estimation with the application of optimization techniques [4]. Such a procedure allows for the choice of optimal values of parameters of the model which accurately reflects clients' expectations. The obtained model is used as a pattern during manufacturing of the product what contributes that the quality of the manufactured product becomes dependant on model quality.

The mathematical model obtained during product design can be used to elaboration of a control method of product quality. Such a method can be applied to assess the product quality in the process of manufacturing and exploitation. The quality control of products in the manufacturing process is done by a producer and aims at the elimination of products, which do not fulfil quality requirements assumed in the design stage. The supervision of quality, which takes place during product exploitation, is directed to detection of deterioration of product quality (or its parameters). In this case it is vital to detect it early enough not to accept the situation where a damaged component can make a breakdown of other product elements.

Unfortunately, there are many reasons which cause that the obtained mathematical model of the product is uncertain so it does not reflect the expected characteristics or behaviour of design product [2]. In order to overcome this problem it is necessary to ensure that the product is insensitive to manufacturing variance, influence of environmental factors during exploitation and lowering quality of components of the product during exploitation. This goal can be obtained by the application of product design and control methods which are robust to disturbances appearing in manufacturing and exploitation processes [1]. The presented product design method [5] based on the Bounded-Error Approach [6] was extended and applied to the product quality control.

2 Sources of Knowledge Applied in Model of Product Design

The most dynamically developing methods of product design are methods which are based on optimization techniques [2, 3, 7, 8]. While the application of such an approach it is necessary to define a model of a product which mathematically

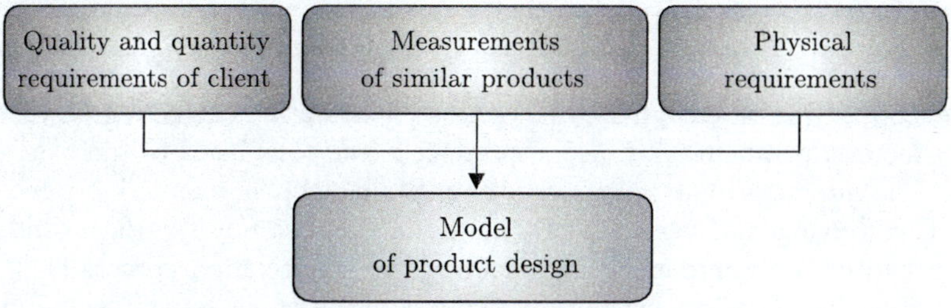

Fig. 1 Sources of knowledge used in model design (Adapted from [6, 9])

reflects the relationship between the inputs and product trait. The model of the product is usually created on the basis of known physical relations and parameters describing characteristics or mechanism of product functioning which is depicted on Fig. 1.

Knowing regulations and rules according to which the product functions it is possible to predict precisely its behaviour and characteristics in a moment of change of particular parameters. It allows to design the product in such a manner that it will meet specific criteria. Unfortunately, in most of contemporary products physical requirements, which describe mechanism of their function, are not always known. Furthermore, even if they are identified they are many a time too compound to apply them in quality design. Designers of products often have only knowledge about the behaviour of an expected product. This knowledge can be usually represented as a pair of data $\{\boldsymbol{r}(k), y(k)\}$ which represent a state of input factors $\boldsymbol{r}(k) = [r_1(k), r_2(k), \ldots, r_{n_r}(k)]$ influencing the product and corresponding them product trait $y(k)$. The examples of input factors can be: force influencing product, the amount of supplied fuel and the level of voltage. The input factors $\boldsymbol{r}(k)$ can change in the subsequent moments k causing the change of state of product trait $y(k)$ such as: angular speed, electrical power and efficiency. Thus, the product behaviour is defined as the set of all possible trajectories of the pairs $\boldsymbol{r}(k)$ and $y(k)$ for moments k which can occur for a given product during its use [10]. The range of changes of $y(k)$ depends on the changes of values of $\boldsymbol{r}(k)$ and the values of $\boldsymbol{P}p \in \mathbb{R}^{n_p}$:

$$y(k) = f(\boldsymbol{r}(k), \boldsymbol{p}) = f(r_1(k), r_2(k), \ldots, r_{n_r}(k), p_1, p_2, \ldots, p_{n_p}), \tag{1}$$

where $\boldsymbol{p}$ are parameters representing physical traits of a manufactured product such as dimension, physical characteristics of product components or chemical composition [1]. All these values manufactured while the process of production are under the influence of control factors $\boldsymbol{s}$. Therefore, quality of product depends on the precision of its manufacturing expressed as the tolerance interval:

$$\boldsymbol{p}^m \leq \boldsymbol{p} \leq \boldsymbol{p}^M, \tag{2}$$

where $\boldsymbol{p}^m$ and $\boldsymbol{p}^M$ are minimum and maximum permissible values of product parameters.

3 Design Robust to Disturbances

The concept of robust design methods relies on an optimal selection of values of control factors, parameters of design product $\boldsymbol{p}$ and tolerances of their manufacturing. The products and processes designed in such a manner respond with insignificant changes of values of product traits $y(k)$ even by significant influence of disturbances in the product manufacturing and exploitation process [11].

Disturbances can affect in both the manufacturing stage as well as the exploitation stage of the product. The disturbances are uncontrollable factors which have a negative influence on the product and process quality. The occurrence of them is an undesirable phenomenon which should be taken into account and limited in the stage of product design.

The disturbances occurring in the manufacturing stage can be divided into resource and object disturbances. As for object disturbances, they involve changeability of environment in the technological process, labour division and machines breakdowns. Resource disturbances relate to energetic disturbances, lack of financial resources, material defects or shortage of raw materials and errors in production planning and control. The individual influence of particular disturbing factors, being in the technological processes, is relatively small, but their amount in the manufacturing process may be so great that the whole process may occur to be completely incapable of manufacturing product of planned quality. The disturbances of exploitation conditions are the next group of factors which influence product during its exploitation. Most of the products are exposed to environmental disturbances among which humidity, pressure, pollution and temperature can be differentiated. Apart from the mentioned above disturbances, products also depend on electrical and mechanical disturbances such as inaccuracy of control system, transient voltage, unbalanced voltage, voltage fluctuations, poor mounting, mechanical over load and pulsating load [5, 12, 13].

For these reasons methods of process and product design robust to disturbances were elaborated [14]. The practical implication of robust design methods enables improvement of product quality by making it robust against to:

- manufacturing variance in the technological process,
- influence of environmental factors appearing during the product exploitation,
- using up of product components in the product exploitation [1].

4 Parameter Estimation and Model Uncertainty

One of the basic reasons of occurrence of model uncertainty of the designed product follows from the fact that in the most of applied methods of model parameter estimation such as Least Mean Square [6] it is assumed that the disturbances affecting process and product are negligible small or can be modelled as realization of independent random variables with known distribution. However, the more realistic assumption is that the values of disturbances are in a certain limited interval. Such an assumption leads to estimation algorithms with Bounded-Error Approach (BEA) [6]. In the Bounded-Error Approach a model of a designed product can be created on the basis of a set $k = 1, \ldots, n_D$ of a pair of data $\{\boldsymbol{r}(k), y(k)\}$ representing requirements for the designed product. Firstly, a linear form of the designed product is assumed:

$$y(k) = \boldsymbol{p}\boldsymbol{r}^T(k) + \varepsilon(k), \tag{3}$$

where $\varepsilon(k)$ stands for the disturbances affecting a product trait, in particularly, the disturbances appearing in the manufacturing process and the disturbances of the exploitation conditions. Furthermore, the BEA method is differentiated from other estimation methods by the fact that the values of the disturbances $\varepsilon(k)$ affecting $y(k)$ are bounded in a certain limited interval:

$$\varepsilon^m(k) \leq \varepsilon(k) \leq \varepsilon^M(k) \tag{4}$$

The limitations $\varepsilon^m(k) < 0$ and $\varepsilon^M(k) > 0$ are assumed a'priori by the designer based on the known values of the disturbances of the manufacturing process and exploitation conditions. Obviously, there might exist disturbances which are unknown. However, they do not influence the presented design method if their values are included in the interval (4).

On the basis of the defined data set $\{\boldsymbol{r}(k), y(k)\}$ by the designer strips limited by pairs of hyperplanes, which define unknown parameters of the designed product, are obtained according to the Eq. (5):

$$\mathbb{S}(k) = \{\ y(k) - \varepsilon^M(k) \leq \boldsymbol{p}\boldsymbol{r}^T(k) \leq y(k) - \varepsilon^m(k)\}, \quad k = 1, \ldots, n_D. \tag{5}$$

The principle of the operation of the BEA relies on a calculation of a feasible parameter set $\mathbb{P}$, in the parameters area:

$$\mathbb{P} = \{\boldsymbol{p} \in \mathbb{R}^{n_p} \mid y(k) - \varepsilon^M(k) \leq \boldsymbol{p}\boldsymbol{r}^T(k) \leq y(k) - \varepsilon^m(k)|\}, k = 1, \ldots, n_D \tag{6}$$

which is obtained by the intersection of n_D pairs of hyperplanes:

$$\mathbb{P} = \bigcap_{k}^{n_D} \mathbb{S}(k). \tag{7}$$

The centre of the feasible parameter set $\mathbb{P}$ (for $n_p = 2$) represents the solution of the parameter estimation task for the model of the designed product. Knowing minimal and maximum values of the parameters:

$$p_i^{\min} = \arg \min_{p \in \mathbb{P}} p_i, \tag{8}$$

$$p_i^{\max} = \arg \max_{p \in \mathbb{P}} p_i, \tag{9}$$

it is possible to calculate the values of parameter estimates:

$$\hat{p}_i = \frac{p_i^{\min} + p_i^{\max}}{2}, \quad i = 1, \cdots, n_p. \tag{10}$$

The values of the calculated parameter estimates are used as nominal ones in the manufacturing process. Nevertheless, each vector $\hat{\boldsymbol{p}}$, which is in the region $\mathbb{P}$, is a valid estimate of the product parameters $\boldsymbol{p}$ that is consistent with data concerning product requirements. Theoretically, it is possible to use them in the manufacturing process. Unfortunately, as a result of a limited precision of the method and many other disturbances affecting the manufacturing process, it is not possible to manufacture the product which parameters are identical to the calculated parameter estimate. It follows that admitting parameter estimates, which are in the centre of the feasible parameter set, enables keeping the maximal acceptable tolerance region in the process production [5]. Such a solution allows to minimize the number of products which do not fulfil designer's requirements. The acceptable tolerance of manufacturing parameters may be calculated directly on the basis of the distance of minimal and maximal value (10) of parameters estimates from the centre of feasible parameter set $\mathbb{P}$.

5 Method of Quality Control Based on the Product Parameter Estimation

In order to overcome the uncomfortable and expensive disassembly of the product a control method based on parameter estimation can be applied. The concept of this method relies on the application of the Bounded-Error Approach to calculate a parameter estimation $\hat{\boldsymbol{p}}'$ and feasible parameter set $\hat{\mathbb{P}}'$ on the basis of data containing input factors and the trait of the controlled product $\{\boldsymbol{r}(k), y(k)\}$ (Fig. 2).

If the calculated parameters and feasible parameter set of the controlled product in the process of estimation differ from the nominal parameters and the feasible parameter set of the manufactured product calculated in the product design, it means that the product is faulty. The mechanism of functioning of the proposed method of quality control robust to uncertainty of model and disturbances based on parameter estimation of the controlled product with the application of the BEA is presented in (Fig. 3)

6 Application of the BEA into Quality Control of Product

In order to show the effectiveness of the proposed design method the model of the brushless DC motor [15] was implemented in the Matlab Simulink. The mathematical model of the motor can be divided in electrical and mechanical subsystems. The equation describing the electrical subsystem is the following:

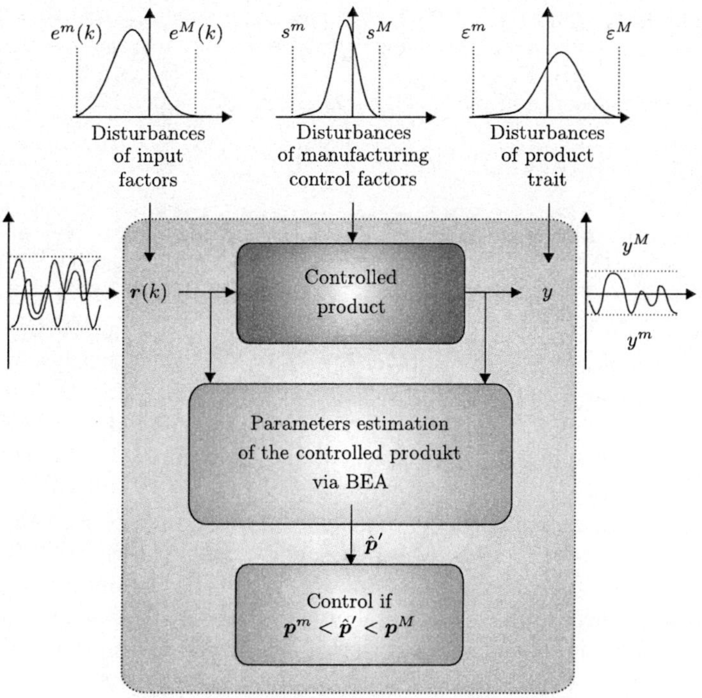

Fig. 2 The concept of quality control method robust to model uncertainty and disturbances based on parameter estimation of the controlled product

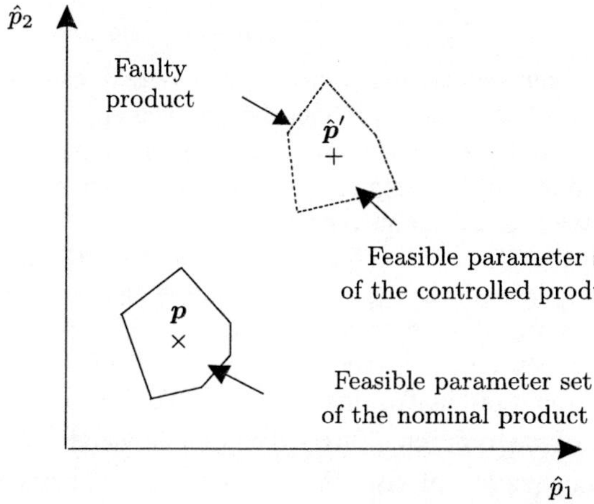

Fig. 3 Feasible parameter sets in quality control method robust to model uncertainty and disturbances

$$i(k) = \frac{1}{R}u(k) + \frac{k_e}{R}\omega(k), \tag{11}$$

and the equation describing the mechanical subsystem is defined as:

$$\omega(k) = \frac{k_T}{J}i(k) + \exp^{(-Tc_v/J)}\,\omega(k-1). \tag{12}$$

Voltage $u(k)$ and rotor's angular velocity $\omega(k)$ are the part of vector of input factors $\boldsymbol{r}(k) = [u(k)\omega(k)]$, of electrical subsystem. The output factor is current $i(k)$. The nominal values of model parameters are back electromotive force constant, which is k_e=0.186 Vs/rad, and coil resistance which amounts to R=4Ω. However, the vector of input factors $\boldsymbol{r}(k) = [i(k)\omega(k-1)]$ of mechanical subsystem consists of current $i(k)$ and delayed rotor's angular velocity $\omega(k-1)$. In this case the output factor is rotor's angular velocity $\omega(k)$. The nominal values of model parameters are the following and equal: torque constant k_T=0.186Vs/rad, rotor interia J=0.015kgm^2 and viscose friction coefficient c_v=0.2Nms/rad.

The assumption of nominal values of parameters of motor implemented in Matlab allowed to generate $n_{\mathcal{D}} = 30$ pair of data $\{[u(k)\omega(k)], i(k)\}$ and $\{[i(k)\omega(k-1)], \omega(k)\}$, which enabled to carry out the design of motor and verification of the obtained results. During real process of design the nominal parameters are unknown. Thus in the next step, it was assumed that the disturbances affecting $u(k)$, $\omega(k)$ and $i(k)$ are generated according to $\boldsymbol{e}(k) \in \mathcal{N}(-0.1, 0.2)$ and $\varepsilon(k) \in \mathcal{N}(-0.05, 0.2)$. Following the elaborated method, assuming that the nominal parameters are unknown, only on the basis of data parameters estimates and feasible parameter set were calculated with the application of the BEA. The parameters estimates calculated via the BEA method amounted to $\hat{\boldsymbol{p}} = [0.2498, 0.04644]$ for electrical subsystem and $\hat{\boldsymbol{p}} = [12.4026, 0.9862]$ for mechanical subsystem and very close to nominal values $\boldsymbol{p} = \left[\frac{1}{R}, \frac{k_e}{R}\right] = [0.25, 0.0465]$ and $\boldsymbol{p} = \left[\frac{k_T}{J}, \exp^{\left(\frac{-Tc_V}{J}\right)}\right] = [12.4, 0.9868]$ assumed during data generation. The obtained results confirm effectiveness of the elaborated method in product design. The application of the BEA allows to obtain unbiased parameters estimates despite of the occurrence of disturbances. The calculated feasible parameter set constitutes an acceptable tolerance of manufacturing parameters in the manufacturing process.

The model of the brushless DC motor implemented in the Matlab was also applied to generate the data containing two faults in the electrical and mechanical subsystems. The first fault, in the electrical subsystem, relied on manufacturing a stator of the motor, which coil resistance amounted to 50% nominal values i.e. R=2Ω. The second fault simulated in the mechanical subsystem, relied on simultaneous change of the rotor interia from the nominal value $\hat{J} = 0.015$ kgm^2 to $\hat{J} = 0.016$ kgm^2 and increasing of viscose friction coefficient from nominal value $\hat{c}_v = 0.2$ Nms/rad to $\hat{c}_v = 0.7$ Nms/rad. Such type of damage can be caused by imprecise manufacture of mechanical parameters of rotor or bearings.

The proposed control approach was applied to parameters and its uncertainty estimation for data containing the faults simulated in the electrical and mechanical

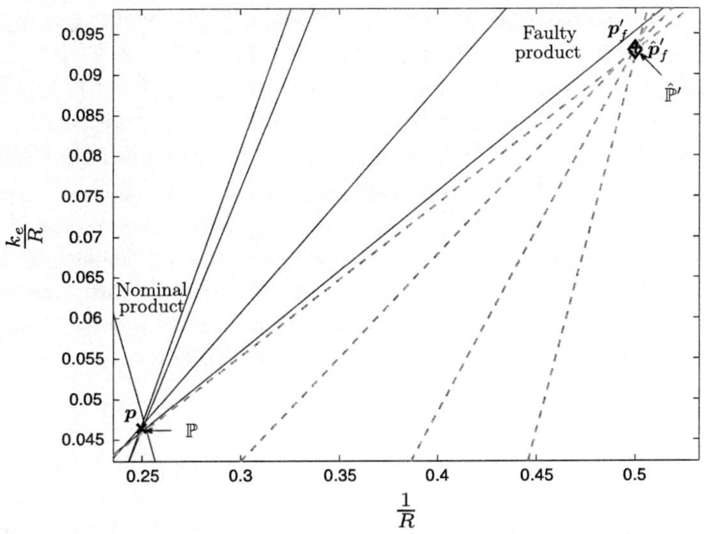

Fig. 4 Identification of decreased coil resistance via the proposed method

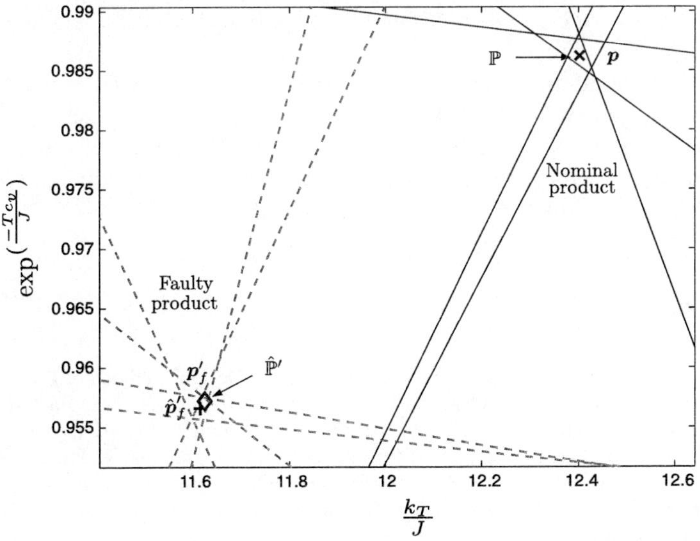

Fig. 5 Identification of change rotor inertia and increase of viscose friction coefficient in the movement rotary via the proposed method

subsystems. In the case of the electrical subsystem (Fig. 4), the estimate of parameters of the controlled motor $\hat{\boldsymbol{p}}'_f = [0.5003, 0.0932]$ obtained via the BEA were significantly different from the nominal parameters $\boldsymbol{p} = [0.25, 0.0465]$, and similar to parameters $\boldsymbol{p}'_f = [0.5, 0.093]$ assumed during simulation of the fault in the motor. On the basis of the obtained parameters estimate it is possible to calculate the estimates $\hat{R}_f = 1.9988\Omega$ and $\hat{k}_e = 0.1863$ Vs/rad.

In the mechanical subsystem, the estimate of parameters $\hat{\boldsymbol{p}}'_f = [11.6156, 0.9566]$ also were significantly different from the nominal parameters, and similar to parameters $\boldsymbol{p}'_f = [11.6177, 0.9569]$ assumed during simulation of the fault (Fig. 5). On the basis of the obtained parameters estimate it is possible to calculate the estimates $\hat{J} = 0.016$ kgm^2 and $\hat{c}_v = 0.7$ Nms/rad of the faulty motor.

The developed approaches were successfully applied to design and control of the brushless DC motor. The presented results confirm that the proposed robust control method allow to detect the faults and also identify precisely the value of the faulty parameter in spite of occurrence of disturbances.

7 Conclusions

The growing market competiveness makes that most of producers conduct systematic research to improve quality of manufactured products. The products of better quality, which can be used in a wide scope of exploitation conditions, contribute to improvement of market and economic position of production companies. In order to achieve it, it is suitable to apply optimization techniques in the process of product design and its control. In the paper the method of design and quality control based on model uncertainty, which can be applied in practice in the manufacturing process, was presented. The effectiveness of the research was considered on the basis of the brushless DC motor. The research results indicated that the proposed method is efficient as it allowed to detect not only faulty products but also product parameters which were responsible for it.

References

1. Mrugalska B., Kawecka-Endler A.: An Analysis of the Design Methods of Products Robust to Disturbances. In: XI Inter. Conf. on Human Aspects of Advanced Manufacturing: Agility and Hybryd Automation, Value Stream Activities Management, 117–124 Poznań (2007).
2. Mrugalska B.: Product Design Robust to Disturbances in Technological Process. In: Fertsch M., Grzybowska K., Stachowiak A., (eds.) Production Management and Logistics - Concepts, Methods and Practical Solutions, pp. 460–468. Poznań University of Technology, Poznań (2006) (in Polish).
3. Doltsinis, I., Kang, Z.: Robust Design of Structures Using Optimization Methods. Computer Methods in Applied Mechanics and Engineering **193**, 2221–2237 (2004)
4. Hendrix, E.M.T., Mecking, C.J., Hendriks, T.H.B.: Finding Robust Solutions for Product Design Problems. European Journal of Operational Research. **92**(1), 28–36 (1996)
5. Mrugalska, B., Kawecka-Endler, A.: Practical Application of Product Design Method Robust to Disturbances. Human Factors and Ergonomics in Manufacturing & Service Industries. **22**, 121–129 (2012)
6. Walter, E., Pronzato, L.: Identification of Parametric Models from Experimental Data. Springer, Berlin (1997)

7. Robinson, T.J., Borror, C.M., Myers, R.H.: Robust Parameter Design: A Review. Quality and Reliability Engineering International **20**(1), 81–101 (2004)
8. Vuchkov, I.N., Boyadjieva, L.N.: Quality Improvement with Design of Experiments: A Response Surface Approach. Kluwer Academic Publishers, Boston (2002)
9. Soderstrom, T., Stoica, P.: System Identification. Prentice Hall, Upper Saddle River (1989)
10. Blanke, M., Kinnaert, M., Lunze, J., Staroswiecki, M. (eds.): Diagnosis and Fault-Tolerant Control. Springer-Verlag, New York (2003)
11. Hamrol, A., Mantura, W.: Quality Management. Theory and Practice, PWN, Warsaw (2002). (in Polish)
12. EN 60079–0: Electrical Apparatus for Explosive Gas Atmospheres. Part 0: General Requirements (2004).
13. Singh, G.K., Kazzaz, A.S.: Induction Machine Drive Condition Monitoring and Diagnostic Research - A Survey. Electric Power Systems Research. **64**(2), 145–158 (2003)
14. Phadke, S.M.: Quality Engineering Using Robust Design. Prentice Hall, Englewood Cliffs, New York (1989)
15. Moseler, O., Isermann, R.: Application of Model-Based Fault Detection to a Brushless DC Motor. IEEE Transactions on Industrial Electronics **47**(5), 1015–1020 (2000)

Dynamic Business Model Analysis for Strategic Foresight in Production Networks

Hans-Christian Haag and Meike Tilebein

Abstract In today's uncertain business environments, many small and medium-sized firms organize their production resources and processes within networks. As a consequence, these firms are increasingly connected through a complex and dynamic pattern of inter-organizational flows of material and information. Within such networks, approaches focusing on a single firm's perspective are inadequate strategic foresight. However, approaches, that explicitly consider the network perspective to enhance single firm perspectives, help firms to align flexibility with uncertainty to achieve greater robustness of their strategies. This chapter combines the introduction of a methodological approach with the practical experience from the application within a research project consisting of different application partners. Thereby, it shows an unconventional way of applying system dynamics within a strategic foresight approach in production networks. By demonstrating the application of the approach within an illustrative example, important modeling steps are shown and crucial tasks are evaluated.

Keywords Strategic foresight · System dynamics · Network modeling · Scenario development

H.-C. Haag (✉) · M. Tilebein
University of Stuttgart, Institute for Diversity Studies in Engineering, Stuttgart, Germany
e-mail: hans-christian.haag@ids.uni-stuttgart.de

M. Tilebein
e-mail: meike.tilebein@ids.uni-stuttgart.de

K. Windt (ed.), *Robust Manufacturing Control*, Lecture Notes in Production Engineering, 507
DOI: 10.1007/978-3-642-30749-2_37,

1 Introduction

Since the 1970s, different approaches have been developed to help single firms cope with environmental changes [1]. Ansoff's concept of weak signals marks the introduction of strategic foresight in literature, as a primary step in the strategic management process for firms [2]. In the past two decades a large amount of literature focusing on the performance of strategic foresight has been published [3–6]. Moreover, many firms adapted their strategies by cooperating within business networks [7, 8]. Firms within networks focus on their core competencies and build dynamic capabilities to address fast changing global market demands [9]. Networks, e.g. production networks are complex and dynamic structures which often evade a single firm's strategic perspective [10]. A firm can understand the influence of the business environment on its strategy through evaluation of dependencies between other network actors. This evaluation process, however, is difficult and requires a methodological sophisticated approach [11]. Over the past few years various ideas to successfully integrate different foresight methods e.g. the integration of scenario analysis with road mapping, have been explored [12]. Whereby, the role of doubting and learning in foresight activities is very important [13, 14]. Therefore, the research question of this sequel is, how strategic foresight for firms in production networks can be performed in a profound approach, systematically integrating existing methods e.g. system dynamics. In the following three sections, first a concept how to perform strategic foresight in production networks is introduces. Second, an example is given, illustrating the application of the approach. Third, results are discussed and an outlook given.

2 Concept of a Dynamic Business Model Analysis in Production Networks

2.1 Insights From a Single Firm Perspective

There are various foresight processes described in literature. Horton distinguishes between three steps to gain a profound output for corporate strategy development: Input generation, foresight activities and output generation for further activities [5]. Accordingly, Voros integrates these steps to a generic foresight process with three foresight tasks: Analysis, interpretation and prospection [15]. The strategic foresight time horizon is thereby fixed and much longer than in other approaches e.g. from the field of supply chain management. This chapter evaluates the three tasks in depth and identifies their contribution to the network perspective.

At the beginning of performing strategic foresight, environmental scanning helps to identify change drivers [6]. Thereby, change drivers from the political, economic, social, technological, legal and ecological field are identified [16]. PESTLE-Analysis is a rather well-known method and will therefore not be further

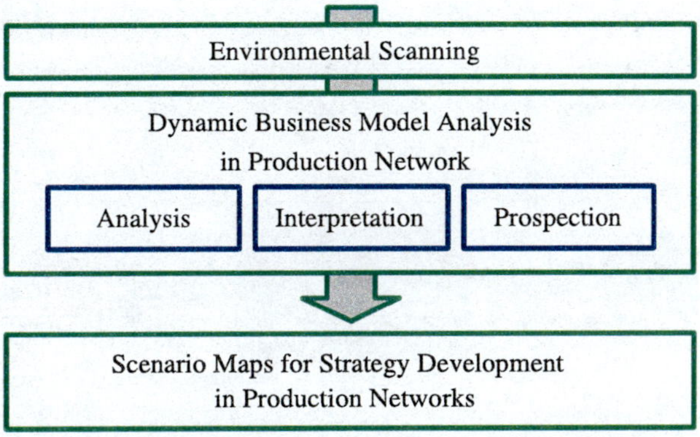

Fig. 1 Strategic foresight in production networks

explained. Based on the change drivers identified during environmental scanning, a dynamic business model analysis in production networks can be performed to transform change drivers to a profound prospection for future strategies (Fig. 1). This enables consideration of the prevailing complexity and dynamic interrelations within production networks.

1. **Analysis**: Input information gained from the previous step is analyzed. Specifically its relevance for the firm is evaluated. This evaluation identifies the impact of the external environment on a firm's business unit.
2. **Interpretation**: A deep understanding of the underlying system structure is important in strategic foresight. The correct interpretation of the impact of external effects on the corporate strategy is crucial for successful prospection.
3. **Prospection**: Creating coherent future scenarios based on the interpretation of changes on a corporate strategy is the last task to develop a profound future prospection for a firm. Thereby, different possible futures are developed based on the interpretation of the input signals as part of a distinct set of future values.

A concept to perform these three generic tasks during a dynamic business model analysis for firms operating in production networks is presented in the following three sections. Each section further describes one generic task of the dynamic business model analysis for production networks.

2.2 *Analysis From a Network Perspective*

A strategic network model is developed to perform an impact analysis within the network. Key factors are identified as a basis for interpretation. In order to deduce key factors with an important impact on a firm's business model, the structure of the production network is modeled. Network actors are identified and connected through information and product flows. A qualitative impact analysis is carried out based on the network model. Each change driver (d) has a chain of impacts through the network along the network relations (Fig. 2, following [11]). For

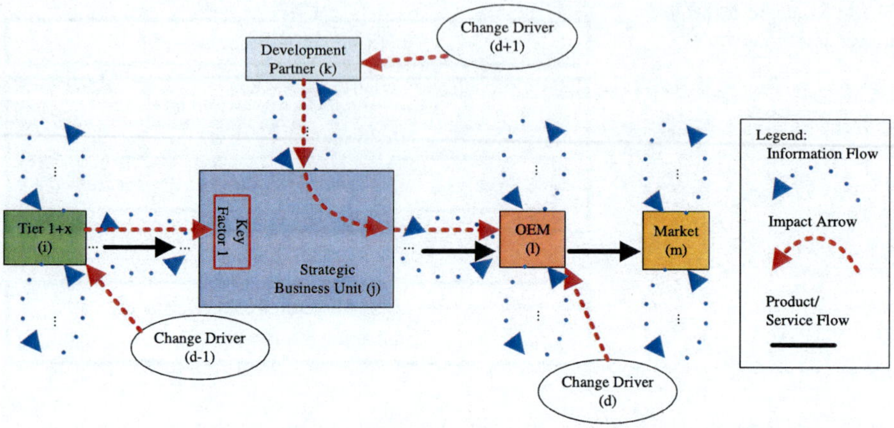

Fig. 2 Analysis from a network perspective

change drivers ending at the strategic business unit, a set of key factors is defined (represented by key factor 1 in Fig. 2).

This set of all key factors is compared to a generic business model to ensure that all relevant areas for the future course of a business unit are included. The generic business model from Johnson et al. includes the four elements (Fig. 3): Customer value proposition, profit formula, key resources and key processes [17].

The key factors are named according to the rules for variable naming which are well known from system dynamics literature [18]. At the end of the analysis task a set of key factors exist. These key factors are used as variables in the following step to set up a system dynamics model.

2.3 *Interpretation by the Use of System Dynamic*

Modeling the impact of the identified key factors on the business model of a firm helps to evaluate and interpret future scenarios. Therefore, appropriate assumptions on the business model need to be made. However, developing a model for such an uncertain and complex issue is rather challenging. This is especially true, since users often tend to question assumptions made during the modeling process. Therefore, the method system dynamics [18] is used in a group model building process [19] to ensure visualization of the underlying mental models. Here different representatives of the firm are integrated in a joint effort to learn about the future development of a business unit.

The development of a system dynamics model within a group model building process consists of two main steps. First, a qualitative model is built. This is done through the use of group model building concepts, such as those summarized in [19] e.g. group memories, workbooks etc. The model purpose is to visualize future

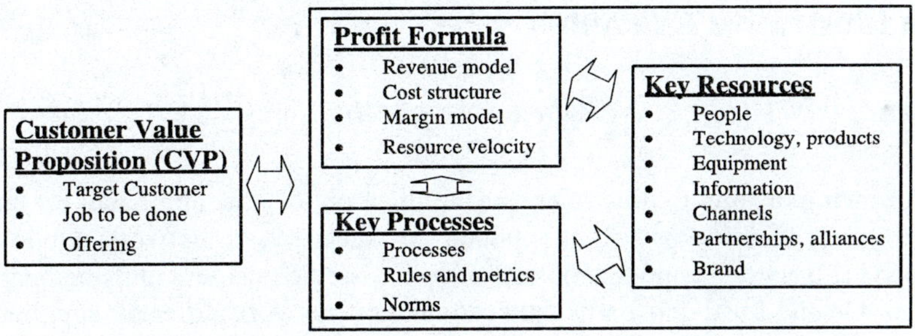

Fig. 3 Generic business model [17]

changes of the business model resulting from the impact of change drivers transformed by the network or directly affecting the business unit of the firm. Dynamic system analysis makes the various mental models of strategists and decision makers on the future course of the firm available. Therefore, it contributes to a firm-wide discussion about it. To ensure an efficient model building process, a so called preliminary model can be used. A possible preliminary model for setting up a qualitative system dynamics might be given by the spreading fixed cost model [18]. In the model the share from spreading fix costs is a reinforcing loop. This is true for a market, where an increase of product attractiveness is automatically followed by an increase of market share. Second, a quantitative model is built. Every relation between two variables needs to be considered and evaluated. For qualitative relations table functions are considered, where simple pairs of values can be inter- or extrapolated.

2.4 Prospection with Respect to Future Scenarios

Prospection aims at building and evaluating different future scenarios. This is accomplished through scenario development based on the given key factors, and through simulation of discrete value sets of the impact factors within a scenario.

There are different approaches available within literature on how to develop future scenarios for single firms [20], including quantitative scenarios. Since, developing quantitative scenarios is well-known from literature, no further explanation is given at this point.

Simulation of scenarios within the system dynamics model visualizes the mental model of the group. Thereby, assumptions on the future course can be interpreted, while discussing the simulation runs of the different scenarios.

By the use of the network models and the simulation models, verbal scenario descriptions and interpretations are completed to a comprehensive view on the meaning for the organization. These network scenario maps enable a firm to use the resulting scenarios of the approach and to consider the underlying models of the network structure during strategy making.

3 An Illustrative Example

3.1 Analysis From a Network Perspective

The following example illustrates the application of the approach for a firm producing machine tools. As explained in Sect. 2.2, a network model was developed (Fig. 4). The model consists of two strategic business units belonging to the firm. On the upstream part of the production network different suppliers are connected over a complex pattern of material and information flows [10]. On the downstream part of the production network, three different classes of direct customers are serving various end customer markets.

In addition to the network structure, the impacts of four selected change drivers on the network are shown in Fig. 4. These change drivers were identified as relevant for the firm. For a later interpretation, business unit II was selected, although the same approach could also be applied to another business unit of the firm.

Table 1 shows the mapping of these four selected change drivers to key factors within the business model of business unit II which can be explained as follows: A possible increase of consolidations and joint ventures activities of suppliers could increase the market power of these suppliers over the firm. A possible increase of technological improvements by the suppliers could directly increase the product attractiveness of specific product features and is therefore relevant for the business unit. The further integration of customers in the product development process is directly represented in the model by a ratio key factor. The demand for system solutions is represented by a key factor representing the share between system solutions and 'normal' product features. The interpretation of the identified key factors follows in next section.

3.2 Interpretation by the Use of System Dynamics

A qualitative system dynamics model has been developed for the selected business unit based on the preliminary model referred to in Sect. 2.3. Figure 5 depicts a simplified version of the original model developed with the firm in a group model building process. In the centre of the model, an improved version of the preliminary model can be seen. At the outer edge of the model the key factors are connected. The corresponding change drivers are visualized by dotted arrows (which are not part of the system dynamics syntax given within literature). The model shows that an increase of market power of the suppliers (first key factor) will lead to an increase in purchasing costs for the firm.

The second key factor, attractiveness of product features, positively correlates with the overall product attractiveness. While an increase of the third key factor, ratio of cooperative development projects with customers, leads to an increase of the product portfolio attractiveness. While the product attractiveness only affects the market share, the portfolio attractiveness affects the market size as well. This

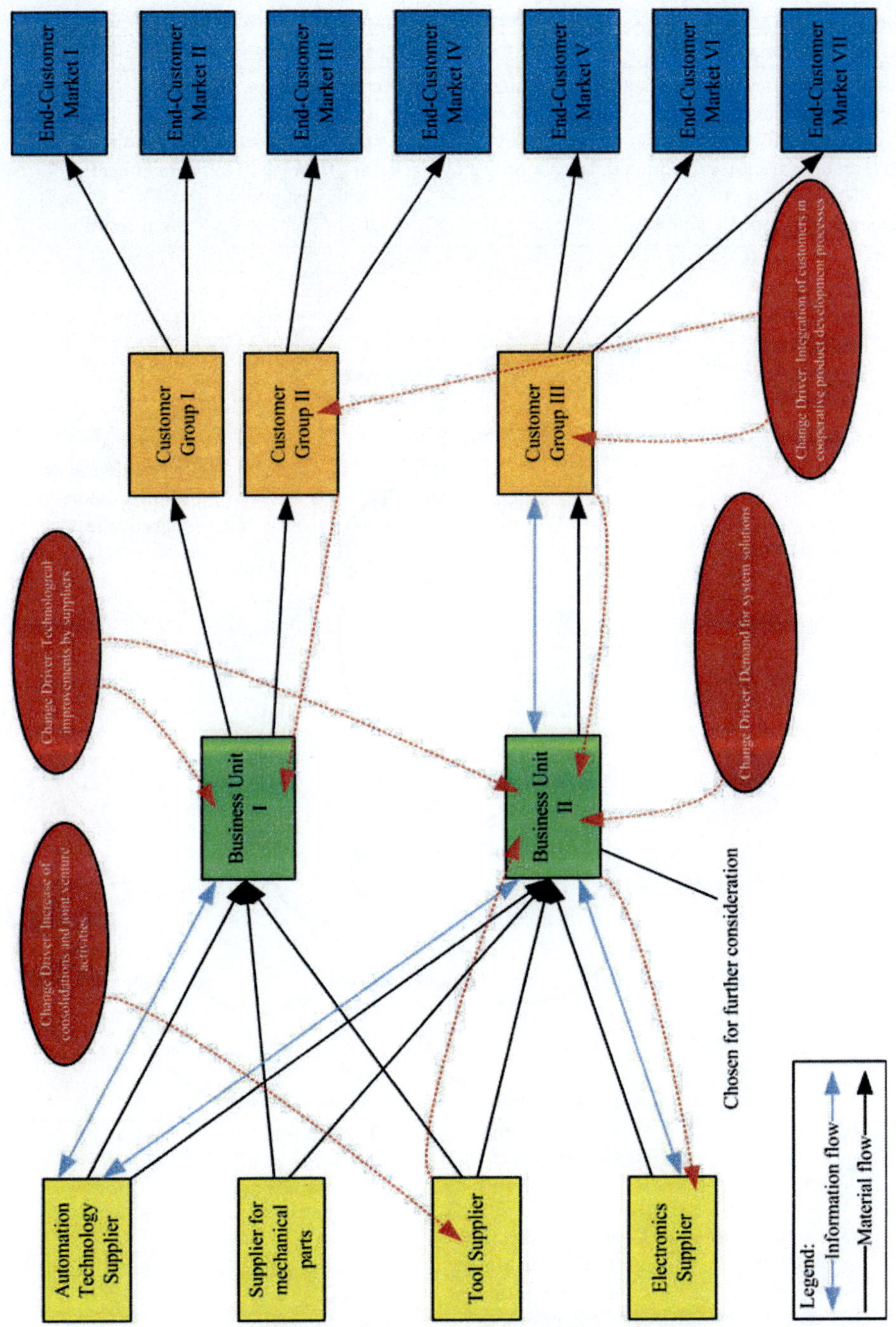

Fig. 4 Illustrative example for performing an analysis from a network perspective

Table 1 Mapping of change drivers on key factors relevant for business unit II

Change driver	Key factor
Increase of consolidations and joint venture activities	Market power of suppliers
Technological improvements by suppliers	Attractiveness of product features
Integration of customers in cooperative product development processes	Ratio of cooperative development projects with customers
Demand for system solutions	Share of demand for system solutions

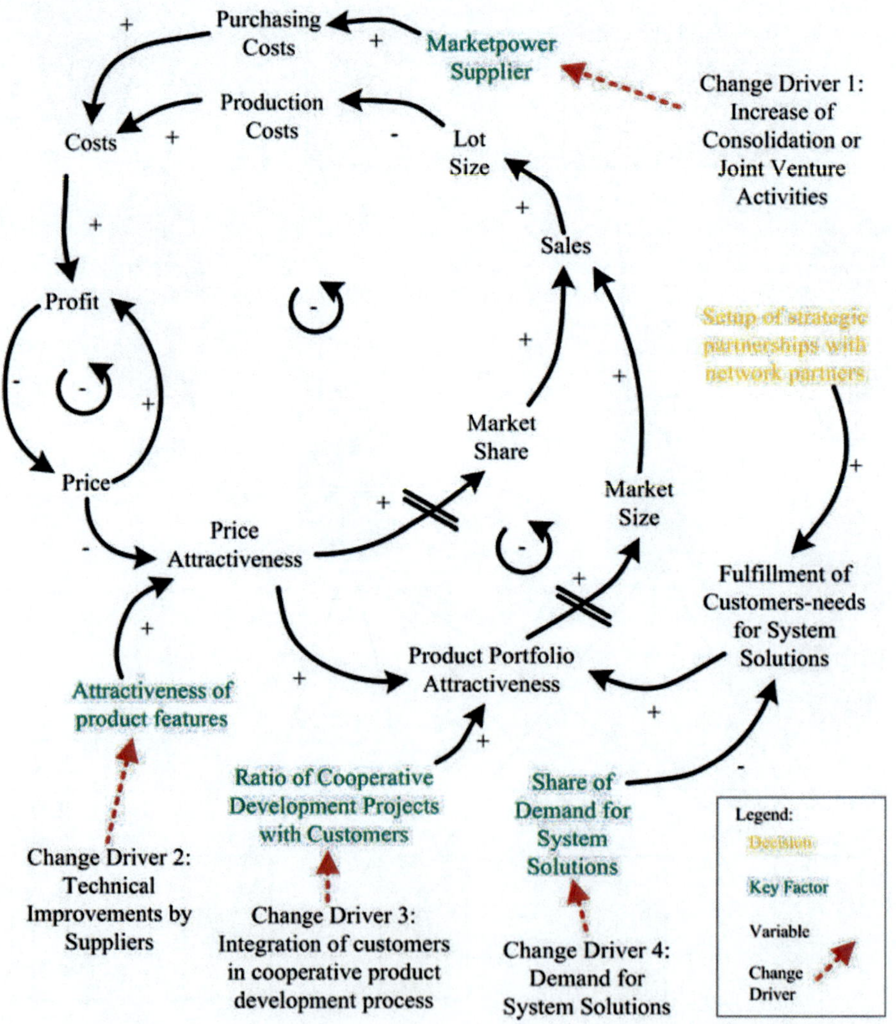

Fig. 5 Illustrative example for a system dynamics model to simulate dynamic changes of a business model

results, because innovative products may address new market areas. The fourth key factor, share of demand for system solutions, decreases the fulfillment of customers' needs for system solutions. This decrease may, however, be diminished by setting up strategic partnerships with other network partners, which would increase the overall system solutions performance of the products of the business unit.

Using a recursive modeling process, a quantitative model was developed based on the previously explained qualitative model. Every relation was quantified, and its value ranges and dimensions defined. The resulting simulation model evaluates the impact of different sets of future values of key factors on the business model. The simulation model can be used to evaluate the future course of the business unit developed in the following section.

3.3 Prospection with Respect to Future Scenarios

In order to develop an appropriate prospection of the business unit's future, two steps need to be fulfilled. First, scenarios based on the key factors identified during analysis with the corresponding value sets defined during interpretation have been developed. Second, these scenarios have been interpreted by the use of the system dynamics model developed during interpretation.

For the given example a consistency matrix combining each possible value of a key factor with each other has been developed. Based on the results, commercial scenario development software selected five different value sets. Simulation of the scenarios showed the impact of the scenarios on the business model of a firm. Different variables, like the profit or the market share, were chosen for a direct comparison of the scenarios. As reference a pre-defined base run was simulated from today's input values.

By simulating the future scenarios, the mental model of the group has been explicitly visualized. Future assumption could be challenged and a profound knowledge base for strategy development has been set up. The greater benefit of applying the approach was rather the participatory group model process itself, than the resulting simulation model whose accuracy stayed short due to high uncertainty and complexity.

4 Results and Discussion

The chapter presents a strategic foresight approach for firms in production networks. This approach builds on earlier published generic foresight processes by addressing the dynamic complexity firm are facing within production networks. Network analysis enables firms to focus on the network perspective. In addition the development of a system dynamics model contributes to a comprehensive interpretation of the interaction between network changes and corporate strategy. This aids in determining the robustness of strategies for different future scenarios.

Finally, the chapter provides an illustrative example of the application of the approach and demonstrates the contribution to the strategy process of a single firm operating in production networks.

The introduced approach has been applied in four different firms within a research project over a period of one and a half years. Feedback from the application partners reveals that the approach is still too complicated for being applied without external advise. Therefore, further research is needed to make the application more intuitive. Developing a simple system dynamics reference model might be an appropriate step, to improve practicability. Also part of further research should be a larger case study in order to examine the limitation of the presented approach. Limitation might be for example the size of a firm, the industrial sector or special attributes of the network like the degree of complexity. Combining the results of the case study with a detailed guide to apply the approach might help to further spread the application of the approach.

Acknowledgments This research is supported by the German Federal Ministry of Education and Research (BMBF) within the framework "Research for the Production of Tomorrow" and coordinated by the Project Management Agency Karlsruhe (PTKA) from the Karlsruhe Institute of Technology.

References

1. Duncan, R.B.: Characteristics of organizational environments and perceived environmental uncertainty. ASQ **17**, 313–327 (1972)
2. Ansoff, H.I.: Managing strategic surprise by response to weak signals. CMR **18**, 21–33 (1975)
3. Frishammar, J.: Characteristics in information processing approaches. IJIM **22**, 143–156 (2002)
4. Tsoukas, H., Shepherd, J.: Coping with the future: developing organizational foresightfulness. Futures **36**, 137–144 (2004)
5. Horton, A.: A simple guide to successful foresight. Foresight **1**, 5–9 (1999)
6. Slaughter, R.A.: Futures studies as an intellectual and applied discipline. ABS **42**, 372–385 (1998)
7. Jarillo, C.J.: On strategic networks. SMJ **9**, 31–41 (1988)
8. Jarillo, C.J.: Strategic Networks, Creating the Borderless Organization. Butterworth-Heinemann, Oxford (1993)
9. Teece, D.J., Pisano, G., Shuen, A.: Dynamic capabilities and strategic management. SMJ **18**, 509–533 (1997)
10. Zheng, L., Possel-Dölken, F.: Cooperation between production companies. In: Zheng, L., Possel-Dölken, F. (eds.) Strategic Production Networks, pp. 7–44. Springer, Berlin (2002)
11. Haag, H.-C., Tilebein, M.: Strategic foresight in production networks—a structured approach. In: Proceedings of the International Conference on Production Research (ICPR) (2011)
12. Rohrbeck, R., Gemünden, H.G.: Corporate foresight: its three roles in enhancing the innovation capacity of a firm. Technol. Forecast. Soc. Change **78**, 231–243 (2011)
13. Blackman, D.A., Henderson, S.: How foresight creates unforeseen futures: the role of doubting. Futures **36**, 253–266 (2004)
14. Vecchiato, R.: Environmental uncertainty, foresight and strategic decision making: an integrated study. Technol. Forecast. Soc. Change **79**, 436–447 (2012)
15. Voros, J.: A generic foresight process framework. Foresight **5**, 10–21 (2003)
16. Johnson, G., Scholes, K.: Exploring Corporate Strategy. Prentice Hall Europe, London (2000)

17. Johnson, M.W., Christensen, C.M., Kagermann, H.: Reinventing your business model. HBR, 78–93 (2008)
18. Sterman, J.D.: Business Dynamics, Systems Thinking and Modeling for a Complex World. Irwin/McGraw-Hill, Boston (2000)
19. Vennix, J.A.: Group Model Building: Facilitating Team Learning Using System Dynamics. Wiley, Chichester (1996)
20. van Heijden, K.D.: Scenarios, The Art of Strategic Conversation. Wiley, Chichester (2007)

Dynamic Capabilities in Manufacturing Processes: A Knowledge-based Approach for the Development of Manufacturing Flexibilities

Philip Cordes and Michael Hülsmann

Abstract Manufacturing systems are subject to the risk of path dependencies and resulting lock- in situations. In order to avoid and cope with them, they need manufacturing flexibilities. A management approach that triggers strategic flexibilities is the concept of dynamic capabilities. Therefore, a knowledge-based conceptualization of dynamic capabilities—i.e. knowledge codification, transfer, abstraction and absorption—is taken and analyzed regarding its contributions and limitations to flexibilize manufacturing systems on the basic or component as well as the system and the aggregated level.

Keywords Dynamic capabilities · Knowledge management · Manufacturing flexibility · Path dependency

1 Introduction

As any other organization, manufacturing systems are confronted with the risk of lock-in situations due to path dependent developments [1, 2]. This risk is associated with the risk of being unable to alter elements on the component or basic level, on the system level and on the aggregated level of manufacturing systems, when it is required [3]. In other words, they need manufacturing flexibilities in order to avoid or leave such inefficient system states.

P. Cordes (✉) · M. Hülsmann
Jacobs University Bremen, Systems Management, Bremen, Germany
e-mail: p.cordes@jacobs-university.de

M. Hülsmann
e-mail: m.huelsmann@jacobs-university.de

K. Windt (ed.), *Robust Manufacturing Control*, Lecture Notes in Production Engineering,
DOI: 10.1007/978-3-642-30749-2_38, © Springer-Verlag Berlin Heidelberg 2013

Arafa and Elmaraghy (2011) speak of such a manufacturing flexibility as dynamic capabilities. With respect to the notion that the most important resource of today's business organization is knowledge (e.g. [4, 5]), Burmann [6] developed a deeper conceptualization and operationalization of dynamic capabilities by defining two dimensions: Replication ability (codifying and transferring knowledge) and reconfiguration ability (abstracting and absorbing knowledge).

Adopting this approach, the following question arises: **How do knowledge codification, transfer, abstraction and absorption in manufacturing systems affect their flexibilities on the component or basic level, on the system level as well as on the aggregated level?**

Therefore, after an introduction in Sects. 1 and 2 aims to describe risks of path dependencies for the flexibility of manufacturing systems. Section 3 aims to depict and adapt the dynamic capabilities approach in order to make it applicable to manufacturing systems. Section 4 discusses the effects of knowledge-based dynamic capabilities—i.e. knowledge codification, transfer, abstraction and absorption—on the different levels of manufacturing flexibilities. In order to do so, associated contributions and limitations will be identified and opposed to each other. In Sect. 5 finally, further research requirements and managerial implications will be deduced.

2 Risks of Path Dependencies for Manufacturing Flexibilities

2.1 General Characteristics of Path Dependencies and Lock-In Situations

According to path dependency theory, today's decisions influence the scope of managerial decision alternatives and might reduce its amount over time (e.g. [7, 8]). The underlying principle is called **'history matters'** [9]. Accordingly, decisions can have formative character for subsequent decisions [7, 8]. Organizational processes are based on decisions of many kinds and hence have an essential historic character [7]. This implies that process developments or at least parts of them are irreversible [10]. Therefore, David [7] calls historical events that determine the future development in an undesired way 'historical accidents' that determine the circumstances and managerial options in the future. Van Driel and Dolfsma [2] point out that organizations are generally sensitive to initial conditions—i.e. a moment or event in the past that is to be seen as a starting point for a path dependent development.

Consequently, it cannot be assumed that economic actors have a free choice. Instead, their scope of available decision alternatives is restricted by decisions made in the past. This argument counts all the more, when **'increasing returns'** (also termed as self-reinforcing effects) occur—the second main characteristic of path dependent processes. They occur when an increase of a certain variable leads to a further increase of the same variable in the next time step [11].

The third main characteristic is finally the evolvement of a **'lock-in situation'**, which is to be seen as the main critical result of path dependent developments. In its original meaning a lock-in situation describes a situation in which users adopt one particular technology although other technologies are superior. However, due to path dependent developments—i.e. historical events and increasing returns—the inferior becomes the quasi standard. Common examples are the QWERTY-typewriter keyboard [7] and the VHS format [12], which are respectively were both technology standards in their respective industry although superior alternatives exist. The main characteristic of suchlike technological lock-ins is that the users are not flexible enough to change the technologies they currently use although others might be superior [7].

Authors like Sydow et al. [9] transferred the underlying mechanisms to the behavior of organizations and their selections of managerial options. Accordingly, self-reinforcing effects that emanate from the selection of a certain option or a type of option can result in a decrease of the amount of options that is generally available . Since combinations of such managerial options, which aim to create a certain industry position, are regarded as corporate strategies [13], it can be deduced that such path dependent developments can reduce the amount of strategies generally available. Consequently, companies that face such a reduction of potential strategic alternatives loose their ability to react flexible to environmental changes. Therefore, one main characteristic of organizational lock-in situations is strategic inflexibility. Hence, with recourse to the problem depicted in Sect. 1, the question arises, in how far manufacturing systems are subject to such risks of lock-ins and resulting inflexibilities.

2.2 *Risks of Path Dependencies and Lock-In Situations in Manufacturing Systems*

Whereas path dependency was mainly used for analyses on the level of entire organizations [9, 14], industries [15] or even nations [16], it was much less used on the micro levels of organizations, such as manufacturing systems [2]. Van Driel and Dolfsma [2] e.g. apply it to the evolvement of the Toyota production system and identify initial conditions and lock-in characteristics of the development of just-in-time production strategies. Morrey at al. [1] find in a case study on a construction company that the process of implementation of a lean culture exhibits fundamental path depende0nt characteristics . Dean and Snell [17] examine organizational inertia, which have strong similarities to lock-in situations, that arise from integrated manufacturing concepts. Hence, manufacturing systems are generally exposed to path dependent developments. Thereby, the risk of lock-in situations for manufacturing systems occurs on three levels:

First, on a technological level, the decision for a particular technology in manufacturing affects subsequent decisions through determining the scope of technologies that are compatible with the selected one. Hence, companies might choose subsequent technologies that are not the best on the market, to either ensure compatibility with the already existing one or to be able to use existing knowledge and capabilities of employees to handle the technology in order to avoid sunk-costs and additional costs for training of employees or technology upgrades. Hence, increasing returns occur: the selection of a certain technology increases the probability that a similar or at least a compatible technology is selected again in the next time step. Finally, a lock-in situation occurs if there is no possibility left to choose another technology than the current one in use although there are superior ones available on the market.

Second, on a managerial level, the decision for a certain manufacturing strategy [18] affects the subsequent decisions through determining the remaining scope of potential manufacturing strategies. E.g. a manufacturing strategy that aims to achieve economies of scale and low cost production through the creation of giant plants and docile work forces [19] cannot be reversed without any efforts into a strategy that focuses on flexible and individual customer-tailored production lines. Hence, in order to avoid related sunk costs (e.g. marketing efforts for positioning as a cost leader) and the loss of an already established market position, subsequent strategic decisions are self-reinforced by the previous ones. Therewith, the risk of an institutional lock-in situation occurs that would reduce the amount of decision alternatives on manufacturing strategies to a restricted field of options.

Third, there are strong interrelations between technological and managerial path dependencies. When the selected manufacturing strategy is based on the investment in a certain manufacturing technology that enables e.g. mass production but no customer-tailored production, the increasing returns on the technological level trigger also increasing returns on the management level. The other way round, if a certain strategy has been chosen that involves the use of a particular technology, increasing returns on the institutional level trigger also increasing returns on the management level.

Consequently, path dependent developments endanger the ability of manufacturing systems to react flexible on internal and external changes. Therefore, inflexibility can be regarded on the one hand as an outcome of lock-in situations. On the other hand, the key to avoid or to cope with path dependencies in manufacturing systems might be manufacturing flexibility.

2.3 Flexibility in Manufacturing Systems

From a general perspective Arafa and ElMaraghy (2011) define flexibility in manufacturing systems *"[...] as the ability of a system or facility to adjust to the changes in its internal or external environment with little penalty in time, effort, cost, or performance*"[20, p. 508]. However, this general perspective seems to need a concretization in order to be applied in particular contexts. Sethi And Sethi [3] propose a classification of flexibility types that distinguish between **component or basic flexibilities**, **system flexibilities** and **aggregate flexibilities** (other classifications have been developed e.g. by Gupta and Goyal [19] based on [21]).

Component or basic flexibilities describe possibilities to change the machines, the material handling and the operations. Thereby, **machine flexibility** *"[...] refers to the various types of operations that the machine can perform without requiring a prohibitive effort in switching from one operation to another."*[3, p. 297]. **Material handling flexibility** describes the *"[...] ability to move different part types efficiently for proper positioning and processing through the manufacturing facility it serves"*[3, p. 300]. **Operation flexibility** refers to the ability of a part *"[...] to be produced in different ways"*[3, p. 301].

System flexibilities reflect possibilities to change the manufacturing system's processes, routings and volumes and to expand. Thereby, **process flexibility** *"[...] relates to the set of part types that the system can produce without major setups"*[3, p. 302]. **Routing flexibility** *"[...] of a manufacturing system is its ability to produce a part by alternate routes through the system"*[3, p. 305]. **Product flexibility** *"[...] is the ease with which new parts can be added or substituted for existing parts"*[3, p. 304]. **Volume flexibility** *"[...] of a manufacturing system is its ability to be operated profitably at different overall output levels"*[3, p. 307]. **Expansion flexibility** *"[...] is the ease with which its capacity and capability can be increased when needed"*[3, p. 309].

Aggregated flexibilities finally reflect possibilities to change elements on a higher level within an organization—i.e. the program, production and the market. **Program flexibility** is *"[...] the ability of the system to run virtually untended for a long enough period"*[3, p. 310]. **Production flexibility** *"[...] is the universe of part types that the manufacturing system can pro- duce without adding major capital equipment"*[3, p. 310]. **Market flexibility** *"[...] is the ease with which the manufacturing system can adapt to a changing market environment"*[3, p. 312].

Considering these different types of manufacturing flexibilities, the question arises, how they enable manufacturing systems to avoid or to cope with path dependencies and resulting lock-in situations?

2.4 Manufacturing Flexibilities as Enablers to Avoid and Cope with Lock-In Situations

On the level of the components or the manufacturing system's basics, the different types of flexibilities ensure or even increase the availability of certain amounts of different decision-alternatives regarding the operations machines can perform, the materials the system can handle and the ways in which parts of products are produced. Hence, the occurrence of initial events that determine the paths on which the manufacturing system develops can be prevented when a great amount of varying decision alternatives is ensured in the first place. Therewith, the risk of lock-ins decreases.

The same argument counts for the flexibilities on the entire system's level. Maintaining the availability of certain amounts of different decision-alternatives e.g. regarding the manufacturing system's processes or the volumes of produced parts or products avoids that initial events can reduce these alternatives to only one remaining one.

Finally, these flexibilities form on the aggregated level. When lock-ins can be avoided on the prior levels, the risk that they occur decreases also on the aggregated level. Furthermore, in order to ensure market flexibility e.g., it is necessary to maintain different decision-alternatives regarding the products that are produced in the manufacturing system.

Subsuming, lock-ins that are based on technological and managerial path dependencies can be avoided through ensuring and maintaining manufacturing flexibilities on the components and basic level as well as on the entire system's and the aggregated level. Therefore, the question arises, how these flexibilities can be developed and maintained. Arafa and ElMaraghy [20, p. 508] draw an interconnection to the general management approach of dynamic capabilities of organizations. Accordingly, *"from a manufacturing perspective the dynamic capability for enterprise organizations is known as manufacturing flexibility"*. Considering the above-mentioned contributions of manufacturing flexibilities to avoiding and coping with lock-ins, this notion conforms with O'Reilly and Tushman [22, p. 187] who state that overcoming *"[...] inertia and path dependencies is at the core of dynamic capabilities"*. Therefore, the following questions arise: *First*, what exactly can be understood of dynamic capabilities from the perspective of manufacturing systems? *Second*, how do such dynamic capabilities of manufacturing systems contribute or limit the development of flexibilities on the components or basic level, on the system's as well as on the aggregate level?

3 Dynamic Capabilities in Manufacturing Systems

3.1 A General Understanding of Knowledge-based Dynamic Capabilities

The dynamic capabilities approach picks up the main assumptions of the resource-based view (e.g. [23]). Accordingly, competitive advantages can be gained and sustained, if companies possess resources that fulfill the so-called VRIN-criteria—i.e. when they are valuable, rare, inimitable and not substitutable (e.g. [24]). However, Katkalo et al. [25, p. 1176] argue that *"[...] even the VRIN-est of resources can lead to little benefit, when managed by incompetent indivudals [...]"*. Therefore, the dynamic capabilities approach also incorporates the underlying assumptions of the competence-based view that traces competitive advantages back to individual and organizational competences (e.g. [26]). However, since both fail to explain, why companies can perform substantially different from others, although they are equipped with the same resources and competences [27], Teece and Pisano [28] challenged the underlying core-notion by stating that the essential capability is to be able to alter organizational resources and competences over time and under consideration of environmental changes. However, the question arises: What are concrete dimensions of dynamic capabilities that can be applied in manufacturing systems?

One concretization was developed by Burmann [6] who conceptualized dynamic capabilities as the ability to **replicate** and **reconfigure** organizational resources (including competences) through managerial and organizational processes. Furthermore, as a multitude of authors have stated, the most essential resource of an organization to gain and maintain competitive advantages is assumed to be **knowledge** (e.g. [4, 5]). According to Burmann [6], all resources that fulfill the VRIN-criteria are based on an advance in knowledge. Even the essential resource knowledge can only be acquired—i.e. organizations learn through knowledge accumulation—if there is knowledge that can be acquired and exchanged [29]. Daniels and Bryson [30, p. 977] observe a particular importance of knowledge for manufacturing systems since *"[...] there is an important shift away from production that is dependent upon material resources to production that utilizes knowledge as the key source of competitiveness and innovation"*. Consequently, a concretization of dynamic capabilities as replication and reconfiguration of organizational resources should focus on the resource knowledge.

3.2 Knowledge-based Dynamic Capabilities in Manufacturing Systems

Following Burmann [6] organizational flexibilities and hence dynamic capabilities through knowledge replication and reconfiguration occur if organizational knowledge...

- ...can be identified and externalized through **knowledge codification**
- ...can be made available to the entire organization without an unintended diffusion to competitors through internal **knowledge transfer**
- ...can be devolved to new fields of appliances respectively markets through **knowledge abstraction**
- ...can be combined with new organization-external knowledge through **knowledge absorption.**

Consequently, the four dimensions of dynamic capabilities in this knowledge-oriented perspective are knowledge codification, transfer, abstraction and absorption.

3.2.1 Knowledge Codification

Grant and Gregory [31] examine the transferability of manufacturing knowledge and state that *"[...] the transfer of know-how, some of which may be tacit and hard to transfer, is clearly critical for learning"*. This involves the notion that tacit knowledge on manufacturing processes has to be converted somehow into a form that allows to circulate and to exchange it [29]. Hence, manufacturing knowledge has to be codified so that *"[...] knowledge managers and users can categorize knowledge, describe it, map and model it, stimulate it, and embed it in rules and recipes"*[32, p. 80]. Consequently, knowledge codification in manufacturing systems refers to processes of transforming implicit knowledge on manufacturing processes into explicit knowledge through representations in symbolic forms.

3.2.2 Knowledge Transfer

The transfer of knowledge is an essential process within manufacturing companies in order to enable a strategic manufacturing alignment. Accordingly, manufacturing performance can be increased through process changes and knowledge creation, based on knowledge transfer [30]. Consequently, knowledge transfer in manufacturing systems refers to processes of devolving knowledge from one application place to another. Such a transfer can be made either within one manufacturing system or between a manufacturing system and other organizations that are in cooperation with each other (based on [6]).

3.2.3 Knowledge Abstraction

Knowledge abstraction latter refers to disengaging knowledge from its original context and making it applicable in other contexts [6]. Hence, it is necessary to reveal causal relations [33] that underlie a company's manufacturing processes. Not until then, knowledge on manufacturing processes can be devolved to new fields of appliances—i.e. to other manufacturing processes in different circumstances.

3.2.4 Knowledge Absorption

Cohen and Levinthal [34, p. 128] speak about a firm's absorptive capacity as *"[...] the ability of a firm to recognize the value of new external information, assimilate it, and apply it to commercial ends"*. For manufacturing systems it is therefore necessary to be able to identify knowledge about manufacturing processes from its environment, to evaluate its potential contribution when being applied and finally to apply it as appropriate in similar or totally different forms within the own manufacturing processes (based on [35]).

Having identified these four dimensions of a manufacturing system's dynamic capabilities, the question arises how they contribute or limit the component/basic flexibilities, the system's flexibilities as well as the aggregated flexibilities of manufacturing systems.

4 Effects of Dynamic Capabilities on the Flexibility of Manufacturing Systems

4.1 Effects of Knowledge Codification on Manufacturing Flexibilities

Knowledge codification leads to both contributions as well as limitations of a manufacturing system's development of flexibilities on the component or basic level, the system level and the aggregated level.

One exemplary **contribution** arises from the fact that codifying knowledge requires forming a mental model of it [36]. Hence, manufacturing knowledge has to be brought to mind by the codifying employees, which facilitates generating new ideas on how to change existing routines and identifying strengths and weaknesses. Hence, knowledge codification does not only reduce the tacitness of manufacturing knowledge. It does also increase the deepness with which manufacturing knowledge is anchored within the employees, which in turn enlarges the scope of possibilities regarding the appliance of certain knowledge.

For the component or basic level of manufacturing systems, this means that knowledge e.g. on how to flexibilize machines, on how to enable them to handle material and on how to ensure an operational flexibility is enlarged and converted into symbolic forms. These forms, in turn, can be transferred and hence diffused within the entire manufacturing systems, so that other machines can learn from the machine, material handling and operation flexibilities of other machines. The same argument counts for the system flexibilities. For example, if knowledge on how to flexibilize the manufacturing volume of certain machines is deepened, then codified and hence made available to other members of the manufacturing system, this knowledge can be applied to other machines. In consequence, the volume flexibility of the entire manufacturing system can increase. Finally, Sethi and Sethi [3] show the interrelations between the system and the aggregated flexibilities. Accordingly, if knowledge codification increases the elements of the system flexibilities, it affects positively the flexibilities on the aggregated level. For example, increasing product flexibility enables a manufacturing system to offer a wider production range. Hence, new options are created that enlarge the scope of actions within manufacturing systems and hence, increases flexibilities on all three manufacturing system levels.

One exemplary **limitation** of knowledge codification results from the risk of a solidification of learned routines: In an empirical study García-Muiña et al.

[37, p. 144] showed that *"[...] an excessive presence of codified knowledge, strongly institutionalized in the heart of the company, can put a serious brake on the creativity, intuition and employees' radical improvisation skills that major innovative activity requires"*, (see also [38–41]). Consequently, the codification of manufacturing knowledge increases the risk that routines within a manufacturing system—e.g. how to program machines or how to design manufacturing processes—are solidified. The reason is that most of the existing codification tools create guidelines for the execution of tasks in the future or for other employees [36]. Such guidelines however, hinder employees to find new and innovative solutions for occurring problems or even to find any solutions for problems that are not covered in the respective guidelines. This problem occurs on all three levels of manufacturing flexibilities: Guidelines for machines—e.g. on how to handle certain material—hinder them to develop material handling flexibility. Guidelines for an entire manufacturing system e.g. on the volume that is intended to be produced hinders the system to deviate from this volume if necessary und hence to develop volume flexibilities. Finally, guidelines that determine e.g. the market for which products are manufactured hinder the manufacturing system on an aggregated level to change their target markets if appropriate.

Additionally, codification of manufacturing knowledge requires efforts that have to be undertaken (e.g. development of manuals or other process-specific tools [36]). Hence, investments in information technologies are necessary that enable a search for knowledge that is worth to be codified and knowledge management systems have to be maintained over time [42]. Furthermore, alternate costs occur since the respective personnel is not able to conduct their usual tasks while codifying their manufacturing knowledge [6, 32]. In sum, knowledge codification is expensive and requires time [32]. Consequently, financial resources are bounded to processes of codifying manufacturing knowledge that cannot be used to make investments that might increase manufacturing flexibilities. One example on the component or basic level is an investment in different machines in order to widen the amount of different markets for which products can be manufactured and thus to increase market flexibility on an aggregated level. Another example on the system level is an investment in an extension of the manufacturing capacities in order to ensure volume flexibility [43]. Hence, the costs that are associated with the codification of manufacturing knowledge can lead to the necessity to resign flexibility investments, which in turn might reduce manufacturing flexibilities on all three levels.

4.2 Effects of Knowledge Transfer on Manufacturing Flexibilities

Knowledge transfer does also lead to both contributions as well as limitations of manufacturing flexibilities on all three levels—the component or basic, the system and the aggregated level.

One exemplary **contribution** is that knowledge transfer enables the combination of existing knowledge with internally transferred and received knowledge—so-called combinative capabilities. On this basis, new knowledge is created [44]. On a component or basic level the combination of manufacturing knowledge (e.g. general knowledge on how to modify machines with special knowledge on particular machine functions) leads to additional decision alternatives and hence, to a wider scope of actions. Therewith, flexibilities on the component or basic level increase. The same argument counts for the system as well as the aggregated flexibility level: If knowledge on how to expand a manufacturing system's capacities is combined with knowledge on how to serve different markets, a chain of flexibility effects is triggered that results in a higher manufacturing flexibilities.

Furthermore, the transfer of manufacturing knowledge can also lead to a higher efficiency of manufacturing processes. The underlying assumption is that most of the knowledge that is transferred within a company is based on the transfer of best practices – hence, knowledge that has already proven to lead e.g. to higher efficiency [45]. An increasing efficiency instead leads to redundant potentials in the manufacturing systems that can be used elsewhere. One example is a machine that is working to full capacity anymore since knowledge has been transferred on how to run the machine more efficient. The redundant capacity can be used e.g. to ensure volume or expansion flexibility. The same argument counts for the component or basic as well as the aggregated flexibility level: E.g. redundant capacities might result in new options to handle material or to change the manufacturing system's program.

An exemplary **limitation** result from the transfer of 'locked-in' knowledge. Zollo and Winter [36] argue that codified and transferred knowledge consists mainly of the creation and diffusion of guidelines, that have already proven successful in the past. However, as it has been shown in Sect. 2, such knowledge or beliefs of what might work because it has been working in the past might be subject to path dependencies and hence, to lock-in situations. Hence, the risk occurs that lock-ins are created or even solidified due to a diffusion of best practices. On the basic or component level of manufacturing systems, this risk refers to a diffusion of knowledge on how to handle machines and conduct manufacturing operations, which is solidified as a result. Hence, machine, material handling and operation flexibilities are reduced. On a system level, knowledge on processes, routing, products or the volume and capacity expansion can be solidified that results in an inability of the manufacturing system to react to changing circumstances that require approaches that are different from the ones that were successful in the past. Finally, as Sethi and Sethi [3] show, there are strong interrelations between the levels, wherefore also the aggregated flexibilities can be negatively influenced.

Additionally, as for the codification of manufacturing knowledge, its transfer requires efforts that result in costs. Correspondingly to knowledge codification costs, costs for knowledge transfer might include a necessary employment of additional personnel, the implementation of additional tools and technological equipment that enable a transfer of manufacturing knowledge as well as alternate costs because the personnel cannot be utilized in order to conduct their usual tasks during the time required for the transfer of manufacturing knowledge. Consequently, financial resources are bounded that could otherwise be utilized in order to invest in manufacturing flexibilities on all three levels. Furthermore, Paliszkiewicz's [32] prefigure that a transfer of codified knowledge increases the risk that knowledge is transferred to the wrong participants—on purpose or involuntary. Such an unintended leak out of manufacturing knowledge might weaken the knowledge's fulfillment of the VRIN criteria, since it might enable competitors to imitate certain manufacturing abilities. Hence, this risk can result in a loss of competitive advantages, which in the end might lead to financial losses. These in turn would decrease the manufacturing system's ability to invest in flexibilities on the three observed levels of a manufacturing system.

4.3 Effects of Knowledge Abstraction on Manufacturing Flexibilities

The third dimension of knowledge-based dynamic capabilities—knowledge abstraction—does also contribute and limit manufacturing flexibilities on the component or basic, the system as well as the aggregated level.

One exemplary **contribution** results from an increased scope of application areas of existing manufacturing knowledge: According to Burmann [6], knowledge abstraction enables the organization members to identify new fields of appliances of the underlying causal relations of their knowledge. Consequently, abstraction of manufacturing knowledge on the basic or component level, the system level as well as the aggregated level leads to a wider scope of possibilities where this knowledge can be applied. For instance, process-related knowledge that is decontextualized from the particular manufacturing processes from which it origins can be used in order to flexibilize other manufacturing processes that do not have that much in common so that a simple knowledge transfer would enable an applicability. Such underlying process- related knowledge could also be applied on other levels such as the processes of the machines' material handlings or the processes of identifying new markets for which the manufacturing system can produce. Hence, decontextualizing manufacturing knowledge enables manufacturing systems to develop new flexibilities on all three levels.

One exemplary flexibility-based **limitation** of abstracting knowledge is that it requires time and hence financial resources without a coercive flexibility-benefit. Although Burmann [6] assumes that the abstraction costs are relatively small in comparison to codification and transfer costs, they are not negligible. Accordingly, the main share of costs that is associated with knowledge abstraction processes account for alternate costs of employees that are decontextualizing their knowledge. While doing so, they are usually not able to fulfill their usual tasks. However, according to Smith et al. [46], knowledge abstraction requires expertise regarding the knowledge that is abstracted . Expertise in turn, e.g. in the form of cognitive flexibility [47] to find new fields of appliances of manufacturing knowledge can be assumed to be costly. Higher educated people have stronger abilities to abstract manufacturing knowledge from their original context but demand also higher salaries [48]. Hence, again, financial resources are bounded to knowledge abstraction processes that cannot be utilized for investments in manufacturing flexibilities, such as an expansion of the manufacturing system's capacity.

4.4 *Effects of Knowledge Absorption on Manufacturing Flexibilities*

There are also both contributions and limitations of the fourth dimension of knowledge-based dynamic capabilities—knowledge absorption—to the development of manufacturing flexibilities on all three observed levels.

The main **contribution** can be assumed to be an increased innovativeness. Cohen and Levinthal [34] argue that a firm's absorptive capacity contributes largely to its innovative capabilities. Empirical validations of this assumption have been conducted e.g. by [49]. The reason is the acquisition of external information and its combination with internally existing knowledge. For a manufacturing system, this refers to the acquisition of external knowledge on machines, material handling or operations as well as processes, routines, products and possibilities to expand the manufacturing system's capacity. Additionally, knowledge on the aggregated level of manufacturing systems can be internalized through knowledge absorption. A combination with such knowledge with internally existing knowledge can lead to a widening of the scope of decision alternatives on all three levels, e.g. how to handle material, how to design manufacturing processes or knowledge on certain market characteristics for which products are manufactured. Hence, when manufacturing systems absorb external knowledge and combine it with internally existing knowledge, their manufacturing flexibilities can be increased on all three levels.

An associated **limitation** however is that an information overload can occur [50]. The underlying notion is that an internalization of a large amount of external knowledge that has not proven its contribution to the manufacturing system's flexibility yet might paralyze the manufacturing system. Hence, its abilities to change the elements on the basic or component, the system as well as the aggregated level of manufacturing systems are hindered through an overload with information on how they could be changed. If the amount of incoming information is too high, the manufacturing system cannot evaluate its benefits and might utilize the 'wrong' knowledge that can lead to contradicting results. Consequently, too much knowledge absorption might result in lower manufacturing flexibilities.

5 Conclusion

The research question of this paper is how knowledge-based dynamic capabilities—i.e. knowledge codification, transfer, abstraction and absorption—in manufacturing systems affect their flexibilities on the component or basic level, on the system level as well as on the aggregated level.

Thereby, both contributions as well as limitations of the dimensions of knowledge-based dynamic capabilities of manufacturing systems to the development of manufacturing flexibilities have to be considered. E.g. knowledge transfer enables on the one hand to combine existing with newly received knowledge and hence to create entirely new manufacturing knowledge that leads to manufacturing flexibilities on the component or basic as well as the system and the aggregated level of manufacturing systems. On the other hand, the risk occurs that knowledge is transferred and applied elsewhere that has proven successful in the past but is not appropriate anymore in new conditions. Hence, manufacturing flexibilities can be limited through an internal transfer of such 'locked-in' knowledge.

Nevertheless, there is a chance that the demonstrated contributions to manufacturing flexibilities predominate the associated limitations. Hence, while designing and managing manufacturing systems, the positive effects of knowledge codification, transfer, abstraction and absorption should be stimulated in order to benefit from their positive effects on manufacturing flexibilities on all three observed levels.

However, neither do the exemplary limitations and contributions provide a complete picture of positive and negative effects, nor are they weighted. Hence, a net-effect could not be identified, wherefore future research is necessary that aims to quantify the influences of knowledge-based dynamic capabilities on the different elements of manufacturing flexibilities.

References

1. Morrey, N., Pasquire, C., Dainty, A: . The Impact of Path Dependencies on Lean Implementation Within a Construction Company. *Lean Construction Journal,* pp. 86-96 (2010)
2. Van Driel, H., Dolfsma, W.: Path dependence, initial conditions, and routines in organizations: the toyota production system re-examined. J. Organ. Change Manage. **22**(1), 49–72 (2009)
3. Sethi, A.K., Sethi, S.P.: Flexibility in manufacturing: a survey. Int. J. Flex. Manuf. Syst. **2**(4), 289–328 (1990)
4. Drucker, P.F.: Post-capitalist society. Harper, New York (1993)
5. Quinn, J.B.: The intelligent enterprise a new paradigm. Executive **6**(4), 48–63 (1992)
6. Burmann, C.: Strategische Flexibilität und Strategiewechsel als Determinanten des Unternehmenswertes. Dt. Univ.-Verl, Wiesbaden (2002)
7. David, P.A.: Clio and the economics of QWERTY. Am. Econ. Rev. **75**, 332–337 (1985)
8. Liebowitz, S.J., Margolis, S.E.: Path dependence, lock-in, and history. J. Law Econ. Organ. **11**(1), 205–226 (1995)
9. Sydow, J., Schreyögg, G., Koch, J.: Organizational Paths: Path Dependency and Beyond, 21st EGOW Colloquium, 30 June–2 July, 2005, Berlin, Germany. Subtheme 1: Path Dependence and Creation Processes in the Emergence of Markets, Technologies and Institutions. Convenors: Callon, Michel; Garud, Raghu; Karnoe, Peter, Berlin (2005)

10. Ackermann, R.: Die Pfadabhängigkeitstheorie als Erklärungsansatz unternehmerischer Entwicklungsprozesse. In: Schreyögg, G., Sydow, J. (eds.) Strategische Prozesse und Pfade, pp. 225–256. Gabler, Wiesbaden (2003)
11. Arthur, W.B.: Competing technologies, increasing returns, and lock-in by historical events. Econ. J. **99**(394), 116–131 (1989)
12. Arthur, W.B.: Positive feedbacks in the economy. Sci. Am. **262**(2), 92–99 (1990)
13. Porter, M.E. Wettbewerbsstrategie : Methoden zur Analyse von Branchen und Konkurrenten = (Competitive strategy), 10., durchges. und erw. Aufl edn, Campus-Verl., Frankfurt/Main u.a (1999)
14. David, P.A.: Why are institutions the 'carriers of history'?: path dependence and the evolution of conventions, organizations and institutions. Struct. Change Econ. Dyn. **5**, 205–220 (1994)
15. Cordes, P.: Non- physische Tonträger im Verdrängungswettbewerb mit physischen Tonträgern—Eine Analyse institutioneller Pfade in der Musikindustrie, 1st edn. Universität Bremen FB Wirtschaftswissenschaften, Bremen (2008)
16. Pierson, P.: Increasing returns, path dependence, and the study of politics. Am. Polit. Sci. Rev. **94**(2), 251–267 (2000)
17. Dean J.W Jr., Snell, S.A.: Integrated manufacturing and job design: moderating effects of organizational inertia. Acad. Manage. J. **34**(4), 776–804 (1991)
18. Gerwin, D.: Manufacturing flexibility: a strategic perspective. Manage. Sci. **39**(4), 395–410 (1993)
19. Gupta, Y.P., Goyal, S.: Flexibility of manufacturing systems: concepts and measurements. Eur. J. Oper. Res. **43**(2), 119–135 (1989)
20. Arafa, A., ElMaraghy, W.: Manufacturing strategy and enterprise dynamic capability. CIRP Ann. Manuf. Technol. **60**, 507–510 (2011)
21. Browne, J., Dubois, D., Rathmill, K., Sethi, S.P., Stecke, K.E.: Classification of flexible manufacturing systems. FMS Mag. **2**(2), 114–117 (1984)
22. O'Reilly, C.A., Tushman, M.L.: Ambidexterity as a dynamic capability: Resolving the innovator's dilemma. Stanford University Graduate School of Business Research Paper No. 1963. Available at SSRN: http://ssrn.com/abstract=978493, vol. 28, pp. 185–206
23. Selznick, P.: Leadership in Administration: A Sociological Interpretation. Harper, New York (1957)
24. Barney, J.B.: Firm resources and sustained competitive advantage. J. Manage. **17**(1), 99–120 (1991)
25. Katkalo, V.S., Pitelis, C.N., Teece, D.J.: Introduction: on the nature and scope of dynamic capabilities. Ind. Corp. Change **19**(4), 1175–1186 (2010)
26. Prahalad, C.K., Hamel, G.: The core competence of the corporation. Harvard Bus. Rev. **68**(3), 79–91 (1990)
27. Freiling, J.: A competence-based theory of the firm. Manag.-Rev. **15**, 27–52 (2004)
28. Teece, D.J., Pisano, G.: The dynamic capabilities of firms: an introduction. Ind. Corp. Change **3**(3), 537–556 (1994)
29. Ancori, B., Bureth, A., Cohendet, P.: The economics of knowledge: the debate about codification and tacit knowledge. Indus. Corp. Change **9**(2), 255–287 (2000)
30. Carrillo, J.E., Gaimon, C.: Improving manufacturing performance through process change and knowledge creation. Manage. Sci. **46**(2), 265–288 (2000)
31. Grant, E.B., Gregory, M.J.: Tacit knowledge, the life cycle and international manufacturing transfer. Technol. Anal. Strat. Manage. **9**(2), 149–162 (1997)
32. Paliszkiewicz, J.O.: Knowledge codification and organisational performance in small and medium enterprises. Int. J. Manage. Enterp. Dev. **6**(1), 80–87 (2009)
33. Nahapiet, J., Ghoshal, S.: Social capital, intellectual capital, and the organizational advantage. Acade. Manage. Rev. **23**(2), 242–266 (1998)
34. Cohen, W.M., Levinthal, D.A.: Absorptive capacity: a new perspective on learning and innovation. Adm. Sci. Q. **35**(1), 128–152 (1990)
35. Zahra, S.A., George, G.: Absorptive capacity: a review, reconceptualization, and extension. Acad. Manage. Rev. **27**(2), 185–203 (2002)
36. Zollo, M., Winter, S.G.: Deliberate learning and the evolution of dynamic capabilities. Organ. Sci. **13**(3), 339–351 (2002)

37. García-Muiña, F.E., Pelechano-Barahona, E., Navas-López, J.E.: Knowledge codification and technological innovation success: empirical evidence from Spanish biotech companies. Technol. Forecast. Soc. Chang. **76**(1), 141 (2009)
38. Casper, S., Whitley, R.: Managing competences in entrepreneurial technology firms: a comparative institutional analysis of Germany, Sweden and the UK* 1. Res. Policy **33**(1), 89–106 (2004)
39. Crossan, M.M., Bedrow, I.: Organizational learning and strategic renewal. Strateg. Manag. J. **24**(11), 1087–1105 (2003)
40. Hedlund, G.: A model of knowledge management and the n-form corporation. Strateg. Manag. J. **15**, 73–90 (1994)
41. Vera, D., Crossan, M.: Improvisation and innovative performance in teams. Organ. Sci. **16**(3), 203–224 (2005)
42. Yao-Sheng, L.: The effects of knowledge management strategy and organization structure on innovation. Int. J. Manage. **24**(1), 53–60 (2007)
43. De Toni, A., Tonchia, S.: Manufacturing flexibility: a literature review. Int. J. Prod. Res. **36**(6), 1587–1617 (1998)
44. Kogut, B., Zander, U.: Knowledge of the firm, combinative capabilities, and the replication of technology. Organ. Sci. **3**(3), 383–397 (1992)
45. Van Wijk, R., Jansen, J.J.P., Lyles, M.A.: Inter- and intra-organizational knowledge transfer: a meta-analytic review and assessment of its antecedents and consequences. J. Manage. Stud. **45**(4), 830–853 (2008)
46. Smith III, J.P., Disessa, A.A., Roschelle, J.: "Misconceptions reconceived: a constructivist analysis of knowledge in transition. J. Learn. Sci. 3(2), 115-163 (1993)
47. Sanchez, R.: Understanding competence-based management—Identifying and managing five modes of competence. J. Bus. Res. **57**(5), 518–532 (2004)
48. Psacharopoulos, G., Anthony, H.: Returns to investment in education: a further update. In: Education Economics 12(2), 111–134
49. Fosfuri, A., Tribó, J.A.: Exploring the antecedents of potential absorptive capacity and its impact on innovation performance. Omega **36**(2), 173–187 (2008)
50. Hülsmann, M., Grapp, J., Li, Y.: Strategic adaptivity in global supply chains—Competitive advantage by autonomous cooperation. Int. J. Prod. Econ. **114**(1), 14–26 (2008)

Evaluation Model for Robustness and Efficiency Trade-offs in Production Capacity Decisions

Max Monauni, Mirja Meyer and Katja Windt

Abstract In volatile market environments with complex production processes, robustness against fluctuations is gaining in importance. Yet robustness can be seen as conflicting with efficiency, e.g., when excess capacities need to be provided. We suggest a model to assess the trade-off between robustness and efficiency for capacity investment decisions from a strategic as well as from an operations perspective. We argue that one needs to consider both perspectives in order to ensure the sustainability of a company.

Keywords Capacity management · Manufacturing · Trade-offs · Robustness

1 Introduction

Capacity management in production requires long-term, strategic decisions in which the amounts of machines and human resources have to be determined according to the anticipated product output. Usually capacity decisions are taken from an efficiency point of view, meaning that idle times of machines and excess

M. Monauni (✉)
Graduate School of Excellence Advanced Manufacturing Engineering,
University of Stuttgart, Stuttgart, Germany
e-mail: max.monauni@gsame.uni-stuttgart.de

M. Meyer · K. Windt
Jacobs University Bremen, Global Production Logistics,
Bremen, Germany
e-mail: mi.meyer@jacobs-university.de

K. Windt
e-mail: k.windt@jacobs-university.de

K. Windt (ed.), *Robust Manufacturing Control*, Lecture Notes in Production Engineering,
DOI: 10.1007/978-3-642-30749-2_39,

capacities should be avoided as they incur fixed costs. Yet in today's uncertain and turbulent market environments, production systems need to cope with fluctuating factors such as non-linear demand. In the past, research on capacity investment and management has focused on rendering production systems more flexible in order to cope with factors such as uncertainty or fluctuating demand [1–3]. Although increased flexibility can result in a better performance of the manufacturing system, one also needs to consider the influence of excess capacity on the system behavior. If production systems lack excess capacities (e.g., buffers) and the ability to adapt to demand changes, peak demands cannot be served and potential revenues are missed out [4]. In addition to that, due to the increasing complexity of production systems, failures are more likely to propagate through the system of capacity resources if no excess capacities are provided, which can negatively influence the performance (e.g., due date reliability) [5]. However, excess capacities do not only render production systems more robust, they also cause potential idle time costs [6]. Therefore capacity decisions have to take into account the trade-off between efficient and robust design of capacities and companies have to identify their business-specific optimal position within this target conflict of goals. Yet in order to facilitate capacity decisions with regards to overall system robustness and efficiency, the benefits (e.g. higher revenue due to fulfillment of all demands) and costs (e.g. increased costs due to idle time costs) of different production capacity settings have to be quantifiable and comparable. However, strategic and operational approaches to determine optimal amounts of capacity differ in the ways that they define and measure capacity and performance. While strategically oriented approaches to determine manufacturing capacities focus on measuring the firm's performance in terms of financial indicators, such as Return on Sales and Return on Assets [7], operationally oriented approaches focus on logistics performance targets, such as due date reliability or throughput time [8].

Therefore, in this paper we suggest a systematic forecast model that on the one hand allows companies to find an ideal balance between efficiency and robustness at which their business model is most sustainable and on the other hand links strategic and operational performance indicators. The remainder of the paper is structured as follows. In the second section, an overview on capacity investment and management literature as well as on trade-offs between robustness and efficiency in capacity management in production is given. The third and fourth sections present the strategically and operationally oriented measuring models. Section five connects the two complementary approaches and outlines the gains of combining them.

2 Existing Strategic Capacity Dimensioning Approaches

The capacity management in a manufacturing company includes tactical (short-term) and strategic (long-term) decisions on how the capacity of a manufacturing system should be sized, when capacity changes are most suitable, what types of capacity should be added, and where additional capacity should be added [9]. Regarding operational models for capacity management, numerous approaches have been published in recent years that use different modeling methods and each focus on specific aspects of capacity management. Luss et al. define that capacity expansion determines the size, timing, and location of buying additional capacity and provide a detailed survey of literature that uses methods from operations research [10]. In Wu et al. [11], an industry-specific review focusing on analytical approaches for capacity planning and management in high-tech industries is given. Paraskevopoulos et al. use a non-linear programming model to determine a capacity expansion solution in which robustness to likely errors under demand uncertainty is given [12]. While this approach focuses on a single machine scenario, further research also considers capacity expansion for multi-product and multi-machine manufacturing systems with stochastic demands [13].

Considering more strategically oriented research, VanMieghem [14] summarizes approaches for strategic management of capacity under uncertainty as well as approaches from economics for risk aversion in capacity investment. Especially such approaches that explicitly consider capacity management under uncertainty share commonalities with the idea of robustness to fluctuations and disturbances. More strategically oriented approaches focus rather on the behavior of financial performance indicator of the firm. For instance, in Hendricks et al. [15], the effect of operational slack (excess capacity) on the stock market reaction to supply chain disruptions is investigated. Furthermore, Modi and Mishra [7] suggest a model that analyzes the influence of resource efficiency and resource slack on the financial performance of the firm. No matter if looking at strategically or operationally oriented approaches, capacity decisions are always taken with regard to a certain target. This can be maximization of profits, flexibility or sustainability. The trade-off that we focus on is between robustness and efficiency.

According to the modern portfolio theory, production capacity can be perceived as a risk-return-combination [16]. Capacity holds an expected return on investment, in terms of its specific input–output-efficiency within a certain degree of utilization. The efficiency of the entire company's production capacity is therefore measured as ratio of output (EBIT) and input (capacity's fixed costs, as an enabler for adding value). The risk of the capacity portfolio is displayed in its robustness, the insensitivity of the production system against possible variation from planed conditions, e.g. fluctuating utilization. In this context, Gutenberg proclaimed the benefits of capacity to bear non-optimum utilization without additional costs [17]. In accordance with Markowitz, above-average returns are only accessible through the acceptance of additional risks, while capacity robustness, which is reducing capacity risk, induces higher costs and therefore reduces returns [18]. In this goal conflict between efficiency and robustness, companies have to find their ideal position between idle cost risk and capacity efficiency, based on their individual business model, market volatility and risk aversion of the shareholders [19, 20].

Figure 1 To implement the desired efficiency-robustness position, different methods have been established which combine internal and external capacity instruments that increase robustness, yet also lead to higher costs and therefore

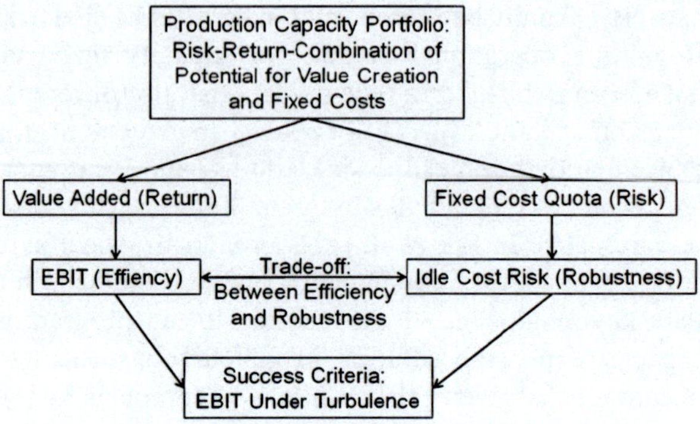

Fig. 1 Production capacity portfolio and trade-off between efficiency and robustness

reduce efficiency [21]. For instance the intended usage of temporary employment, sale-and-lease-back or flexible working times variabilize the company's cost structure [22]. From the idle cost perspective, the variable costs have no influence on a company's efficiency [23]. While company profits will decrease when sales drop, the return of sales (ROS) of a company that only has variable costs will remain equal.

Therefore, capacity management can be perceived as fixed cost management. Schmalenbach propagated the avoidance of idle costs by questioning the fixed cost quota (FCQ), which is an important element of robustness [24]. This is quantified by Dudenhöfer, who argued that for capacity intensive industries like the automotive sector a decrease of utilization by 15 % will result in a decline of ROS of 5.3 %-points, which is typically the entire profit margin [25]. Similar findings arise from the PIMS-study (Profit Impact of Market Strategies), which indicates a strongly negative correlation between return on investment (ROI) and capital investment intensity [26]. It can be stated that production capacity is both an essential factor for value creation and a considerable risk factor. In this context, FCQ is not per definition negative, but because it implies additional equity to cope with the idle costs risk, additional FCQ have to be legitimated with extra profits to compensate capital commitment.

3 Strategic Modeling Approach: The Idle-Cost-Model

To cope with the above described trade-off between efficiency and robustness, the so called idle-cost-model is subsequently designed. It confronts the specific fixed cost quota of a certain business model with its returns [27]. Thereby high value added business models, which require big capacities and above-average FCQ are questioned regarding their risk adequate returns [28]. So FCQ, which is indicating the idle cost risk, is tied to the return on sales, which identify to what extent the company-specific cost risks are justified by appropriate returns. The aim of the idle-cost-model is exclusively the detection of the company-specific, optimum position for production capacity in this goal conflict, not the execution of the identified position.

Fig. 2 Risk-return-coordinate system

		Efficiency (ROS)	
		High (10%)	Low (5%)
Robustness (FCQ from Sales)	High (10%)	Company 1	Company 2
	Low (50%)	Company 3	Company 4

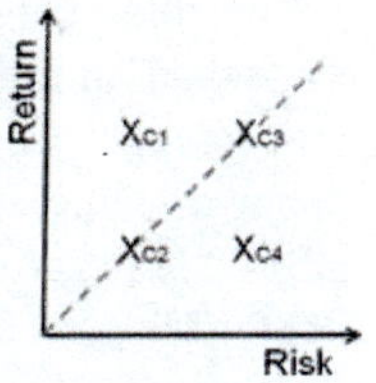

Starting-point for the idle-cost-model is Markowitz's efficiency frontier [16], which represents the best possible risk-return-combination. As high production capacity forms a critical element for value and profits creation, its efficient usage is essential. On the other side the production's fixed costs are lowering the robustness against market break-ins. In this goal conflict between robustness and efficiency the optimum capacity portfolio cannot exceed the efficiency frontier (dashed line in Fig. 2) [19]. Within this context, an output-oriented model is needed to forecast the consequences of occurring turbulences for future profits. This will quantitatively state how strong the influence of losses in sales will be on a company's profits. This approach allows to review the discussed trade-offs between the competing goals efficiency and robustness, since the results of the risk-return-capacity-portfolio are visualized.

For illustrative purposes, Fig. 2 shows the positioning of sample companies in the risk-return-coordinate-system: while company 1 is a theoretical, real-world impossible concept of a "perfect portfolio", company 4 represents an inefficient combination, because the same returns are possible with smaller risks (company 2), respectively higher returns are available within same risk aversion (company 3). Company 2 and 3 display two different characteristics of efficient capacity portfolios. Therefore, the final position for the optimal capacity portfolio depends on the companies' business sector, market volatility and risk aversion of the shareholders, but is always placed on the efficiency frontier.

To forecast the outcomes of different capacity portfolios, the companies undertake a stress testing based on their actual cost and results situation. The methodical realization of this idle-cost-model arises from forecasting the potential earnings before interest and taxes (EBIT) and thereby operationalizing the consequences of different sales trends. By observing the company's development in critical situations of market break-ins, the company-specific capacity risk is quantified by its profit reduction, which indicates the present ratio of companies' efficiency and robustness. Thereby the idle-cost-model executes an enhancement of the break-even-analysis through a dynamic examination proportional to the company's actual financial position instead of a single extrapolation of sales and EBIT. In contrast to the calculation of an abstract KPI comparable with the operating leverage, an absolute value (the quantifiable, potential EBIT under turbulence) is used as success criterion. This approach is comparable with Nink, who defines risk as the existence of unstableness, which can result in additional costs [29]. The company-specific capacity risk is quantified by the probability of turbulence multiplied by the amount of losses.

The idle-cost-model considers market break-ins as key indicator of EBIT losses. Based on the multiple future theory, the model forecasts impact of sales losses [30]. This is put into action by the quantification of consequences from different scenarios. Therefore, the command variable of potential EBIT after turbulence (EBIT t1) is introduced as a KPI for efficiency and robustness. While the current EBIT indicates the company's efficiency, a slight difference between the current EBIT and EBIT (t1) after a sales collapse indicates high robustness.

Objective target of the idle-cost-model is therefore the identification of the company-specific ideal capacity portfolio position, which provides a strong EBIT after potential losses in sales. The relevant parameters for the success criterion are the ROS, as an indicator of operating efficiency, and the fixed cost quota, which indicates the company's robustness as the fixed costs in proportion to sales, the FCQ. Low fixed costs enable companies to cost reduction on short notice when sales decrease. Therefore, fixed costs are an indicator for minor transformation ability [31]. Wages and salaries are not classified as variable costs, contrary to the traditional view. In addition to the increasing relevance of knowledge work, the employment law of many industrial nations inhibits dismissals on short notice [32]. Strong legal constraints obviate mass layoffs even aiming to generally deny these [33]. On this base wages and salaries are currently questioned in their definition as rapid changeable variable costs [34] or are even categorized as fixed costs [35]. As a result the dismissal of production staff as an adjustment of capacity is not realizable without losses, therefore wages and salaries are accounted as (periodic) fixed costs. The utilization of production capacity is determined as the key indicator of EBIT losses by the idle cost stress test [36]. This can be illustrated by means of the economic equation:

$$\text{EBIT} = \text{Sales} - \text{Costs (fixed and variable)} \tag{1}$$

$$\text{Gross Value Added} = \text{Sales} - \text{Intermediate Consumption} \tag{2}$$

Under the assumption that wages and salaries are (periodic-) fixed costs as well as the depreciation and amortization of fixed assets, the Intermediate Consumption (IC) equals the variable costs. Therefore, the Value Added equals the fixed costs plus EBIT. This corresponds to the definition of Value Added from the British Department for Innovation, which defines Value Added as EBITDA plus wages and salaries [37]. With the implementation of the auxiliary variable Turbulence Coefficient (TC), which is defined as future sales (t1) divided by actual sales (t0), the following conditions apply:

$$\text{EBIT (t0)} = \text{Sales (t0)} - \text{Costs (t0)} \tag{3}$$

$$\text{EBIT (t1)} = \text{Sales (t0)} * \text{TC} - \text{IC (t0)} * \text{TC} - \text{VA (t0)} + \text{EBIT (t0)} \tag{4}$$

Due to the fact that Value Added contains only fixed costs, it is per definition irreducible in the short term and therefore not influenced by market changes,

Fig. 3 Results idle-cost-model

Company	1	2	3	4
Sales	100	100	100	100
FQC	10%	50%	10%	50%
EBIT	10	10	5	5
ROS (t0)	10%	10%	5%	5%
TC = 0,95				
EBIT (t1)	9	7	4,25	2,25
ROS (t1)	9,5%	7,4%	4,5%	2,4%
TC = 0,85				
EBIT (t1)	7	1	2,75	-3,25
ROS (t1)	8,2%	1,2%	3,2%	-3,8%

represented by the Turbulence Coefficient. Through mathematical conversion the above EBIT (t1) equation can be converted into the final idle-cost-formula:

$$\text{EBIT (t1)} = \text{EBIT (t0)} + \text{VA (t0)} * (\text{TC} - 1) \qquad (5)$$

This equation contains the relevant influence drivers for efficiency and robustness due to the company's cost structure. The first figure (EBIT t1) indicates the company's rate of return on the basis of its production capacity, which results from three factors. First, the current state efficiency of production capacity (EBIT t0), which indicates the success of the present business model. Second, the maximum, potential amount of idle cost (IC t0), which occur at a sales drop to zero, based on the total fixed costs. Third, the TC-1, which specify the utilized capacity portfolio taking into account the company-, industry- and market-specifics which the cooperation is subject to. The turbulence factor is, in contrast to the other two pure quantitative figures, a qualitative risk weighting, similar to the weighted average cost of capital (WACC) or the industry specific ß-factor based on the capital asset pricing model (CAPM) [38].

With a Turbulence Coefficient of 0.95 respectively 0.85, representing market break-ins of 5 % respectively 15 %, using the idle-cost-model results in the ROS values for the four above displayed companies depicted in Fig. 3. It becomes obvious that the value-added-oriented capacity portfolio of company 2 is favorable in markets with little volatility (e.g. market-break-ins of 5 %), while the risk-averse capacity-portfolio of company 3 outperforms company 2 at TC < 0.89.

Although the prognosis of market turbulence is neither accurately determinable nor developed by the idle-cost-model, the quantification of the EBIT (t1) break-even-point identifies the steady-state capacity portfolio costs. Under these conditions, the idle-cost-model enables companies to quantify the idle cost risk as a result of its current EBIT, the fixed cost quota and the current market volatility, by enhancing the operating leverage model by an explicit consideration of the company's value added structures. The gain is the future EBIT projection under turbulence in total numbers.

In accordance with the present state of knowledge, wages and salaries are classified as (periodic-) fixed costs. In contrast to existing, models based on the

Gaussian distribution, the idle-cost-model considers the absolute, possible maxima of idle costs risk contrasted with the company's returns. Potential worst case scenarios are contemplated. The idle-cost-model enables management to setup a warning and control system for the early detection of financial distress, as it is postulated in the current regulations, e.g. the German KontraG (Control and Transparency Business Act) [39]. In this context, the idle-cost-model provides no methods for the elimination of turbulence, but forecasts the turbulence impacts and thereby implicates needs for additional robustness.

4 Operational Modeling Approach: Robust Manufacturing Performance

The manufacturing system of a company is operated and controlled according to certain targets, depending on the company's products and specific market conditions. For instance, in the process industry where machines or workstations are costly investments, a high utilization of machines usually is the overarching goal. These targets for manufacturing are described by target systems, such as the performance targets of production according to Westkämper [40], where the conflicting targets are grouped into time, quality and cost targets. A second target model that focuses more on manufacturing aspects are the four production logistics targets described by Gutenberg [17], which are a high due date reliability, short throughput times, high utilization and low inventory. These targets can be divided into logistics performance and logistics cost targets and they are contradictory, meaning that if the logistics performance (e.g., higher due date reliability) is increased this usually results in increased logistics costs (e.g., higher inventory) [41]. This is generally described as a trade-off situation and has been investigated for different contradicting performance values in the past [27, 42]. Since not all targets can be equally achieved at the same time, a company has to decide which targets to emphasize, which is called logistical positioning.

With complexity and fluctuating influences (e.g., demand, supply rates) in manufacturing systems rising, failures are more likely to propagate through the systems, easily causing performance decreases. Robustness as a system characteristic which "enables the system to maintain its functionalities against external and internal perturbations" [43], is therefore gaining importance for manufacturing systems. In order to achieve performance robustness against fluctuations and disturbances in manufacturing (e.g., demand, supply rates), different measures, such as excess capacity (resources) need to be installed. Since these usually incur costs, the relationship between performance robustness and resource efficiency can be described as a trade-off. For instance, increasing the flexibility of a machine (e.g., making several different operations possible) can lead to improved robustness of the manufacturing system, but at the same time might incur costs for investing into new and flexible resources. In a market scenario where flexibility is required due to rapidly changing demands this increased flexibility might be beneficial and thus increase robustness. Therefore, when capacity adjustments are necessary and thus capacity investment decisions have to be made, the trade-off between efficient or robust design should be taken into consideration.

In order to model the trade-off, we first need to define how robustness and efficiency can be measured. We argue that a key element of achieving robustness is redundancy. In the context of reliability engineering, redundancy describes the adding of identical components in order to design systems or components fail safe. In previous works, we have depicted a manufacturing system as a network with nodes and edges and suggested a path analytical approach to examine the influence of redundancy on system robustness [44]. As measures to increase the redundancy are connected to increased costs, we suggest the machine capacity used (i.e. the redundancy in the system) as a measure for cost efficiency. If a new capacity setting (*cs*) is evaluated with regard to its efficiency, the efficiency *e* can be measured as a ratio of the machine capacity *C* (in hours) needed for the new scenario *n* and the machine capacity needed for an initial scenario *i*.

$$e_{cs} = Ci \,/\, Cn \tag{6}$$

A higher capacity (i.e. percentage of machines used) is considered to be less cost efficient (as they incur cost for maintenance, resource investment), whereas a lower capacity is considered to be more cost efficient.

On the other hand, we propose the logistics performance values (e.g., due date reliability) as a measure for system robustness under fluctuations and disturbances. In case of a disturbance, a manufacturing system can be described as robust if the disturbance does not negatively affect the performance. To evaluate the robustness r of a capacity setting, one can compare the initial performance (e.g. due date reliability, measured in days) p_i to the performance of a new capacity setting p_n.

$$r_{cs} = p_n / \, p_i \tag{7}$$

Using feedback data from manufacturing systems, these measurements are suitable to be used in a simulation study, where the system behavior for different capacity settings and different fluctuating factors (e.g., supplier rates, machine breakdowns) can be tested. In this way, the relationship between redundancy and efficiency can be investigated for different turbulence scenarios and depicted in a graph as shown in Fig. 4.

5 Model Combination and Discussion

We have presented a strategic and operational model to analyze the trade-off between efficiency and robustness. The necessity of defining a manufacturing strategy and aligning it with other functions in the company, such as production operations, has long been claimed [45]. For this purpose, modeling approaches that

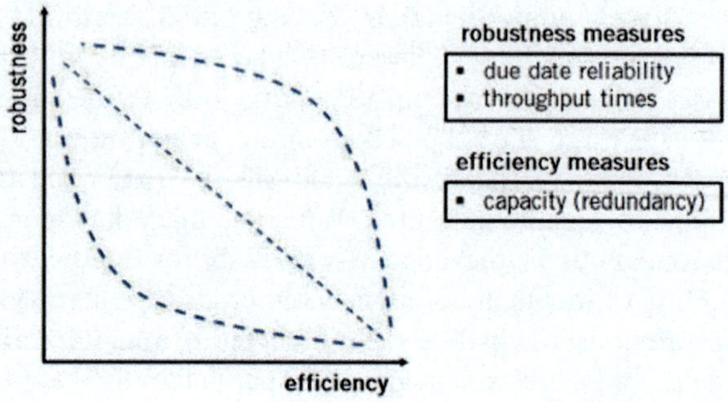

Fig. 4 Possible relationships between robustness and efficiency in manufacturing systems

link the perspectives from manufacturing strategy and sales volatility to operations planning have been suggested [46]. It has further been shown that manufacturing performance substantially contributes to business performance and thus manufacturing and business strategy should be aligned [47].

Looking at the summarized variables of our previously introduced models in Fig. 5, it becomes evident that the models have different areas of application. The operational model allows for the control of the manufacturing system, which is on short-term basis. Contrary to this, the strategic model is preoccupied with long-term financial values and thus focuses on risk aversion aspects. Yet the purpose of both models is the same: according to the key performance figures of the respective areas of application, an optimal capacity configuration is to be found. Furthermore, both models identify an existing trade-off between efficiency and robustness in capacity investment decisions, which has to be addressed systematically.

We thus argue that when it comes to capacity decisions, a company needs to position itself in the trade-off between robustness and efficiency, and they need to consider both the operational and strategic perspective in order to guarantee the overall success and sustainability of a company. Firstly, the operational model enables them to determine adequate manufacturing capacity for an ideal ratio between robustness to fluctuations or disruptions and process efficiency. Secondly, the strategic model allows to test whether the chosen capacity portfolio display a sustainable business model within the industry-specific market volatility.

Yet such a combination of two models of different application areas surely also has its shortcomings. For one, a large difficulty lies in obtaining the necessary data to fill both models for the same company, since they are of a substantially different nature. Thus applying this model requires that the analyst is familiar with both environments (i.e. the operational shop floor level and the strategic business level) and their respective key performance indicators. Future research will be concerned with applying both models to different financial and feedback data from manufacturing companies in order to test their validity.

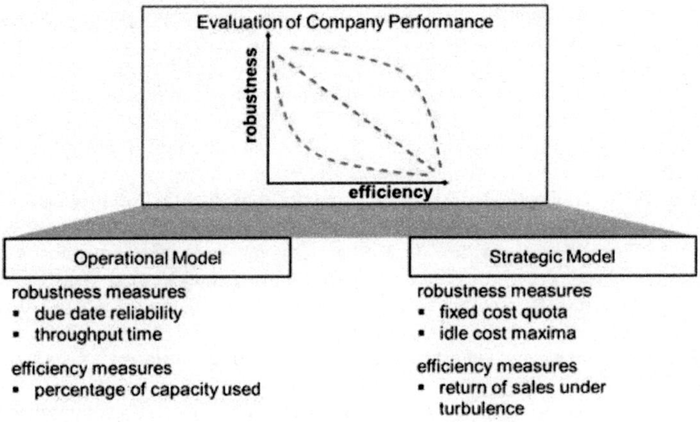

Fig. 5 Comparison of operational vs. strategic model

Acknowledgments The research of Prof. Dr.-Ing. Katja Windt is supported by the Alfried Krupp Prize for Young University Teachers of the Krupp Foundation.

References

1. Yang, T., Peters, B.A.: Flexible machine layout design for dynamic and uncertain production environments. Eur. J. Oper. Res. **108**, 49–64 (1994)
2. Alp, O., Tan, T.: Tactical capacity management under capacity flexibility in make-to-stock systems. IIE Trans. **40**, 221–237 (2006)
3. Zaeh, M.F., Müller, N.: A modeling approach for evaluating capacity flexibilities in uncertain markets. Int. J. Flex. Manuf. Sys. **19**, 151–172 (2007)
4. Caputo, M.: Uncertainty, flexibility and buffers in the management of the firm operating system. Prod. Plann. Control **7**, 518–528 (1996)
5. Battini, D., Persona, A., Regattieri, A.: Buffer size design linked to reliability performance—a simulative study. Comput. Ind. Eng. **56**, 1633–1641 (2009)
6. Karmarkar, U., Kekre, S.: Manufacturing configuration, capacity and mix decisions considering operational costs. J. Manuf. Syst. **4**, 315–324 (1987)
7. Modi, S.B., Mishra, S.: What drives financial performance—resource efficiency or resource slack—evidence from US based manufacturing firms from 1991 to 2006. J. Oper. Manage. **29**, 254–273 (2011)
8. Feyzioğlu, O., Pierreval, H., Deflandre, D.: A simulation-based optimization approach to size manufacturing systems. Int. J. Prod. Res. **43**, 247–266 (2005)
9. Hopp, W.J., Spearman, M.L.: Factory Physics. Mcgraw-Hill, New York (2008)
10. Luss, H.: Operations research and capacity expansion problems: a survey. Oper. Res. **30**, 907–947 (1982)
11. Wu, D., Erkoc, M., Karabuk, S.: Managing capacity in the high-tech industry—a review of literature. Eng. Econ. **50**, 125–158 (2005)
12. Paraskevopoulos, D., Karakitsos, E., Rustem, B.: Robust capacity planning under uncertainty. Manage. Sci. **37**, 787–800 (1991)
13. Zhang, F., Roundy, R., Cakanyildirim, M., Huh, W.T.: Optimal capacity expansion for multi-product, multi-machine manufacturing systems with stochastic demand. IIE Trans. **36**, 23–36 (2004)

14. Van Mieghem, J.A.: Capacity management, investment, and hedging: review and recent developments. Manuf. Serv. Oper. Manage. **5**, 269–302 (2003)
15. Hendricks, K.B., Singhal, V.R., Zhang, R.: The effect of operational slack, diversification, and vertical relatedness on the stock market reaction to supply chain disruptions. J. Oper. Manage. **27**, 233–246 (2009)
16. Markowitz, H.: Portfolio Selection: Efficient Diversification of Investments. Blackwell, Oxford, UK (1991)
17. Gutenberg, E.: Grundlagen der Betriebswirtschaftslehre. 1.Bd.: Die Produktion, Springer, Berlin (1951)
18. Nagel, M.: Flexibilitätsmanagement—Ein systemdynamischer Ansatz zur quantitativen Bewertung von Produktionsflexibilität. Deutscher Universitäts-Verlag, Wiesbaden (2003)
19. Adler, P.S., Goldoftas, B., Levine, D.I.: Flexibility versus efficiency? a case study of model changeovers in the toyota production system. In. Organ. Sci. **1**, 43–68 (1999)
20. Zhou, S.: Faites votre jeu: Unternehmensplanung unter quantitativer Risiko-Betrachtung. risknews, 2, 42–47 (2005)
21. Monauni, M.: Fixkostenmanagement—strategischer Ansatz zur Flexibilisierung von Produktionskapazitäten. Eul Verlag, Köln (2011)
22. Dekkers, R.: Strategic capacity management: meeting technological demands and performance criteria. Int. J. Prod. Res. **40**, 3895–3911 (2002)
23. Wildemann, H.: Fixkostenmanagement—Leitfaden zur Anpassung von Kostenstrukturen an volatile Märkte. TCW, München (2009)
24. Schmalenbach, E.: Die Betriebswirtschaftslehre an der Schwelle der neuen Wirtschaftsverfassung. In: Funk, J., Hax, H., Potthoff, E. (eds) Kapazitätsrisiken und Unternehmenspolitik, pp. 91-101, Handelsblatt GmbH, Düsseldorf (1984)
25. Dudenhöffer, F.: Automobilindustrie hat noch Flexibilitätsreserven. VDI Nachrichten 1, 10 (2010)
26. Schoeffler, S., Buzzel, R., Heany, D.: Impact of strategic planning on profit performance. Harvard Bus. Rev. **53**, 137–145 (1974)
27. da Silveira, G.: Improving trade-offs in manufacturing: method and illustration. Int. J. Prod. Econ. **95**, 27–38 (2005)
28. Krcal, H.: Strategische Implikationen einer geringen Fertigungstiefe in der Automobilindustrie. zfbf: Schmalenbachs Zeitschrift für betriebswirtschaftliche Forschung 8, 778–808 (2008)
29. Nink, J.: Strategisches Fixkostenmanagement—Konzeption und ausgewählte Instrumente zur Bestimmung von Fixkostenstrategien. Cuvillier, Göttingen (2002)
30. Zahn, E., Bullinger, H.-J., Gagsch, B.: Führungskonzepte im Wandel. In: Handbuch Unternehmensorganisationpp. pp. 109–125, Springer, Berlin (2008)
31. Westkämper, E., Zahn, E.: Wandlungsfähige Unternehmensstrukturen—Das Stuttgarter Unternehmensmodell. Springer, Berlin (2008)
32. Mayer, R.: Kapazitätskostenrechnung—Neukonzeption einer kapazitäts- und prozeßorientierten Kostenrechnung, Franz Vahlen, München (1998)
33. Brox, H., Rüthers, B., Henssler, M.: Arbeitsrecht, 17th edn. Kohlhammer, Stuttgart (2007)
34. Rade, K.: Angemessene Herstellungskosten nach BilMoG – Keine Irrelevanz der Abgrenzung von Einzel- und Gemeinkosten, DStR 28, 1334–1338 (2011)
35. Westkämper, E.: Die Digitale Fabrik—Kontinuierliche und Partizipative Planung. In: Milberg, J., Schuh G. (eds.): Erfolg in Netzwerken, pp.246-259, Springer, Berlin (2002)
36. Banker, R.D., Chang, H., Majumdar, S.K.: A Framework for analyzing changes in strategic performance. Strateg. Manag. J. **17**, 693–712 (1996)
37. Department for Innovation, Universities and Skills, The 2008 Value Added Scoreboard. Crown Copyright (2008)
38. Gleißner, W.: Wertorientierte Unternehmensführung und risikogerechte Kapitalkosten, Risikoanalyse statt Kapitalmarktdaten als Informationsgrundlage. Controlling 3, 165–171 (2011)

39. KontraG Nr. 29, Art. 1 §91.2: „Der Vorstand hat geeignete Maßnahmen zu treffen, insbesondere ein Überwachungssystem einzurichten, damit den Fortbestand der Gesellschaft gefährdende Entwicklungen früh erkannt werden
40. Westkämper, E.: Einführung in die Organisation der Produktion. Springer, Berlin (2006)
41. Wiendahl, H.-P.: Betriebsorganisation für Ingenieure. Hanser, München (2008)
42. Jodlbauer, H., Altendorfer, K.: Trade-off between capacity invested and inventory needed. Eur. J. Oper. Res. **203**, 118–133 (2010)
43. Kitano, H.: Biological robustness. Nat. Rev. Genet. **5**, 826–837 (2004)
44. Windt, K., Meyer, M., Hütt, M.-T.: A modeling approach to analyze redundancy in manufacturing systems. In: ElMaraghy, H. (ed.) Enabling Manufacturing Competitiveness and Economic Sustainability. Springer, Berlin (2012). (in print)
45. Wheelwright, S.C.: Manufacturing strategy: defining the missing link. Strateg. Manag. J. **5**, 77–91 (1984)
46. Olhager, J., Rudberg, M., Wikner, J.: Long-term capacity management: Linking the perspectives from manufacturing strategy and sales and operations planning. Int. J. Prod. Econ. **69**, 215–225 (2001)
47. Sun, H., Hong, C.: The alignment between manufacturing and business strategies: its influence on business performance. Technovation **22**, 699–705 (2002)

Index

K. Windt (ed.), *Robust Manufacturing Control*, Lecture Notes in Production Engineering
DOI: 10.1007/978-3-642-30749-2, © Springer-Verlag Berlin Heidelberg 2013